Electrical Engineering Fundamentals

Electrical Engineering Fundamentals

Maryam A. Al-Othman, John H. Cole, and Dimitrios Peroulis

Purdue University Press, Lafayette, Indiana

Cataloging-in-Publication Data on file at the Library of Congress.

978-1-62671-098-6 (hardcover)
978-1-62671-099-3 (epdf)

Contents

Preface

This is the textbook for ECE 20001: Electrical Engineering Fundamentals I at Purdue University. Before ECE 20001, we used to have ECE 201: Linear Circuits Analysis I. ECE 201 served as the first required circuits course for students from at least six majors at Purdue University (computer, electrical, mechanical, multidisciplinary, industrial, and nuclear engineering). Besides being the first course, ECE 201 was also the only required circuits course for all the above-mentioned majors (except electrical and computer engineering). The traditional ECE 201 syllabus was focused on the fundamentals of linear circuits including DC, first-order, second-order, and AC circuits.

It was in 2009 that the senior author of this book started working on creating ECE 20001 and the first notes for this book. In Fall 2009, Prof. Peroulis offered, for the first time, a small section of ECE 201 with a revised syllabus that, for the first time, included a three-week segment on semiconductors, diodes, and transistors. It was offered as an honors version to 44 students (about 10% of the total enrollment in ECE 201 at the time). This experimental offering conveyed, for the first time, the idea that students need a basic understanding of the fundamental semiconductor technology that fuels all aspects of modern living. The COVID-19 pandemic underlined the importance of semiconductor chips for practically everyone worldwide. But in 2009 this was not as obvious, even in educational settings. We can still recall faculty conversations questioning the ability of sophomore students to handle semiconductors, band theory, and nonlinear devices. This first offering was indeed successful, and Prof. Peroulis repeated the honors offerings in the fall semesters of 2010 (36 students), 2011 (47 students), and 2012 (50 students). Every one of these offerings included improvements in content coverage and delivery based on valuable student feedback. All offerings included semiconductors.

In 2016, a small committee in the School of Electrical and Computer Engineering at Purdue University volunteered to work on a comprehensive curriculum reform. The committee was appointed by the then ECE Head, Prof. Venkataramanan (Ragu) Balakrishnan and was chaired by the then Associate Head of Education, Prof. Mike Melloch. Prof. Peroulis was a member of the committee. The committee focused on reforming the required circuits courses. The committee addressed the question of which concepts were needed to be covered in the school's courses and which ones should be retired. After several months of hard work, in November 2016 the committee recommended that two new courses be formally developed: ECE 20001 and ECE 20002. Prof. Peroulis led the ECE 20001 development as an improvement of the aforementioned ECE

201 honors version. The first offering was in Fall 2017. It was at about that time that Drs. Al-Othman and Cole joined the team with unparalleled enthusiasm to create a high-quality textbook for this class. Semester after semester, the team worked hard to create the book you are reading now.

All three coauthors agreed that this textbook should be offered to everyone for free. Moreover, we created a set of many dozens of videos that complement this book. These recordings are available through Purdue University. We hope you enjoy this book and that it inspires you to further explore the fascinating world of electrons, bits, and atoms.

We are very thankful to all of our colleagues and students whose encouragement, energy, wisdom, and corrections made this project possible. We would also like to thank Purdue President Dr. Mung Chiang and Provost Dr. Patrick Wolfe, who created the Purdue University Books Initiative that made the production of this book possible. We extend our heartfelt appreciation to Professor Beth McNeil, Dean of Purdue University Libraries, for her critical support, invaluable encouragement, and wise advice at various stages. Additionally, we offer our deep thanks to Katherine Purple, who managed this project at Purdue University Press. Katherine's positive energy and meticulous attention to detail were crucial to the success of this project.

As we conclude this short preface, we would like to particularly thank our families for their constant support and encouragement during this process.

We hope you will permit us to include a couple more words from Prof. Peroulis: I owe a profound debt of gratitude to my wife, Prof. Alina Alexeenko, whose love has guided me on a journey like no other. Her unwavering belief in me and her boundless inspiration have been the cornerstone of this endeavor.

Chapter 1

The Basic Trio: Power, Current, and Voltage

Introduction

We are used to, as students, being given problems to solve every time we learn something new. Many of us find the word "problem" to be intimidating, and why not? The word itself implies an attempt to solve something that might not even have a solution or might require innovation. In this book, we prefer the word "exercises" since they do all have a solution and you will be given the tools needed to solve them.

As an engineer, you will encounter real problems that need a solution. There will always be something that needs to be made more efficient, faster, lighter, or brighter, which will require you to come up with a solution that does not exist in the back of a book—a solution that will not be graded by a teacher but by your ability to achieve the desired goal.

For this reason, we will try to provide you with as many tools as possible during your journey as an engineering student to aid you in finding your future solutions. You will see many concepts such as modeling, simulation, testing, and heuristics. We advise you to think of your classes as opportunities to add more tools to your tool belt instead of as a series of exams to be solved and grades to be taken.

In this course you will be introduced to many electrical and electronic engineering fundamentals—basic tools and concepts that you will build on as you progress in this class and in your future classes. These tools and concepts are not only critical to electrical and computer engineers but are helpful in many other engineering disciplines as well.

Some concepts like power calculations, Ohm's law, and Kirchhoff's voltage and current laws (KVL and KCL, respectively) are necessary for anyone dealing with electricity. An understanding of these concepts can help even nonengineers resolve common electrical dilemmas.

Imagine yourself in the following scenario: you are in your bedroom with your computer and TV on and are playing a game on your game console. All of the devices are connected to one power strip that is connected to your bedroom outlet. This outlet connects to a 15 ampere (A) circuit breaker in the house's breaker box, as shown in Figure 1.1. The current draw of each device is also shown in the figure. Notice that a hair dryer is connected to the power strip in addition to your other devices but that it is connected through a switch that is off. The switch is included to indicate that the hair dryer is not always on; it will draw no current while it is off but draws 2 A when you switch it on to use it.

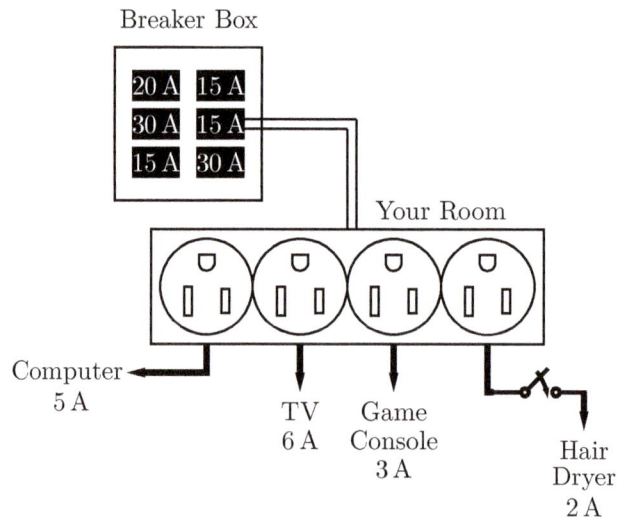

Figure 1.1: Common household electronics connected to a single circuit breaker through a power strip.

Figure 1.2: A photo of a circuit breaker box.

Now suppose you left everything on and went to take a shower. After your shower you are about to dry your hair, but when you turn your hair dryer on, all of your devices shut down and your hair dryer stops working. Hopefully you had your game and all of the work on your computer saved.

How can we explain this shutdown? We start by graphing the electrical connections of each device to make a circuit diagram like that shown in Figure 1.3. With some basic electrical engineering knowledge, you can quickly identify the cause of the failure from this circuit diagram.

All of the devices connected to your power strip are connected in parallel and, using Kirchhoff's current law (KCL), the current through the circuit breaker is the sum of the currents drawn by each device. So when the hair dryer is off, the total current, (i), drawn through the circuit breaker is calculated as

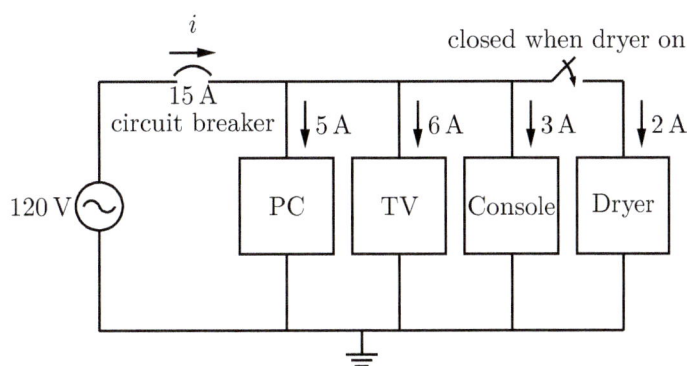

Figure 1.3: A circuit diagram showing common household electronics connected through a circuit breaker, and their associated current draws.

$$i = 5\,\text{A} + 6\,\text{A} + 3\,\text{A} = 14\,\text{A}.$$

Since this is less than the circuit breaker's 15 A rating, everything works. On the other hand, when you turn the hair dryer on, the total current draw becomes

$$i = 5\,\text{A} + 6\,\text{A} + 3\,\text{A} + 2\,\text{A} = 16\,\text{A}.$$

The resulting current is larger than the circuit breaker's limit, so it opens the circuit, halting all current flow.

If your devices remain plugged in and on, the circuit breaker will not allow you to reclose the circuit. So what is the solution? If air drying your hair is not an option, you can connect your hair dryer to an outlet that is connected to a *different* circuit breaker, or you can shut down or unplug one of the other devices. Another solution would be to replace your 15 A circuit breaker with a higher-capacity breaker. The problem with this last solution, though, is that we cannot exceed the current rating of the wires connecting the wall outlet to the circuit breaker. If the wire is only rated for 15 A, a higher current could result in melting the wire, or worse, lead to an electric fire. While this is a very simple example, hopefully it motivates you to learn more.

Application 1.0.1

For the authors of this book, a washer and dryer are among the most important amenities to have in a house. While hunting for an apartment together, two of us found a place with a full laundry room! We asked the owner, who was also the civil engineer of the building, if the two outlets in the laundry room were connected to separate circuit breakers. He replied, "Yes, of course." We decided to rent the place and before long we started doing laundry. As soon as we had the washer and dryer running at the same time, they both abruptly stopped working. When we checked the breaker box, we found that one 15 A breaker had tripped yet both the washer and dryer stopped. What does this tell us about how they are connected?

- Since both the washer and dryer lost power even though they were connected to separate outlets but only one circuit breaker tripped, we know that both outlets are connected to the same circuit breaker as shown in Figure 1.4.

The washer and dryer were both rated to draw 15 A, so when both devices were turned on at the same time, the total current drawn could be as high as

$$i_{\text{total}} = \text{washer current} + \text{dryer current} = 15\,\text{A} + 15\,\text{A} = 30\,\text{A}$$

which was the reason for the circuit breaker tripping.

Since using only one unit at a time was not a good option for us, we called the owner, who arrived along with his electrician. The electrician said it was "impossible" to move one of the outlets to another circuit breaker, so he suggested changing the existing 15 A breaker to a 30 A breaker. As electrical engineers, the first thing we had to ask was, "Are your wires rated for 30 A currents? We could have a fire if they are not!" The owner replied, "You live on the first floor—why would you be afraid of a fire? You can just jump out of the window!"

We sincerely hope that with the aid of this book, you will learn enough about the general principles of electricity that you will never give a response as sad as this.

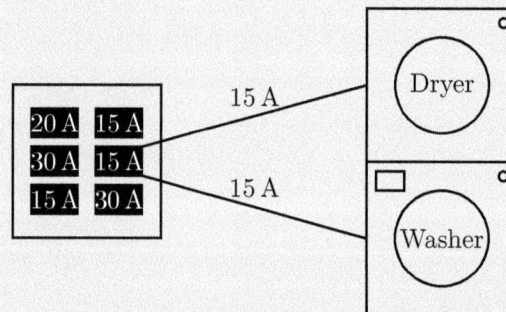

Figure 1.4: A washer and dryer unit connected to a single 15 A circuit breaker for application 1.0.1.

1.1   What Matters the Most

Since the most important subject (quantity) in the world (scratch that; we mean in the universe) is energy, it is only appropriate to start with it. We are all familiar with the **law of conservation of energy**, which states that:

> Energy can neither be created nor destroyed; rather, it transforms from one form to another.

We are also aware that energy is "expensive," so the ultimate goal is to minimize the amount of energy it takes to do something useful. For example, we used to need 100 watts to power an incandescent light bulb that produced 1600 lumens (a measure of visible light). Now, with light-emitting diode (LED) technology, we need only about 17 watts to produce the same amount of light.

Let us get into some definitions. Recall from physics that power is the rate of change in energy (work)

$$p \triangleq \frac{dw}{dt}. \tag{1.1}$$

Where p is in watts, w is in joules (J), and t is in seconds. Since we are talking about electricity, we will also be talking about electric power. For this course, the type of power we are interested in is electrical power. In order to calculate electrical power, we first need to introduce two basic quantities: current and voltage. We will discuss each of these separately in the next two sections and then return to our discussion of power in section 1.1.3.

1.1.1 Current

A potential difference (voltage) applied across a conductor forces electrons to stop moving randomly and instead follow the general direction imposed by the applied field. Figure 1.5 provides an approximate illustration of this idea.

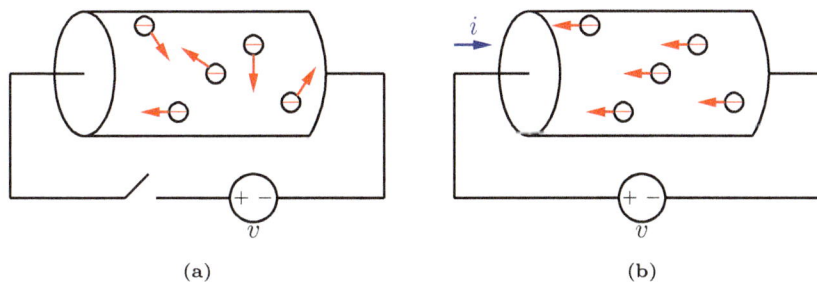

(a)   (b)

Figure 1.5: The random movements of each electron result in zero net current flow until an external voltage is applied.

In Figure 1.5a, we can see a few representative electrons in a conductor. While the switch connecting the voltage source is open, the motions of the electrons are random and the net current flow is zero. Closing the switch allows for a potential difference to be applied to the conductor, resulting in a net movement of negatively charged electrons toward the positive terminal of the voltage source. Although we know that electrons are the charges that actually move, we take the direction of the current to be the direction that equivalent positive charges would have followed, as seen in the figure. This is due to historical reasons when current was thought to be due to movement of positive charges. Thankfully, it makes no difference in our calculations, so the electrical engineering community has maintained this convention.

The flow rate of the free electrons in the conductor is referred to as the electric current, and is defined mathematically as:

$$\boxed{i(t) \triangleq \frac{dq(t)}{dt},} \tag{1.2}$$

where q is the charge measured in coulombs (C) and t is the time in seconds (s). The units of electric current are amperes (A) where $1\,\mathrm{A} = 1\,\mathrm{C/s}$.

Example 1.1.1

Find the current $i(t)$ for the following charge flow

$$q(t) = \frac{t^2 + t + 2}{t^2 + 2}\ \mathrm{C}.$$

From (1.2), we know that the current $i(t)$ can be calculated by taking the derivative of the charge

$$i(t) = \frac{\mathrm{d}q(t)}{\mathrm{d}t}.$$

To calculate the derivative of the given charge flow, we will use the chain rule. To make things easier, we can rewrite $q(t)$ as

$$q(t) = f_1(t)f_2(t),$$

where

$$f_1(t) = t^2 + t + 2$$

and

$$f_2(t) = \frac{1}{t^2 + 2}.$$

Then

$$i(t) = \frac{\mathrm{d}f_1(t)}{\mathrm{d}t}f_2(t) + \frac{\mathrm{d}f_2(t)}{\mathrm{d}t}f_1(t).$$

We now find the derivatives of $f_1(t)$ and $f_2(t)$ as

$$\frac{\mathrm{d}f_1(t)}{\mathrm{d}t} = 2t + 1,$$

and

$$\frac{\mathrm{d}f_2(t)}{\mathrm{d}t} = \frac{\mathrm{d}}{\mathrm{d}t}(t^2 + 2)^{-1} = \frac{-2t}{(t^2 + 2)^2}.$$

Getting back to the current, we have

$$i(t) = \frac{2t + 1}{t^2 + 2} + \frac{-2t(t^2 + t + 2)}{(t^2 + 2)^2} = \frac{2t + 1}{t^2 + 2} - \frac{2t^3 + 2t^2 + 4t}{(t^2 + 2)^2}.$$

This can be simplified to yield

$$i(t) = -\frac{t^2 - 2}{(t^2 + 2)^2}\ \mathrm{A}.$$

If we already have the current, we can find the total charge flow between two times t_0 and t as

$$\boxed{\ q_{\text{total}}(t) \triangleq \int_{t_0}^{t} i(t)\,\mathrm{d}t.\ } \tag{1.3}$$

Note: We hope you will not be confused by the use of t as the integral limit and the integration variable. If this is confusing, you may write this equation as:

$$q_{\text{total}}(t) \triangleq \int_{t_0}^{t} i(x)\,\mathrm{d}x. \tag{1.4}$$

Example 1.1.2

Find the charge flow $q(t)$ for the following current over the interval $0 < t < t_f$, given the current

$$i(t) = (6t + 3)\cos(t^2 + t)\,\mathrm{A}.$$

The total charge that flows between time 0 and t_f can be found by integrating the current over the given interval

$$q(t_f) = \int_0^{t_f} i(t)\,\mathrm{d}t = \int_0^{t_f} (6t + 3)\cos(t^2 + t)\,\mathrm{d}t.$$

We can use integration by substitution here as follows:

$$u = t^2 + t.$$

Then

$$\mathrm{d}u = (2t + 1)\,\mathrm{d}t.$$

Remember that we will have to change the limits of the integral at

$$t = 0 \rightarrow u = 0$$

and at

$$t = t_f \rightarrow u = t_f^2 + t_f.$$

To make things more clear, we can rearrange the integral as

$$q(t_f) = 3\int_0^{t_f} \cos(t^2 + t)(2t + 1)\,\mathrm{d}t.$$

Now we can substitute u and $\mathrm{d}u$ in the integral to get

$$q(t_f) = 3\int_0^{t_f^2 + t_f} \cos(u)\,\mathrm{d}u.$$

Completing this simple integration results in

$$q(t_f) = 3\sin(t_f^2 + t_f)\,\mathrm{C}.$$

Circuit diagrams are drawn using branches and nodes. The circuit's elements are drawn on the diagram's branches and the connections between the elements form the nodes. Current runs through circuit branches, and we represent current flow in the diagram with an arrow that points in the direction of positive current flow. We must always specify the current's value and its direction. Failure to pay attention to both the current's value and its direction is the cause of many sign errors. Figure 1.6 shows two equivalent ways to represent current flowing in the branch between nodes a and b. We can say either a positive current i

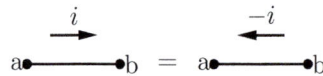

Figure 1.6: Equivalent current specifications for current through a branch between nodes a and b.

flows from node a to node b, or that the negative value of current i (i.e., $-i$) flows from node b to node a.

When trying to find an unknown current, it is practically always necessary to assume its direction. You should not feel afraid to assume (or assign) a direction to an unknown current. If you assign a direction to a current in your analysis and the result ends up being a negative value, then the actual direction of the current is the opposite of what you assumed. In this case, you can change the direction of the current and omit the negative value or just keep the negative sign and the assumed direction. Either way is fine; as long as you maintain consistency, there is nothing to worry about. The following example illustrates this.

Example 1.1.3

For the circuit in Figure 1.7, find the currents I_1, I_2, and I_3.

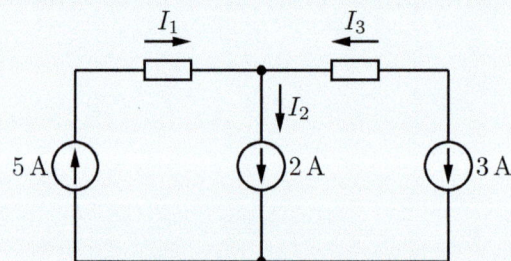

Figure 1.7: Example 1.1.3.

The assumed current directions are shown in the figure. The current I_1 is in the same direction as the 5 A current source, so

$$I_1 = 5\,\text{A}.$$

The current I_2 is in the same direction as the 2 A current source, so

$$I_2 = 2\,\text{A}.$$

Meanwhile, the current I_3 is in the opposite direction of the 3 A current source, so

$$I_3 = -3\,\text{A}.$$

Again, the negative sign does not indicate a wrong answer.

1.1.2 Voltage

Let us look again at what happens inside a conductor when a battery is connected across it (Figure 1.8). The battery creates an electric field $\mathscr{E}$ inside the conductor that exerts a force $F = q\mathscr{E}$ on any charge q inside it. If $q > 0$, the force is in the direction of the applied electric field. Thus, electrons for which $q < 0$ are pushed in the opposite direction of the applied electric field: from the negative pole toward the positive pole of the battery. Since ions do not move in typical solid conductors, all of the electric current is due to the motion of electrons.

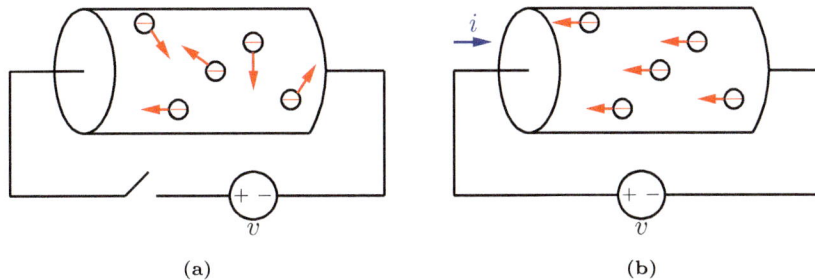

(a)   (b)

Figure 1.8: Figure 1.5 is repeated here for convenience.

Intuitively, we know that a stronger battery results in a higher current. A battery is defined by many parameters (e.g., volume, shape, stored energy), but a critical one is the voltage difference between its poles. A battery with a voltage difference of 10 volts (V) signifies a battery 10 times stronger than one with a difference of 1 V (for the same current). This is because a voltage difference between two points is defined as the difference in electric (potential) energy between these points per unit charge. Therefore, a 10 V voltage difference is equal to 10 J/C. Now let us visualize an electron in such a case. To move this electron against the electric field (from the positive to the negative pole), we will need to consume $10\,\text{V} \times 1.6 \times 10^{-19}\,\text{C} = 1.6 \times 10^{-18}\,\text{J}$.

More generally, we define the voltage difference, or potential difference, or simply "the voltage," between two points as the work necessary to move a unit charge from one point to the other. Mathematically, we can write this as

$$v_{12} \triangleq \frac{dw}{dq}. \tag{1.5}$$

where v is in volts, w is the work in joules, and q is charge in coulombs.

One thing we also need to clarify here is the meaning of the word "voltage." Voltage can mean two things, and the correct meaning is often derived from the context it is used in.

- Voltage may simply represent a shorthand notation for voltage difference.

- Voltage may also refer to the voltage difference between a point and a reference point with an assigned voltage of zero volts. We can only talk about absolute voltage of a point if some other point is assigned to have zero voltage. The reference point is also often called "ground."[1] Any point can be assigned to be a reference point. Sometimes the actual physical ground is taken as our reference point. When this is not possible or obvious, the lowest point in a circuit schematic is often assigned to be the reference point. For example, in Figure 1.3, the bottom edge of the circuit schematic is used as the reference point.

Now let us see how we can represent voltage in a circuit. Figure 1.9 shows a circuit element that is connected to some circuit at nodes a and b. Since we are talking about potential difference across elements, one node will have a higher potential than the other. This is often indicated by assigning a polarity between nodes a and b. In the figure we have assigned the polarity as shown. Please note that assigning polarity when defining a voltage difference is critical; without a polarity, it is impossible to properly define the voltage between the nodes a and b. With the assigned polarity, we are saying that the node a is at a higher potential than the node b.

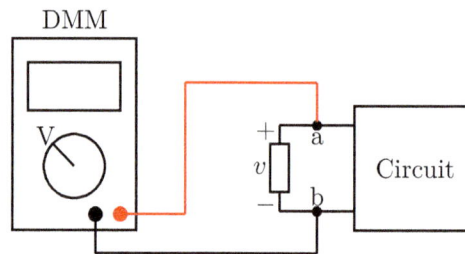

Figure 1.9: Reading the potential difference across an element using a digital multimeter (DMM).

If we connect the positive lead of a digital multimeter (DMM) to node a and the negative lead to node b (as seen in the figure), we can get a voltage reading. If the reading is a positive number, then indeed node a is at a higher potential than node b. If we get a negative reading, then node b is at a higher potential than node a. We can always assign polarity to our circuit elements however we see fit and then perform our analysis. Negative calculation results only mean that the true polarity is the opposite of the assumed one; it is not an indication of a mistake. In practice, we don't know the voltage (including its polarity) before we start solving a problem. Instead, we have to *assume* a voltage polarity, solve the problem, and accept whatever sign comes out of the calculations. It does not make sense to spend time trying to guess the correct polarity ahead of our calculations.

Let us see the same thing in a different way. Consider the two circuits shown in Figure 1.10. In this figure we see two different assumed polarities for the voltage V_x. In the circuit of Figure 1.10a, point a is assumed to be at a higher voltage than point b. For the given excitation, this clearly means that $V_x = 5\,\text{V}$. If we pick the opposite polarity and take a reading as in Figure 1.10b, we will have a negative reading of voltage V_x, meaning $V_x = -5\,\text{V}$.

Before we move on, we need to talk about some notation. Voltage may or may not be a function of time. For example, a 1.5 V battery (theoretically)

---

[1]This is similar to mechanical potential energy: we often assume that the potential energy on the ground level is zero.

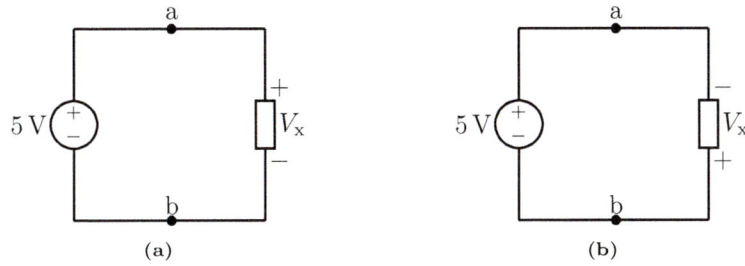

Figure 1.10: Voltage polarity.

provides 1.5 V of constant voltage until it runs out. We call this kind of voltage "DC voltage," where DC stands for "direct current" (don't let the word "current" mislead you; we will talk more about this later). We use uppercase italic letters for DC voltages (and currents), so the voltage that the 1.5 V battery provides can be written as

$$V_{\text{battery}} = 1.5\,\text{V}. \tag{1.6}$$

Other types of voltages can change in time. For example, a voltage can typically be sinusoidal like the outlet voltage, and other functions of time (e.g., square or triangular waveforms) are also quite common. In time-dependent cases, we use lowercase italic symbols as v. This notation can also be used when we are just talking in general about voltage or writing equations that are general and are true for any type of voltage.

Now that we are a little bit more comfortable with the concept of voltage difference, let us examine Figure 1.9 again, but this time with numbers. Assume you have the DC circuit in Figure 1.11 and you are asked to find the voltage V across the element placed between nodes a and b.

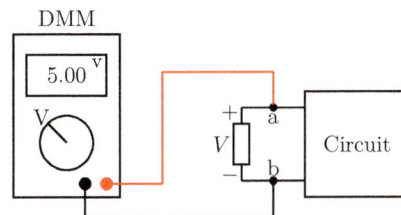

Figure 1.11: Voltage polarity and DMM reading.

Since you are following the polarity asked for in the figure, you connect your positive (red) probe to node a and negative (black) probe to node b, and the number you get on the LED screen is 5.00 V. Basically, we are taking the difference between the voltage at node a (V_a) and the voltage at node b (V_b). Notice that we are using uppercase italic now for the voltage because we know it is a DC circuit. If we want to write it in an equation form, we would say

$$V = V_a - V_b = 5.00\,\text{V}. \tag{1.7}$$

The voltage in (1.7) can be also referred to as the voltage V_{ab}.

$$V = V_{ab} = V_a - V_b = 5.00\,\text{V}. \tag{1.8}$$

What if you were asked to find the voltage V_{ba} instead? Since we know that

$$V_{ba} = V_b - V_a \tag{1.9}$$

and since we already know V_{ab}, we can rewrite (1.9) as

$$V_{ba} = V_b - V_a = -(V_a - V_b) = -V_{ab} = -5.00\,\text{V}. \tag{1.10}$$

Note that in this example we have not explicitly defined our reference (ground) node. If we connect node b in Figure 1.11 to ground (i.e., we set $V_b = 0$), then the reading we get on the digital multimeter (DMM) is the voltage V_a and it is 5 V.

The voltage difference equation can be written in general for any type of voltage as

$$\boxed{v_{ab} = v_a - v_b = -v_{ba}.} \tag{1.11}$$

Example 1.1.4

For the circuit in Figure 1.12, $V_{s_1} = 10\,\text{V}$, $V_d = 0\,\text{V}$, $V_b = 8\,\text{V}$, and $V_{cb} = -5\,\text{V}$. Find V_{ad}, V_a, V_{ab}, V_{bd}, V_c, and V_{dc}.

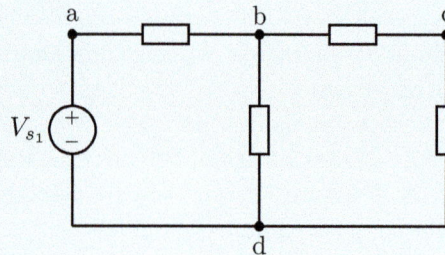

Figure 1.12: Example 1.1.4.

The first thing we notice is that the voltage V_{ad} is the voltage provided by the source V_{s_1}

$$V_{ad} = V_{s_1} = 10\,\text{V}.$$

We also know V_d and we know from (1.11) that

$$V_{ad} = V_a - V_d.$$

So we can write

$$V_a = V_{ad} + V_d = 10\,\text{V} + 0\,\text{V} = 10\,\text{V}.$$

Now we can find V_{ab}

$$V_{ab} = V_a - V_b = 10\,\text{V} - 8\,\text{V} = 2\,\text{V}.$$

We can find V_{bd} the same way:

$$V_{bd} = V_b - V_d = 8\,\text{V} - 0\,\text{V} = 8\,\text{V}.$$

Now, for V_c, we know the value of V_{cb} and V_b, so

$$V_{cb} = V_c - V_b = -5\,\text{V}.$$

$$V_c = V_{cb} + V_b = -5\,\text{V} + 8\,\text{V} = 3\,\text{V}.$$

Once we have V_c, we can find V_{dc}:

$$V_{dc} = V_d - V_c = 0\,\text{V} - 3\,\text{V} = -3\,\text{V}.$$

Now, let us use the circuit diagram in Example 1.1.4 again, but with changes to some of the numbers. See if you can work out the voltages in the exercise below.

Exercise 1.1.1

For the circuit in Figure 1.13, $V_{s_1} = 15\,\text{V}$, $V_d = 10\,\text{V}$, $V_b = 8\,\text{V}$, and $V_{cb} = -5\,\text{V}$. Find V_{ad}, V_a, V_{ab}, V_{bd}, V_c, and V_{dc}.

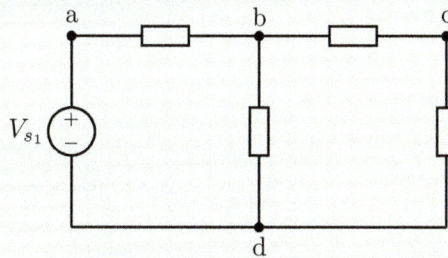

Figure 1.13: Exercise 1.1.1.

$V_{ad} = 15\,\text{V}$, $V_a = 25\,\text{V}$, $V_{ab} = 17\,\text{V}$, $V_{bd} = -2\,\text{V}$, $V_c = 3\,\text{V}$, and $V_{dc} = 7\,\text{V}$.

1.1.3 Back to Power

If we multiply the definition of voltage in (1.5) by the definition of current in (1.2), we get

$$vi = \frac{\mathrm{d}w}{\mathrm{d}q}\frac{\mathrm{d}q}{\mathrm{d}t} = \frac{\mathrm{d}w}{\mathrm{d}t} = p, \tag{1.12}$$

where p is power. As expected, voltage and current are the two main quantities we need to calculate power, which explains our long journey of trying to define and calculate voltages and currents. As you will be learning new methods that can yield voltages and currents in a circuit and at any point in time, it is important to remember the reason for doing this: computing power (and energy).

Let us start with a single element as seen in Figure 1.14. Assume that we know both the voltage across that element and the current through it. In Figure 1.14, we have two orientations for the current. The current and voltage are assumed to be positive for both cases. Let us examine the element in Figure 1.14a first. Notice that the current enters the device through the positive voltage node. This is called the **passive sign convention**, and the device is often labeled as a **load** or **passive device**. The meaning behind these names is that the load absorbs or consumes positive power as given by

$$p_{\mathrm{abs}} = vi. \tag{1.13}$$

Of course, we can equivalently say that it delivers negative power as given by

$$p_{\mathrm{del}} = -vi. \tag{1.14}$$

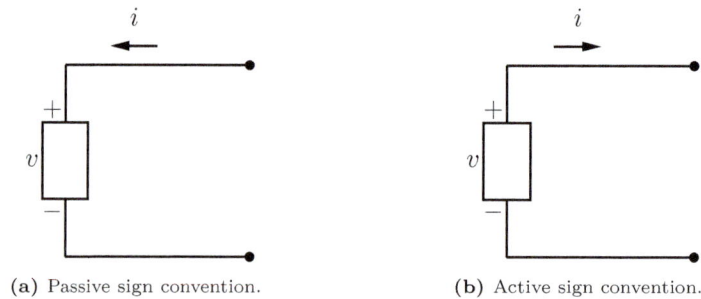

(a) Passive sign convention.   **(b)** Active sign convention.

Figure 1.14: Sign conventions.

It is critical to understand why the product vi yields consumed power when calculated under the passive sign convention. Let us think about the following argument. As we discussed, current direction indicates movement of positive charges. A positive charge dq entering the "+" node and leaving through the "−" node will *lose* energy equal to $dw = v\,dq$. This is the energy consumed by the load. Hence,

$$vi = \frac{dw}{dq}\frac{dq}{dt} = \frac{dw}{dt} = p. \tag{1.15}$$

> **Load (passive device)**: As depicted in Figure 1.14a, a load is a circuit element that absorbs (consumes) positive power. A load has positive current entering its positive (higher-potential) terminal, which is called passive sign convention. Examples of common loads include your fridge, lights, game console, washers, and dryers.

All loads, such as your phone, computer, and lights, need to consume energy in order to operate. This is similar to your body needing food to function. Our loads and bodies need a source of energy. The energy that is needed to be consumed has to be generated by something. For us, food is the source of energy, but for our devices, a battery, generator, or solar cell are often the sources of energy. Sources are also called active devices. These sources provide the energy required to make our devices function. For a device to be considered a **source** or active device, it must deliver (and not absorb) positive power. Based on the aforementioned arguments, this will be the case if the current enters the device from its negative voltage terminal. We call this notation **active sign convention**, and it is shown in Figure 1.14b. Consequently, the element in Figure 1.14b delivers positive power given by

$$p_{\text{del}} = vi. \tag{1.16}$$

Equivalently, we can say that it absorbs negative power equal to

$$p_{\text{abs}} = -vi. \tag{1.17}$$

> **Source (active element):** As depicted in Figure 1.14b, a source is a circuit element that delivers positive power. The element delivers the energy by current leaving the positive (higher-potential) terminal; this is called active sign convention. Batteries, generators, and solar cells are examples of common sources.

As you already know, power should always be balanced in a closed system. For us, this means that what gets delivered as electric power has to be absorbed as electric power in our circuits (we will assume our circuits are ideal closed systems). In the real world, some of the provided electric power gets dissipated as heat.

Example 1.1.5

For the circuit in Figure 1.15, find the current I. Which elements are responsible for delivering power?

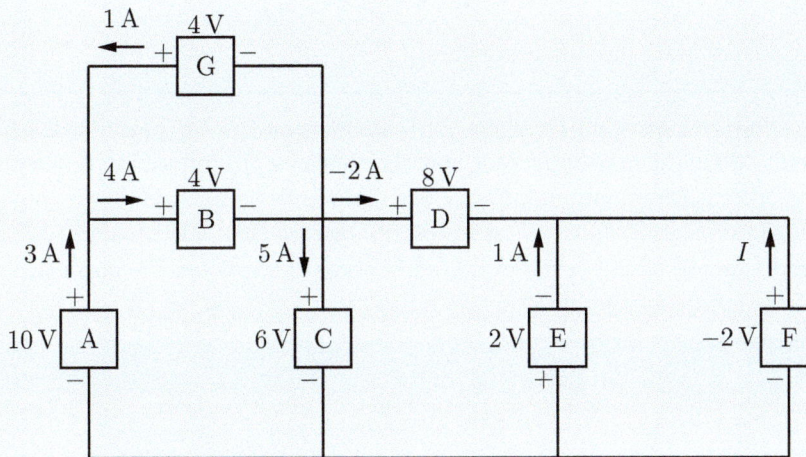

Figure 1.15: Example 1.1.5.

So far, conservation of energy is the only tool we have. We can use $\sum P_{\text{abs}} = 0\,\text{W}$, and since we already need to state which elements are responsible for delivering power, let us try to do this in a table.

Power Absorbed	
Symbol	(W)
P_A	-30
P_B	16
P_C	30
P_D	-16
P_E	2
P_F	$2I$
P_G	-4

For the circuit to be valid, we will need

$$P_A + P_B + P_C + P_D + P_E + P_F + P_G = 0\,\text{W}$$

Simplifying, we get

$$P_E + P_F + P_G = 2\,\text{W} + 2\,\text{V} \cdot I - 4\,\text{W} = 0\,\text{W}.$$

Solving for I, we find that

$$I = \frac{2\,\text{W}}{2\,\text{V}} = 1\,\text{A}.$$

From the table, we can see that elements A, D, and G are power-delivering elements (i.e., sources of power).

As we can see from Example 1.1.5, it does not matter how "complex" the circuit schematic may appear, or how the elements are connected, or even what type of elements are present. Power in a closed system must always be balanced (i.e., sum to zero). This is a very powerful feature that will allow us to analyze and design our circuits even in this early stage of the book. We make this more clear in the next example.

Example 1.1.6

For the circuit in Figure 1.16, elements A and B are loads. The power absorbed by each load is shown in the figure. Find the current I and the voltages V_1 and V_2.

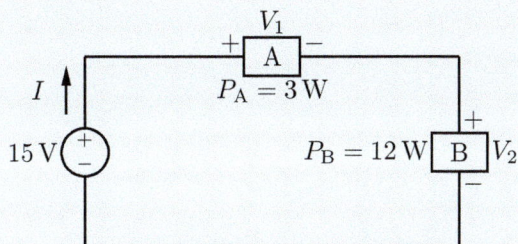

Figure 1.16: Example 1.1.6.

We can see that with what we have so far, the only way to solve this circuit is by using power. We know that the power generated by the 15 V source is absorbed by loads A and B,

$$P_{15\,\text{V}} = P_\text{A} + P_\text{B},$$

and the power generated can be found by

$$P_{15\,\text{V}} = VI = 15\,\text{V} \cdot I.$$

We can now find the total current I as

$$15\,\text{V} \cdot I = 3\,\text{W} + 12\,\text{W} \implies I = 1\,\text{A}.$$

Now that we know the total current and know the power absorbed by each load, we can find the voltages.

$$V_1 = \frac{P_\text{A}}{I} = \frac{3\,\text{W}}{1\,\text{A}} = 3\,\text{V}$$

and

$$V_2 = \frac{P_\text{B}}{I} = \frac{12\,\text{W}}{1\,\text{A}} = 12\,\text{V}.$$

Example 1.1.7

For the circuit in Figure 1.17, the elements A, B, C, D, and E are loads, and their absorbed powers are shown in the circuit. Find the voltage V.

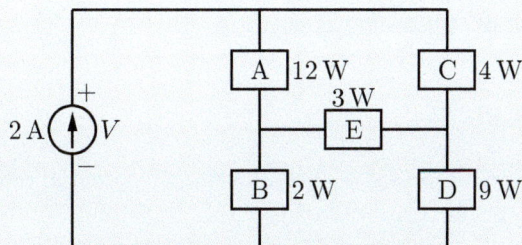

Figure 1.17: Example 1.1.7.

The only element that delivers power is the current source, so

$$P_{\text{source}} = P_A + P_B + P_C + P_D + P_E,$$

and

$$P_{\text{source}} = VI = V \cdot 2\,\text{A}.$$

Using both equations, we get

$$V \cdot 2\,\text{A} = 12\,\text{W} + 2\,\text{W} + 4\,\text{W} + 9\,\text{W} + 3\,\text{W} = 30\,\text{W}.$$

Then

$$V = \frac{30\,\text{W}}{2\,\text{A}} = 15\,\text{V}.$$

Exercise 1.1.2

For the circuit in Figure 1.18, the powers absorbed by each load are shown in the circuit. Find the voltage V.

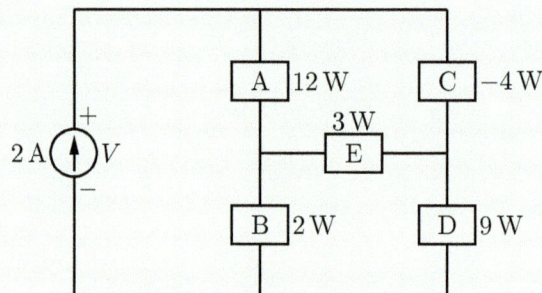

Figure 1.18: Exercise 1.1.2.

$V = 11\,\text{V}.$

Before we finish this section, it is critical to clarify that a device may act as a source in one circuit and a load in another. For example, a battery may deliver power in one circuit while it absorbs power in another. Consider what happens when you plug in an electronic device (such as your phone) to charge. Your battery has been working to deliver power all day, but when you plug it in to charge, it starts absorbing power.

1.2 Sources

By now we are convinced that we need a voltage (potential difference) to pump current through our devices. We achieve this by adding sources to our devices such as batteries or generators. Most circuits that are practically important include sources. These circuits are modeled and analyzed before they are physically built. In this book, we will learn how to analyze some of these circuits. In circuit analysis we often have two types of sources: independent and dependent sources. We will talk about independent sources first.

1.2.1 Independent Sources

As the name implies, an independent source provides the circuit with a steady voltage or current independent of its connection or other elements in the circuit. For example, a 5 V battery will provide the circuit with 5 V whether it is connected to one load or more.

> **Independent source**: A source whose value is unaffected by the circuit it is connected to.

An independent source can either be constant (DC) or vary with time. It can also be a current or a voltage source. Independent sources are typically represented by circles, as seen in Figure 1.19.

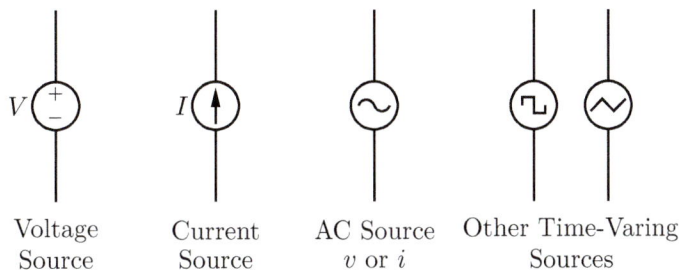

| Voltage Source | Current Source | AC Source v or i | Other Time-Varing Sources |

Figure 1.19: Independent source representation.

Since batteries are such an important part of practical circuits, they deserve a dedicated symbol. Figure 1.20 shows the symbol used for single- or multicell batteries. Note that the polarity is determined by the line length: the negative terminal is on the end with the shorter dash.

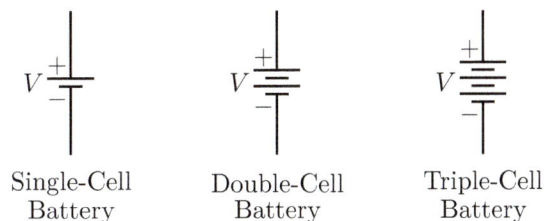

Single-Cell Battery   Double-Cell Battery   Triple-Cell Battery

Figure 1.20: Representation of DC voltage sources.

We typically model good-quality sources as ideal elements. There are two critical attributes of ideal sources. First, their value remains unchanged indefinitely. Second, they can deliver any amount of power needed by their loads. This means that an ideal voltage source can deliver any amount of current needed. Consequently, you can never determine the current through an ideal voltage source unless you know its load. Similarly, the voltage drop across an ideal current source is determined by the load connected to it. Practical sources have power limitations. For example, while a coin-cell battery can power a watch, it cannot turn on an automobile. We will see later that we can model this imperfection or limitation with other circuit elements. We will try to illustrate this concept in Example 1.2.1.

Example 1.2.1

Consider the circuits (a) and (b) in Figure 1.22. Both circuits contain the same independent source with a DC value of 5 V. We attach two different loads to our voltage source: load A, which draws 5 W, and load B, which draws 25 W.

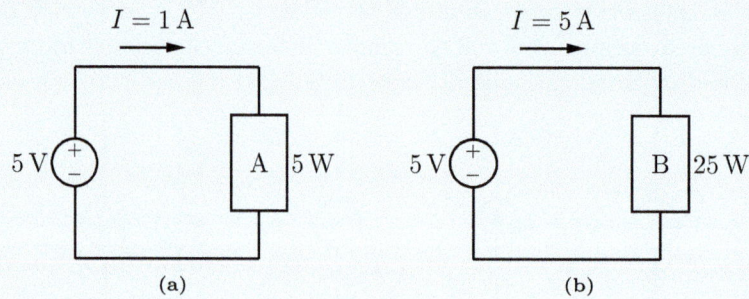

Figure 1.22: Example 1.2.1.

For DC circuits, we remember that

$$P = VI.$$

Thus, if the voltage cannot change, then the amount of current pumped out of the source has to change in order to fulfill the load's power requirement. Hence in the first case, the source has to deliver 1 A of current, while in the second, the current needs to be increased to 5 A. Ideal sources can do this. In real life, the maximum current (or more generally power) a source can provide will be limited by its type and/or fabrication technology.

Exercise 1.2.1

The circuits in Figure 1.24 each contain a 5 A independent current source. Circuit (a) contains a 5 W load and circuit (b) contains a 25 W load. Knowing that the current will not change, can you find the voltages V_1 and V_2 across each load?

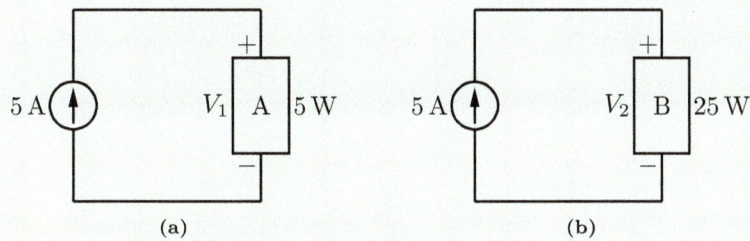

Figure 1.24: Current source example.

$V_1 = 1\,\text{V}$ and $V_2 = 5\,\text{V}$.

1.2.2 Dependent Sources

The concept of a dependent source can be a bit confusing at first because it appears to model unrealistic cases. At this point, it is easier for us to find real-life approximations for independent sources (e.g., batteries or solar cells) than for dependent ones. Soon though we will be able to change this. Let us look at the definition of a dependent source first.

> **Dependent source:** A source whose value is determined by other quantities in a circuit.

A good example to illustrate this definition is the induction circuit shown in Figure 1.25. Even though the two coiled wires (inductors) are not physically touching each other, a changing current (an alternating current (AC) in this case) applied to the left inductor results in an induced voltage across the neighboring inductor, which also results in a current flow that lights the bulb. Thus the current flow through the bulb (right circuit) depends on the excitation of the left circuit. This is an effect we could readily model with a dependent source on the right circuit. We will learn more about this later.

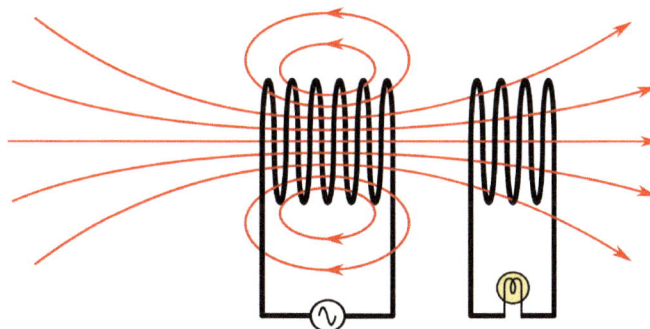

Figure 1.25: Induction circuit.

A dependent source's value can depend on a current or voltage in any part of the circuit. A dependent source is represented with a diamond as seen in Figure 1.26.

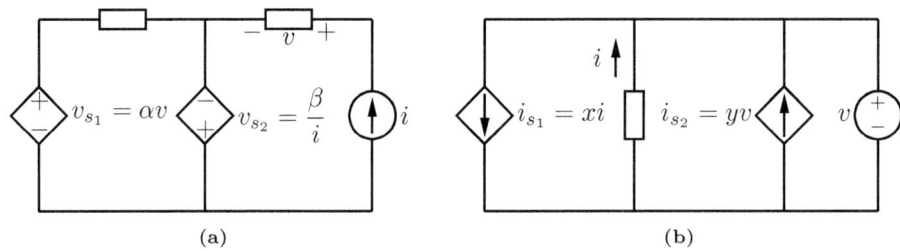

Figure 1.26: Dependent source representation.

In Figure 1.26, we have two circuits: (a) and (b). Circuit (a) contains the two possible types of dependent voltage sources. The dependent source v_{s_1} is a voltage-controlled (dependent) voltage source (VCVS). That means that the voltage supplied by the source v_{s_1} is dependent on another voltage in the circuit. In this case, v_{s_1} depends on the voltage across an element in the circuit. In order to find the voltage v_{s_1}, we have to find the voltage v and then multiply it by the constant α. The other dependent voltage source, v_{s_2}, is a current-controlled (dependent) voltage source (CCVS). The voltage supplied by the source v_{s_2} is inversely proportional to the current i supplied by the independent current source in the circuit. Note that α and β are just constants. They do, however, carry different units. The constant α has to be unitless in order for the resulting αv to have the unit of volts. The constant β has to have the unit of volt-amperes (VA) for β/i to be volts.

Circuit (b) contains the two possible types of dependent current sources. The first current source i_{s_1} is a current-controlled (dependent) current source (CCCS). We will have to find the value of the current i in the circuit to find the value of i_{s_1}. The second source i_{s_2} is a voltage-controlled (dependent) current source (VCCS), where the current is proportional to the voltage supplied by the independent voltage source. Again, the variables x and y are just constants but have different units depending on the situation. The constant x has to be unitless to keep the resulting source in amps. The unit of the constant y has to be A/V for the result to be in amps.

We would like to emphasize again that these sources are not off-the-shelf components as described so far. They are simplified models of practical devices. Let us see an example of dependent sources modeling amplifiers. The circuit in Figure 1.27 shows a simple example of a voltage amplification circuit.

In this figure, we have drawn two equivalent representations. The one on the left connects the two ground points to emphasize the common ground of the circuits. The one on the right does not have this connection. However, this does not affect the electrical performance. The voltage of the ground point is 0 V for both cases. It is important to emphasize, though, that $i_z = 0$. If this were not the case, there would need to be a voltage difference between the two ground points, which is obviously not the case. Before we leave this discussion, let us say that while most engineers would prefer the schematic on the left, both are common.

The source v_s could represent the voltage output of a sensor. This voltage is applied across a device that most engineers would call the input resistance of the amplifier. The voltage across this device, v_i, would be amplified by the constant

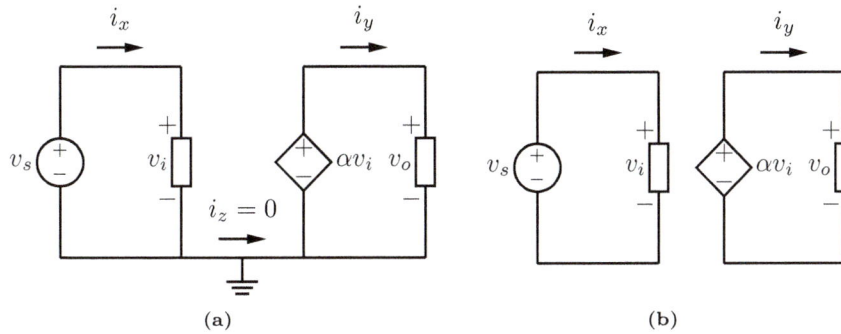

Figure 1.27: A simple example of a voltage amplification circuit. Both circuits are equivalent. The one on the left (a) emphasizes the common ground.

$\alpha > 1$ and produce the output voltage αv_i. This voltage would normally be seen across another device usually called (yes, you guessed it) the output resistance. You need not worry if you are not familiar with the term "resistance" yet, as we will discuss this later.

Please keep in mind that even though dependent sources are controlled by other voltages or currents in the circuit, they are still treated like ideal sources. The difference between dependent and independent sources is that we have to calculate the value of dependent sources by solving the circuit. Let us look at some examples to better understand this.

Example 1.2.2

Figure 1.29 contains two circuits: (a) and (b). Both circuits include the same current-controlled voltage source (CCVS). The only difference between the two circuits is the load value. What is the voltage value supplied by the dependent voltage source in each circuit?

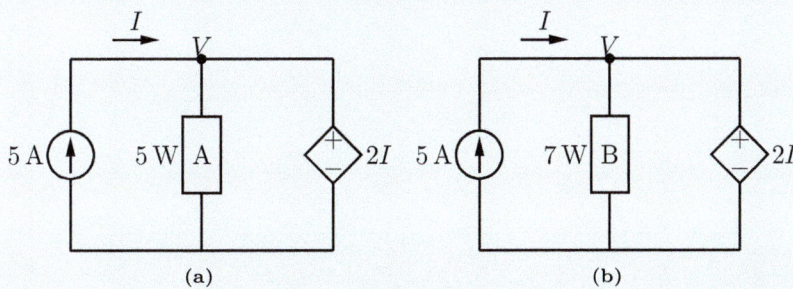

Figure 1.29: Example 1.2.2.

For circuit (a), we notice that the dependent source is controlled by the current I. The current I is clearly the same current generated by the 5 A current source. Therefore, the voltage supplied by the dependent voltage source is:

$$V = 2I = 2\,\mathrm{V\,A}^{-1} \cdot 5\,\mathrm{A} = 10\,\mathrm{V}.$$

Now, since the controlled source value does not depend on anything but the current I, and the current I is the same in both circuits, the value of the controlled source voltage in circuit (b) will be the same as the value in circuit (a). You may want to continue this example by finding the currents through the loads. While the voltage drops across them are the same, their currents will not be since they absorb different powers.

Now let us look at a circuit where the source value *is* affected by the change of load.

Example 1.2.3

For the circuit in Figure 1.31, find the value of the voltage supplied by the voltage-controlled voltage source (VCVS) in both circuits (a) and (b).

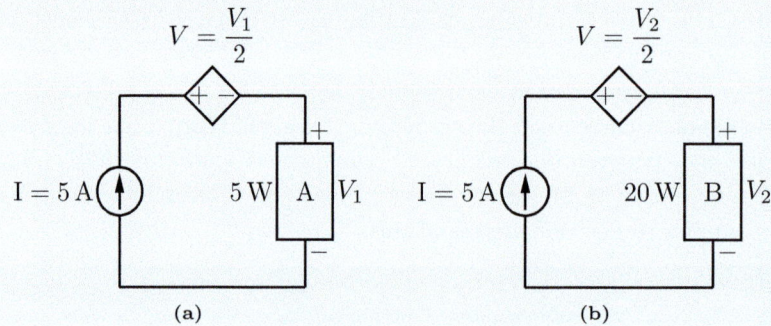

Figure 1.31: Example 1.2.3.

For circuit (a), we need to first find voltage V_1. Since we know the current that goes through load A, we can find the voltage

$$V_1 = \frac{P}{I} = \frac{5\,\text{W}}{5\,\text{A}} = 1\,\text{V}.$$

The voltage generated by the VCVS can now be found as

$$V = \frac{V_1}{2} = \frac{1\,\text{V}}{2} = 0.5\,\text{V}.$$

For circuit (b), the load is different, so the voltage across load B has to be found as
$$V_2 = \frac{P}{I} = \frac{20\,\text{W}}{5\,\text{A}} = 4\,\text{V}.$$

The voltage generated by the VCVS can now be found as

$$V = \frac{V_2}{2} = \frac{4\,\text{V}}{2} = 2\,\text{V}.$$

Example 1.2.4

The circuit in Figure 1.32 contains a switch that is initially connected to the 15 V voltage source. After one second (at $t = 1\,\text{s}$), the switch disconnects from the voltage source and connects to the 2 A current source. Find the value of the current I_2 before switching ($t < 1\,\text{s}$) and after switching ($t > 1\,\text{s}$).

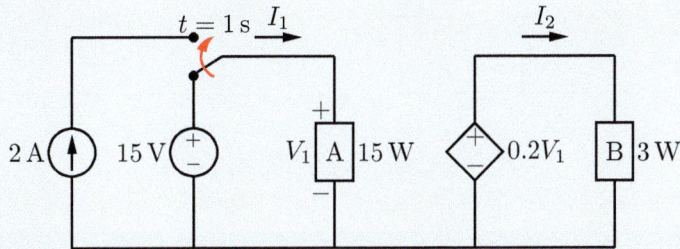

Figure 1.32: Example 1.2.4.

Before switching ($t < 1\,\text{s}$), we have the 15 V source connected to the circuit. We need to find V_1 first in order to find I_2. In this case, V_1 is equal to the voltage source—that is,

$$V_1 = 15\,\text{V}.$$

The current I_2 can now be calculated using the power absorbed by element B and the voltage across element B, which is equal to the voltage provided by the voltage-controlled voltage source (VCVS):

$$I_2 = \frac{P_\text{B}}{0.2V_1} = \frac{3\,\text{W}}{0.2(15\,\text{V})} = 1\,\text{A}.$$

After switching ($t > 1\,\text{s}$), the 15 V source is disconnected and the 2 A current source is connected. To find the new V_1, we use its current and power:

$$V_1 = \frac{P_\text{A}}{I_1} = \frac{15\,\text{W}}{2\,\text{A}} = 7.5\,\text{V}.$$

Then I_2 is calculated the same way as before but now with a new value for the dependent source's voltage:

$$I_2 = \frac{P_\text{B}}{0.2V_1} = \frac{3\,\text{W}}{0.2(7.5\,\text{A})} = 2\,\text{A}.$$

1.2.3 Open and Short Circuits

The terms "open circuit" (or "open") and "short circuit" (or "short") are among the most important ones in electrical engineering. An open circuit is an indication that current is not flowing. For example, the circuit in Figure 1.33a will not function. The circuit is open between the source and rest of the devices, which means current cannot flow from the source to the elements in this circuit.

In Figure 1.33b, the open circuit is just above element b. This open circuit will only restrict current from flowing through element b; current can still flow through the remaining components.

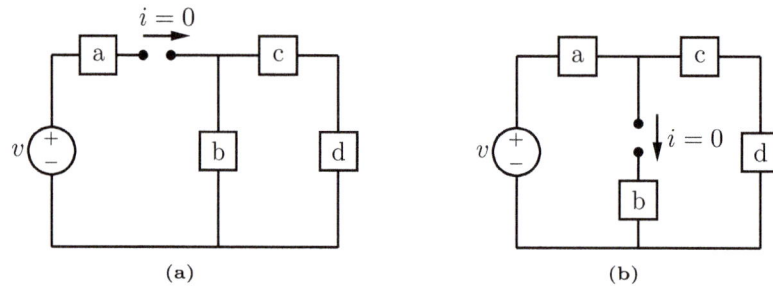

Figure 1.33: Open circuit examples.

Example 1.2.5

For the circuit in Figure 1.34, find the power absorbed by elements a, b, c, and d.

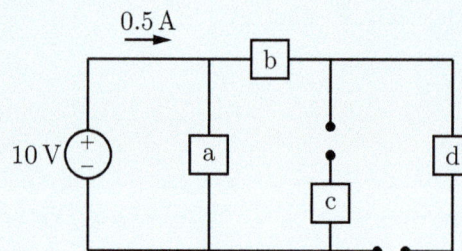

Figure 1.34: Example 1.2.5.

We have two open circuits that prevent current from flowing to elements b, c, and d, and so $P_b = P_c = P_d = 0\,\mathrm{W}$. The only element that has power delivered to it is element a, and its power absorbed is given by

$$P_a = 10\,\mathrm{V} \cdot 0.5\,\mathrm{A} = 5\,\mathrm{W}.$$

A short circuit can be used to bypass an element as in Figure 1.35a. By adding a wire across element a (shorting), we provide an easier path, or to be more accurate, a zero-resistance path for the current to flow through. So by shorting element a, no current will pass through this element. Instead, the current passes through the provided short circuit and, of course, the rest of the circuit elements.

In Figure 1.35b, we have shorted the voltage source. This should not be done in real life, as it may result in very high current, causing injury or damage to the equipment. Due to the short circuit, current will not flow through any of the circuit elements. Short or open circuits can be intentionally implemented (e.g., by opening or closing switches) or by accident. Because of this, it is often important to make sure that our designs are safe even under accidentally occurring opens or shorts.

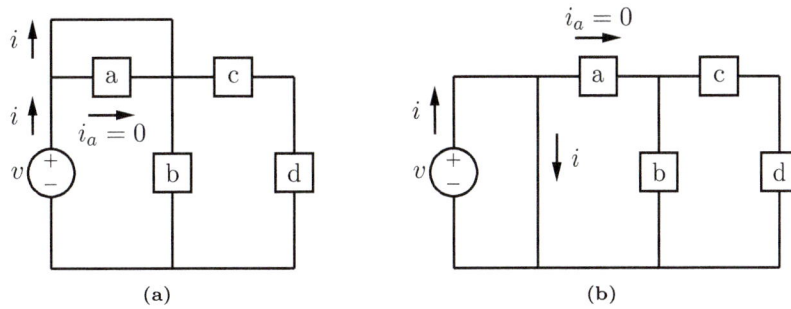

Figure 1.35: Short circuit connection examples.

Example 1.2.6

For the circuit in Figure 1.36, find the power absorbed by elements a, b, c, and d.

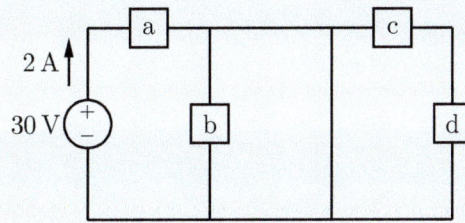

Figure 1.36: Example 1.2.6.

The short circuit will prevent current from reaching elements b, c, and d, and so $P_b = P_c = P_d = 0\,\text{W}$. The only element left to absorb power is element a, and its power is given by

$$P_a = 30\,\text{V} \cdot 2\,\text{A} = 60\,\text{W}.$$

Turning Off Sources

Another place that we use the terms "open circuit" and "short circuit" is when we turn sources off. When turned off, a voltage source is represented by a short circuit, as seen in Figure 1.37. Why do we represent a turned-off voltage source with a short circuit? Because a short circuit ensures a zero voltage drop across it, and zero volts means that the source is off. Notice in Figure 1.37 that element d is now completely bypassed because of the short circuit across it.

What about turning off current sources? Turning off a current source means having zero current through it. Of course, zero current requires an open circuit. So we replace a turned-off current source with an open circuit, as seen in Figure 1.38.

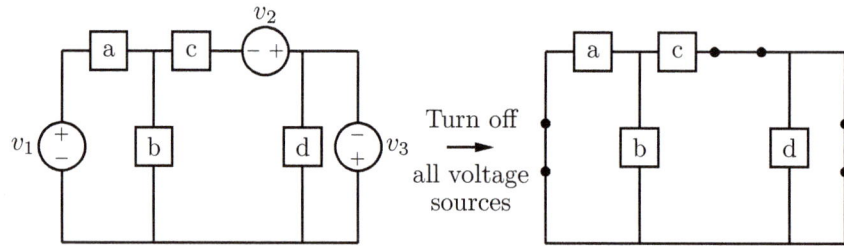

Figure 1.37: Turning off a voltage source is accomplished by replacing it with a short circuit.

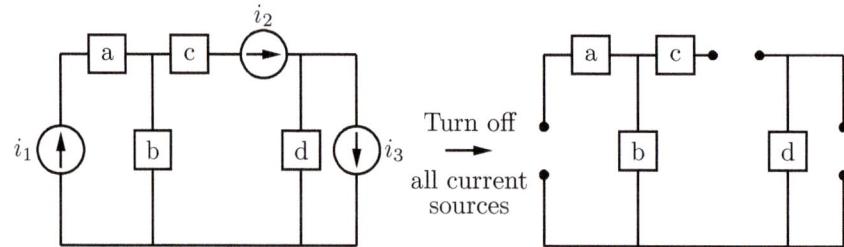

Figure 1.38: Turning off a current source is accomplished by replacing it with an open circuit.

Example 1.2.7

The power absorbed by each load is shown in the circuit in Figure 1.39.

1. Find I_1 if the 25 V source is turned off.

2. Find I_2 if the 100 V source is turned off.

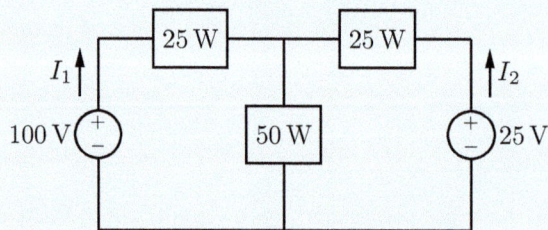

Figure 1.39: Example 1.2.7.

When we turn off the 25 V source, we can replace it with a short circuit, as seen in Figure 1.40.

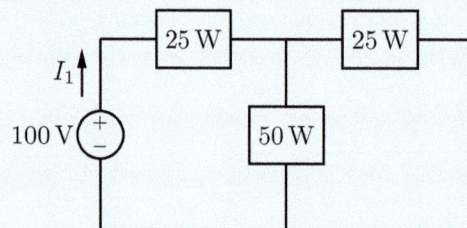

Figure 1.40: Example 1.2.7 with the 25 V source turned off.

We can now calculate the current I_1 using the total power absorbed by the circuit:

$$I_1 = \frac{P_T}{V} = \frac{25\,\text{W} + 25\,\text{W} + 50\,\text{W}}{100\,\text{V}} = 1\,\text{A}.$$

Similarly, turning off the 100 V source will result in the circuit shown in Figure 1.41.

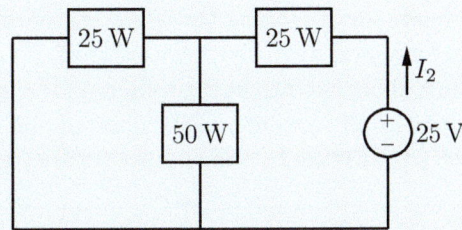

Figure 1.41: Example 1.2.7 with the 100 V source turned off.

The current I_2 can be calculated in the same way by using the total power absorbed by the circuit:

$$I_2 = \frac{P_T}{V} = \frac{25\,\text{W} + 25\,\text{W} + 50\,\text{W}}{25\,\text{V}} = 4\,\text{A}.$$

1.2.4 Circuit Connections

Now that we have identified multiple circuit elements (loads and sources), we will consider the two most common types of circuit connections: the series connection and the parallel connection. Other types of circuit connections, such as delta connections and wye connections, are discussed in more advanced texts.

A **series** connection is defined by the fact that the *same* current passes through all devices (Figure 1.42a). While the same current flows through all devices connected in series, their voltage drops may be different. On the other hand, circuit elements connected in **parallel** will necessarily have the *same* voltage value across their terminals, as seen in Figure 1.42b. Current through parallel-connected elements will not be, in general, the same. The appropriate connection depends on the design goals and application. For instance, current flow will cease through all components connected in series if an accidental open occurs. This is a disadvantage if the series-connected elements are

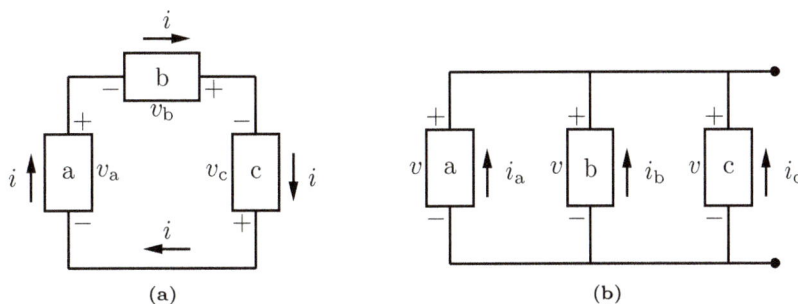

Figure 1.42: Circuit elements connected (a) in series, and (b) in parallel.

light bulbs. A burnt-out light bulb is an open circuit and it will stop the current from flowing, which will result in total darkness. On the other hand, it is the appropriate thing to do when a fuse is employed to protect a circuit. The role of a fuse is to stop current flow if the current exceeds a maximum limit in order to protect more valuable devices.

Analogous comments can be made for parallel connections. For example, light bulbs are often connected in parallel. If one of them is burnt out and results in an open circuit, the remaining bulbs will still function. Failing as an open circuit is typical for a light bulb. On the other hand, if a component fails as a short (i.e., results in a short circuit after it is damaged), a parallel connection may not be appropriate, as all elements will be shorted out. In practice we use a combination of such connections to achieve specific design goals.

In-series connection: Connected elements share the same current.
In-parallel connection: Connected elements share the same voltage drop.

Ideal voltage sources *cannot* be connected directly in parallel, but they can be combined in series. Why? Let's look at Example 1.2.8.

Example 1.2.8

Consider the parallel connection of sources in Figure 1.43a. By definition, an ideal voltage source cannot change value. Thus, by connecting them in parallel we have an obvious contradiction. On the other hand, the series connection in Figure 1.43b is totally valid and does not contradict the definition of an ideal source. We can even go one step further and lump both sources in Figure 1.43b into one voltage source with a 9 V value. More on this will follow.

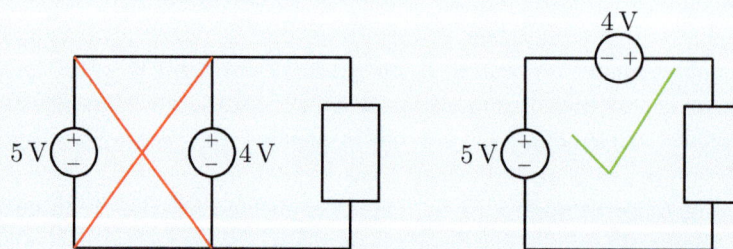

(a) Incorrect connection of voltage sources.   (b) Correct connection of voltage sources.

Figure 1.43: Example 1.2.8.

Ideal current sources *cannot* be connected in series, but they can be combined in parallel. Connecting the current sources in series would again contradict the definition of the ideal source: two different currents cannot occupy the same path.

Example 1.2.9

In this example, we can see from Figure 1.44a that current sources cannot be connected in series. Meanwhile, the connection in Figure 1.44b is valid and we can even go one step further by lumping the two sources together into one current source with a 9 A value.

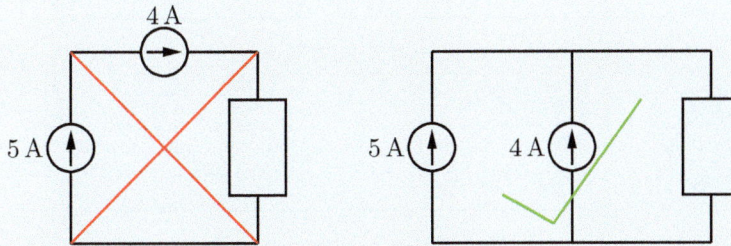

(a) Incorrect connection of current sources.　(b) Correct connection of current sources.

Figure 1.44: Example 1.2.9.

Maybe it is time for some exercises to solidify these concepts. In each exercise, state whether the source connection is valid and explain why.

Exercise 1.2.2

In Figure 1.45, we have a current-controlled current source in series with an independent source. Is this connection valid?

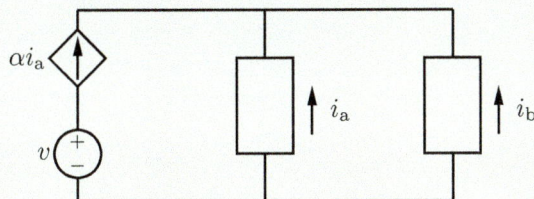

Figure 1.45: Exercise 1.2.2.

Connecting a current source in series with a voltage source breaks no rules. The only thing we have to keep in mind is that now the current through the voltage source will be αi_a.

Exercise 1.2.3

In Figure 1.46, we have a current-controlled voltage source in parallel with a voltage-controlled voltage source. Is this connection valid?

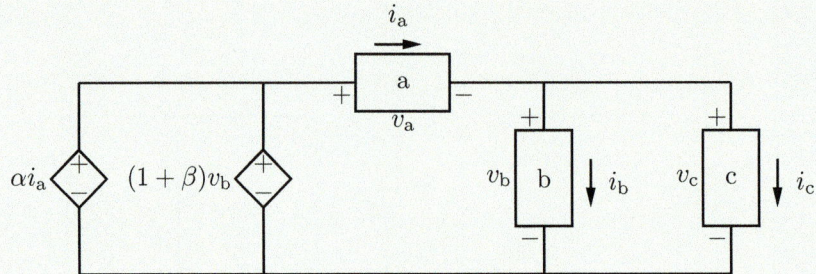

Figure 1.46: Exercise 1.2.3.

Connecting two voltage sources in parallel is invalid. The only way this connection can be valid is if

$$\alpha i_{\mathrm{a}} = (1 + \beta)v_{\mathrm{b}}.$$

1.3 Exercises

Subsection 1.1.1

1.1. For the charge flows given in Exersises a–d, find the current $i(t)$.

  a) $q(t) = 7t^2 + 5t + 2$ C.

  b) $q(t) = 2t^3 - 6t^2 + 12t$ C.

  c) $q(t) = 24\cos(3t + a)$ C.

  d) $q(t) = \dfrac{t^2 + t}{t^2 + 2}$ C.

1.2. For the currents given in Exersises a–c, find the charge flow $q(t)$ with $t_0 = 0$ s.

  a) $i(t) = 6t^2 + 4t + 2$ A.

  b) $i(t) = t^3 - 4t$ A.

  c) $i(t) = 4t\sin(t^2)$ A.

Subsection 1.1.2

1.3. For the circuit in Figure 1.47, find

  a) V_{ac}.   b) V_{cb}.   c) V_{bc}.   d) V_{ab}.

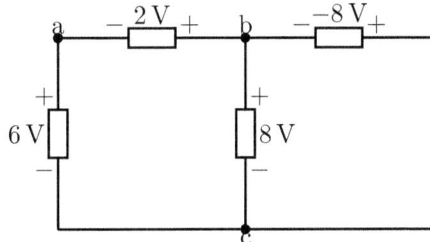

Figure 1.47

Subsection 1.1.3

1.4. For the circuit in Figure 1.48, identify which elements deliver power (i.e., act as sources) and which elements absorb power (i.e., act as loads).

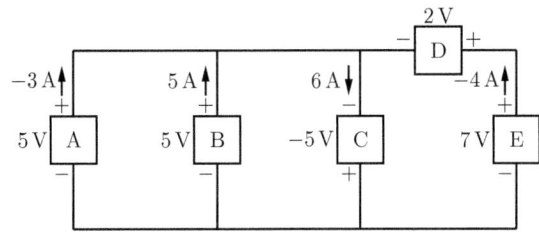

Figure 1.48

1.5. For the circuits in Figure 1.49a and Figure 1.49b, the power absorbed by each element is shown in the schematics. Determine whether the circuit is valid or not. (Hint: Use conservation of energy.)

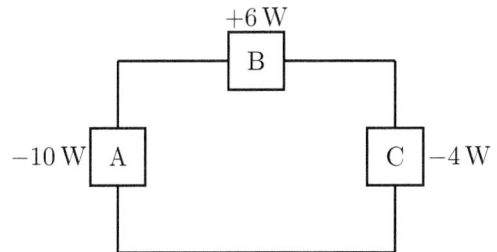

(a)

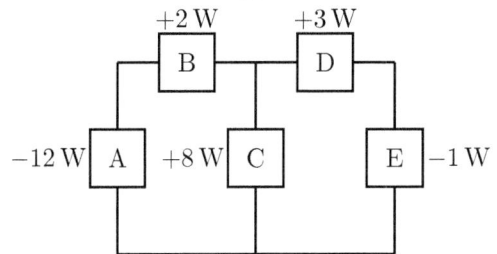

(b)

Figure 1.49

1.6. For the circuit in Figure 1.50, find the current I. (Hint: Use conservation of energy.)

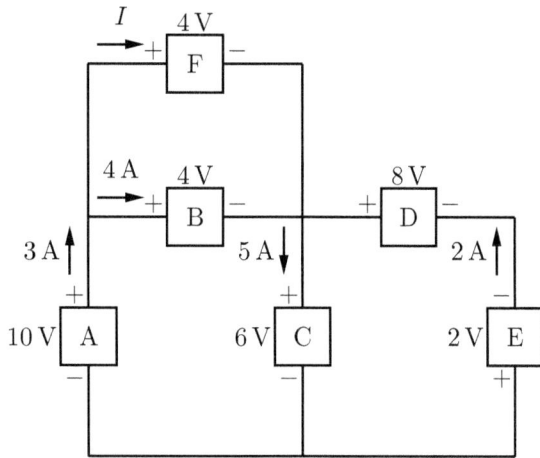

Figure 1.50

Section 1.2

1.7. For the circuit in Figure 1.51, elements A and B are loads. The power absorbed by each load is shown in the figure. Find the currents I, I_A, and I_B. (Hint: $V_{ab} = 15\,\text{V}$.)

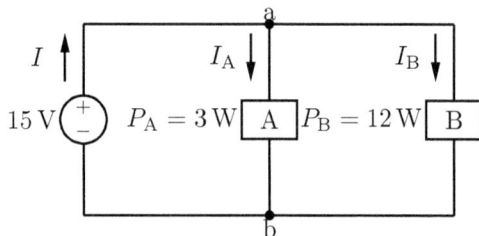

Figure 1.51

1.8. For each of the circuits in Figure 1.52, find the current I. Note that the power absorbed by each element is shown in each circuit.

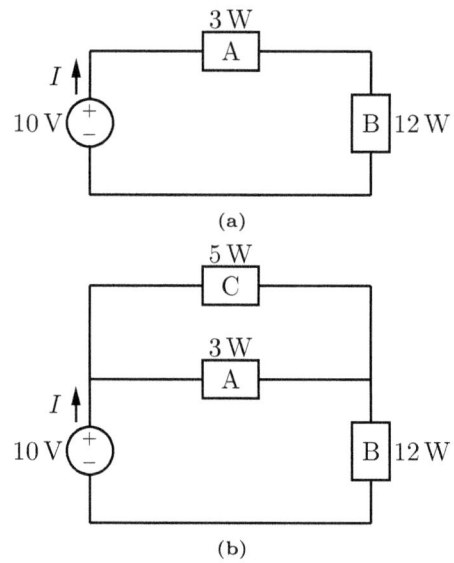

(a)

(b)

Figure 1.52

1.9. For each of the circuits in Figure 1.53, find the current I. Note that the power absorbed by each element is shown in each circuit.

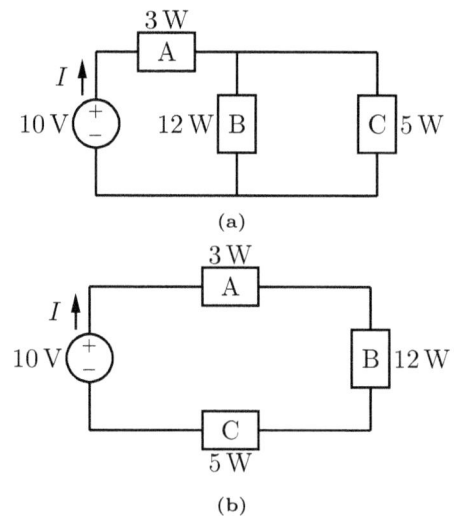

(a)

(b)

Figure 1.53

1.10. For the circuit in Figure 1.54, find the voltages V_{ab}, V_{bc}, and V_{ca}.

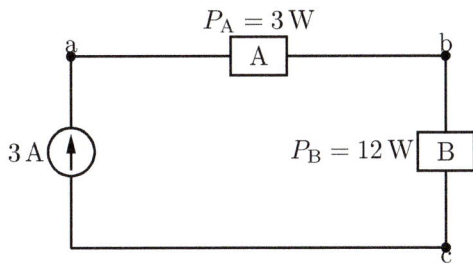

Figure 1.54

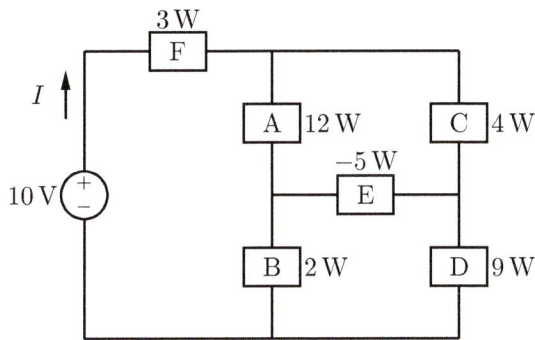

Figure 1.57

1.11. For the circuit in Figure 1.55, the power absorbed by each element is shown in the circuit. Find the current I.

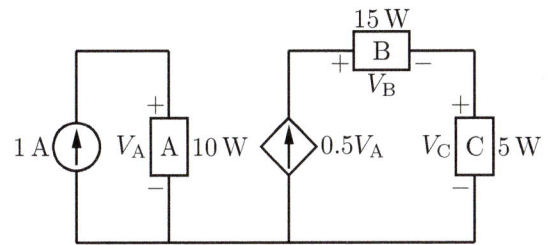

Figure 1.55

1.12. For the circuit in Figure 1.56, the power absorbed by each element is $P_A = 15\,\mathrm{W}$ and $P_B = 5\,\mathrm{W}$. Find the current I_2 before flipping the switch ($t < 3\,\mathrm{s}$) and after flipping it ($t > 3\,\mathrm{s}$).

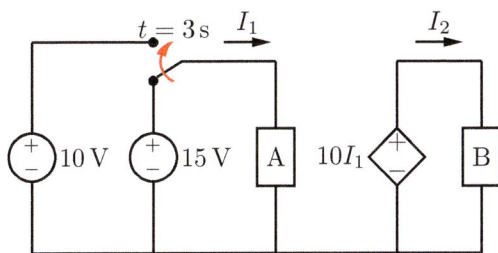

Figure 1.56

1.13. For the circuit in Figure 1.57, the power absorbed by each element is shown in the circuit. Find the voltages V_B and V_C.

1.14. The power absorbed by each load is shown in the circuit in Figure 1.58.

a) Find the current I if the 2 A source is turned off.

b) Find voltage V if the 50 V source is turned off.

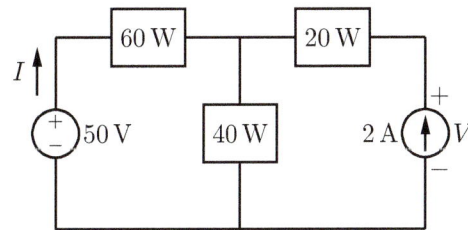

Figure 1.58

Chapter 2

Fundamental Laws

2.1 The Simplest Circuit We Can Build

A very simple circuit can be constructed by touching your tongue (a resistor) to a 9 V battery (DC voltage source) as seen in Figure 2.1a. Although it will not hurt your tongue, we do not recommend you try this, as the battery may not be clean. If the battery is fully charged, your tongue will connect the positive and negative terminals of the battery, closing the circuit and allowing current to flow. The amount of current flow will depend on the resistance of your tongue.

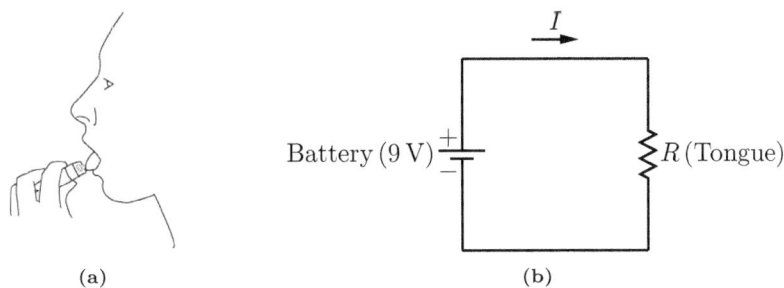

Figure 2.1: A simple circuit: (a) tongue touching a 9 V battery; (b) circuit model of tongue touching a 9 V battery.

The circuit created by your tongue (Figure 2.1b) is governed by one of the most fundamental laws of electrical engineering. This law is called Ohm's law. The law only applies to resistive circuits, but that does not make it any less powerful. The DC form of Ohm's law is written as

$$\text{Voltage} = \text{Current} \times \text{Resistance}$$

or

$$V = IR. \tag{2.1}$$

From Ohm's law, we find out that the voltage drop (V) across your tongue (the resistor) results in a current flow (I) that can be calculated as

$$I = \frac{V}{R}.$$ (2.2)

Many real-life situations can be modeled as resistive circuits. For example, a resistor can represent an electric toaster, stove, or heater. While practical circuits are often more complicated, you will find out soon that we can model many circuits with only a resistor and a source. And we are not limited to just DC circuits. Ohm's law can be written more generally as

$$\boxed{v(t) = i(t)R}$$ (2.3)

where lowercase letters v and i are used to represent voltages and currents that vary with time. We continue discussing Ohm's law (a bit more thoroughly) in section 2.3.

2.2   Resistivity and Resistance

Before we talk about resistive circuits or resistors, we need to talk about resistivity. From an electrical point of view, we often view elements as conductors, semiconductors, or insulators based on their resistivity. **Resistivity** is a natural material property that quantifies its ability to impede current flow through it. Some materials have higher resistivity than others. Conductors, as their name implies, have a much lower resistivity than insulators, which block practically all current flow. Table 2.1 contains the resistivity of some common materials.

Material	Resistivity ρ ($\Omega\,$m)
Copper	1.68×10^{-8}
Aluminum	2.65×10^{-8}
Silicon (pure)	2.5×10^{3}
Glass	1×10^{9} to 1×10^{13}
Rubber	1×10^{13} to 1×10^{15}

Table 2.1: Resistivity table of some materials.

To better understand resistance and resistivity, let us examine a typical cylindrical piece of wire as seen in Figure 2.2. The wire is made out of a uniform material with resistivity ρ ($\Omega\,$m), length l (m), and a cross-sectional area A (m^2). We recall from basic physics that resistance can be calculated using

$$R = \rho\frac{l}{A}.$$ (2.4)

Consequently, resistance depends on both the material of the wire and its geometry.

A resistor is a device designed to have a specific resistance. It might seem a little counterintuitive to want to make and use resistors since they reduce current flow. However, it helps to keep in mind that they are useful in limiting

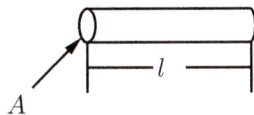

Figure 2.2: Cylindrical wire.

or regulating current flow for a given potential difference. Also, resistors turn electrical energy into heat, such as in hair dryers and toaster ovens.

It is worth noting that resistivity is solely a material property (for given environmental conditions) while resistance is a property of an object. For example, while the resistivity of copper at given conditions is constant, we can manufacture resistors with different resistances out of copper by varying their geometries. In practice, resistors come in many different shapes and form factors, as needed to satisfy diverse requirements such as power handling, cost, and manufacturing tolerance.

Resistance is measured in ohms (Ω). Most of the time, we use the symbol in Figure 2.3a to represent a resistor, but sometimes we use a rectangle as seen in Figure 2.3b.

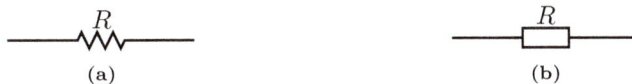

Figure 2.3: Common resistor symbols in circuit diagrams.

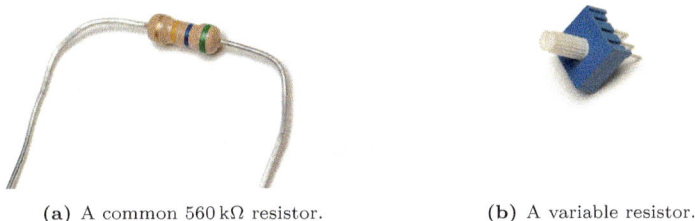

(a) A common 560 kΩ resistor.   **(b)** A variable resistor.

Figure 2.4: Pictures of widely used resistors.

2.2.1   Conductance

A material's ability to conduct electricity, the inverse of resistivity, is called conductivity. **Conductivity** is defined as $\sigma = 1/\rho$. Similarly, we can use the inverse of an object's resistance, called conductance, which is calculated as

$$G = \frac{1}{R} \tag{2.5}$$

The unit for conductance is siemens (S) and was mhos ($\mho$) in older books. We use the same symbol to represent conductors as we use for resistors, as seen in Figure 2.5.

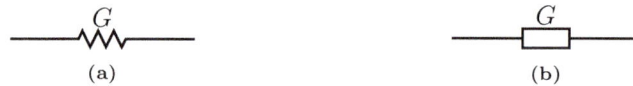

$$
\text{(a)} \qquad\qquad\qquad\qquad \text{(b)}
$$

Figure 2.5: Resistors can also be represented by their conductivity.

2.2.2   Hydraulic Analogy

We can improve our intuitive understanding of circuits through a hydraulic analogy. Take a look at the radiator hydraulic system shown in Figure 2.6a. We are trying to pump water from a water tank through the radiator. Water and water flow are analogous to charge and charge flow (current) respectively. The pump is analogous to a voltage source. The radiator in our small hydraulic system is analogous to a resistor. Why? The radiator consists of very small-diameter pipes that wind around. This increases resistivity to water flow. An equivalent electrical circuit diagram is shown in Figure 2.6b.

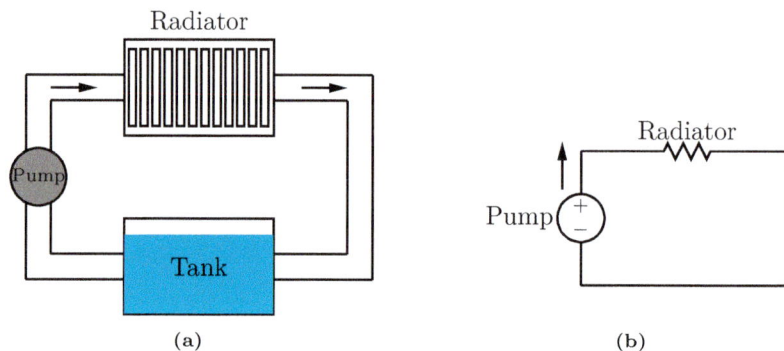

Figure 2.6: A simple hydraulic system and its equivalent circuit diagram.

2.3   Ohm's Law

If we connect the wire in Figure 2.2 to a varying voltage source $v(t)$ and measure the current $i(t)$ at different voltages, we will notice that (for an ideal resistor)

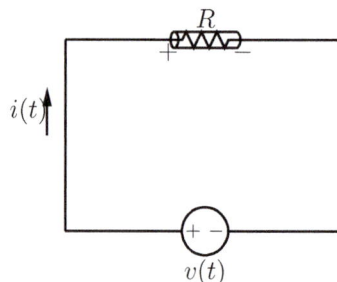

Figure 2.7: This circuit is used to define Ohm's law. Notice that the current enters the resistor from the positive (high) voltage node. This is called "passive sign convention" (section 1.1.3).

the voltage is linearly proportional to the current. Expressed mathematically, that is

$$v(t) = \alpha i(t). \tag{2.6}$$

The proportionality constant α is the resistance of the wire as defined by the German physicist Georg Simon Ohm (1789-1854)

$$\alpha = R = \frac{v(t)}{i(t)}. \tag{2.7}$$

It is *critical* to note that $v(t)$ and $i(t)$ *must* be in the passive sign convention for the previous two equations to be valid. In other words, the voltage drop $v(t)$ across a resistor R is given by the product $i(t)R$, provided that $i(t)$ *enters* the resistor from the positive voltage node.

Ohm's law states that the voltage drop across an ideal resistor is linearly proportional to its current by a constant called the resistance as seen in (2.8):

$$\boxed{v(t) = i(t)R.} \tag{2.8}$$

Ohm's law can be used for any (ideal) resistor, but we will find out later that not every material (or object) obeys Ohm's law.

Let us examine (2.8) more closely. What happens in Figure 2.7 if the resistance R of the wire is very large—so large that it approaches infinity? Also, what happens if the resistance is zero? For the first scenario, we rearrange Ohm's law as

$$i(t) = \frac{v(t)}{R}. \tag{2.9}$$

If the resistance is approaching infinity, then the ratio $\frac{v(t)}{R}$ will go to zero, which means the current $i(t)$ will be zero. Since no current can pass, this represents an open circuit. For the second scenario, when the resistance is zero, the voltage will be zero as well and the current will go to infinity. In other words, we will have a short circuit. Figure 2.8 shows these equivalent circuit representations.

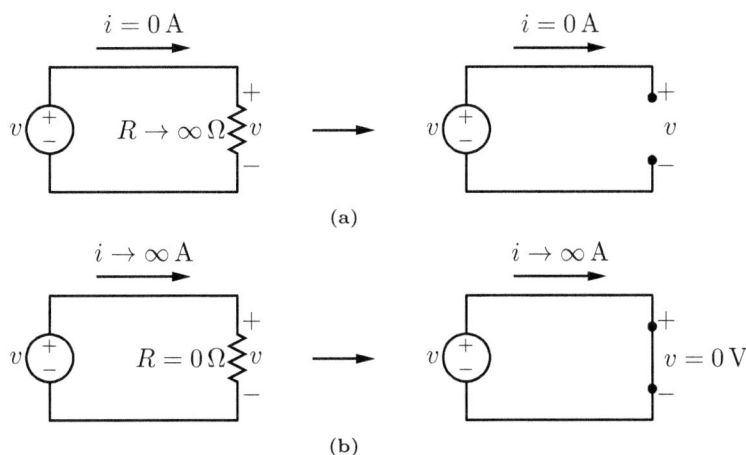

Figure 2.8: Open and short circuit equivalent representations of (a) very large resistance and (b) very low resistance.

Example 2.3.1

Consider the circuit in Figure 2.9. Find the voltage V and current I for $R = 0.5\,\Omega, 5\,\Omega$, and $\infty\,\Omega$.

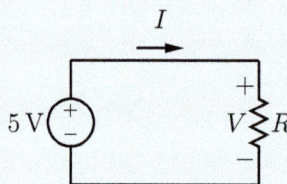

Figure 2.9: Example 2.3.1.

We know that for $R = 0.5\,\Omega$, $5\,\Omega$, and $\infty\,\Omega$, $V = 5\,\text{V}$. The current, on the other hand, will have to be calculated using (2.9).

1. $R = 0.5\,\Omega \rightarrow I = \dfrac{5\,\text{V}}{0.5\,\Omega} = 10\,\text{A}$

2. $R = 5\,\Omega \rightarrow I = \dfrac{5\,\text{V}}{5\,\Omega} = 1\,\text{A}$

3. $R = \infty\,\Omega \rightarrow I = \dfrac{5\,\text{V}}{\infty\,\Omega} = 0\,\text{A}$

Ohm's law can also be written, in general, in terms of conductance as

$$i(t) = Gv(t), \tag{2.10}$$

or as

$$v(t) = \frac{i(t)}{G}. \tag{2.11}$$

Since conductance is the inverse of resistance, the open and short circuit representations of conductance are the opposite from those of resistance, as seen in Figure 2.10.

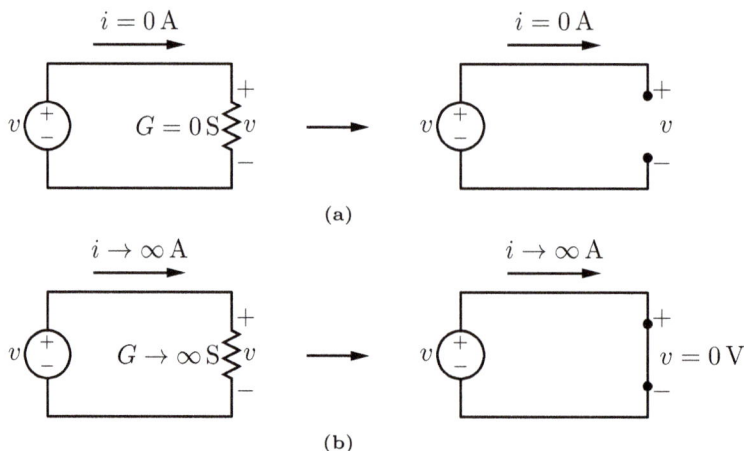

(a)

(b)

Figure 2.10: Open and short circuit equivalent representations using conductance.

Example 2.3.2

Consider the circuit in Figure 2.11. Find the current I for $G = 0.5\,\text{S}$, $5\,\text{S}$, and $0\,\text{S}$. Also find the resistance in each case.

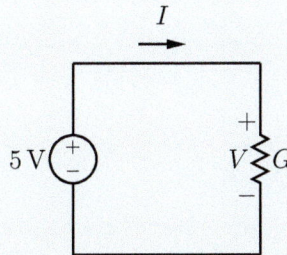

Figure 2.11: Example 2.3.2.

We will calculate the current using (2.10) and $V = 5\,\text{V}$.

1. $G = 0.5\,\text{S} \rightarrow I = (5\,\text{V})(0.5\,\text{S}) = 2.5\,\text{A}$; $R = \dfrac{1}{G} = \dfrac{1}{0.5\,\text{S}} = 2\,\Omega$

2. $G = 5\,\text{S} \rightarrow I = (5\,\text{V})(5\,\text{S}) = 25\,\text{A}$; $R = \dfrac{1}{G} = \dfrac{1}{5\,\text{S}} = 0.2\,\Omega$

3. $G = 0\,\text{S} \rightarrow I = (5\,\text{V})(0\,\text{S}) = 0\,\text{A}$; $R = \dfrac{1}{G} = \dfrac{1}{0\,\text{S}} = \infty\,\Omega$

2.4 Resistive Power

We recall from section 1.1.3 that

$$p = vi \tag{2.12}$$

results in consumed (absorbed) power if v and i are defined in the passive sign convention. Now, if we use Ohm's law to substitute for the current in (2.12), in terms of voltage and resistance we get

$$p = vi = v\left(\frac{v}{R}\right), \tag{2.13}$$

which leads to

$$\boxed{p = \frac{v^2}{R}.} \tag{2.14}$$

Or we can find the power in terms of current by using Ohm's law to substitute for the voltage in (2.12) in terms of current and resistance to get

$$p = vi = (iR)i, \tag{2.15}$$

which leads to

$$\boxed{p = i^2 R.} \tag{2.16}$$

The term "resistive losses" is often used to describe (2.16). From what we have learned so far, we know that no matter how conductive the material is, we will always have some resistivity. In practice, even a highly conductive wire has a nonzero resistance, which means that some of the connected power will be dissipated as heat in the wire itself. Let us look at the simple circuit in Figure 2.12.

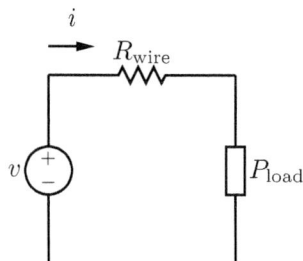

Figure 2.12: Simple circuit with a load and a nonideal wire.

For this circuit, the total power generated by the source can be calculated by

$$P_{\text{del}} = i^2 R_{\text{wire}} + P_{\text{load}}.$$

Now, if we were in a perfect world, the resistance of the wire would be zero and the power delivered by the source would be the same as the power required by the load. However, in practice, the nonzero wire resistance means that the source will have to generate more power than will be absorbed by the load.

We can also write power in terms of voltage and conductance—that is,

$$p = vi = v(Gv), \tag{2.17}$$

which leads to

$$\boxed{p = v^2 G.} \tag{2.18}$$

Or we can find the power in terms of current and conductance

$$p = vi = \left(\frac{i}{G}\right)i, \tag{2.19}$$

which leads to

$$\boxed{p = \frac{i^2}{G}.} \tag{2.20}$$

2.5   Kirchhoff's Laws

Although Ohm's law is one of the most important and basic concepts in electrical engineering, Kirchhoff's laws deserve (at least) equal attention. Why are Kirchhoff's laws so important? Because they apply to *any* and *every* circuit.[1]

---

[1] Kirchhoff's laws are derived from more fundamental equations (Maxwell's equations) for DC circuits. As a result, strictly speaking, Kirchhoff's laws are only valid in DC circuits. In time-varying circuits, they are only approximately true. However, this approximation is very accurate until we reach very high-frequency circuits. You can learn more about these in more advanced courses or books. In this book, we will assume that Kirchhoff's laws are always valid.

It does not matter what elements are connected to the circuit or what state it is in; Kirchhoff's laws are always valid.

In order to talk about Kirchhoff's laws, we need to first set up our environment. We talked a little bit about circuit connections in section 1.2.4, and now we need to talk a little bit more about nodes and connections before we can dive into Kirchhoff's laws.

2.5.1 Is Every Node Essential?

What is a node? A node is a junction or point where two or more circuit elements are connected or joined. There is also a special type of node we call an essential node. An essential node is a node where three or more elements are joined together. Why is it special? When we have three or more elements connected to a node, the current starts to split. Think of this like connecting pipes. Connecting two water pipes together end to end does not divert the water; it just continues along the same path. But if, for example, we add a T connector (a three-way split), then the flow of water can split. Figure 2.13 illustrates this concept. In the figure, the nonessential nodes are drawn with green circles and the essential nodes are in red. Notice how the current splits at the essential nodes.

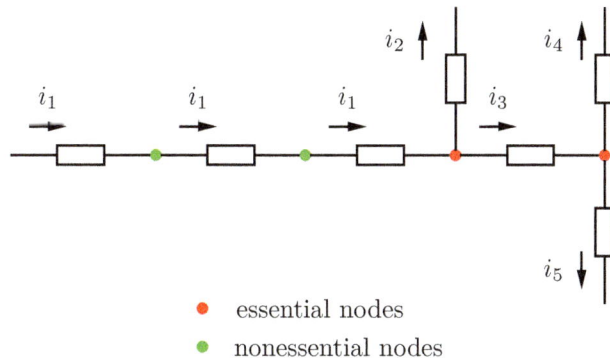

Figure 2.13: Essential versus nonessential nodes.

It is important to emphasize here that the wire length in a circuit schematic is not electrically important. A wire does not separate nodes; in fact, it combines them into one equivalent electrical node. Nodes connected directly by wire have the same potential and, therefore, can be, combined into a single node. Let us take a look at Example 2.5.1 to better understand this.

Example 2.5.1

The circuit in Figure 2.14 has four circuit elements and six points that can be labeled as nodes. We will examine each node and decide whether it is essential or not.

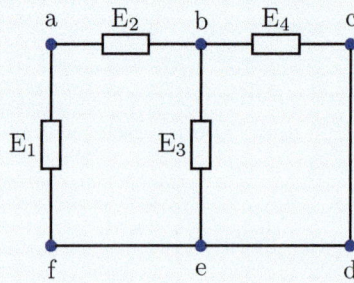

Figure 2.14: Example 2.5.1.

1. Node a connects elements E_1 and E_2. The current is the same through both elements, so node a is a nonessential node.

2. We can clearly see that node b connects three elements: E_2, E_3, and E_4. The current does split, which makes node b an essential node.

3. Node c is directly connected to node d with a wire, which makes node c the same as node d. We can say the same about node e and node f. Thus, we can combine nodes c, d, e, and f into a single node. This combined node connects three elements—E_1, E_3, and E_4—so it is an essential node (see Figure 2.15).

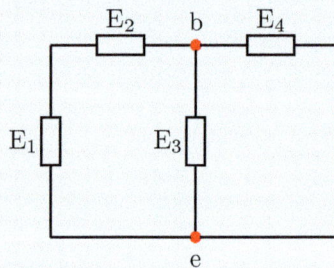

Figure 2.15: Example 2.5.1 with essential nodes only.

Now that we understand nodes, let us discuss Kirchhoff's current law (KCL) first, and then Kirchhoff's voltage law (KVL).

2.6   Kirchhoff's Current Law (KCL)

KCL is applied to nodes. While Kirchhoff's Current Law (KCL) can be applied to both essential and nonessential nodes, it is most beneficial when applied to essential nodes. Sometimes it is easier to look at KCL under the prism of our water-current fluid analogy, as seen in Figure 2.16a. The figure shows a section of a pipe with two junctions (essential nodes). The main pipe branches into three pipes at one junction and then the three pipes join again at another junction. If all pipes are the same and there is no leakage, the water will split evenly through all available paths. We also expect that the outward flow is equal to the inward flow.

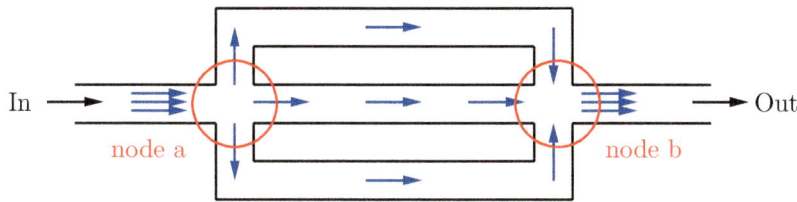

(a) KCL in fluid form.

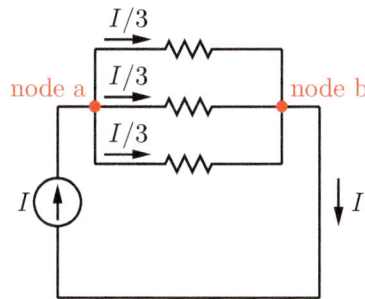

(b) The circuit representation of Figure 2.16a.

Figure 2.16: Analogous representations of KCL.

We can draw an equivalent electrical circuit to the system in Figure 2.16a, as shown in Figure 2.16b. To do so, we use a current source to represent the total amount of water flow entering the system. Similar to the water flow, some of the current I flows through each branch. We know that the pipes have some resistance to the flow of water, so we represent each branch with a resistor. We claimed earlier that the pipe in each branch is identical, so in the circuit representation, we use equal resistors. This ensures that, like the water, the current will split evenly between the branches. Mathematically, we represent this with the following equation for the current at node a:

$$I = \frac{I}{3} + \frac{I}{3} + \frac{I}{3}. \tag{2.21}$$

Equation (2.21) is essentially KCL for node a. There are many different ways one can state KCL, but the basic message remains unchanged. This basic message is based on a fundamental physics principle stating that charge can neither be created nor destroyed (charge conservation principle). Consequently, if a current carries several coulombs per second toward a node, some other current must carry the same rate of charge away from that node. If this were not true, that node would have to be a source or sink of charge, and this is not possible. Consequently, we can state KCL as follows:

KCL: The sum of all currents leaving a node is equal to zero.

In math form, KCL can be stated as

$$\sum_{n=1}^{N} i_n(t) = 0 \, \text{A}. \tag{2.22}$$

In the above statement, every current leaving the node is taken as positive, while every current entering the node is considered negative. Alternatively, we could state KCL as: the sum of all currents entering a node is equal to zero. In that case, every current leaving the node would be taken as negative, while every current entering the node would be considered positive. Obviously, the two statements are equivalent. Yet another way to state KCL would be: the sum of all currents entering a node should equal the sum of all currents leaving that node.

We will choose the first of the three possible expressions to remain consistent in our discussion. Therefore, looking at Figure 2.17, the sum of all currents leaving the node is zero. Of course, some of the currents will have a negative value in order for the sum to be zero. This is analogous to building an accounting system that leads to a budget: outward currents are positive while inward currents are taken as negative. Let us look at a numerical example.

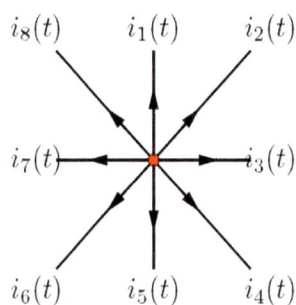

Figure 2.17: Graphical representation of KCL.

Example 2.6.1

In Figure 2.18 we have a single node with four branches. We have the magnitude and direction for three out of the four currents and we need to find the value of the fourth current. In simple cases, it may be tempting to choose any of three possible KCL equations. However, it is important to train our minds to always apply KCL consistently in the form shown by (2.22). This will be very valuable in more complex situations.

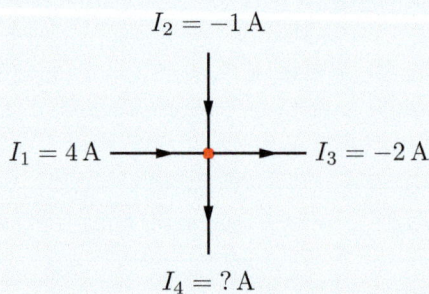

Figure 2.18: Example 2.6.1.

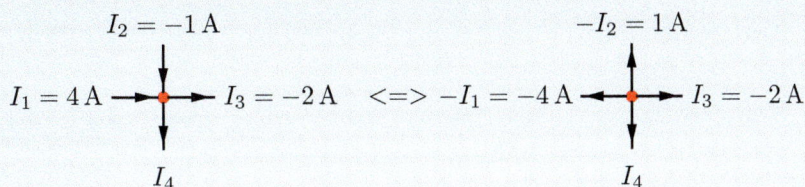

Figure 2.19: Redrawing Example 2.6.1 in a form convenient for applying KCL per (2.22).

Summing the current leaving the node (Figure 2.19), we get

$$-I_1 - I_2 + I_3 + I_4 = 0\,\mathrm{A}$$
$$-4\,\mathrm{A} + 1\,\mathrm{A} - 2\,\mathrm{A} + I_4 = 0\,\mathrm{A}$$
$$I_4 = 5\,\mathrm{A}.$$

As you become more experienced, you will not necessarily have to draw Figure 2.19. Simply accounting outward currents as positive and inward currents as negative is effective and sufficient. Try to write KCL here without Figure 2.19 to practice.

Example 2.6.2

Find the value of the current I in Figure 2.20.

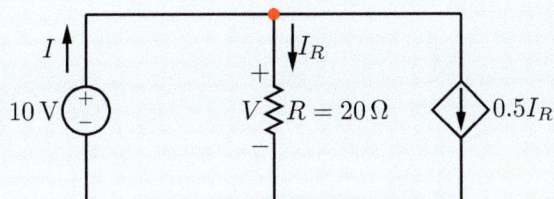

Figure 2.20: Example 2.6.2.

KCL at our essential node results in

$$-I + I_R + 0.5I_R = 0\,\mathrm{A},$$

or

$$I = I_R + 0.5I_R = 1.5I_R.$$

We can find the current I_R using Ohm's law:

$$I_R = \frac{V}{R} = \frac{10\,\mathrm{V}}{20\,\Omega} = 0.5\,\mathrm{A}.$$

Then

$$I = 1.5(0.5\,\mathrm{A}) = 0.75\,\mathrm{A}.$$

Example 2.6.3

Find the value of the current I in Figure 2.21.

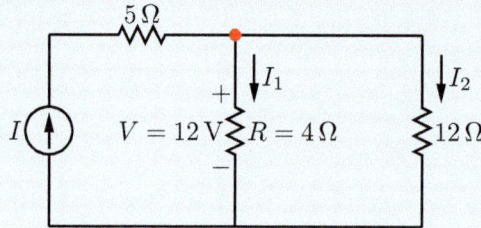

Figure 2.21: Example 2.6.3.

We can use Ohm's law to find the currents I_1 and I_2, and then we can use KCL at the essential node shown in red to find the current I. We write

$$I_1 = \frac{V}{R} = \frac{12\,\text{V}}{4\,\Omega} = 3\,\text{A}$$

and

$$I_2 = \frac{12\,\text{V}}{12\,\Omega} = 1\,\text{A}.$$

KCL applied to the essential node gives

$$-I + I_1 + I_2 = 0\,\text{A},$$

$$I = I_1 + I_2 = 4\,\text{A}.$$

Example 2.6.3 can also be solved by simplifying the circuit diagram. Specifically, the $4\,\Omega$ and $12\,\Omega$ resistors are in parallel and can be combined into one equivalent resistor, $R_{eq} = 3\,\Omega$, and then the current I can be easily calculated as

$$I = \frac{12\,\text{V}}{3\,\Omega} = 4\,\text{A}.$$

So now let us look at how to combine resistors in parallel by applying KCL.

2.6.1   Resistors in Parallel

In section 1.2.4, we discussed different types of connections and saw how we can combine sources connected in series or parallel. We will examine these connections again for resistors, starting with parallel connections.

Resistors in parallel share the *same* nodes and necessarily have the *same* voltage drop across them. Checking if the voltage drop across two or more resistors is the same is the only safe way to determine whether they are connected in parallel or not. Current, on the other hand, will be divided among the resistors in a manner that we will discover later. Intuitively speaking, from what we know so far about current and resistance, current always picks the least resistive path to go through. As a result, small resistors will have more current go through them than larger ones. We will soon prove this mathematically.

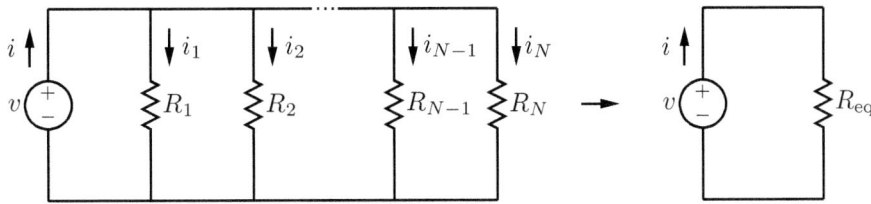

Figure 2.22: Resistors connected in parallel.

Figure 2.22 shows N resistors connected in parallel. How do we know they are in parallel? They all share the same top node and the same bottom node. Thus the voltage across all resistors is the same. We denote this by v. Let us navigate through this circuit step by step. Say we want to find the current i. We can use KCL to sum the currents:

$$-i + i_1 + i_2 + \cdots + i_{N-1} + i_N = 0$$
$$i = i_1 + i_2 + \cdots + i_{N-1} + i_N. \tag{2.23}$$

Now we can use Ohm's law since we know the voltage across the resistors:

$$
\begin{aligned}
i &= \quad i_1 + \quad i_2 + \cdots + \quad i_{N-1} + \quad i_N \\
i &= \quad \frac{v}{R_1} + \frac{v}{R_2} + \cdots + \frac{v}{R_{N-1}} + \frac{v}{R_N} \\
i &= v \left(\frac{1}{R_1} + \frac{1}{R_2} + \cdots + \frac{1}{R_{N-1}} + \frac{1}{R_N} \right).
\end{aligned}
\tag{2.24}
$$

$$\tag{2.25}$$

Rearranging this yields

$$\frac{i}{v} = \frac{1}{R_{\text{eq}}} = \sum_{n=1}^{N} \frac{1}{R_n}, \tag{2.26}$$

or

$$\boxed{R_{\text{eq}} = \frac{1}{\sum_{n=1}^{N} \frac{1}{R_n}}.} \tag{2.27}$$

Consequently, the **equivalent resistor** is defined as a resistor that can replace the given circuit such that the $v - i$ relationship remains unchanged. This is a powerful and general concept that, as we will see later, applies to all circuits, not just the ones that contain resistors connected in parallel.

There is a way to make (2.27) more visually appealing. Remember conductance G? Conductance is the reciprocal of the resistance, which means (2.25) can be rewritten as

$$
\begin{aligned}
i &= v \left(G_1 + G_2 + \cdots + G_{N-1} + G_N \right) \\
i &= v G_{\text{eq}}.
\end{aligned}
\tag{2.28}
$$

As a result, for parallel combination the conductances can simply be summed together:

$$G_{\text{eq}} = \sum_{n=1}^{N} G_n. \tag{2.29}$$

Some Interesting Facts and Special Cases

As a result of summing the reciprocal of resistance, the resulting equivalent resistance will always be smaller than the smallest resistor.

Example 2.6.4

If $R_1 = 6\,\Omega$, $R_2 = 3\,\Omega$, and $R_3 = 2\,\Omega$ are connected in parallel, find the equivalent resistance. From (2.27), we get

$$R_{\text{eq}} = \frac{1}{\frac{1}{6\,\Omega} + \frac{1}{3\,\Omega} + \frac{1}{2\,\Omega}} = 1\,\Omega.$$

The equivalent resistance is smaller than the smallest resistor.

Another special case is to have identical resistors connected in parallel. The equivalent resistance is their common value divided by the number of resistors. In math form for N identical resistors with a value R,

$$R_{\text{eq}} = \frac{R}{N}.$$

Example 2.6.5

If $R_1 = R_2 = R_3 = R_4 = 8\,\Omega$ are connected in parallel, find the equivalent resistance. From the above discussion, we have

$$R_{\text{eq}} = \frac{8\,\Omega}{4} = 2\,\Omega.$$

Also notice that equivalent resistance is smaller than each of the resistors.

If we have only two resistors in parallel, the equivalent resistor can be calculated using

$$R_{\text{eq}} = \frac{R_1 R_2}{R_1 + R_2}. \tag{2.30}$$

If the two parallel resistors are equal, $R_1 = R_2 = R$, then

$$R_{\text{eq}} = \frac{R}{2}. \tag{2.31}$$

Also notice again that the resulting R_{eq} is smaller than each of the resistors.

Example 2.6.6

Find the power delivered by the voltage source in Figure 2.23.

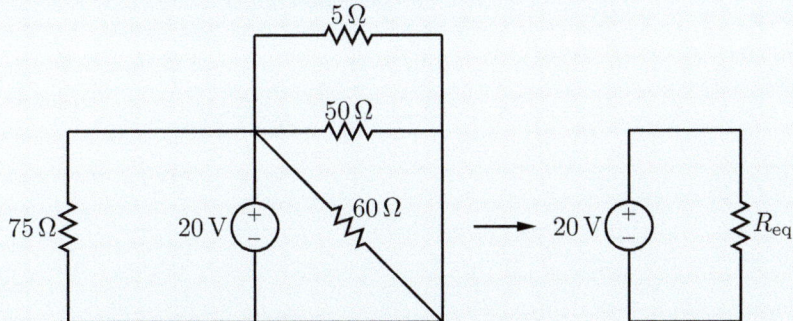

Figure 2.23: Example 2.6.6.

We need to find the equivalent resistance using (2.26):

$$\frac{1}{R_{eq}} = \frac{1}{75\,\Omega} + \frac{1}{60\,\Omega} + \frac{1}{50\,\Omega} + \frac{1}{5\,\Omega} = \frac{1}{4\,\Omega}.$$

Thus,

$$R_{eq} = 4\,\Omega.$$

The power delivered by the voltage source is

$$P = \frac{V^2}{R_{eq}} = \frac{(20\,\text{V})^2}{4\,\Omega} = 100\,\text{W}.$$

2.6.2 Current Division

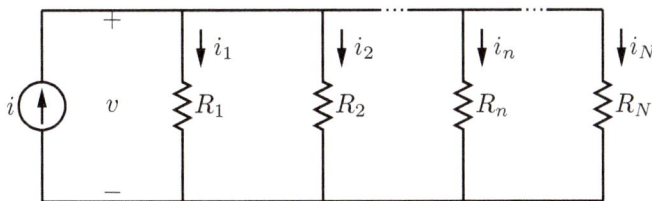

Figure 2.24: Resistors connected in parallel.

Figure 2.24 is the same circuit as in Figure 2.22, except we have replaced the voltage source with a current source, so the voltage across the resistors is not known directly. What if we would like to find a current, i_n, in the circuit? The concept of current division can be used in this situation. Let us first attempt finding the current i_n using what we know so far. From Ohm's law,

$$i_n = \frac{v}{R_n},$$

and to find v we can use

$$v = iR_{\text{eq}}.$$

From (2.27),

$$R_{\text{eq}} = \frac{1}{\frac{1}{R_1} + \frac{1}{R_2} + \cdots + \frac{1}{R_n} + \cdots + \frac{1}{R_N}},$$

and so

$$v = \frac{i}{\frac{1}{R_1} + \frac{1}{R_2} + \cdots + \frac{1}{R_n} + \cdots + \frac{1}{R_N}},$$

which yields

$$i_n(t) = \left(\frac{\frac{1}{R_n}}{\frac{1}{R_1} + \frac{1}{R_2} + \cdots + \frac{1}{R_n} + \cdots + \frac{1}{R_N}} \right) i(t). \tag{2.32}$$

Equation (2.32) is the formula for finding the current through parallel connected resistors. Why do we need it? It simplifies our solution, which in turn minimizes errors. We can also write the current division equation in terms of conductances, which makes it very pleasing to the eyes.

$$i_n(t) = \left(\frac{G_n}{G_1 + G_2 + \cdots + G_n + \cdots + G_N} \right) i(t). \tag{2.33}$$

The current division equation for two resistors in parallel can be written as

$$i_1 = \left(\frac{R_2}{R_1 + R_2} \right) i \tag{2.34}$$

and

$$i_2 = \left(\frac{R_1}{R_1 + R_2} \right) i. \tag{2.35}$$

Can you derive (2.34) and (2.35) from (2.32)? Try it out! To help you memorize this special but practically important case, look at the following expression:

$$i_{1,2}(t) = \left(\frac{\text{Other resistor}}{\text{Sum of resistors}} \right) i(t).$$

Example 2.6.7

The power delivered by the voltage source in Figure 2.25 is $P_s = 40\,\text{W}$. Find currents I_1 through I_6.

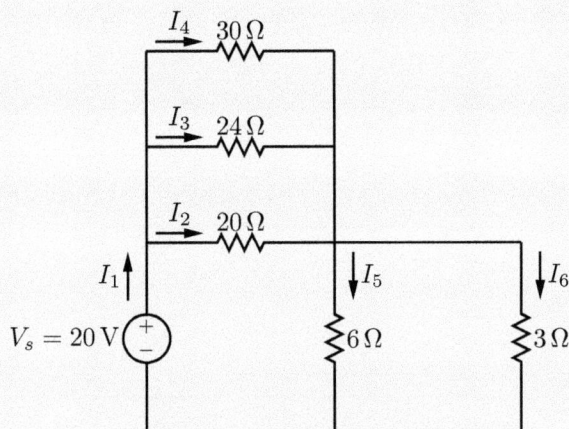

Figure 2.25: Example 2.6.7.

Since we want to find the currents in every branch, it does not help to start combining resistors, and this is where current division can shine. First, the total current I_1 delivered by the source can be found since we know the power delivered by the source:

$$I_1 = \frac{P_s}{V_s} = \frac{40\,\text{W}}{20\,\text{V}} = 2\,\text{A}.$$

Now, I_1 branches out into I_2, I_3, and I_4. Notice also that I_1 branches out into I_5 and I_6. We can use (2.32) to find the currents, repeated here for convenience:

$$i_n = \left(\frac{\frac{1}{R_n}}{\frac{1}{R_1} + \frac{1}{R_2} + \cdots + \frac{1}{R_n} + \cdots + \frac{1}{R_N}} \right) i.$$

Since the $20\,\Omega$, $24\,\Omega$, and $30\,\Omega$ resistors are in parallel,

$$I_2 = \frac{\frac{1}{20\,\Omega}}{\frac{1}{20\,\Omega} + \frac{1}{24\,\Omega} + \frac{1}{30\,\Omega}} I_1 = \frac{4}{5}\,\text{A}$$

$$I_3 = \frac{\frac{1}{24\,\Omega}}{\frac{1}{20\,\Omega} + \frac{1}{24\,\Omega} + \frac{1}{30\,\Omega}} I_1 = \frac{2}{3}\,\text{A}$$

$$I_4 = \frac{\frac{1}{30\,\Omega}}{\frac{1}{20\,\Omega} + \frac{1}{24\,\Omega} + \frac{1}{30\,\Omega}} I_1 = \frac{8}{15}\,\text{A}.$$

To check, we can see that the addition of I_2, I_3, and I_4 is equal to the total current I_1. Since the $6\,\Omega$ and $3\,\Omega$ resistors are in parallel, we can use the same equation, or we can use (2.34) and (2.35) as

$$I_5 = \left(\frac{3\,\Omega}{6\,\Omega + 3\,\Omega} \right) I_1 = \frac{2}{3}\,\text{A},$$

and

$$I_6 = \left(\frac{6\,\Omega}{6\,\Omega + 3\,\Omega}\right) I_1 = \frac{4}{3}\,\text{A}.$$

Again, the sum of I_5 and I_6 is equal to I_1.

Example 2.6.8

The power delivered by the current source in Figure 2.26 is 40 W. Find currents I_1 and I_2.

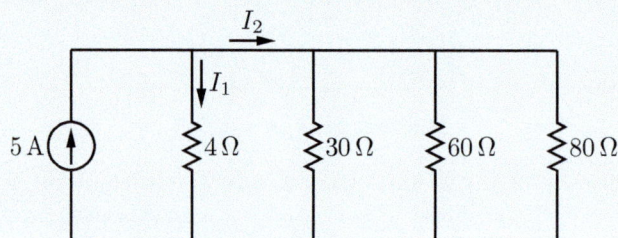

Figure 2.26: Example 2.6.8.

We can first find the equivalent resistance for the current I_2. Then we will have the current source in parallel with the $4\,\Omega$ resistor, and the equivalent resistance as seen in Figure 2.27.

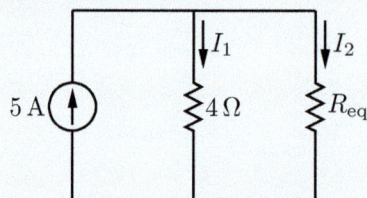

Figure 2.27: The equivalent circuit of Example 2.6.8.

We write

$$R_{\text{eq}} = \frac{1}{\frac{1}{30\,\Omega} + \frac{1}{60\,\Omega} + \frac{1}{80\,\Omega}} = 16\,\Omega.$$

Now that the $4\,\Omega$ and $16\,\Omega$ resistors are in parallel, we can use (2.34) and (2.35) as

$$I_1 = \left(\frac{16\,\Omega}{4\,\Omega + 16\,\Omega}\right) 5\,\text{A} = 4\,\text{A}$$

and

$$I_2 = \left(\frac{4\,\Omega}{4\,\Omega + 16\,\Omega}\right) 5\,\text{A} = 1\,\text{A}.$$

2.7   Kirchhoff's Voltage Law (KVL)

Kirchhoff's Voltage Law (KVL) is applied to closed loops (paths). What counts as a closed loop? The reality is that some paths can be more obvious than others. So let us make sure we can identify closed loops before we state KVL. A closed path is any path where you start and end at the same point, regardless of whether there are wires or devices connecting all points of the path. In other words, KVL applies to all closed paths regardless of whether electrical connections exist between the points of the considered closed path. Figure 2.28 can help illustrate the point.

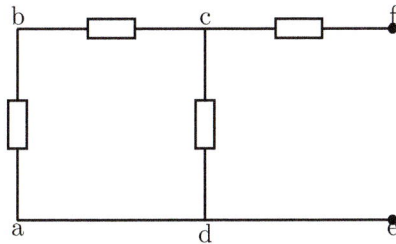

Figure 2.28: Closed loop identification.

The path (a,b,c,d back to a) probably makes the most obvious closed loop. The path (d,c,f,e back to d) makes another closed loop, and the path (a,b,c,f,e,d back to a) is a third possible closed loop. The fact that the second and third loops are not physically closed did not stop us from considering them as closed paths. Now we can state KVL.

KVL: The sum of all voltages around a closed path is zero.

In math form, KVL can be stated as

$$\sum_{n=1}^{N} v_n(t) = 0\,\text{V}. \tag{2.36}$$

Let us redraw Figure 2.28 with assigned voltages. Remember that you may assign any voltage polarity you wish. In fact, this is often necessary at the beginning of a problem, as polarities are, in general, unknown. If the true polarity of a voltage drop ends up being the opposite of what you assign, the result will simply end up being negative for that voltage drop.

We have assigned a polarity to each voltage we want in Figure 2.29. This is the first important step to applying KVL. To be consistent in every loop (this helps tremendously in avoiding mistakes):

- We start from the lowest left corner of each loop.

- We then move clockwise around a loop.

- Finally, we use the first voltage sign we encounter.

Let us now apply these steps to the following paths.

1. Path (a,b,c,d back to a):

$$-v_1 - v_2 + v_4 = 0\,\text{V}$$
$$v_4 = v_1 + v_2.$$

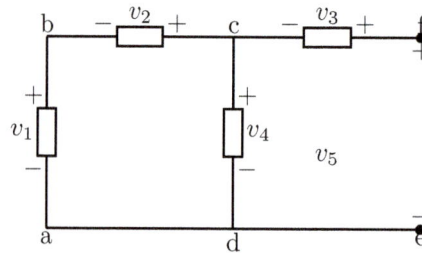

Figure 2.29: Figure 2.28 with assigned voltages.

2. Path (d,c,f,e back to d):

$$-v_4 - v_3 + v_5 = 0\,\text{V}$$

$$v_5 = v_3 + v_4.$$

3. Path (a,b,c,f,e,d back to a):

$$-v_1 - v_2 - v_3 + v_5 = 0\,\text{V}$$

$$v_5 = v_1 + v_2 + v_3.$$

When we apply KVL systematically to solve a circuit, we find out that we do not really need to find all possible loops; we only need to solve for the ones that contain our unknowns with the fewest number of equations. Example 2.7.1 should help solidify this idea.

Example 2.7.1

Find the voltage V_x in Figure 2.30 using KVL.

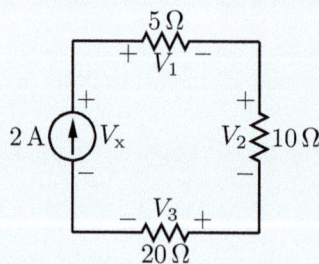

Figure 2.30: Example 2.7.1.

This is a simple example of how to use KVL in a resistive circuit. We can find the voltage across the current source by systematically applying KVL in the single loop of this circuit. We have then

$$-V_x + V_1 + V_2 + V_3 = 0\,\text{A}$$

or

$$V_x = V_1 + V_2 + V_3.$$

We can find the voltages across each resistor using Ohm's law since the current is known:

$$V_x = 2\,\text{A}(5\,\Omega) + 2\,\text{A}(10\,\Omega) + 2\,\text{A}(20\,\Omega) = 70\,\text{V}.$$

Example 2.7.2

Find the voltage V_y in Figure 2.31 using KVL.

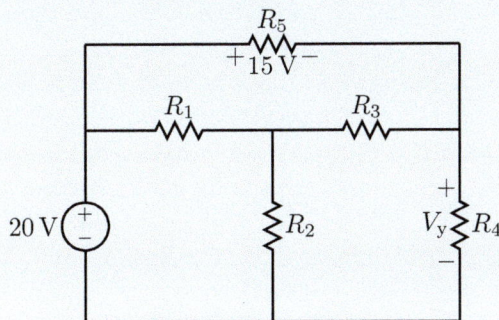

Figure 2.31: Example 2.7.2.

Although it may look like we don't have enough information, if we look closely we can see that a KVL around the outside loop of the circuit will give us what we need:

$$-20\,\text{V} + 15\,\text{V} + V_y = 0\,\text{V}$$

or

$$V_y = 20\,\text{V} - 15\,\text{V} = 5\,\text{V}.$$

Example 2.7.3

Find the voltages V_x, V_y, V_z, and V_w in Figure 2.32 using KVL.

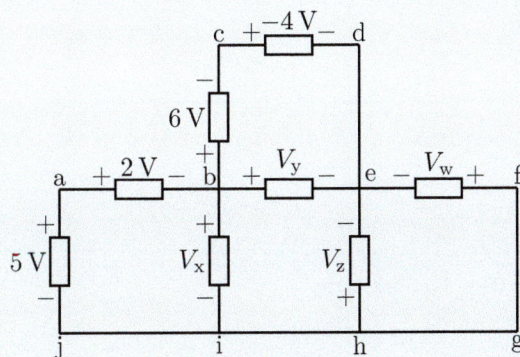

Figure 2.32: Example 2.7.3.

In order to minimize the number of steps to solve this problem, we need to take a good look at the circuit, what is given, and what is needed and then devise a plan.

- Solution plan:

  Loops (i,b,e,h,i) and (h,e,f,g,h) are useless to start with since every single voltage is unknown in these loops. Thus we definitely will not start with them. On the other hand, loops (j,a,b,i,j) and (b,c,d,e,b) contain one unknown per loop, and using KVL around these loops will result in finding V_x and V_y. Once we have these two voltages, loop (i,b,e,h,i) looks attractive again since applying KVL around it will solve for V_z. However, there is an even more attractive option that does not rely on the previous two loops. Loop (j,a,b,c,d,e,h,i,j) contains only V_z as an unknown variable and does not depend on V_x and V_y. A potential error in V_x or V_y will not, therefore, impact finding V_z. Besides, after finding V_z directly, we can use loop (h,e,f,g,h) or (j,a,b,c,d,e,f,g,h,i,j) to solve for V_w.

- Solution:

  Now that we have a plan, let us start solving. Taking the clockwise direction,

  - Path (j,a,b,i,j):

    $$-5\,\text{V} + 2\,\text{V} + V_x = 0\,\text{V} \rightarrow V_x = 3\,\text{V}.$$

  - Path (b,c,d,e,b):

    $$6\,\text{V} + (-4\,\text{V}) - V_y = 0\,\text{V} \rightarrow V_y = 2\,\text{V}.$$

  - Path (j,a,b,c,d,e,h,i,j):

    $$-5\,\text{V} + 2\,\text{V} + 6\,\text{V} + (-4\,\text{V}) - V_z = 0\,\text{V} \rightarrow V_z = -1\,\text{V}.$$

  - Path (h,e,f,g,h):

    $$V_z - V_w = 0\,\text{V} \rightarrow V_w = V_z = -1\,\text{V}.$$

Just as KCL was used in simplifying resistors in parallel, we will examine how KVL is used in finding the equivalent of an in-series combination of resistors.

2.7.1   Resistors in Series

In-series connection is another method of connecting resistors. As a reminder, recall that the same current *must* flow through all components connected in series. By the word "*must*," we mean that there are no possible alternatives for the current: the connection is such that the current has nowhere else to go but to necessarily flow through all in-series-connected components. Asking this question (i.e., is it necessarily true that the current is the same?) is the safest way to understand whether two or more components are connected in series. While for pedagogical purposes we tend to draw components connected in series very clearly when we first introduce this concept, this will not always be true in real-life circuits. Thus, trusting the circuit schematic without checking the current may lead to mistakes.

Figure 2.33 shows N resistors connected in series. The back-to-back connection from one component to the next without having any other components branching out from the connecting nodes guarantees that the current is the same through all resistors. From what we have learned so far, we can use KVL and Ohm's law to find the current i:

$$-v + v_1 + v_2 + \cdots + v_{N-1} + v_N = 0$$
$$v = v_1 + v_2 + \cdots + v_{N-1} + v_N. \tag{2.37}$$

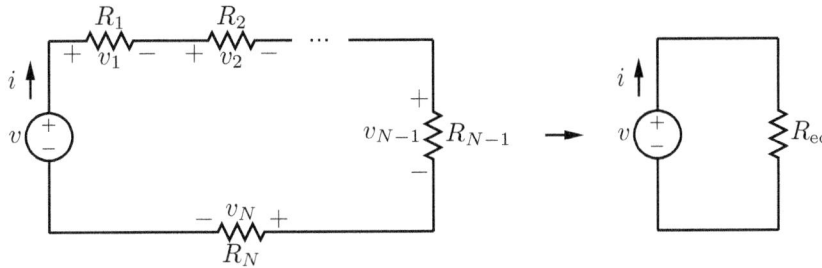

Figure 2.33: Resistors connected in series.

We can now use Ohm's law:

$$v = v_1 + v_2 + \cdots + v_{N-1} + v_N$$
$$v = iR_1 + iR_2 + \cdots + iR_{N-1} + iR_N. \tag{2.38}$$

Since the current i is the same through every resistor, we can rewrite (2.38) as

$$v = i(R_1 + R_2 + \cdots + R_{N-1} + R_N), \tag{2.39}$$

and from that we can calculate the current i:

$$i = \frac{v}{(R_1 + R_2 + \cdots + R_{N-1} + R_N)}. \tag{2.40}$$

This leads us to the conclusion that resistors connected in series can be summed together as one equivalent resistor:

$$\boxed{R_{eq} = \sum_{n=1}^{N} R_n.} \tag{2.41}$$

Equation (2.40) can now be rewritten as

$$R_{eq} = \frac{v}{i}. \tag{2.42}$$

Example 2.7.4

Find the power delivered by the 50 V source and the current I in Figure 2.34.

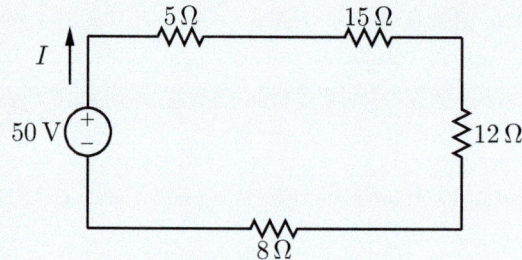

Figure 2.34: Example 2.7.4.

If we find the equivalent resistance, we will easily be able to find the current and the power as

$$P = \frac{(50\,\text{V})^2}{R_{\text{eq}}}$$

and

$$I = \frac{50\,\text{V}}{R_{\text{eq}}}.$$

Since the resistors are in series,

$$R_{\text{eq}} = 5\,\Omega + 15\,\Omega + 12\,\Omega + 8\,\Omega = 40\,\Omega.$$

Then

$$P = \frac{(50\,\text{V})^2}{R_{\text{eq}}} = \frac{(50\,\text{V})^2}{40\,\text{V}} = 62.5\,\text{W}$$

and

$$I = \frac{50\,\text{V}}{R_{\text{eq}}} = \frac{50\,\text{V}}{40\,\Omega} = 1.25\,\text{A}.$$

2.7.2   Voltage Division

Now that we have learned how to calculate the current in a circuit with resistors connected in series, we can ask ourselves: What if we want to find the voltage drop across one of the resistors in Figure 2.33? For example, what will v_2 be? Since we have the current from (2.42), we can use Ohm's law to find the voltage:

$$v_2 = iR_2. \tag{2.43}$$

Let us replace i with its equivalent from (2.42).

$$v_2 = \left(\frac{v}{R_{\text{eq}}}\right) R_2. \tag{2.44}$$

Rearrange (2.44) and replace R_{eq} with the sum of resistors to get

$$v_2 = \left(\frac{R_2}{R_1 + R_2 + \cdots + R_{N-1} + R_N} \right) v. \qquad (2.45)$$

This can be repeated for any resistor in Figure 2.33. In general, we have

$$\boxed{v_n(t) = \left(\frac{R_n}{R_1 + R_2 + \cdots + R_n + \cdots + R_{N-1} + R_N} \right) v(t).} \qquad (2.46)$$

Equation (2.46) is called voltage division, a wonderful consequence of connecting resistors in series. The total voltage is distributed among the resistors according to their values. The amount of voltage drop across each resistor depends on its value with respect to the other resistors. The larger the resistor, the larger the voltage drop across it will be. Let us look at the same circuit from Example 2.7.4.

Example 2.7.5

Find the voltage V across the $12\,\Omega$ resistor in Figure 2.35.

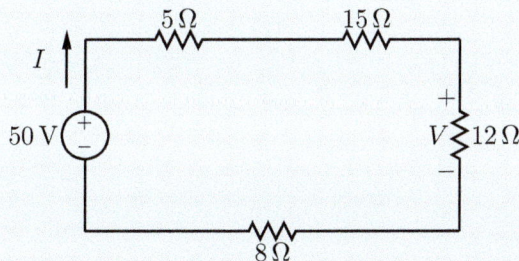

Figure 2.35: Example 2.7.5.

Since we only need the voltage across the $12\,\Omega$ resistor, we can use voltage division:

$$V = \frac{12\,\Omega}{5\,\Omega + 15\,\Omega + 12\,\Omega + 8\,\Omega}(50\,\text{V}) = 15\,\text{V}.$$

Now let us take a look at a common mistake so you can learn to avoid it.

Example 2.7.6

Find the voltage V in Figure 2.36.

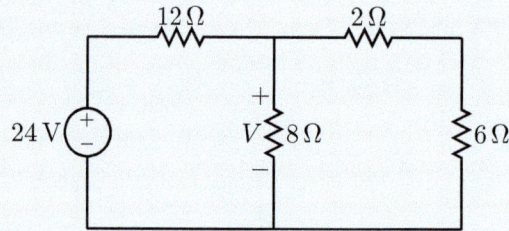

Figure 2.36: Example 2.7.6.

Many might be tempted to apply the voltage division formula for the $12\,\Omega$ and $8\,\Omega$ resistors. But we have to stop and think for a second. The current through the $8\,\Omega$ resistor is not the same as the current that passes through the $12\,\Omega$ resistor, which is the necessary condition to apply the voltage division formula. So what do we do? Notice that the voltage V is the same voltage as the voltage across the sum of the $2\,\Omega$ and $6\,\Omega$ resistors. Figure 2.37 shows the simplified circuit.

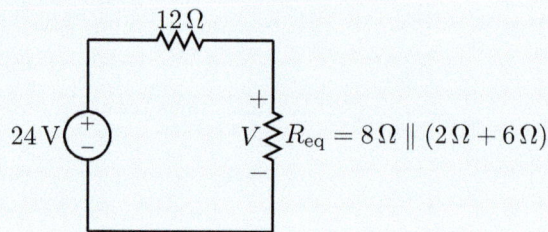

Figure 2.37: Example 2.7.6.

Now we can find V using our voltage division formula,

$$V = \frac{R_{\text{eq}}}{R_{\text{eq}} + 12\,\Omega}(24\,\text{V}),$$

where

$$R_{\text{eq}} = 8\,\Omega \parallel (2\,\Omega + 6\,\Omega) = 4\,\Omega.$$

Then

$$V = \frac{4\,\Omega}{4\,\Omega + 12\,\Omega}(24\,\text{V}) = 6\,\text{V}.$$

Example 2.7.7

Find the voltage V in Figure 2.38.

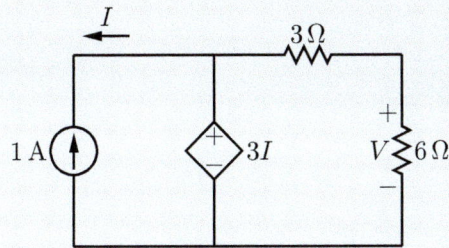

Figure 2.38: Example 2.7.7.

The voltage division can be used to find V using the voltage-controlled source:

$$V = \frac{6\,\Omega}{3\,\Omega + 6\,\Omega}(3I) = 2I.$$

We also have $I = -1\,\text{A}$, which yields

$$V = 2I = -2\,\text{V}.$$

2.8 Parallel and Series Combinations

Now that we know how to deal with resistors connected in series and in parallel, we can move on to circuits with combinations of both parallel and series connections. We will jump into an example right away.

Example 2.8.1

For the circuit in Figure 2.39, find the current I and the voltage V.

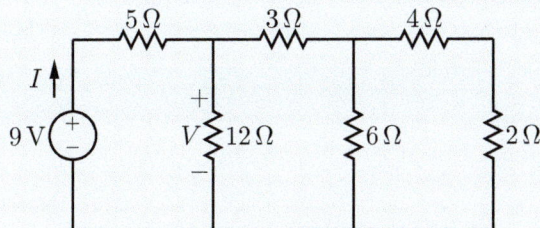

Figure 2.39: Example 2.8.1.

To find I, we can collapse the circuit into a source and an equivalent resistor. We will have to start from the far right of the circuit and work our way to the source. We notice that the 4 Ω and 2 Ω resistors are in series and their combination is in parallel with the 6 Ω resistor. Hence,

$$(4\,\Omega + 2\,\Omega) \parallel 6\,\Omega = \frac{(4\,\Omega + 2\,\Omega)6\,\Omega}{4\,\Omega + 2\,\Omega + 6\,\Omega} = 3\,\Omega.$$

Now the resulting 3 Ω resistor is in series with the middle 3 Ω resistor and their sum is in parallel with the 12 Ω resistor, as seen in the figure below. Thus,

$$(3\,\Omega + 3\,\Omega) \parallel 12\,\Omega = \frac{(3\,\Omega + 3\,\Omega)12\,\Omega}{3\,\Omega + 3\,\Omega + 12\,\Omega} = 4\,\Omega.$$

The resulting 4 Ω resistor is in series with the 5 Ω resistor, as seen in the figure below.

$$R_{\text{eq}} = 5\,\Omega + 4\,\Omega = 9\,\Omega$$

and now we can find the current I as

$$I = \frac{9\,\text{V}}{R_{\text{eq}}} = \frac{9\,\text{V}}{9\,\Omega} = 1\,\text{A}.$$

We can also find the voltage V as

$$V = I(4\,\Omega) = 4\,\text{V}.$$

Notice that the voltage across the 4 Ω resistor is the same as the voltage across the 12 Ω resistor. Why? Because the 4 Ω resistor was a result of a parallel combination.

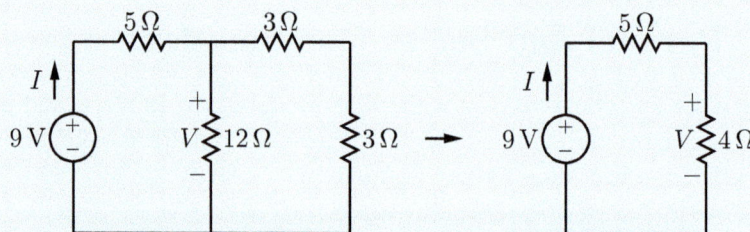

Figure 2.40: Example 2.8.1 after the first (left) and second (right) steps.

In Example 2.8.1, we had to collapse the circuit and find R_{eq}. Let us try another example to see if we can use more of what we have learned so far.

Example 2.8.2

Find the currents I_1 and I_2 in Figure 2.41.

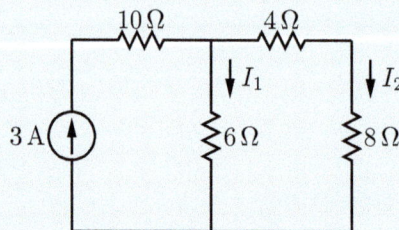

Figure 2.41: Example 2.8.2.

For this example, we could find R_{eq}, but it will be easier if we use current division. We can even use the special case in (2.34). But what about the 10 Ω resistor? The existence of this resistor has no effect on the total current supplied by the current source. The 3 A current is going to be divided between the two branches:

$$I_1 = \frac{4\,\Omega + 8\,\Omega}{6\,\Omega + 4\,\Omega + 8\,\Omega} 3\,\text{A} = 2\,\text{A}.$$

Now we can use current division again to find the second current, but we can also find it using KCL:

$$I_2 = I - I_1 = 3\,\text{A} - 2\,\text{A} = 1\,\text{A}.$$

Try finding I_2 using current division to double-check the result.

Example 2.8.3

Find the voltage V in Figure 2.42.

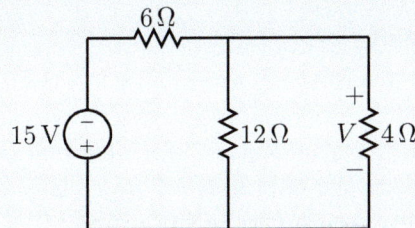

Figure 2.42: Example 2.8.3.

In this example we can clearly use voltage division, but we have to be careful about two things: the first is that voltage division is done between in-series resistors; the second is noting the voltage source polarity.

Since the 4 Ω and 12 Ω resistors are in parallel, the voltage across them is the same and the voltage across their parallel combination will remain the same. The 6 Ω resistor will then be in series with the combined resistors and we can use voltage division. First we write

$$R = 12\,\Omega \parallel 4\,\Omega = 3\,\Omega.$$

Now we can use voltage division (being careful of the polarity):

$$V = \frac{R}{6\,\Omega + R}(-15\,\text{V}) = \frac{3\,\Omega}{6\,\Omega + 3\,\Omega}(-15\,\text{V}) = -5\,\text{V}.$$

2.8.1   Other Types of Connections

There are additional ways to connect resistors, such as a delta connection and a wye connection. There are even ways to convert a delta connection into a wye and vice versa. We will not be discussing these for now, but Figure 2.43 shows the connections.

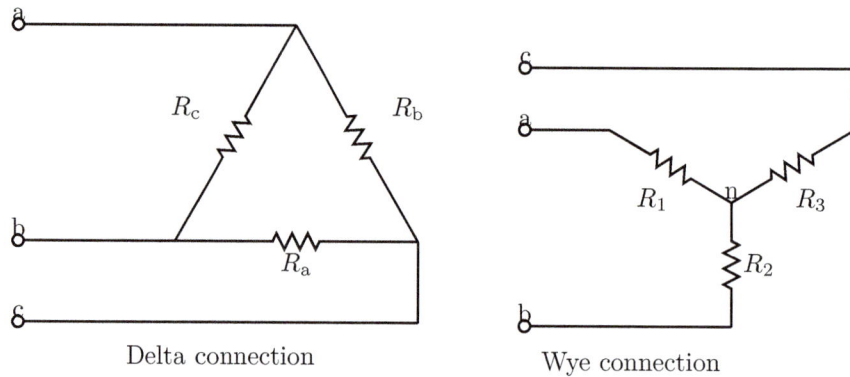

Figure 2.43: Delta and wye connections.

2.9   Finding the Equivalent Resistance of a Circuit

Why do we need to find the equivalent resistance of a circuit? In many situations we are only interested in a certain port or nodes of a circuit and not in all of its details. We would only be interested in the resistance seen from that port. For example, many of our everyday devices have input and output ports (laptops, tablets, cellphones, etc.). We only interact with many such devices through their ports. As a result, we can often greatly simplify things by replacing the entire circuit by an equivalent (fictitious) resistance that maintains the same $v - i$ relationships. Many circuits have several ports with different equivalent resistance at each port. Let us look into this through a simple example.

Example 2.9.1

In Figure 2.44, we have a 25 V source connected in series with a 7 A fuse. The 7-A fuse will pop (open the circuit) when current supplied by the source exceeds 7 A. The fuse is assumed to have negligible resistance. We would like to know which one of the circuits on the right would make the fuse pop. Note that we will connect port a-b from the source to port a-b of the resistive circuit.

Now let us start with circuit a; we need to find the resistance seen by the source. So we need to find R_{eq} as seen from port a-b. We start from the far right, moving our way to the a-b nodes.

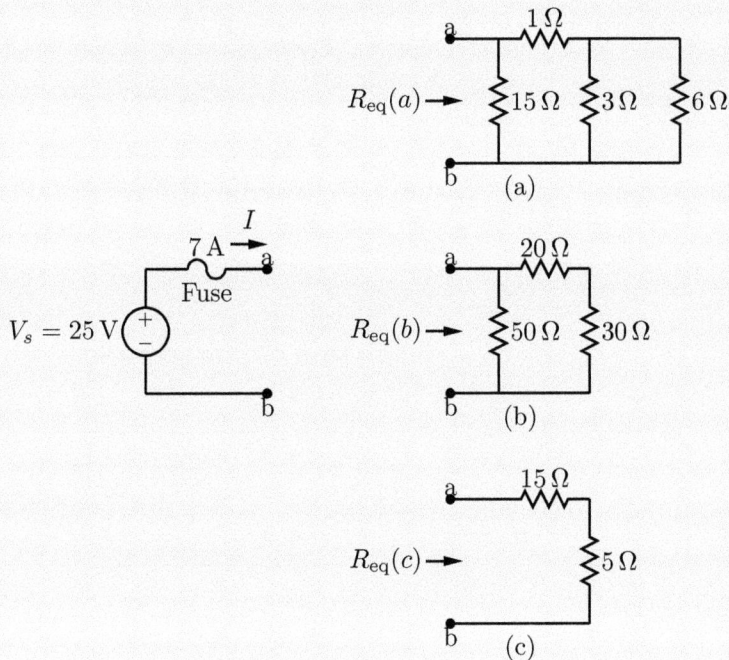

Figure 2.44: Example 2.9.1.

$$R_{eq}(a) = 15\,\Omega \parallel (1\,\Omega + (3\,\Omega \parallel 6\,\Omega)) = 15\,\Omega \parallel (1\,\Omega + 2\,\Omega) = 2.5\,\Omega.$$

Now we check for the current:

$$I = \frac{V_s}{R_{eq}(a)} = \frac{25\,\text{V}}{2.5\,\Omega} = 10\,\text{A}.$$

The current will cause the fuse to pop, so we will not be able to connect circuit (a) to the source as is.

For circuit (b),

$$R_{eq}(b) = 50\,\Omega \parallel (20\,\Omega + 30\,\Omega) = 25\,\Omega.$$

Check the current:

$$I = \frac{V_s}{R_{eq}(b)} = \frac{25\,\text{V}}{25\,\Omega} = 1\,\text{A}.$$

The circuit can be connected without blowing the fuse.

For circuit c,

$$R_{eq}(c) = 5\,\Omega + 15\,\Omega = 20\,\Omega.$$

Check the current:

$$I = \frac{V_s}{R_{eq}(c)} = \frac{25\,\text{V}}{20\,\Omega} = 1.25\,\text{A}.$$

Circuit (c) passes inspection. We can see from this example that finding the equivalent resistance is important. One might have thought that the source would only see the $15\,\Omega$ and $50\,\Omega$ resistance in circuits (a) and (b), but this is clearly not the case.

Let us examine another example to see how the same circuit can have different equivalent resistance depending on which port we are looking from.

Example 2.9.2

Figure 2.45 shows a circuit with multiple ports, and we would like to find the resistance seen from these different ports.

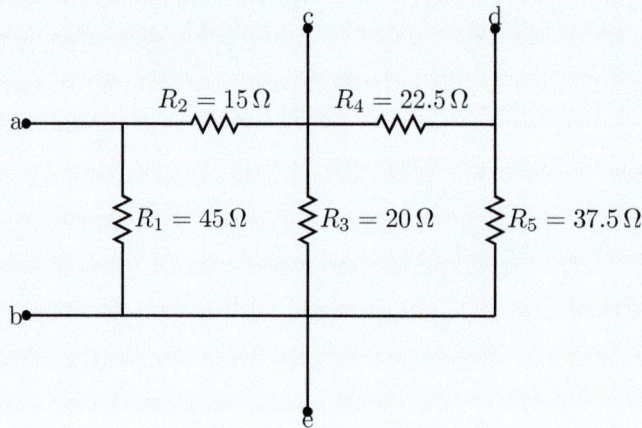

Figure 2.45: Example 2.9.2.

Let us start with the a-b port. If you applied your ohm meter across this port, you would not see R_1, but you would see

$$R_{\text{ab}} = R_1 \parallel [R_2 + (R_3 \parallel (R_4 + R_5))] = 18\,\Omega.$$

When looking at the circuit from port c-d, we have

$$R_{\text{cd}} = R_4 \parallel [R_5 + (R_3 \parallel (R_1 + R_2))] = 15.75\,\Omega,$$

and when we look from port c-e, we find

$$R_{\text{ce}} = R_3 \parallel (R_1 + R_2) \parallel (R_4 + R_5) = 12\,\Omega.$$

Although we have one circuit, we see a different equivalent resistance depending on the port we are looking from. Now that we are convinced that we sometimes need the equivalent resistance and that the equivalent resistance for a circuit can change depending on what port we are looking through, we will turn our attention to calculating the equivalent resistance. There is more than one way to find the equivalent resistance. If we are lucky, the circuit will contain only series and parallel combinations of resistors, which is not a big deal.

Sometimes we will need some elbow grease. We will go through some situations and try to see how to find the equivalent resistance in each.

Circuits with No Sources

A circuit with no sources can easily be reduced to an equivalent resistance if it contains only series and parallel combinations of resistors. A circuit will sometimes contain a delta or a wye connection, and if you knew the formula to transform it, the solution could keep moving forward. If you were not familiar with delta-to-wye transformations, another method could be used. This method is called the $i - v$ test method and is discussed later.

Example 2.9.3

Find the equivalent resistance in Figure 2.46.

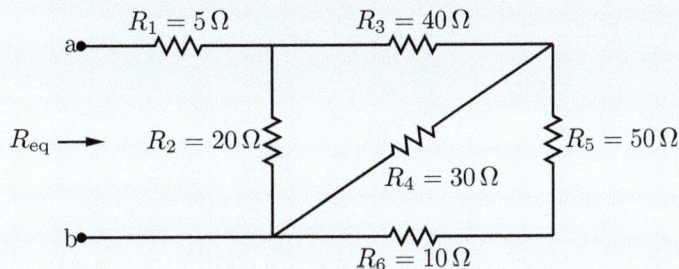

Figure 2.46: Example 2.9.3.

We will have to start with the far right of the circuit. We can see that R_5 and R_6 are in series and their sum is in parallel with R_4. Hence

$$R_{eq_1} = R_4 \parallel (R_5 + R_6) = 30\,\Omega \parallel (50\,\Omega + 10\,\Omega) = 20\,\Omega.$$

The resulting resistance R_{eq_1} is then in series with R_3:

$$R_{eq_2} = R_{eq_1} + R_3 = 20\,\Omega + 40\,\Omega = 60\,\Omega.$$

The resulting resistance R_{eq_2} is in parallel with R_2:

$$R_{eq_3} = R_{eq_2} \parallel R_2 = 60\,\Omega \parallel 20\,\Omega = 15\,\Omega.$$

The resulting resistance R_{eq_3} is then in series with R_1:

$$R_{ab} = R_{veq_3} + R_1 = 15\,\Omega + 5\,\Omega = 20\,\Omega.$$

As you get better with practice, you will be able to write all these steps in one equation:

$$R_{ab} = R_1 + [R_2 \parallel (R_3 + (R_4 \parallel (R_5 + R_6)))].$$

Just remember that there is an order for these operations, so you need to pay attention.

Circuits with Sources

a. Independent Sources Only

In Figure 2.47, we have considered again the circuit in Figure 2.46 from Example 2.9.3 but have added two independent current sources and a voltage source. We would like to find R_{eq} as before. Finding R_{eq} in a circuit that contains independent sources requires *turning off all independent sources first*. This is a general rule and will always apply. Remember what we have learned from section 1.2.3 about turning off sources: voltage sources are replaced by short circuits; current sources are replaced by open circuits. Figure 2.48 shows the circuit with all sources turned off.

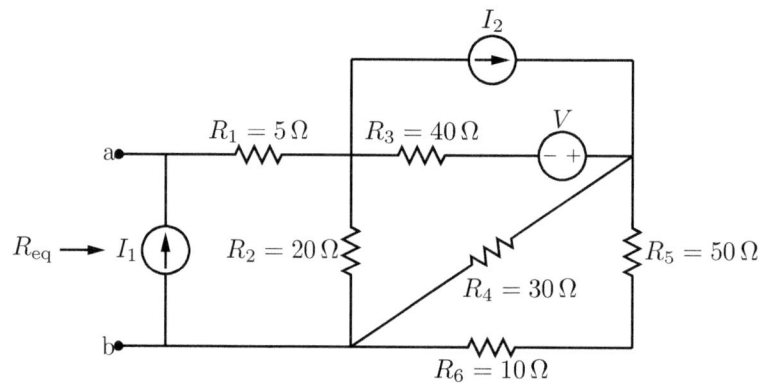

Figure 2.47: Finding the equivalent resistance with independent sources.

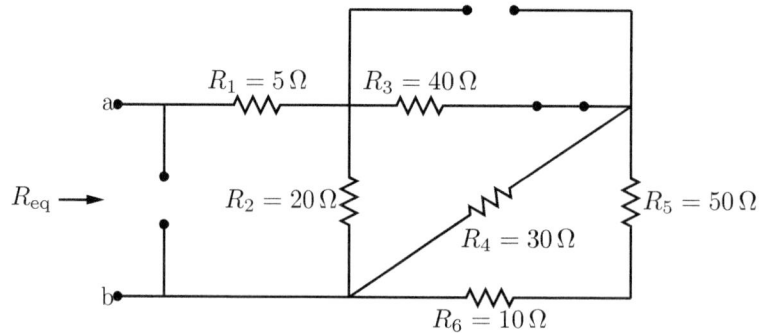

Figure 2.48: Turning off all sources in Figure 2.47.

Now the circuit looks exactly like Figure 2.46 from Example 2.9.3. The equivalent resistance R_{eq} is calculated the same way, and we will have the same result.

$$R_{eq} = R_1 + [R_2 \parallel (R_3 + (R_4 \parallel (R_5 + R_6)))] = 20\,\Omega$$

Let us change the circuit again and see what happens.

Example 2.9.4

Find the equivalent resistance R_{ab} in Figure 2.49.

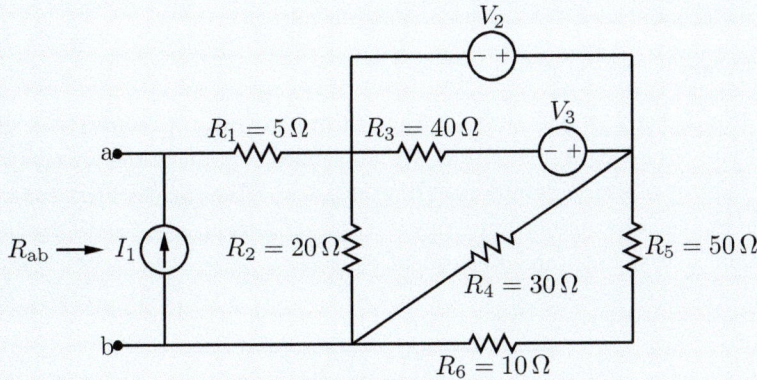

Figure 2.49: Example 2.9.4.

After turning all sources off, R_3 is shorted, which will make $R_2 \parallel R_4 \parallel (R_5 + R_6)$. The circuit with turned off sources is shown in Figure 2.50.

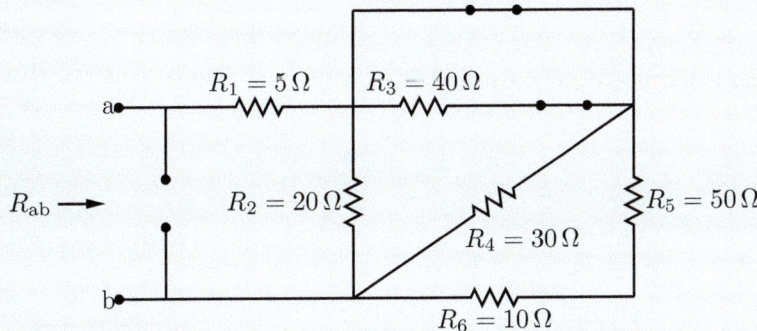

Figure 2.50: Turning all sources off in Figure 2.49.

$$
\begin{aligned}
R_{ab} &= R_1 + [\ R_2 \parallel R_4 \parallel (R_5 + R_6)\ ] \\
&= 5\,\Omega + [\,20\,\Omega \parallel 30\,\Omega \parallel (50\,\Omega + 10\,\Omega)\,] \\
&= 15\,\Omega
\end{aligned}
$$

b. Independent and Dependent Sources

Now that we know how to find the equivalent resistance when we have independent sources only, let us throw in some dependent sources and see what happens. Figure 2.51 shows an example circuit that contains both dependent and independent sources. Finding the equivalent resistance for a circuit that contains a dependent source is a little bit more involved than for circuits with independent sources alone. Dependent sources *remain active* when we find R_{eq}.

Consequently, the only way to find R_{eq} is to perform the $i - v$ test and apply the definition. This means that:

a. We apply a test voltage (or current) source at the required port.

b. We calculate the resulting current (or voltage).

c. We apply the definition $R_{eq} = v/i$.

Of course, the $i - v$ test—that is, the definition of R_{eq}—can always be applied even if the circuit contains only series and parallel connections of resistors. It is just that in this case combining these resistors often yields the correct result faster. On the other hand, applying the R_{eq} definition is the *only* way to find R_{eq} when a circuit contains dependent sources.

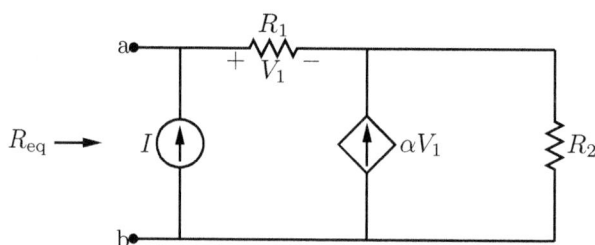

Figure 2.51: Finding the equivalent resistance with independent and dependent sources.

These steps are graphically illustrated in Figure 2.52, and mathematically we get

$$R_{eq} = \frac{V_{test}}{I_{test}}. \tag{2.47}$$

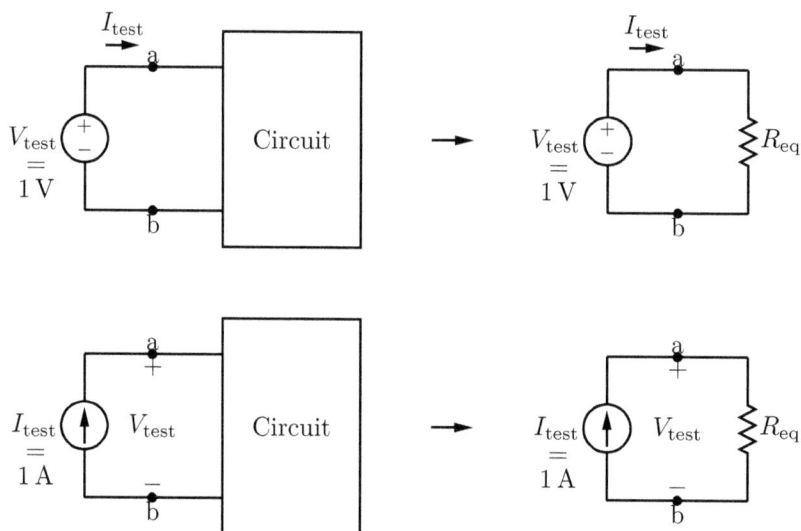

Figure 2.52: Graphic illustration of the $i - v$ test in finding R_{eq}. Notice that all independent sources must be turned off before applying the test voltage or current source.

As shown, we need to add an independent test source at the port (terminal) where we wish to calculate the equivalent resistance. This source can be a voltage source or a current source. The value of the test source can be left as V_{test}, I_{test} or we can give it a value of 1 V or 1 A. Notice that choosing a value of 1 (V or 1 A) for the applied voltage or current may make the calculations easier.

$$R_{\text{eq}} = \frac{V_{\text{test}}}{I_{\text{test}}} = \frac{1\,\text{V}}{I_{\text{test}}}, \tag{2.48}$$

$$R_{\text{eq}} = \frac{V_{\text{test}}}{I_{\text{test}}} = \frac{V_{\text{test}}}{1\,\text{A}}. \tag{2.49}$$

Let us now give the circuit in Figure 2.51 some numbers and apply what we have learned to solve for the equivalent resistance.

Example 2.9.5

Find the equivalent resistance for the circuit in Figure 2.53.

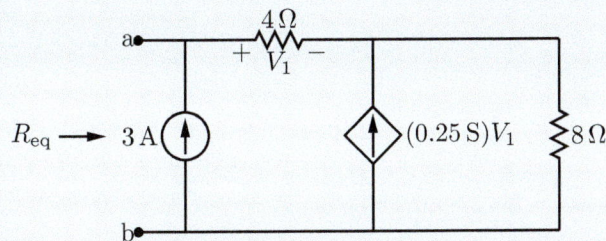

Figure 2.53: Example 2.9.5.

First, we turn off all independent sources. Second, we need to choose whether to apply a test current or voltage source. While most of the time it makes no difference which source we choose, sometimes one or the other option leads to an easier solution. Looking at this circuit, for example, adding a current source makes the solution very easy. We end up knowing the current in two branches and we are able to find the voltage that the dependent source needs (V_1). Try to evaluate the situation if we were to use a test voltage; it would be entirely possible to solve, just a bit longer.

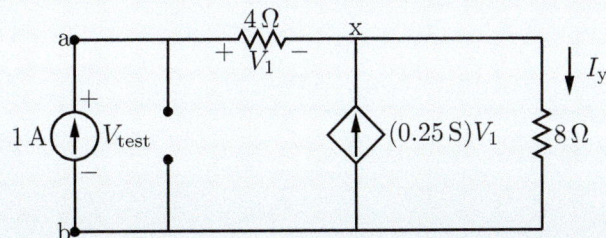

Figure 2.54: Applying $i - v$ test to the circuit in Figure 2.53.

Figure 2.54 shows the circuit with the 3 A source turned off and the added test current source. The goal now is to find V_{test}. Rather than randomly finding currents and voltages in the circuit, it is always a good idea to develop a solution plan. While in the beginning your plans may not be optimal, they will progressively get better as you get more experience. Remember that *a bad solution plan is better than no plan at all*. One plan here is to calculate V_{test} by applying KVL:

$$V_{\text{test}} = V_1 + I_{\text{y}}(8\,\Omega).$$

So now we need to find I_{y}. Since I_{y} flows from node x, we could apply KCL there to find it:

$$I_{\text{y}} = 1\,\text{A} + (0.25\,\text{S})V_1.$$

We can find the voltage V_1 by simply multiplying the test current by the $4\,\Omega$ resistance:

$$V_1 = I_{\text{test}}R = (1\,\text{A})(4\,\Omega) = 4\,\text{V}.$$

So the KCL at node x yields

$$I_{\text{y}} = 1\,\text{A} + (0.25\,\text{S})V_1 = 1\,\text{A} + (0.25\,\text{S})(4\,\text{V}) = 2\,\text{A}.$$

So our KVL for V_{test} becomes

$$V_{\text{test}} = V_1 + I_{\text{y}}(8\,\Omega) = 4\,\text{V} + (2\,\text{A})(8\,\Omega) = 20\,\text{V}.$$

Finally, the equivalent resistance can be calculated as

$$R_{\text{eq}} = \frac{V_{\text{test}}}{I_{\text{test}}} = \frac{20\,\text{V}}{1\,\text{A}} = 20\,\Omega.$$

We can also see here why choosing the 1 A value made the calculations a bit easier. We could have left the value of I_{test} as a variable, and by doing that

$$V_1 = I_{\text{test}} \cdot 4\,\Omega$$

and

$$I_{\text{y}} = I_{\text{test}} + (0.25\,\text{S})V_1 = I_{\text{test}} + (0.25\,\text{S})(I_{\text{test}} \cdot 4\,\Omega) = 2I_{\text{test}}$$

and

$$V_{\text{test}} = V_1 + I_{\text{y}}(8\,\Omega) = I_{\text{test}} \cdot 4\,\Omega + (2I_{\text{test}}) \cdot 8\,\Omega = I_{\text{test}} \cdot 20\,\Omega.$$

Then

$$R_{\text{eq}} = \frac{V_{\text{test}}}{I_{\text{test}}} = \frac{I_{\text{test}} \cdot 20\,\Omega}{I_{\text{test}}} = 20\,\Omega.$$

We see that we get the same result but one might have less chance of error than the other, so it is up to you to choose what you are most comfortable with.

Let us now modify the circuit in Figure 2.53 and make it a bit more interesting.

Example 2.9.6

The circuit in Figure 2.55 has almost the same configuration as the previous one, but we have added a voltage source and changed the controlling voltage of the dependent source.

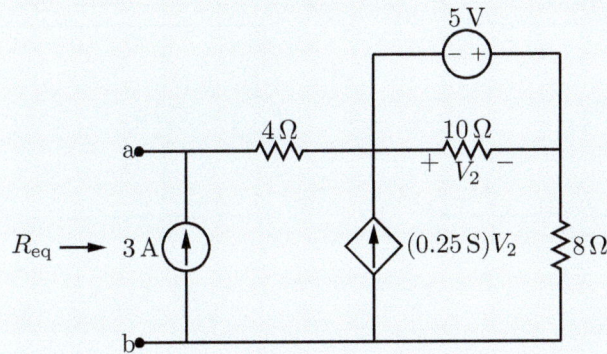

Figure 2.55: Example 2.9.6.

We know the process now: since we have a dependent source, we turn off all independent sources and add a test source. Looking at the circuit, we notice something interesting. We notice that when we turn off the 5 V source we end up shorting the 10 Ω resistor, which in turns makes the voltage $V_2 = 0\,\text{V}$, which leads to zero current from the dependent source. Hence we conclude that the equivalent resistance is

$$R_{\text{eq}} = 4\,\Omega + 8\,\Omega = 12\,\Omega.$$

Finding the equivalent resistance is not intimidating if you have a plan for what to do. You need to carefully evaluate each solution before you start writing equations.

2.10   How to Redraw a Circuit

Circuits are not always presented in the clearest way possible. Our knowledge about nodes and connections will help us detangle circuits if necessary. Let us look at a quick example. The circuits in Figure 2.56 are equivalent but drawn differently. By recognizing (numbering) nodes and keeping track of the elements connected to each node, we can redraw a circuit to make it easier.

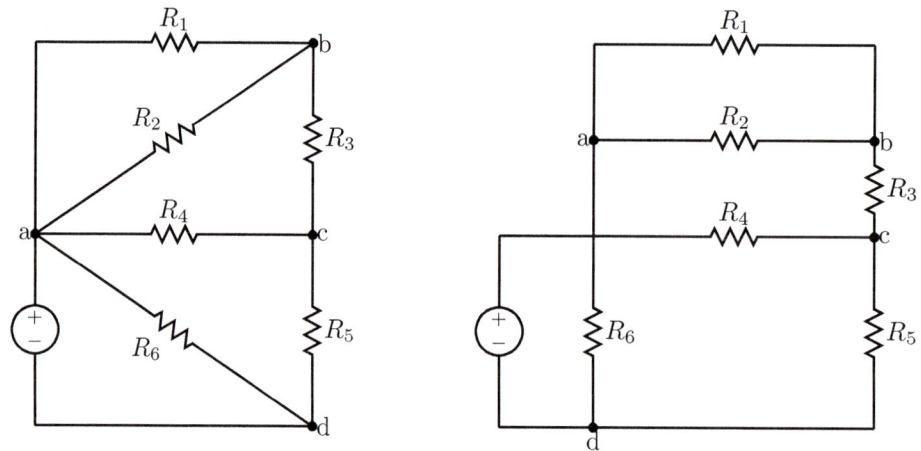

Figure 2.56: The equivalent circuits.

Let us look at the following numerical example to see how redrawing the circuit might make things less complicated.

Example 2.10.1

For the circuit in Figure 2.57, find the voltage V_o after turning off the 10 V voltage source.

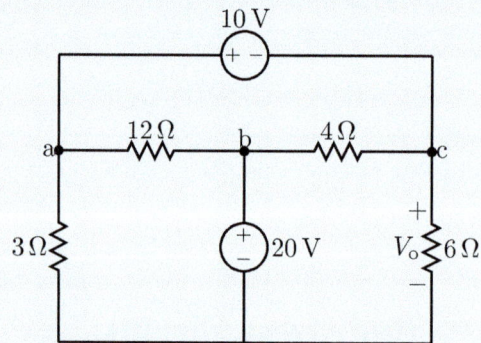

Figure 2.57: Example 2.10.1.

Since we are asked to find the voltage after turning off the 10 V voltage source, we first redraw the circuit with the new condition. Turning off a voltage source is equivalent to replacing it with a short circuit, as seen in Figure 2.58.

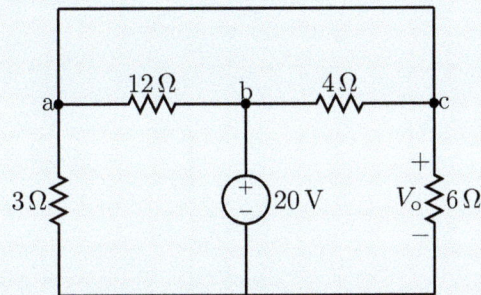

Figure 2.58: Example 2.10.1 with the 10 V voltage source off.

The first thing we notice in Figure 2.58 is that nodes a and c are now the same. We can redraw the circuit one more time to give us a clearer picture of what we need.

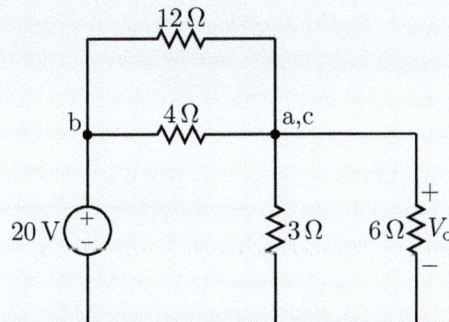

Figure 2.59: Redrawing the circuit in Figure 2.58.

All we need to do now is to apply voltage division to $(12\,\Omega \parallel 4\,\Omega)$ and $(6\,\Omega \parallel 3\,\Omega)$.

$$12\,\Omega \parallel 4\,\Omega = \frac{12\,\Omega \cdot 4\,\Omega}{12\,\Omega + 4\,\Omega} = 3\,\Omega.$$

$$6\,\Omega \parallel 3\,\Omega = \frac{6\,\Omega \cdot 3\,\Omega}{6\,\Omega + 3\,\Omega} = 2\,\Omega.$$

Applying voltage division, we get

$$V_{\mathrm{o}} = \frac{2\,\Omega}{3\,\Omega + 2\,\Omega}(20\,\mathrm{V}) = 8\,\mathrm{V}.$$

2.11   Exercises

Section 2.3

2.1. The circuit in Figure 2.60 contains a variable resistor to control the current flow in the circuit. Find the value of the resistance to obtain 1 A, 0.5 A, and 0.2 A.

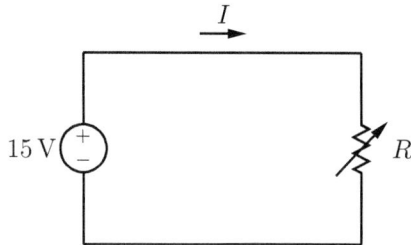

Figure 2.60

2.2. The circuit in Figure 2.61 contains a variable resistor to control the voltage V in the circuit. Find the value of the resistance to obtain 4.5 V, 10 V, and 30 V.

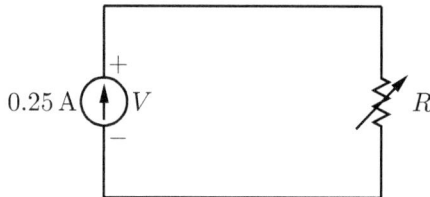

Figure 2.61

2.3. The circuit in Figure 2.62 is equipped with a circuit breaker that opens if the current through it exceeds 0.5 A. Find the value of the smallest resistor we can use.

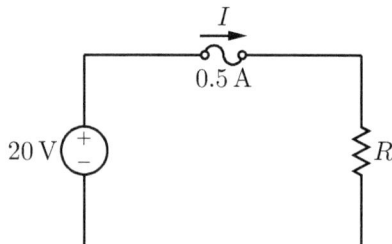

Figure 2.62

2.4. The circuit in Figure 2.63 is equipped with a circuit breaker that opens if the current through

it exceeds 0.5 A. Find the value of the largest conductance we can use.

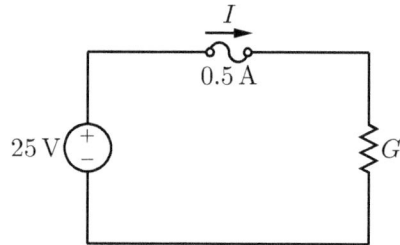

Figure 2.63

Section 2.4

2.5. For the circuit in Figure 2.64, find the voltages V_1 and V_2, the current I_3, and the value of the resistance R_3. (Note that the power absorbed by R_3 is shown in the circuit.)

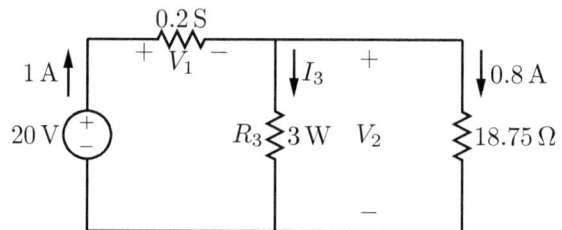

Figure 2.64

2.6. For the circuit in Figure 2.65, The power generated by the 20 V source is 100 W. Find the current I and the power absorbed by each resistor.

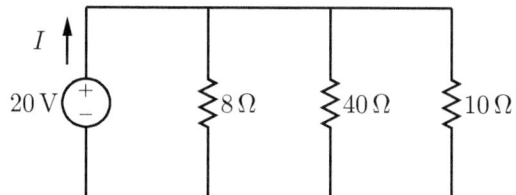

Figure 2.65

Section 2.6

2.7. Find the current I and the voltage V in the circuit in Figure 2.66.

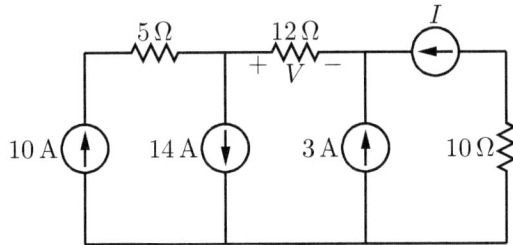

Figure 2.66

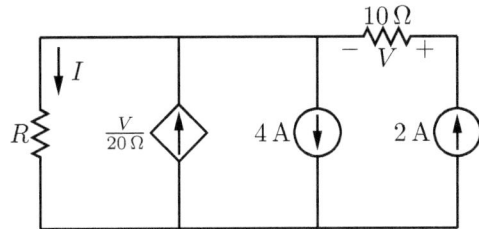

Figure 2.69

2.8. Find the current I and the power delivered by the voltage source in the circuit in Figure 2.67.

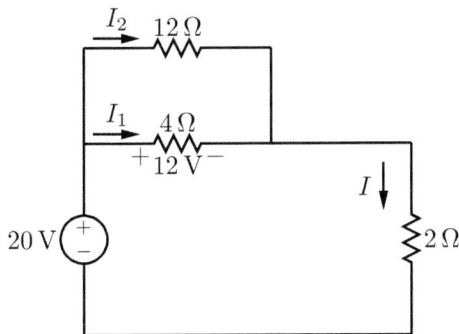

Figure 2.67

2.9. Find the currents I_1 through I_5 and the power delivered by the voltage source in the circuit in Figure 2.68.

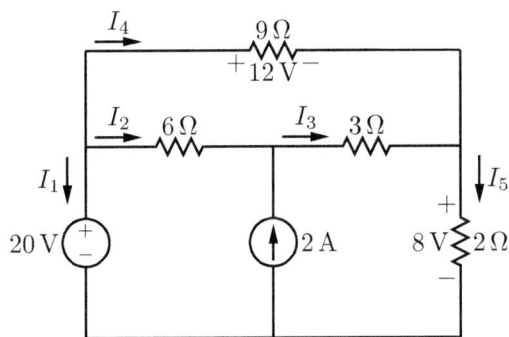

Figure 2.68

2.10. Find the current I and the voltage V in the circuit in Figure 2.69.

Section 2.6.1

2.11. The voltage source in the circuit in Figure 2.70 delivers 24 W. Find the value of the voltage supplied by the source. (Hint: find R_{eq}.)

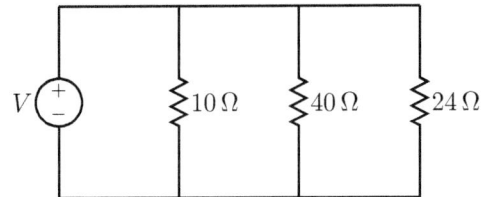

Figure 2.70

2.12. Find the power delivered by the voltage source in Figure 2.71.

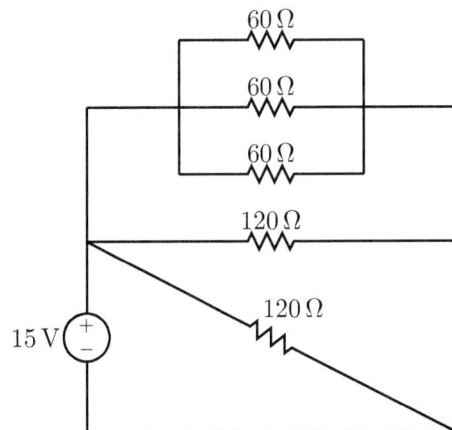

Figure 2.71

Section 2.6.2

2.13. Find the currents I_1 and I_2 in the circuit in Figure 2.72.

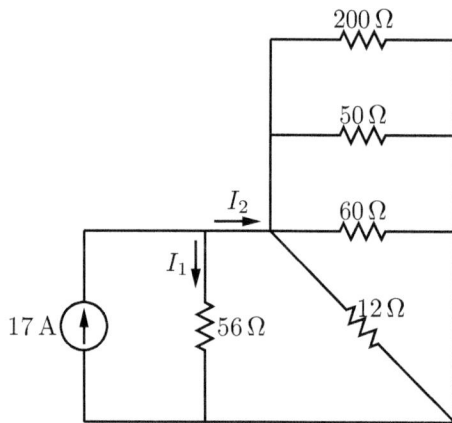

Figure 2.72

2.14. Find the currents I_1 and I_2 in the circuit in Figure 2.73.

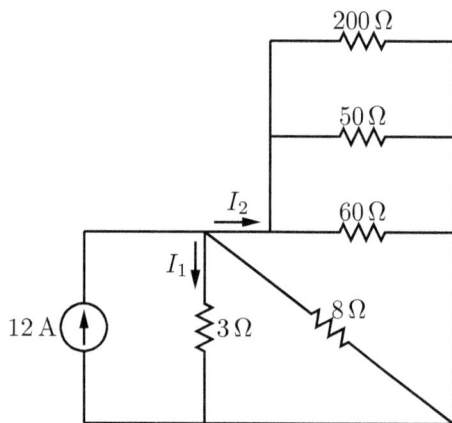

Figure 2.73

2.15. Find the current through the 20 Ω resistor in the circuit in Figure 2.74.

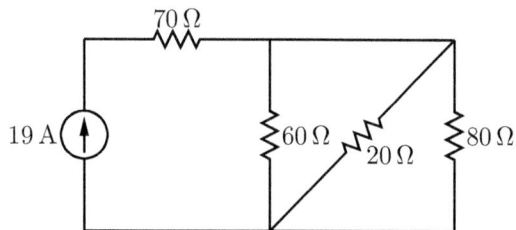

Figure 2.74

Section 2.7

2.16. Find the current I and the voltage V_{ab} in the circuit in Figure 2.75.

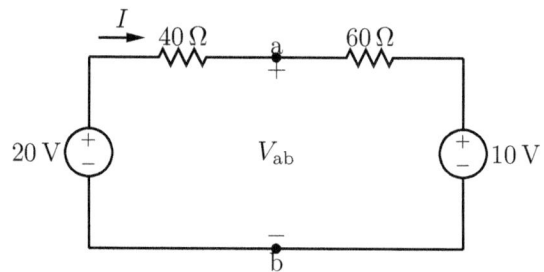

Figure 2.75

2.17. Find the value of the resistor R in the circuit in Figure 2.76.

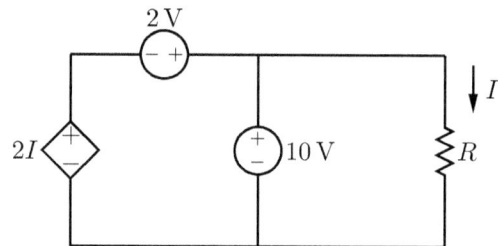

Figure 2.76

Section 2.7.2

2.18. Find the voltages V_1 and V_2 in the circuit in Figure 2.77.

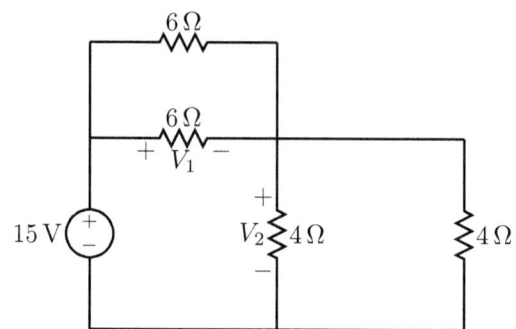

Figure 2.77

2.19. Find the voltages V_1 and V_2 in the circuit in Figure 2.78 if the voltage across the resistor R is given.

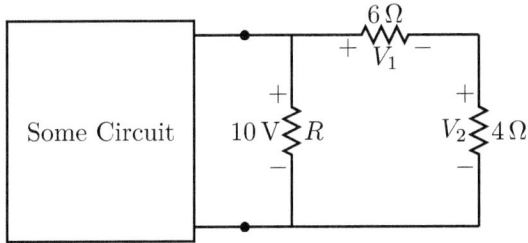

Figure 2.78

2.20. For the circuit in Figure 2.79, the voltage across the resistor R is given. Find whether each of the following statements is true or false.

a) If $R_1 > R_2$, then $V_1 > V_2$.

b) If $R_1 = 0.25R_2$, then $V_1 = 0.25V_2$.

c) If $R_1 = 3R_2$, then $V_1 = \frac{V_2}{3}$.

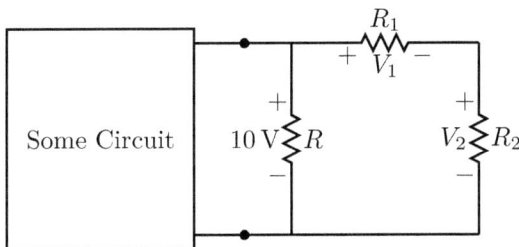

Figure 2.79

Section 2.8

2.21. For the circuit in Figure 2.80, find V.

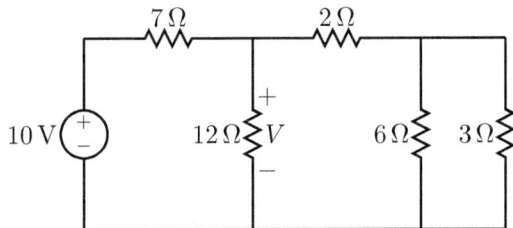

Figure 2.80

2.22. For the circuit in Figure 2.81, find I.

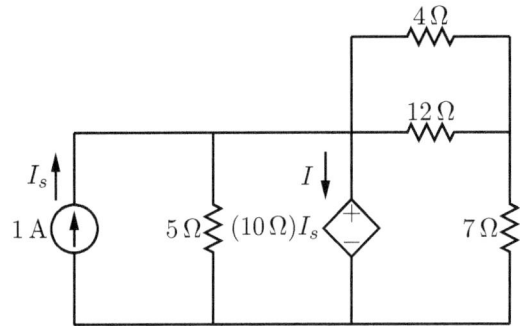

Figure 2.81

2.23. For the circuit in Figure 2.82, find the power delivered by the current source.

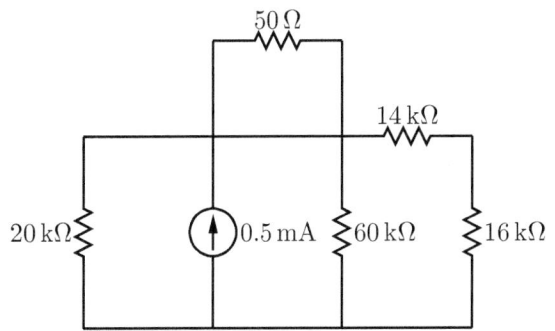

Figure 2.82

2.24. For the circuit in Figure 2.83, find V_1 and V_2.

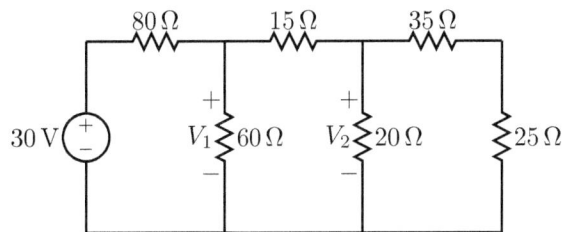

Figure 2.83

Section 2.9

2.25. For the circuit in Figure 2.84, find the equivalent resistance at the port a-b.

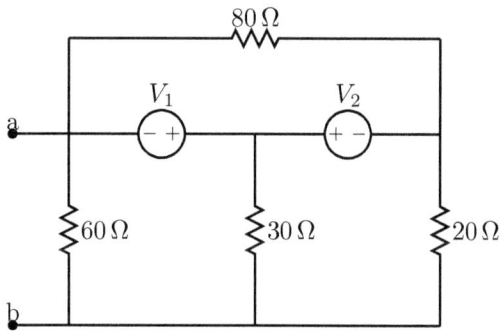

Figure 2.84

2.26. For the circuit in Figure 2.85, find the equivalent resistance at the port a-b.

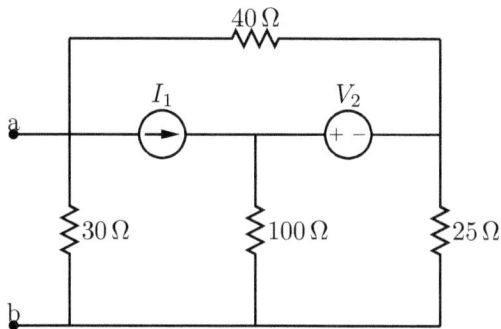

Figure 2.85

2.27. For the circuit in Figure 2.86, find the equivalent resistance at ports a-b and c-d.

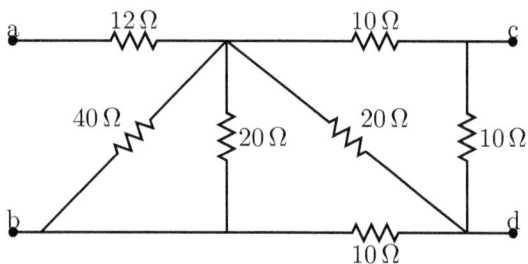

Figure 2.86

2.28. For the circuit in Figure 2.87, find the equivalent resistance seen by the current source.

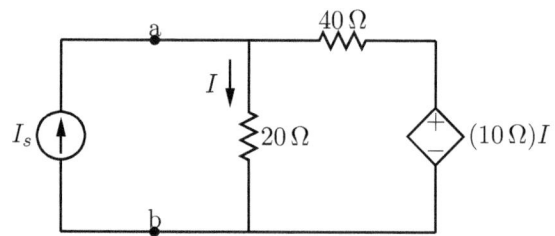

Figure 2.87

2.29. For the circuit in Figure 2.88, find the equivalent resistance at port a-b.

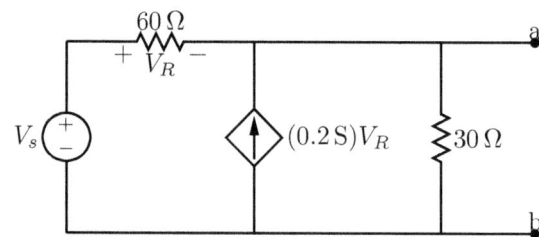

Figure 2.88

2.30. For the circuit in Figure 2.89, find the equivalent resistance at port a-b.

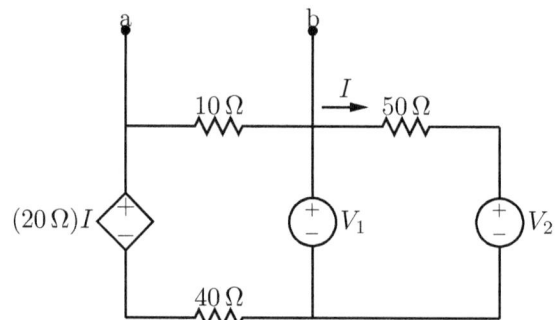

Figure 2.89

Chapter 3

Analysis Techniques Using Kirchhoff's Laws

Introduction

Our focus in this chapter is to learn some basic circuit analysis techniques and theorems for DC resistive circuits. We start with DC resistive circuits because they are the easiest to analyze. We will use these circuits to grasp the concepts and apply them to circuits that are more involved.

As discussed before, a DC circuit is a circuit where current flows in a constant direction and has a constant value. This usually occurs as a result of applying constant voltage or current sources to resistive circuits. There are specific tools and theorems that simplify the process of analyzing circuits, and we will learn them in this chapter. The following is a brief summary.

1. Nodal analysis

   This is a systematic method used to determine the voltage at every (essential) node in a circuit. After picking our reference (ground) node, we systematically apply KCL at every (essential) node in the circuit except, of course, the ground node. By finding every nodal voltage in the circuit, we can find every branch current, which enables us to determine the power absorbed or delivered by every element. Because of its ability to be applied to any circuit, nodal analysis is quite effective in both analytical and numerical approaches.

2. Mesh analysis

   This method is used to determine every loop current in a circuit. We use KVL around meshes (loops) to find the mesh (loop) currents. We can then calculate any voltage or any branch current from the resulting mesh currents. (Basic) mesh analysis has a limitation in that it can only be applied to planar circuits. We will discuss this limitation further in section 3.4.

3.1 Nodal Analysis

So far we have dealt with very small circuits that can be collapsed down to a few elements (a voltage source and a couple of resistors, for example) or with circuits that have many known currents or voltages. Typically, this is not the case, so a general analysis method becomes necessary. Nodal analysis is such a generic method. We will start developing an understanding of the method by starting with an example.

Figure 3.1 shows a circuit with known currents through some branches, and we are asked to find the rest of the unknown currents I_x, I_y, and I_z. We can use KCL at every node to determine the unknown currents.

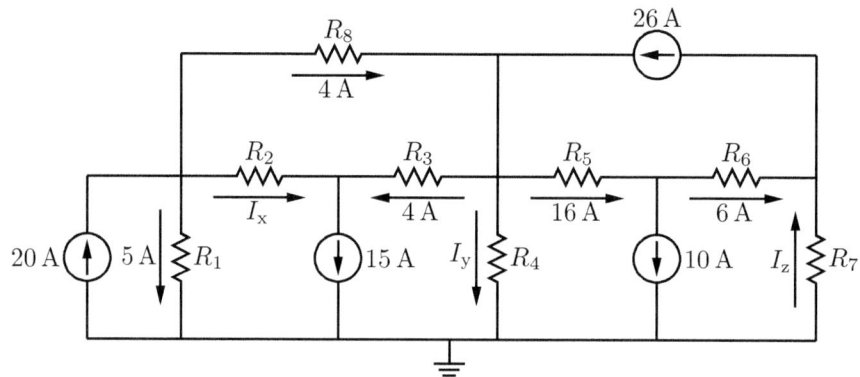

Figure 3.1: Simple multinode circuit.

Starting from the left side of the circuit,

$$-20\,\mathrm{A} + 5\,\mathrm{A} + I_x + 4\,\mathrm{A} = 0\,\mathrm{A},$$

or

$$I_x = 20\,\mathrm{A} - (4\,\mathrm{A} + 5\,\mathrm{A}) = 11\,\mathrm{A}.$$

Applying KCL at the node connecting R_3, R_4, R_5, and R_8 and the 26 A current source to calculate I_y yields

$$4\,\mathrm{A} + I_y + 16\,\mathrm{A} - 26\,\mathrm{A} - 4\,\mathrm{A} = 0\,\mathrm{A},$$

or

$$I_y = 26\,\mathrm{A} + 4\,\mathrm{A} - 16\,\mathrm{A} - 4\,\mathrm{A} = 10\,\mathrm{A}.$$

Applying KCL again at the right side of the circuit to calculate I_z results in

$$-6\,\mathrm{A} - I_z + 26\,\mathrm{A} = 0\,\mathrm{A},$$

or

$$I_z = 26\,\mathrm{A} - 6\,\mathrm{A} = 20\,\mathrm{A}.$$

While this was rather straightforward, what if we were asked to find the same missing currents while knowing only the sources, as in Figure 3.2? At first it may seem hopeless, since a KCL at any node will only yield more unknowns.

This is where nodal analysis shows its power. It is no surprise that, according to many engineers, if only one tool could be given to students, nodal analysis would be the first choice.

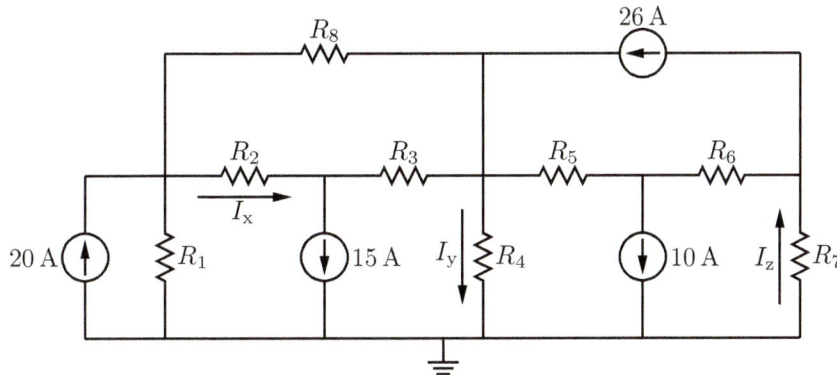

Figure 3.2: Simple multinode circuit.

The key idea is not to focus on the currents flowing through the branches of the circuit but rather on its *nodal voltages*. For the given circuit, we have five nodes (the sixth one is grounded, so the voltage there is zero). These nodal voltages are labeled V_1 through V_5 in Figure 3.3.

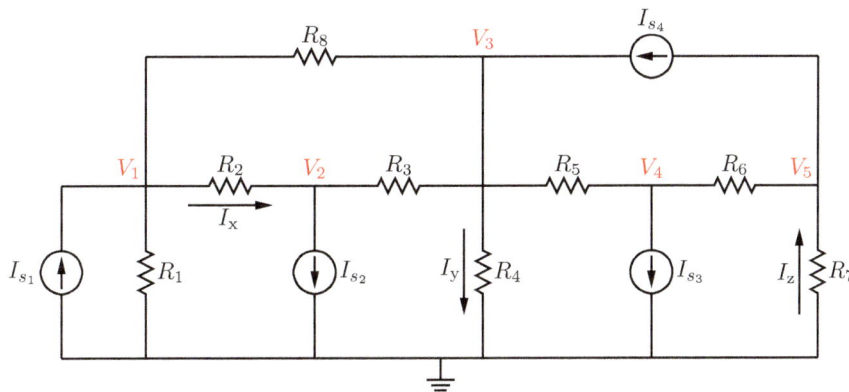

Figure 3.3: Simple multinode circuit.

Notice that if we know V_1, V_2, V_3, V_4, and V_5, we can find our unknown currents and everything else for this circuit. For example, the unknown currents are given by

$$I_\text{x} = \frac{V_1 - V_2}{R_2},$$

$$I_\text{y} = \frac{V_3 - 0\,\text{V}}{R_4},$$

and

$$I_\text{z} = \frac{0\,\text{V} - V_5}{R_7}.$$

Therefore, the key question is how to find these nodal voltages. Since we have five unknown voltages, we need a set of five independent equations. We knew from the beginning that KCL was the answer but did not know how to implement it; now the picture is getting clearer: we have five essential nodes in the circuit and we need to write a KCL equation at each one of the nodes, keeping in mind that all currents have to be represented by the five nodal voltages. Let us do that below:

> Reminder: An **essential node** is a node that connects more than two branches—in other words, a current will split going through that node.

Starting with node 1, we sum all currents, leaving the node as

$$-I_{s_1} + \frac{V_1}{R_1} + \frac{V_1 - V_2}{R_2} + \frac{V_1 - V_3}{R_8} = 0\,\text{A}. \tag{Node 1}$$

Similarly for the other nodes, we get

$$I_{s_2} + \frac{V_2 - V_1}{R_2} + \frac{V_2 - V_3}{R_3} = 0\,\text{A}, \tag{Node 2}$$

$$-I_{s_4} + \frac{V_3}{R_4} + \frac{V_3 - V_4}{R_5} + \frac{V_3 - V_1}{R_8} + \frac{V_3 - V_2}{R_3} = 0\,\text{A}, \tag{Node 3}$$

$$I_{s_3} + \frac{V_4 - V_3}{R_5} + \frac{V_4 - V_5}{R_6} = 0\,\text{A}, \tag{Node 4}$$

and

$$I_{s_4} + \frac{V_5}{R_7} + \frac{V_5 - V_4}{R_6} = 0\,\text{A}. \tag{Node 5}$$

Not so bad, is it? We have fairly well-organized first-order linear equations that only have our nodal voltages as unknowns. If we are to solve them, we will need the equations to be organized better. The first thing that comes to mind is to pull all unknowns on one side and the known quantities on the other. Also, we can group all coefficients of the same nodal voltages together. Let us go ahead and do this.

Node 1:
$$\left(\frac{1}{R_1} + \frac{1}{R_2} + \frac{1}{R_8}\right) V_1 - \left(\frac{1}{R_2}\right) V_2 - \left(\frac{1}{R_8}\right) V_3 = I_{s_1}. \tag{3.1}$$

Node 2:
$$-\left(\frac{1}{R_2}\right) V_1 + \left(\frac{1}{R_2} + \frac{1}{R_3}\right) V_2 - \left(\frac{1}{R_3}\right) V_3 = -I_{s_2}. \tag{3.2}$$

Node 3:
$$-\left(\frac{1}{R_8}\right) V_1 - \left(\frac{1}{R_3}\right) V_2 + \left(\frac{1}{R_3} + \frac{1}{R_4} + \frac{1}{R_5} + \frac{1}{R_8}\right) V_3 - \left(\frac{1}{R_5}\right) V_4 = I_{s_4}. \tag{3.3}$$

Node 4:
$$-\left(\frac{1}{R_5}\right) V_3 + \left(\frac{1}{R_5} + \frac{1}{R_6}\right) V_4 - \left(\frac{1}{R_6}\right) V_5 = -I_{s_3}. \tag{3.4}$$

Node 5:
$$-\left(\frac{1}{R_6}\right) V_4 + \left(\frac{1}{R_6} + \frac{1}{R_7}\right) V_5 = -I_{s_4}. \tag{3.5}$$

Since we are dealing with more than two equations, we employ matrices to solve for our unknowns with ease. We have five equations and five unknowns, which means we will have a 5×5 matrix and two 5×1 vectors. The 5×5 matrix is constructed as follows.

$$\underbrace{\begin{bmatrix} \frac{1}{R_1}+\frac{1}{R_2}+\frac{1}{R_8} & -\frac{1}{R_2} & -\frac{1}{R_8} & 0 & 0 \\ -\frac{1}{R_2} & \frac{1}{R_2}+\frac{1}{R_3} & -\frac{1}{R_3} & 0 & 0 \\ -\frac{1}{R_8} & -\frac{1}{R_3} & \frac{1}{R_3}+\frac{1}{R_4}+\frac{1}{R_5}+\frac{1}{R_8} & -\frac{1}{R_5} & 0 \\ 0 & 0 & -\frac{1}{R_5} & \frac{1}{R_5}+\frac{1}{R_6} & -\frac{1}{R_6} \\ 0 & 0 & 0 & -\frac{1}{R_6} & \frac{1}{R_6}+\frac{1}{R_7} \end{bmatrix}}_{\mathbf{G}} \quad (3.6)$$

The matrix $\mathbf{G}$ is called the conductance matrix because the unit for all elements is siemens. The system of equations can now be written as

$$\underbrace{\begin{bmatrix} \frac{1}{R_1}+\frac{1}{R_2}+\frac{1}{R_8} & -\frac{1}{R_2} & \cdots & 0 \\ -\frac{1}{R_2} & \frac{1}{R_2}+\frac{1}{R_3} & \cdots & 0 \\ \vdots & \vdots & \cdots & \vdots \\ \vdots & \vdots & \cdots & \vdots \\ 0 & 0 & \cdots & \frac{1}{R_6}+\frac{1}{R_7} \end{bmatrix}}_{\mathbf{G}} \underbrace{\begin{bmatrix} V_1 \\ V_2 \\ V_3 \\ V_4 \\ V_5 \end{bmatrix}}_{\vec{V}} = \underbrace{\begin{bmatrix} I_{s_1} \\ -I_{s_2} \\ I_{s_4} \\ -I_{s_3} \\ -I_{s_4} \end{bmatrix}}_{\vec{I}} \quad (3.7)$$

The vector $\vec{V}$ is the vector of unknown nodal voltages, and $\vec{I}$ is the vector of source currents. In short form:

$$\mathbf{G}\vec{V} = \vec{I}. \quad (3.8)$$

All we have to do now is to invert the matrix and solve for the nodal voltages:

$$\vec{V} = \mathbf{G}^{-1}\vec{I}. \quad (3.9)$$

This can be very easily done with modern computer tools in both symbolic and numeric forms.

In a nutshell, this is nodal analysis. Let us review the basic steps here:

1. Number all (essential) nodes of the given circuit.

2. Write KCL for every (essential) node by keeping in mind that only nodal voltages should be used (no currents). If other unknowns are involved (e.g., dependent source equations), express them as a function of the unknown nodal voltages (e.g., by using Ohm's law).

3. Group the resulting equations together in a matrix form.

4. Solve for the unknown nodal voltages by inverting the resulting linear equation.

5. Calculate any quantity of interest (e.g., power consumption) from the known nodal voltages.

In addition, here are a few important notes.

- Always (yes, with no exceptions) apply KCL in the same consistent way no matter what circuit has been given and regardless of how tempting it may be to "invent a new method." The consistent way you should stick with has been presented above and is summarized here: sum all currents, *leaving* a node to be equal to zero. Consistency is key for learning basic skills in life (e.g., for developing a good swing in sports or for learning how to walk) and the same is true for learning nodal analysis.

- Treat each node independently. Notice, for example, that if a current source is between two nodes, it will show up in both equations of these nodes but will carry opposite directions. This makes perfect sense since the current will leave one node and enter the another. This is similar to measuring trade between two states. Exports from one state appear as imports for the other and vice versa.

- Most of the time, we have the freedom to choose our ground node, and a good rule of thumb is to choose the node with the most branches connected to it. Why, you ask? To prevent headaches is a perfectly acceptable answer. Notice how the equations get longer the more branches we have connected to the node, as seen in (3.1) for node 1 and (3.3) for node 3. Typically we draw circuit schematics so the node with the highest number of connected branches appears at the bottom of the circuit. Hence, unless otherwise specified, the bottom node of the circuit is taken as the ground node.

Exercise 3.1.1

For the circuit in Figure 3.4, use nodal analysis to find the final equations in matrix form to solve for the nodal voltages V_1 and V_2.

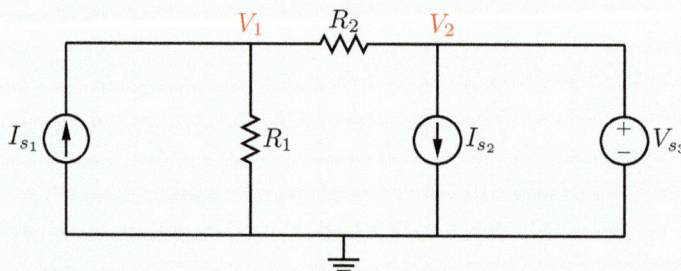

Figure 3.4: Exercise 3.1.1.

$$\begin{bmatrix} \frac{1}{R_1} + \frac{1}{R_2} & -\frac{1}{R_2} \\ 0 & 1 \end{bmatrix} \begin{bmatrix} V_1 \\ V_2 \end{bmatrix} = \begin{bmatrix} I_{s_1} \\ V_{s_3} \end{bmatrix}.$$

3.2 Nodal Analysis: Interesting Cases

In this section, we explore some interesting examples that may present difficulties for some students. Specifically, we look at more involved situations that help us build a higher level of understanding of how nodal analysis works. Just like you can make your exercise more challenging by adding some weights or doing some jumping jacks, adding more elements between our essential nodes provides similar benefits.

3.2.1 Case 1: Dependent Sources

The first interesting case to look at is the existence of a dependent source. Let us jump into an example right away.

Example 3.2.1

We want to find the nodal voltages V_1, V_2, and V_3 in the circuit of Figure 3.5. Notice that we have specified conductances (not resistances) in the schematic.

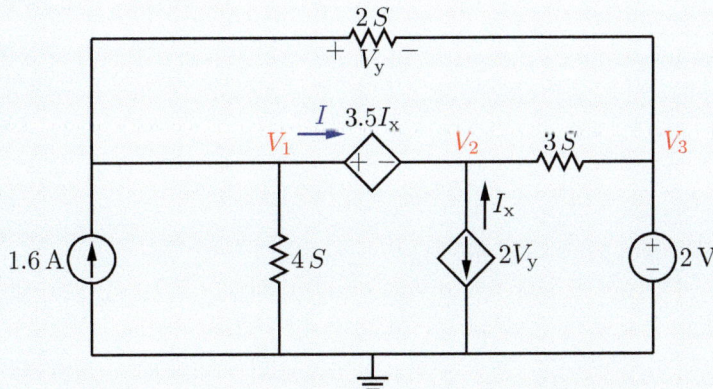

Figure 3.5: Example 3.2.1.

Just glancing quickly through the circuit schematic, we might think that there are three essential node voltages. Consequently, we will have a 3×3 system of equations. Looking at this circuit more closely though reveals that the nodal voltage V_3 is actually known since it is directly connected to a grounded voltage source. Thus we can immediately write the node 3 equation:

$$V_3 = 2\,\text{V}. \tag{3.10}$$

This case occurs quite often in circuits, so keep it in mind next time you see a **grounded** voltage source. Both nodes of such a voltage source have known voltages: the grounded end will be equal to zero, while the other end will have a voltage equal to the voltage source. Of course, this applies to both independent and dependent **grounded** voltage sources.

Now we can shift our attention to the other two nodal voltages. In this circuit, we have two dependent sources: a CCVS between two nodal voltages and a VCCS between a nodal voltage and ground. We have not seen such cases before, so we will have to be extra careful here to avoid making errors.

Let us start with the CCVS. The difficulty here is that this is a **floating voltage source**. If we were to attempt to write the nodal equation for either of its two nodes the way we have done it so far, we would have a problem: the current leaving a node through a voltage source is unknown! To address this problem, we will add a *dummy current variable I*. This dummy variable will become an additional unknown. We will see soon how to find it. With this dummy variable, we can write the nodal equations for both nodes 1 and 2.

Node 1 equation (after cleanup):

$$(4+2)V_1 - 2V_3 = 1.6 - I, \text{or}$$

$$6V_1 - 2V_3 = 1.6 - I. \tag{3.11}$$

Node 2 equation (after cleanup):

$$3V_2 - 3V_3 = -2V_y + I. \tag{3.12}$$

We notice that we can add (3.11) to (3.12) to get rid of our dummy variable.

$$6V_1 + 3V_2 - 5V_3 = 1.6 - 2V_y. \tag{3.13}$$

We now have two equations, (3.10) and (3.13), but four unknowns, V_1, V_2, V_3, and V_y. We will need an additional equation for the voltages at nodes 1 and 2 as well as an additional equation for V_y.

The equation that relates the node 1 and 2 voltages can be readily found by realizing that there is a voltage source between them. Hence

$$V_1 - V_2 = 3.5I_x. \tag{3.14}$$

The current I_x is given by

$$I_x = -2V_y. \tag{3.15}$$

Plugging (3.15) into (3.14) yields

$$V_1 - V_2 = -7V_y. \tag{3.16}$$

Now let us focus on V_y since both (3.13) and (3.16) depend on it. We can readily express V_y with respect to the nodal voltages as

$$V_1 - V_3 = V_y. \tag{3.17}$$

Then plugging (3.17) into (3.13), we get

$$\boxed{8V_1 + 3V_2 - 7V_3 = 1.6,} \tag{3.18}$$

and plugging (3.17) into (3.16) gives us

$$\boxed{8V_1 - 1V_2 - 7V_3 = 0.} \tag{3.19}$$

We now have three equations with three unknowns (the nodal voltages). Specifically, (3.10), (3.18), and (3.19) make up the system of equations to solve. Note that you could easily plug (3.10) into (3.18) and (3.19) to reduce this to a 2×2 system. It is all a matter of preference.

$$\begin{bmatrix} 8 & 3 & -7 \\ 8 & -1 & -7 \\ 0 & 0 & 1 \end{bmatrix} \begin{bmatrix} V_1 \\ V_2 \\ V_3 \end{bmatrix} = \begin{bmatrix} 1.6 \\ 0 \\ 2 \end{bmatrix}.$$

Solving this yields

$$\begin{bmatrix} V_1 \\ V_2 \\ V_3 \end{bmatrix} = \begin{bmatrix} 1.8 \\ 0.4 \\ 2.0 \end{bmatrix} \text{ V.}$$

To summarize, these are the main concepts we learned here:

1. **Grounded voltage source**: A grounded voltage source is great because it immediately reveals the value of a nodal voltage.

2. **Floating voltage source between two essential nodes**: A floating voltage source requires a dummy variable that is needed to write the currents in each of its nodes. By adding the nodal equations with the dummy variable together, you can get rid of it and continue normally.

3. **Dependent source**: You can treat dependent sources the same way as independent ones, except that you will need to add an equation that links its controlling quantity (voltage or current) to the nodal voltages.

3.2.2 Case 2: A Resistor and a Voltage Source Between a Node and Ground

Next let us try to solve the following symbolic Example 3.2.2. It is sometimes beneficial to solve without numbers. Numbers can throw us off and mask some key points that can be extracted and applied in general.

Example 3.2.2

For the circuit in Figure 3.6, find the final nodal equations for the nodal voltages V_1 and V_2 in a matrix form.

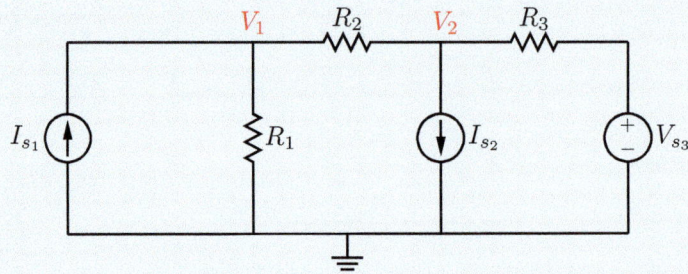

Figure 3.6: Example 3.2.2.

Finding the equation for node 1 is straightforward based on what we have learned so far. Below is the final form of the equation. Try to verify it yourself as practice.

$$\left(\frac{1}{R_1} + \frac{1}{R_2}\right) V_1 - \frac{1}{R_2} V_2 = I_{s_1}.$$

The interesting part starts here. The current leaving node 2 and passing through R_3 can be written as

$$I_{R_3} = \frac{V_2 - V_{s_3}}{R_3}.$$

Now we can use KCL to find the equation for node 2:

$$\frac{V_2 - V_1}{R_2} + I_{R_3} + I_{s_2} = 0.$$

Rearranging variables and cleaning up the equation yields

$$-\frac{1}{R_2} V_1 + \left(\frac{1}{R_2} + \frac{1}{R_3}\right) V_2 = \frac{V_{s_3}}{R_3} - I_{s_2}.$$

The matrix form is

$$\begin{bmatrix} \frac{1}{R_1} + \frac{1}{R_2} & -\frac{1}{R_2} \\ -\frac{1}{R_2} & \frac{1}{R_2} + \frac{1}{R_3} \end{bmatrix} \begin{bmatrix} V_1 \\ V_2 \end{bmatrix} = \begin{bmatrix} I_{s_1} \\ \frac{V_{s_3}}{R_3} - I_{s_2} \end{bmatrix}.$$

We can see why this example was better with symbols. Notice the final node 2 equation. The branch that contains the resistor R_3 and the voltage source V_{s_3} acts as a current source with a value $\frac{V_{s_3}}{R_3}$ that flows into node 2. Now every time you encounter a resistor in series with a voltage source between a node voltage and ground, you might try to convert it to a current source or to just pay attention to the values that will end up in your matrix. We will discuss source transformations in more detail later as well.

3.2.3 Case 3: A Voltage Source and a Resistor Between Two Nodes

This is another example we solve symbolically to show how to deal with a voltage source and a resistor between two nodal voltages.

Example 3.2.3

For the circuit in Figure 3.7, find the final equation (matrix form) to solve for the nodal voltages V_1 and V_2.

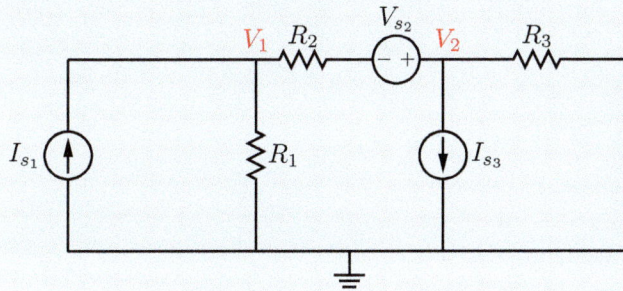

Figure 3.7: Example 3.2.3.

Before we go any further, we have to introduce some dummy variables into the circuit, as seen in Figure 3.8, in order to solve it. The current I_x is a dummy variable we have seen before: it is necessary due to the presence of the floating voltage source. The voltage V_x is a voltage we define in a **nonessential node**. This is necessary because one end of the floating voltage source is connected to this nonessential node. So we need to know its voltage, as will be seen later.

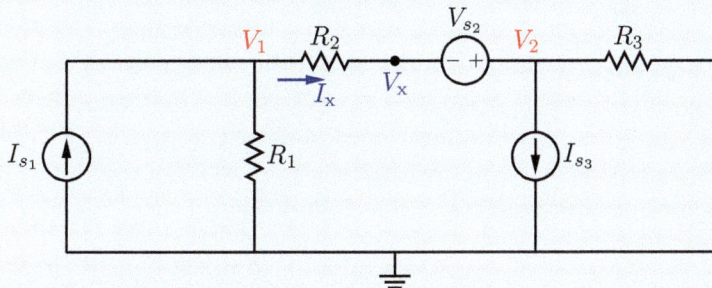

Figure 3.8: Placing dummy variables.

Now we can begin writing the nodal equations starting with node 1.

$$\frac{1}{R_1}V_1 = I_{s_1} - I_x. \tag{3.20}$$

The node 2 equation becomes

$$\frac{1}{R_3}V_2 = -I_{s_3} + I_x. \tag{3.21}$$

We might be tempted right now to just add (3.20) to (3.21) to get rid of I_x, but it can't be that simple if we are dedicating an example to it. The trouble here is that V_1 and V_2 are not directly linked to the floating voltage source as before. So we cannot simply get rid of I_x and then write an additional equation for the two essential nodes and the floating voltage source.

Instead, we will take an alternative route. Let us try to find an equation for I_x:

$$I_x = \frac{V_1 - V_x}{R_2}, \text{or}$$

$$I_x = \frac{1}{R_2}V_1 - \frac{1}{R_2}V_x. \tag{3.22}$$

Now we need an equation for V_x:

$$V_2 - V_x = V_{s_2}, \text{or}$$

$$V_x = V_2 - V_{s_2}. \tag{3.23}$$

Substituting (3.23) into (3.22) will result in an equation for I_x that depends only on the nodal voltages as follows.

$$I_x = \frac{1}{R_2}V_1 - \frac{1}{R_2}V_2 + \frac{V_{s_2}}{R_2} \tag{3.24}$$

Now let us plug (3.24) into the node 1 equation (3.20) and reorganize.

$$\left(\frac{1}{R_1} + \frac{1}{R_2}\right)V_1 - \left(\frac{1}{R_2}\right)V_2 = I_{s_1} - \frac{V_{s_2}}{R_2} \tag{3.25}$$

Let us take a look at the resulting equation (3.25). The first thing to notice is that although the nodal voltage V_2 seems far from node 1, it still affects it. The next thing to notice is that the voltage source contributes with a current whose magnitude is equal to the voltage source value divided by the resistor between node 1 and the source $\left(\frac{V_{s_2}}{R_2}\right)$. This is just like the previous case in Example 3.2.2. We can draw the circuit again to show this result, as seen in Figure 3.9.

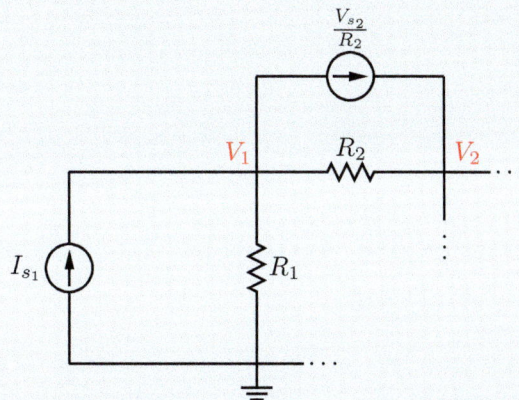

Figure 3.9: Equivalent circuit of the voltage source in series with the resistor.

As mentioned before, this equivalence will become more clear after we discuss source transformations.

We can do the same thing again for the node 2 equation. We will plug (3.24) into the node 2 equation (3.21) and reorganize:

$$-\left(\frac{1}{R_2}\right) V_1 + \left(\frac{1}{R_2} + \frac{1}{R_3}\right) V_2 = \frac{V_{s_2}}{R_2} - I_{s_3}. \tag{3.26}$$

The resulting system of equations can be written as

$$\begin{bmatrix} \frac{1}{R_1} + \frac{1}{R_2} & -\frac{1}{R_2} \\ -\frac{1}{R_2} & \frac{1}{R_2} + \frac{1}{R_3} \end{bmatrix} \begin{bmatrix} V_1 \\ V_2 \end{bmatrix} = \begin{bmatrix} I_{s_1} - \frac{V_{s_2}}{R_2} \\ -I_{s_3} + \frac{V_{s_2}}{R_2} \end{bmatrix}.$$

The following exercise contains both cases of a) a resistor and a voltage source between a node and ground and b) a voltage source and a resistor between two node voltages.

Exercise 3.2.1

For the circuit in Figure 3.10, find the final nodal equations for the nodal voltages V_1 and V_2 in a matrix form.

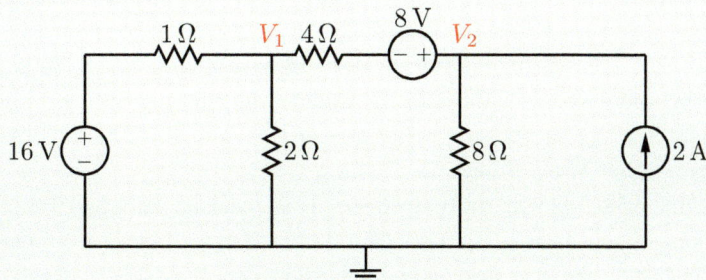

Figure 3.10: Exercise 3.2.1.

$$\begin{bmatrix} \frac{7}{4} & -\frac{1}{4} \\ -\frac{1}{4} & \frac{3}{8} \end{bmatrix} \begin{bmatrix} V_1 \\ V_2 \end{bmatrix} = \begin{bmatrix} 14 \\ 4 \end{bmatrix}.$$

Solving for the voltages yields $V_1 = 10.53$ V and $V_2 = 17.68$ V.

3.3 An Important Concept to Review Before Mesh Analysis

Mesh or **loop** analysis is the second generic circuit solution we will learn. Before doing this, though, we will focus on understanding current and its direction more thoroughly. Figure 3.11a shows a simple circuit with given currents I_a and I_b. If we want to find the value of I_x, we simply use KCL to write

$$I_x = I_a - I_b. \tag{3.27}$$

What if we want to find the current I_y in Figure 3.11b? Again using KCL, we write

$$I_y = I_b - I_a. \tag{3.28}$$

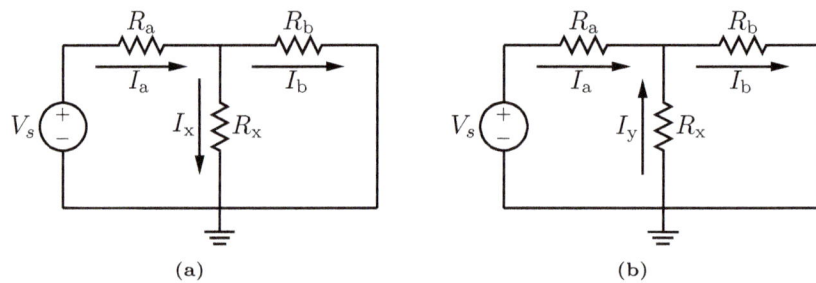

Figure 3.11: Simple circuit with element currents.

The currents I_a, I_b, I_x, and I_y in Figure 3.11 are called **element currents** because they are the currents that flow through individual elements of the circuit. Therefore, the circuit we have here contains three element currents. Of course, we know that we only need to find two of the currents since the third one will be given by KCL. Consequently, it would be suboptimal to develop a method that focuses on finding element currents. Remember that this was the case in nodal analysis as well. In nodal analysis, we did not try to find the element voltage drops but rather the nodal voltages. This resulted in the minimum number of equations.

The same principle will apply to the analysis method (**mesh** or **loop**) we will develop focused on finding currents in a circuit. Instead of directly calculating element currents, we will first calculate other currents, called **mesh** or **loop currents**.

To see how this works, let us redraw the two circuits in Figure 3.11 by adding mesh currents. In Figure 3.12, we have defined two mesh currents, one for each

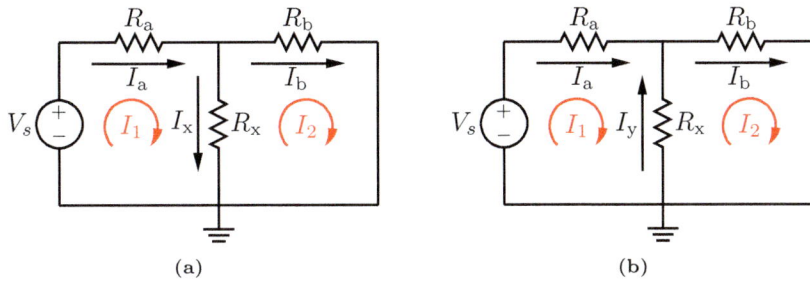

Figure 3.12: Simple circuit with mesh currents.

simple loop of the circuit. The mesh currents are in red with lowercase letters and are surrounded by a loop that indicates their direction—hence the name "loop currents." Someone might ask, "Why do we need these loop currents and how is I_1 different from I_a?" As mentioned before, the currents I_a, I_b, and I_x are element currents. Specifically, the current I_a represents the current that flows through the source V_s and the resistor R_a. The current I_x is the current that flows through the resistor R_x. On the other hand, the loop current I_1 is the current that flows through all elements in its closed path (simple left loop of the circuit). So I_1 flows through elements V_s, R_a, and R_x in the clockwise direction. The same applies for the loop current I_2, which flows through the right simple closed loop. This current passes through both elements R_x and R_b in the clockwise direction, while I_b passes through only the resistor R_b.

Notice that both I_1 and I_2 flow through the resistor R_x but in opposite directions. This is a fact that we also stated in (3.27) in a different way. In other words, we proved that the current I_x is a combination of the currents I_a and I_b. So if we wanted to write an equation for I_x in Figure 3.12a in terms of the mesh currents, we would write

$$I_x = I_1 - I_2. \tag{3.29}$$

Similarly, if we wanted to write an equation for I_y in Figure 3.12b in terms of the mesh currents, we would write

$$I_y = I_2 - I_1. \tag{3.30}$$

We can see now that elements that are not shared among meshes (loops) will carry the same current as the mesh current in their loop—that is,

$$I_a = I_1$$

and

$$I_b = I_2.$$

On the other hand, circuit elements that are included in many loops carry all the currents of the loops to which they belong. Current directions are very important, as demonstrated in (3.29) and (3.30). Consequently, it is critical to clearly and explicitly define both element and mesh currents.

Let us clarify these concepts further by applying KVL to solve the aforementioned circuits. Let us solve the circuit in Figure 3.13 to find the element currents I_a, I_b, and I_x. We will first try to solve it using only element currents. Second, we will try it again with mesh currents. Of course, as we will later learn, we can simply solve this circuit by combining the parallel resistors R_x and R_b, then finding I_a using voltage division, and finally using current division to find I_b and I_x. Nevertheless, we are using this simple circuit to illustrate some techniques that will be used to solve more complex circuits where simply combining resistors does not work.

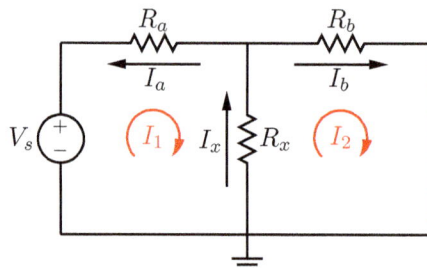

Figure 3.13: Simple circuit with element *and* mesh currents.

First technique (KVL with element currents): starting with the left loop, KVL yields

$$-V_s - R_a I_a - R_x I_x = 0. \tag{3.31}$$

In addition, KVL at the right loop also yields

$$R_x I_x + R_b I_b = 0. \tag{3.32}$$

We have two equations and three unknowns, so we need a third equation (which is KCL):

$$I_x = I_a + I_b. \tag{3.33}$$

Plugging (3.33) into both (3.31) and (3.32), we get

$$-V_s - R_a I_a - R_x(I_a + I_b) = 0$$
$$-(R_a + R_x)I_a - R_x I_b = V_s,$$

and

$$R_x(I_a + I_b) + R_b I_b = 0$$
$$R_x I_a + (R_b + R_x)I_b = 0.$$

We now have two equations with two unknowns that we can easily solve for I_a and I_b.

Second technique (KVL with mesh currents): now let us try to write these equations using mesh currents. Remember that a mesh current is constant along a closed loop. Applying KVL at the first loop yields

$$-V_s + R_a I_1 + R_x (I_1 - I_2) = 0. \tag{3.34}$$

Notice the current through R_x that is influenced by both loops. Applying KVL at the second loop gives

$$R_b I_2 + R_x (I_2 - I_1) = 0. \tag{3.35}$$

We have two equations with two unknowns that we can solve for, I_1 and I_2. Once the mesh currents are known, we will be able to find the element currents as

$$I_a = -I_1.$$
$$I_b = I_2.$$
$$I_x = I_2 - I_1.$$

A mesh current's direction is important. You have the option to choose any direction you wish; after all, the directions are mere variables we create to solve the circuits. Figure 3.14 shows all different mesh current directions for our example circuit. Knowing how to write equations for every mesh current direction is a good practice to help understand how mesh currents work. Consistency, though, is key. Picking the *same* direction and *sticking* with it every time you solve a circuit is an excellent way to minimize calculation errors. In this text we will always define mesh currents in the clockwise direction. Someone might ask, "What if I encounter a question with mesh currents already determined, like in an exam, for example?" With an understanding that reversing directions yields a negative sign, we respond that you are faced with two choices:

1. Solve it as is, keeping in mind that the directions are different now. We will demonstrate how to do this for Figures 3.14b and 3.14c.

2. Armed with confidence, change the directions as you are comfortable with, and then if the question is asking for only the mesh currents, change the sign of any mesh current result that had an opposite sign to your choice. If element currents are needed, you will have to see how the element currents relate to your choice of mesh current direction.

Either way you choose, you have to understand directions of current and how they change voltage drops across elements.

Let us now practice these with our example circuit. We will start by writing the equations for Figure 3.14b. Here both mesh currents are in the same anti-clockwise direction.

Loop 1:

$$R_x(I_1 - I_2) + R_a I_1 + V_s = 0$$
$$(R_a + R_x)I_1 - R_x I_2 = -V_s.$$

Loop 2:

$$R_b I_2 + R_x(I_2 - I_1) = 0$$
$$-R_x I_1 + (R_x + R_b)I_2 = 0.$$

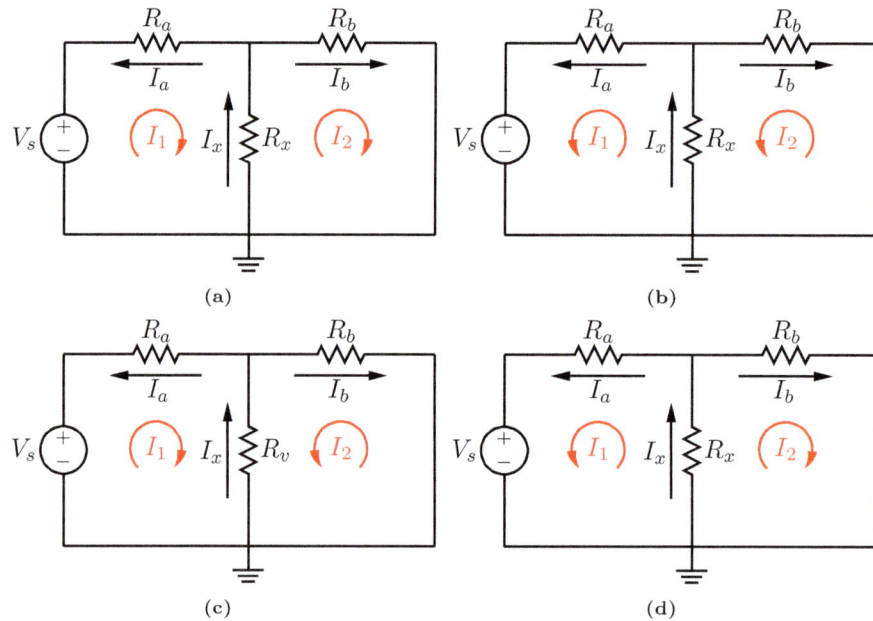

Figure 3.14: Our example circuit with mesh currents defined in all different directions.

After we solve the equations, we can find the element currents as

$$I_a = I_1$$
$$I_b = -I_2$$
$$I_x = I_1 - I_2.$$

Now let us do the same for Figure 3.14c.

Loop 1:

$$-V_s + R_a I_1 + R_x(I_1 + I_2) = 0.$$
$$(R_a + R_x)I_1 + R_x I_2 = V_s.$$

Notice here how the current that goes into the resistor R_x is the sum of both loop currents since they enter the resistor R_x in the same direction.

Loop 2:

$$R_b I_2 + R_x(I_2 + I_1) = 0.$$
$$R_x I_1 + (R_x + R_b)I_2 = 0.$$

Again R_2 gets the sum of both loop currents. After solving the equations, we write

$$I_a = -I_1$$
$$I_b = -I_2$$
$$-I_x = I_1 + I_2.$$

As can be seen from all equations we have solved so far, the direction of the mesh currents is important and staying organized and consistent helps you minimize confusion and errors. It is also important to mention that the choice of meshes and loops is arbitrary, but it is usually better to divide the circuit into its smallest loops to avoid errors and confusion.

3.4 Mesh Analysis

Mesh analysis is an important tool to analyze/solve circuits. Like the node voltages in nodal analysis, we will have mesh currents throughout the circuit, and by solving for them, we will be able to find everything else in the circuit. In mesh analysis, we use KVL around meshes (loops) in the circuit to find the necessary equations. In the previous section, we were able to see that we will have fewer unknowns to solve for if we use mesh currents rather than element currents. Mesh analysis is often regarded as a friendlier tool to use than nodal analysis, but unfortunately mesh analysis can only be used on planar circuits. A planar circuit is a circuit that can be drawn in a two-dimensional plane with no overlap between unconnected elements. Nonplanar circuits are not covered in this text, so both methods can be applied to any circuit in this book. After covering mesh analysis, we will learn how to choose between methods.

3.4.1 Case 1: The Basic Circuit

Examples are often the best way to explain an idea or a method, so let us focus on the circuit in Figure 3.15. Suppose that we were asked to find all element currents in the circuit. We first notice that we cannot solve this circuit by combining resistors to simplify it (more on this later). Second, we notice that we could easily solve this problem with nodal analysis: we have three nodal voltages to solve for and then we can calculate all currents. In fact, it would be a good idea for you to try this out and practice your nodal analysis skills. Third, we see the circuit has seven element currents but only four mesh currents. So, logically, we will solve for mesh currents and then find all element currents after that. The mesh equations, starting with mesh 1, are

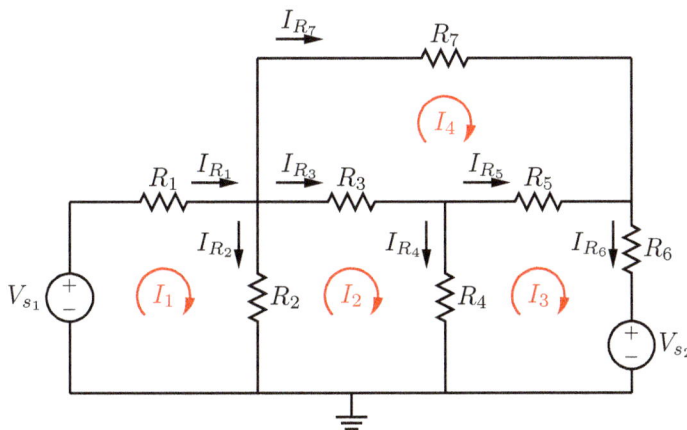

Figure 3.15: Simple circuit with four mesh currents.

Mesh 1:

$$-V_{s_1} + R_1 I_1 + R_2(I_1 - I_2) = 0$$
$$(R_1 + R_2)I_1 - R_2 I_2 = V_{s_1}. \tag{3.36}$$

Mesh 2:

$$R_2(I_2 - I_1) + R_3(I_2 - I_4) + R_4(I_2 - I_3) = 0$$
$$-R_2 I_1 + (R_2 + R_3 + R_4)I_2 - R_4 I_3 - R_3 I_4 = 0. \tag{3.37}$$

Mesh 3:

$$R_4(I_3 - I_2) + R_5(I_3 - I_4) + R_6 I_3 + V_{s_2} = 0$$
$$-R_4 I_2 + (R_4 + R_5 + R_6)I_3 - R_5 I_4 = -V_{s_2}. \tag{3.38}$$

Mesh 4:

$$R_3(I_4 - I_2) + R_5(I_4 - I_3) + R_7 I_4 = 0$$
$$-R_3 I_2 - R_5 I_3 + (R_3 + R_5 + R_7)I_4 = 0. \tag{3.39}$$

Since we have more than two equations to solve for, it is easier to use the matrix form of the equations to avoid calculation errors. We start by building the 4×4 matrix:

$$\underbrace{\begin{bmatrix} R_1 + R_2 & -R_2 & 0 & 0 \\ -R_2 & R_2 + R_3 + R_4 & -R_4 & -R_3 \\ 0 & -R_4 & R_4 + R_5 + R_6 & -R_5 \\ 0 & -R_3 & -R_5 & R_3 + R_5 + R_7 \end{bmatrix}}_{\mathbf{R}} \tag{3.40}$$

where $\mathbf{R}$ is the resistance matrix. The system of equations can now be written as

$$\underbrace{\begin{bmatrix} R_1 + R_2 & \dots & \dots & 0 \\ \vdots & R_2 + R_3 + R_4 & \dots & -R_3 \\ \vdots & \dots & \dots & \vdots \\ 0 & \dots & \dots & R_3 + R_5 + R_7 \end{bmatrix}}_{\mathbf{R}} \underbrace{\begin{bmatrix} I_1 \\ I_2 \\ I_3 \\ I_4 \end{bmatrix}}_{\vec{I}} = \underbrace{\begin{bmatrix} V_{s_1} \\ 0 \\ -V_{s_2} \\ 0 \end{bmatrix}}_{\vec{V}}. \tag{3.41}$$

In short form, this is

$$\mathbf{R}\,\vec{I} = \vec{V} \tag{3.42}$$

where $\vec{I}$ is the vector of the mesh currents and $\vec{V}$ is the vector of source voltages. We can solve for the mesh currents $\vec{I}$ by inverting the resistance matrix $\mathbf{R}$ and multiplying it by the source vector $\vec{V}$:

$$\vec{I} = \mathbf{R}^{-1}\,\vec{V}. \tag{3.43}$$

3.4.2   Case 2: A Known Mesh Current

Let us try another example. Figure 3.16 shows the same circuit as Figure 3.15 but with some changes. Let us see how we can deal with these changes. One change to notice is the mesh that contains a voltage source without a resistor in series on its unshared branch (mesh 4). Is that a problem? Of course not! In mesh analysis, we perform KVLs around closed loops, so any voltage source is welcome to be in the path. The other change is that now we have a current source in an unshared branch in mesh 1. Having a current source in an unshared branch makes life a little bit easier because it tells us that we do not have to solve for the mesh current I_1 anymore! The mesh current I_1 is set to be equal in magnitude to the current source I_{s_1} but will carry a negative sign, indicating that it is going in the opposite direction of the current source.

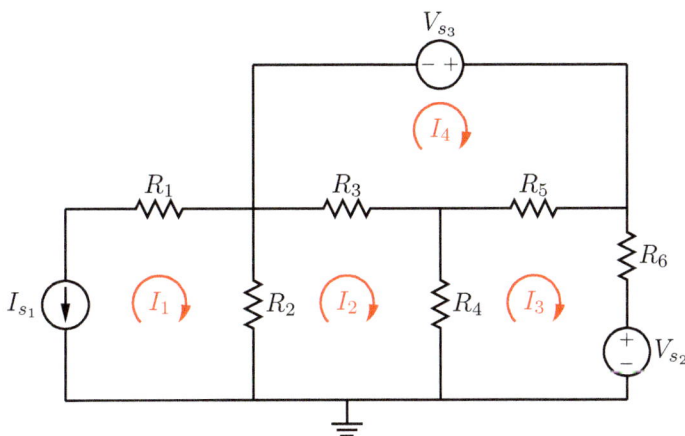

Figure 3.16: A second circuit with four mesh currents.

Now we can start writing our mesh equations.

Mesh 1:

$$I_1 = -I_{s_1} \tag{3.44}$$

Mesh 2:

$$R_2(I_2 - I_1) + R_3(I_2 - I_4) + R_4(I_2 - I_3) = 0$$
$$-R_2 I_1 + (R_2 + R_3 + R_4)I_2 - R_4 I_3 - R_3 I_4 = 0 \tag{3.45}$$

We can plug (3.44) into (3.45) right away or wait until the end; it is a matter of preference. We choose to plug it in now since it is the only time we will have to make this substitution in this example.

$$(R_2 + R_3 + R_4)I_2 - R_4 I_3 - R_3 I_4 = -R_2 I_{s_1}. \tag{3.46}$$

Mesh 3:

$$R_4(I_3 - I_2) + R_5(I_3 - I_4) + R_6 I_3 + V_{s_2} = 0$$
$$-R_4 I_2 + (R_4 + R_5 + R_6)I_3 - R_5 I_4 = -V_{s_2}. \tag{3.47}$$

Mesh 4:

$$R_3(I_4 - I_2) + R_5(I_4 - I_3) - V_{s_3} = 0$$
$$-R_3 I_2 - R_5 I_3 + (R_3 + R_5)I_4 = V_{s_3}. \qquad (3.48)$$

We now have three equations, (3.46), (3.47), and (3.48), and three unknowns, I_2, I_3, and I_4. In matrix form, these equations are expressed as

$$\begin{bmatrix} R_2 + R_3 + R_4 & -R_4 & -R_3 \\ -R_4 & R_4 + R_5 + R_6 & -R_5 \\ -R_3 & -R_5 & R_3 + R_5 \end{bmatrix} \begin{bmatrix} I_2 \\ I_3 \\ I_4 \end{bmatrix} = \begin{bmatrix} -R_2 I_{s_1} \\ -V_{s_2} \\ V_{s_3} \end{bmatrix}. \qquad (3.49)$$

3.4.3   Case 3: A Current Source Between Two Meshes

Let us modify Figure 3.16 again and see if we can make it more complicated. Figure 3.17 shows the new circuit. The circuit now has a current source in place of the resistor R_4, and the voltage source V_{s_2} is now a current-controlled voltage source.

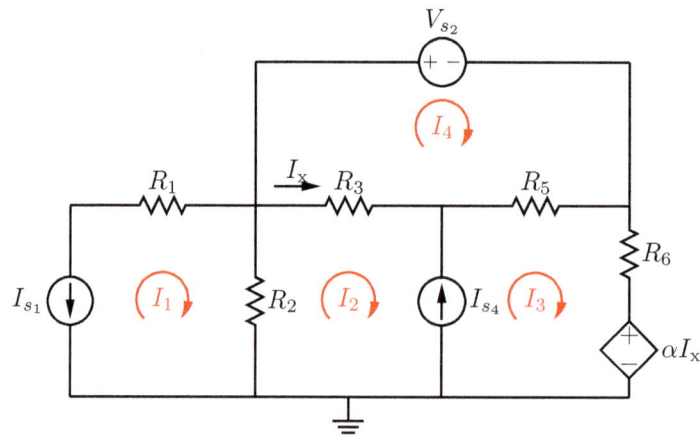

Figure 3.17: A third circuit with four mesh currents.

From looking at the circuit, nothing seems too difficult so far; the controlled voltage source is still a voltage source but we will have to replace its value with what it depends on. Specifically, it depends on the current I_x, which flows through the resistor R_3. The current I_x is between two meshes (2 and 4), but that is not a problem as we already learned how to write a current between two meshes in terms of their mesh currents. The only new problem we can see in this example is the new current source I_{s_4}, as its location is new to us. This is the first time we've seen a current source that belongs to two meshes. We have to think about this carefully. As we already know, in order to write a mesh equation, we use KVL around that mesh, summing the voltages to be equal to zero. But what is the voltage across the current source I_{s_4}? To overcome this difficulty, we must assign a dummy voltage across the current source I_{s_4} and write our equations accordingly. This will end up adding one extra unknown, so we will have to find one more equation to solve for all unknowns.

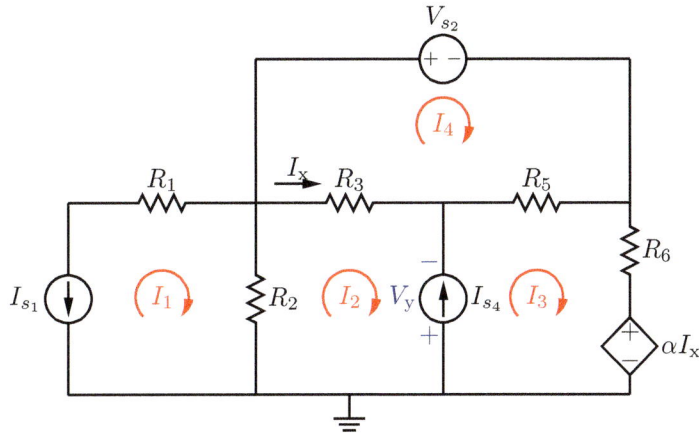

Figure 3.18: Simple circuit with four mesh currents.

Let us now write our mesh equations.

Mesh 1:

$$I_1 = -I_{s_1}. \tag{3.50}$$

Mesh 2:

$$R_2(I_2 - I_1) + R_3(I_2 - I_4) - V_y = 0$$
$$(R_2 + R_3)I_2 - R_3I_4 - V_y - -R_2I_{s_1}. \tag{3.51}$$

Mesh 3:

$$R_5(I_3 - I_4) + R_6I_3 + \alpha I_x + V_y = 0$$
$$(R_5 + R_6)I_3 - R_5I_4 + V_y = -\alpha I_x. \tag{3.52}$$

Mesh 4:

$$R_3(I_4 - I_2) + R_5(I_4 - I_3) + V_{s_2} = 0$$
$$-R_3I_2 - R_5I_3 + (R_3 + R_5)I_4 = -V_{s_2}. \tag{3.53}$$

Let us deal with the dependent voltage term first. Since I_x flows through a branch between mesh 3 and mesh 4, we have

$$I_x = I_2 - I_4. \tag{3.54}$$

The mesh 3 equation now becomes:

$$(R_5 + R_6)I_3 - R_5I_4 + V_y = -\alpha(I_2 - I_4)$$
$$\alpha I_2 + (R_5 + R_6)I_3 - (R_5 + \alpha)I_4 + V_y = 0. \tag{3.55}$$

Now let us deal with the new variable we introduced, V_y. The equation for mesh 2, (3.51), and the cleaned-up mesh 3 equation, (3.55), contain V_y. We can combine these equations to eliminate V_y, but this will also reduce the number of equations. Alternatively, we may choose to keep V_y as a variable in our system.

Either way, we will need to find an additional equation to replace the lost one or to add an extra equation to find V_y.

Let us combine the mesh 2 and mesh 3 equations.

Mesh 2 + mesh 3:

$$(R_2 + R_3 + \alpha)I_2 + (R_5 + R_6)I_3 - (R_3 + R_5 + \alpha)I_4 = -R_2 I_{s_1}. \qquad (3.56)$$

This leaves us with two independent equations, which are:

$$(R_2 + R_3 + \alpha)I_2 + (R_5 + R_6)I_3 - (R_3 + R_5 + \alpha)I_4 = -R_2 I_{s_1} \quad \text{(Mesh 2 + 3)}$$
$$-R_3 I_2 - R_5 I_3 + (R_3 + R_5)I_4 = -V_{s_2}. \qquad \text{(Mesh 4)}$$

Since we have two equations and three unknowns, we need a third equation. Looking at the circuit in Figure 3.18, we see that

$$I_3 - I_2 = I_{s_4}. \qquad (3.57)$$

In matrix form, we write

$$\begin{bmatrix} R_2 + R_3 + \alpha & R_5 + R_6 & -(R_3 + R_5 + \alpha) \\ -R_3 & -R_5 & R_3 + R_5 \\ -1 & 1 & 0 \end{bmatrix} \begin{bmatrix} I_2 \\ I_3 \\ I_4 \end{bmatrix} = \begin{bmatrix} -R_2 I_{s_1} \\ -V_{s_2} \\ I_{s_4} \end{bmatrix}. \qquad (3.58)$$

Another way to look at the problem caused by the current source I_{s_4} is to choose a different path or loop that does not include it in order to avoid summing its voltage. Since KVL is valid for any closed loop, we are free to choose the easiest path. Say we take the dotted path in Figure 3.19. The equations for mesh 1 and mesh 4 will not change, but we don't have a mesh 2 or a mesh 3 equation. Instead, we have an equation for the new path.

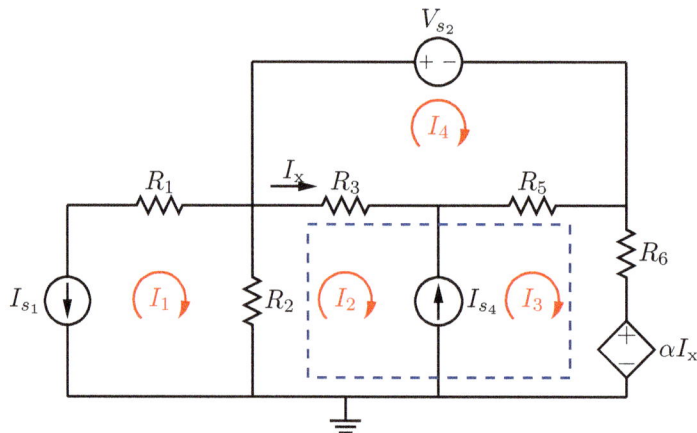

Figure 3.19: Choosing a different path.

Mesh 1 and mesh 4:

$$I_1 = -I_{s_1} \tag{3.59}$$

$$-R_3 I_2 - R_5 I_3 + (R_3 + R_5) I_4 = -V_{s_2}. \tag{3.60}$$

We will follow the new path, but keep in mind that we will use the mesh currents we defined before. Following the dotted path clockwise, starting at the resistor R_2, we have

$$R_2(I_2 - I_1) + R_3(I_2 - I_4) + R_5(I_3 - I_4) + R_6 I_3 + \alpha I_x = 0$$
$$(R_2 + R_3)I_2 + (R_5 + R_6)I_3 - (R_3 + R_5)I_4 = -R_2 I_{s_1} - \alpha I_x. \tag{3.61}$$

Using (3.54) for I_x yields

$$(R_2 + R_3)I_2 + (R_5 + R_6)I_3 - (R_3 + R_5)I_4 = -R_2 I_{s_1} - \alpha(I_2 - I_4)$$
$$(R_2 + R_3 + \alpha)I_2 + (R_5 + R_6)I_3 - (R_3 + R_5 + \alpha)I_4 = -R_2 I_{s_1}. \tag{3.62}$$

Well, well, what do we have here! The resulting equation of the new path is exactly the same as the equation resulting from combining the mesh 2 and mesh 3 equations. All roads lead to Rome! Both using the new path and adding a dummy variable leads us to the same set of independent equations. Let us try to work through a numerical example now.

Example 3.4.1

For the circuit in Figure 3.20, find the final matrix equation to solve for the mesh currents I_1, I_2, and I_3.

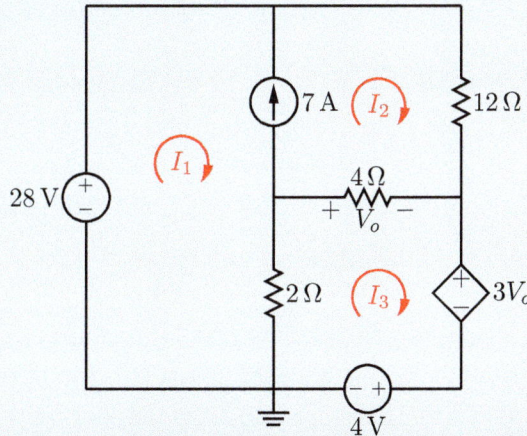

Figure 3.20: Example 3.4.1.

This is a wonderful example for you to study carefully. Why? It has everything that might confuse you: a dependent source and a current source between two meshes. Let us start by assigning the dummy variable, V_x, to the current source, as seen in Figure 3.21.

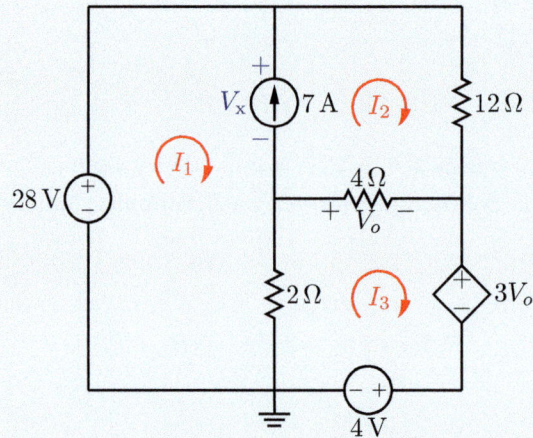

Figure 3.21: Example 3.4.1.

Now we write our mesh equations. For mesh 1, we have

$$2I_1 - 2I_3 = 28 - V_\mathrm{x}.$$

Then the equation for mesh 2 is

$$(12 + 4)I_2 - 4I_3 = V_\mathrm{x}.$$

We can now add both equations to eliminate the dummy variable:

$$\boxed{2I_1 + 16I_2 - 6I_3 = 28.}$$

This is our first equation. Our second equation comes from our 7 A current source. Specifically,

$$\boxed{-I_1 + I_2 = 7.}$$

The equation for mesh 3 is

$$-2I_1 - 4I_2 + 6I_3 = -3V_o - 4.$$

Next we need an equation for V_o in terms of the mesh currents. Using Ohm's law, we have

$$V_o = 4(I_3 - I_2).$$

Combining both equations and cleaning up will result in our third and final equation for our matrix.

$$\boxed{-2I_1 - 16I_2 + 18I_3 = -4.}$$

The final mesh equations in matrix form is

$$\begin{bmatrix} 2 & 16 & -6 \\ -1 & 1 & 0 \\ -2 & -16 & 18 \end{bmatrix} \begin{bmatrix} I_1 \\ I_2 \\ I_3 \end{bmatrix} = \begin{bmatrix} 28 \\ 7 \\ -4 \end{bmatrix}.$$

Solving for the mesh currents, we get

$$\begin{bmatrix} I_1 \\ I_2 \\ I_3 \end{bmatrix} = \begin{bmatrix} -4 \\ 3 \\ 2 \end{bmatrix} \text{ A.}$$

Exercise 3.4.1

For the circuit in Figure 3.22, find the final matrix equation to solve for the mesh currents I_1 and I_2.

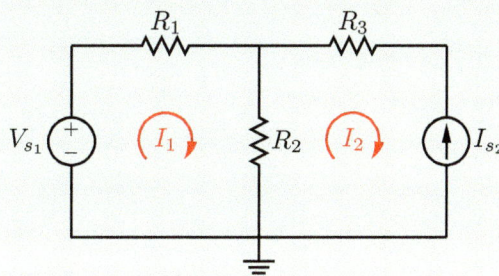

Figure 3.22: Exercise 3.4.1.

$$\begin{bmatrix} R_1 + R_2 & -R_2 \\ 0 & 1 \end{bmatrix} \begin{bmatrix} I_1 \\ I_2 \end{bmatrix} = \begin{bmatrix} V_{s_1} \\ -I_{s_2} \end{bmatrix}.$$

Exercise 3.4.2

For the circuit in Figure 3.23, find the final matrix equation to solve for the mesh currents I_1 and I_2.

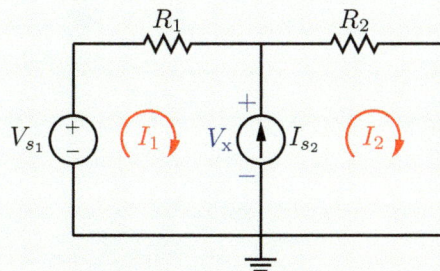

Figure 3.23: Exercise 3.4.2.

$$\begin{bmatrix} R_1 & R_2 \\ -1 & 1 \end{bmatrix} \begin{bmatrix} I_1 \\ I_2 \end{bmatrix} = \begin{bmatrix} V_{s_1} \\ I_{s_2} \end{bmatrix}.$$

3.5   Exercises

Section 3.1

3.1. Using nodal analysis, find the current I and the voltage V in the circuit in Figure 3.24.

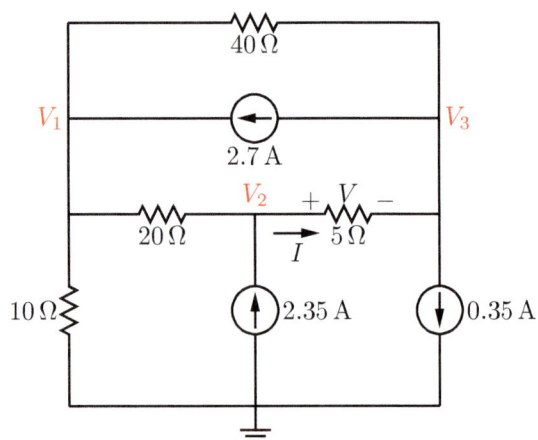

Figure 3.24

3.2. Using nodal analysis, find the nodal voltages V_1, V_2, and V_3 in the circuit in Figure 3.25. Also find the power delivered by the 9 A current source.

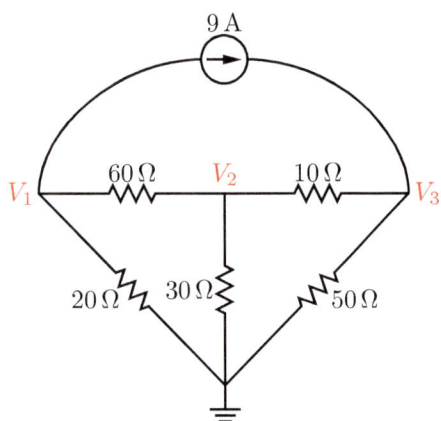

Figure 3.25

3.3. Using nodal analysis, find the nodal voltages V_1, V_2, and V_3 in the circuit in Figure 3.26. Do all sources in the circuit deliver power? If not, identify the ones that absorb power.

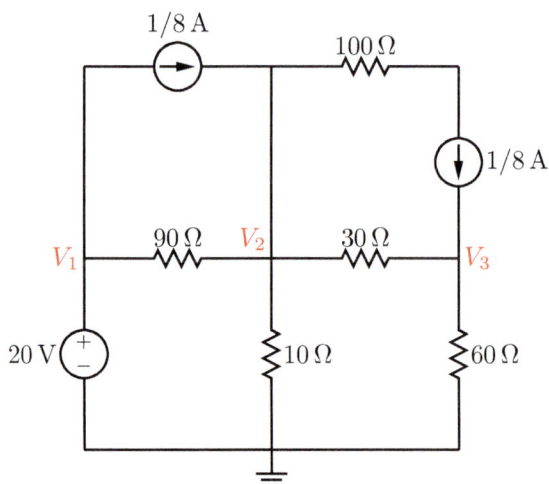

Figure 3.26

3.4. Using nodal analysis, find I_1 and I_2 in the circuit in Figure 3.27.

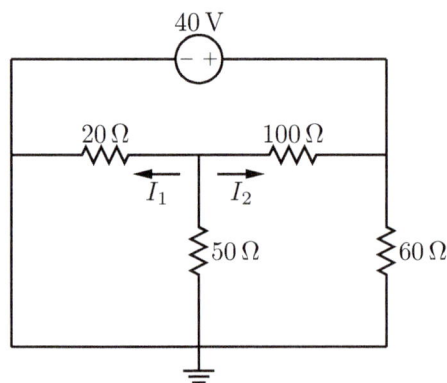

Figure 3.27

3.5. Using nodal analysis, find all nodal voltages indicated in the figure and the total power delivered to passive elements in the circuit in Figure 3.28.

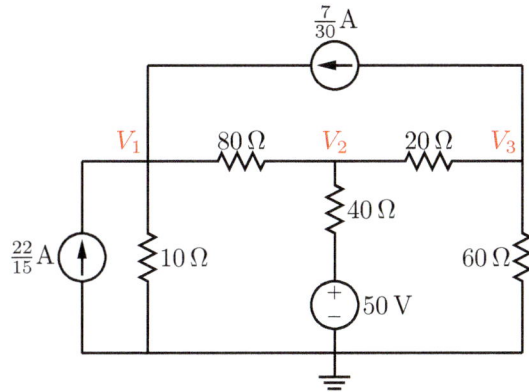

Figure 3.28

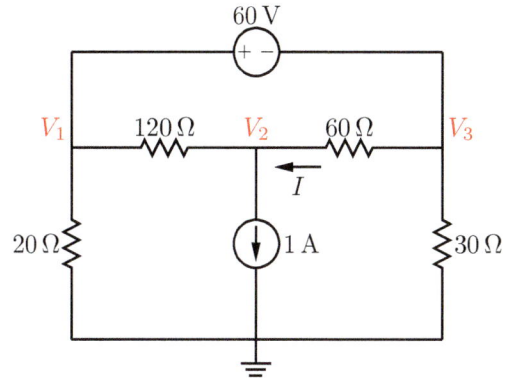

Figure 3.31

3.6. Using nodal analysis, find all nodal voltages indicated in the circuit in Figure 3.29.

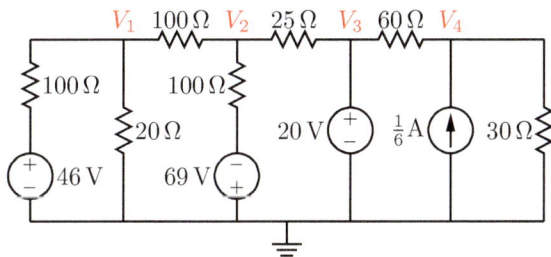

Figure 3.29

3.9. Using nodal analysis, find all nodal voltages indicated in the circuit in Figure 3.32. Also find the power delivered by the voltage source.

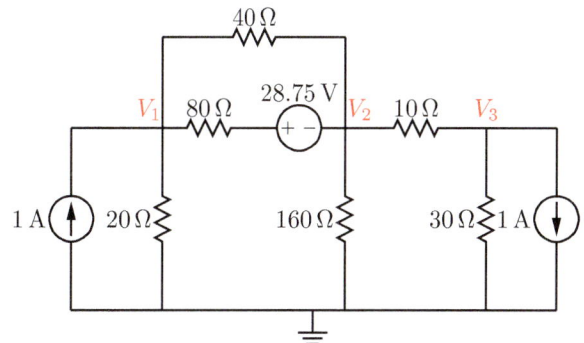

Figure 3.32

3.7. Using nodal analysis, find all the nodal voltages indicated in the circuit in Figure 3.30. Also find the current I.

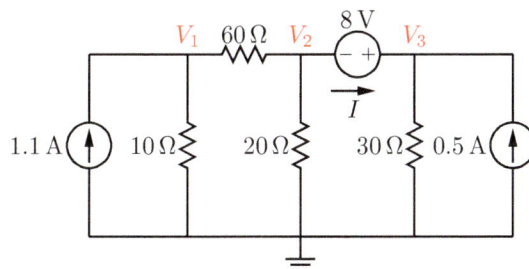

Figure 3.30

3.10. Using nodal analysis, find all nodal voltages indicated in the circuit in Figure 3.33.

3.8. Using nodal analysis, find all nodal voltages indicated in the circuit in Figure 3.31. Also find the current I.

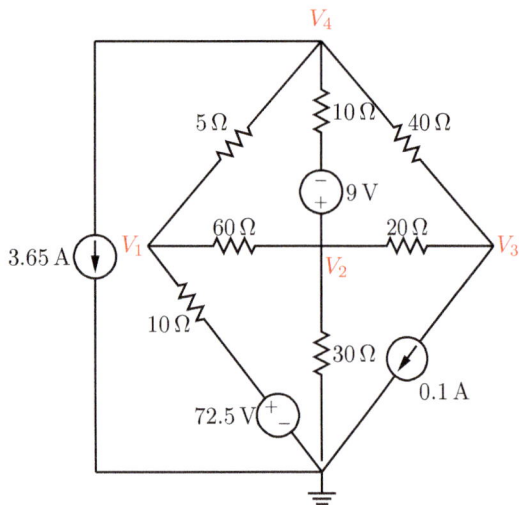

Figure 3.33

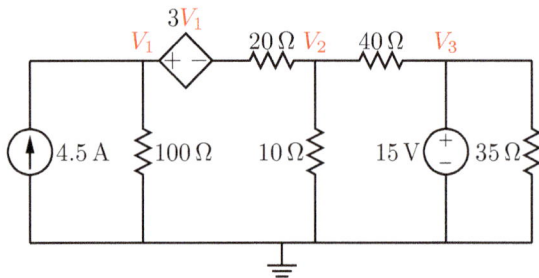

Figure 3.35

3.11. Using nodal analysis, find all nodal voltages indicated in the circuit in Figure 3.34. Also find the power delivered by the independent voltage source.

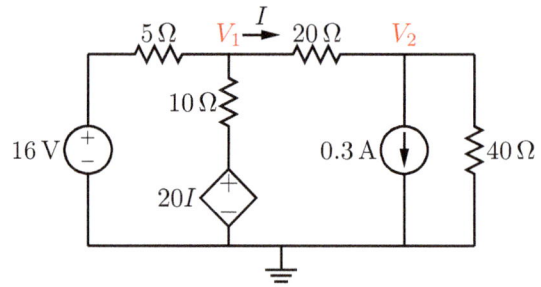

Figure 3.34

3.12. Using nodal analysis, find all nodal voltages indicated in the figure in the circuit in Figure 3.35. Also find the power delivered by the independent voltage source.

3.13. Using nodal analysis, find all nodal voltages indicated in the circuit in Figure 3.36. Also find the power delivered by the dependent voltage source.

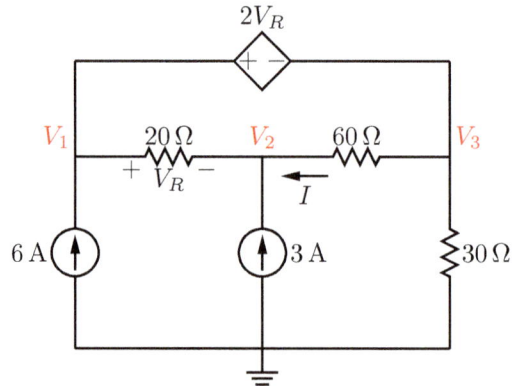

Figure 3.36

3.14. Using nodal analysis, find all nodal voltages indicated in the circuit in Figure 3.37. Also find the current I.

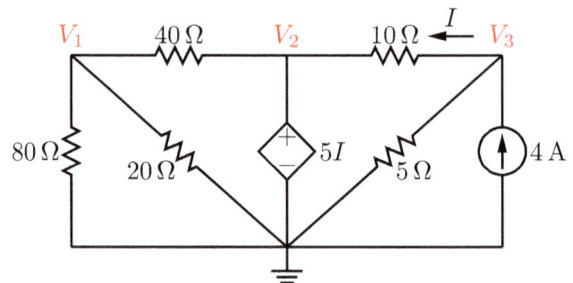

Figure 3.37

3.15. Using nodal analysis, find all nodal voltages indicated in the circuit in Figure 3.38. Also find the power delivered by the voltage source.

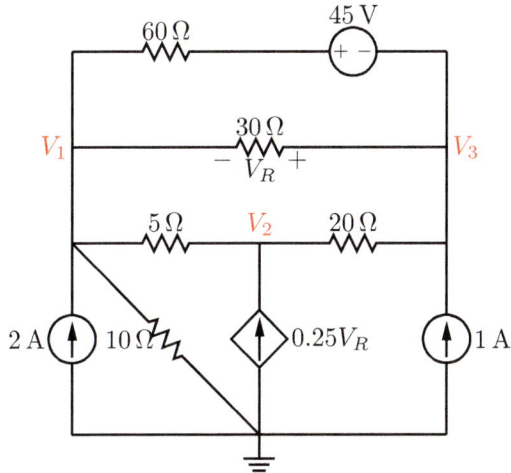

Figure 3.38

3.16. Using nodal analysis, find all nodal voltages indicated in the circuit in Figure 3.39.

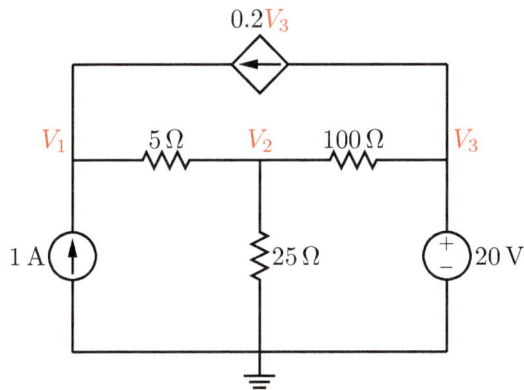

Figure 3.39

3.17. Using nodal analysis, find all nodal voltages indicated in the circuit in Figure 3.40. Also find the current I.

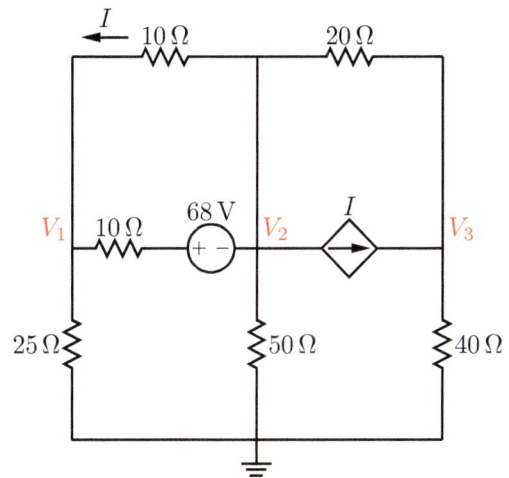

Figure 3.40

3.18. Using nodal analysis, find all nodal voltages indicated in the circuit in Figure 3.41.

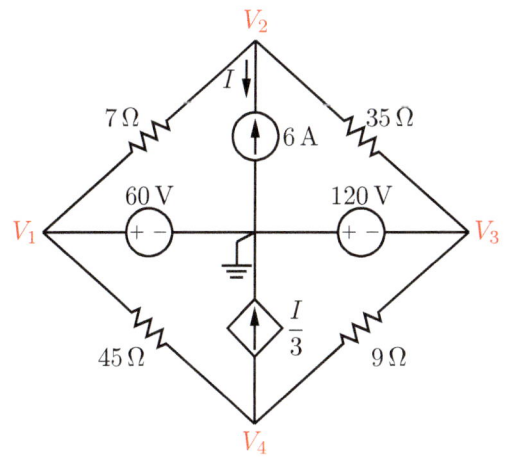

Figure 3.41

Section 3.4

3.19. Using mesh analysis, find all mesh currents indicated in the circuit in Figure 3.42.

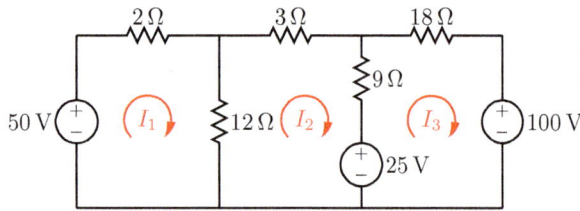

Figure 3.42

3.20. Using mesh analysis, find all mesh currents indicated in the circuit in Figure 3.43. Also find the power absorbed by the independent voltage source.

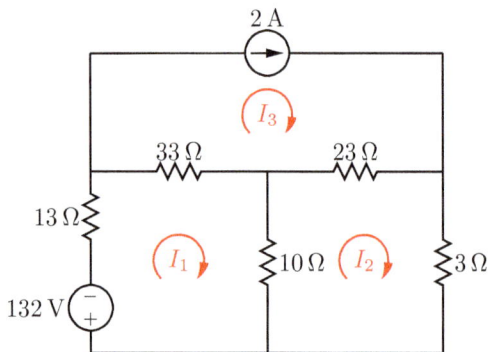

Figure 3.43

3.21. Using mesh analysis, find all mesh currents indicated in the circuit in Figure 3.44. Also find the total power absorbed by the passive elements in the circuit.

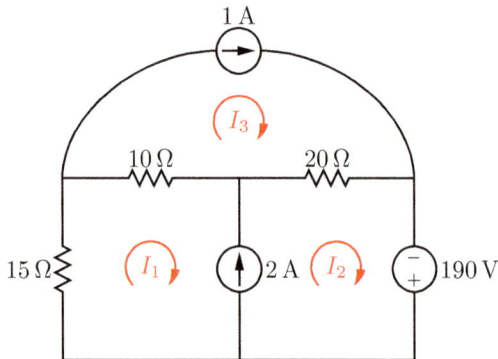

Figure 3.44

3.22. Using mesh analysis, find all mesh currents indicated in the circuit in Figure 3.45. Also find the power delivered by the 47 V voltage source.

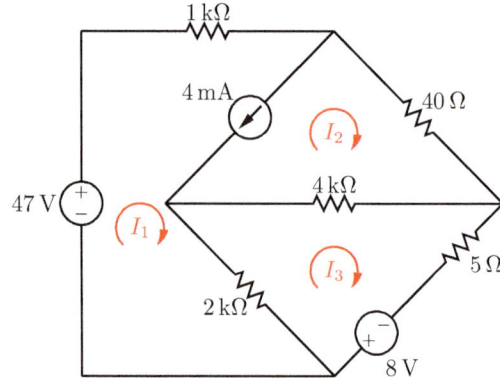

Figure 3.45

3.23. Using mesh analysis, find all mesh currents indicated in the circuit in Figure 3.46. Also find the power absorbed by the independent current source.

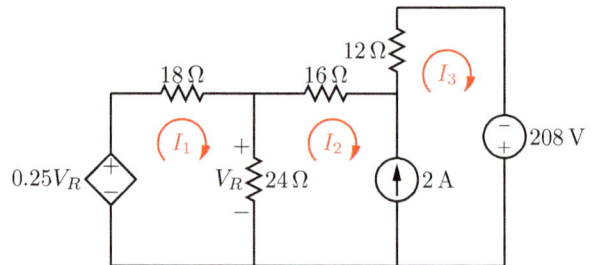

Figure 3.46

3.24. Using mesh analysis, find all mesh currents indicated in the circuit in Figure 3.47. Also find the power absorbed by the dependent voltage source.

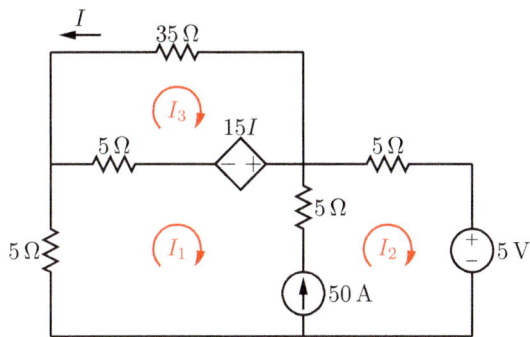

Figure 3.47

3.25. Using mesh analysis, find all mesh currents indicated in the circuit in Figure 3.48.

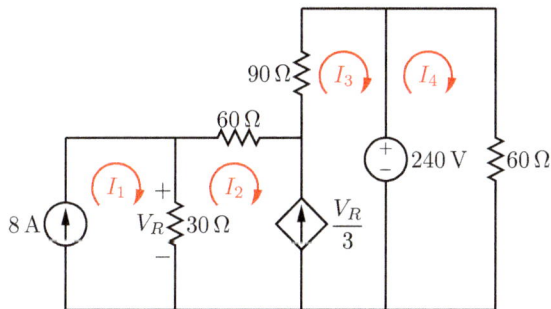

Figure 3.48

3.26. Using mesh analysis, find all mesh currents indicated in the circuit in Figure 3.49. Also find the power delivered by the dependent current source.

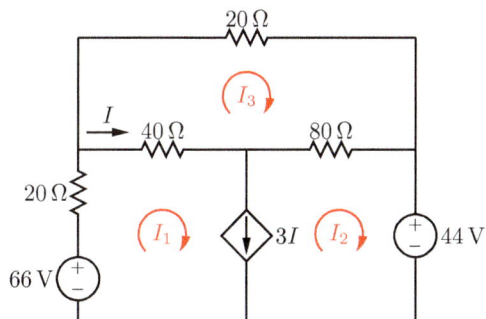

Figure 3.49

Chapter 4

Powerful Circuit Theorems

4.1 Introduction

So far we have learned how to use nodal and mesh analyses to solve for any voltage and current in resistive circuits. In this chapter we will learn about some powerful circuit theorems that will aid us even more in analyzing and understanding our circuits. Just like nodal and mesh analyses, these theorems will be applied first to DC circuits but they apply to any linear circuit. Here is a brief summary of these theorems.

1. Linearity

   Linearity is a mathematical principle that implies additivity and scalability. Using this property will save us from re-solving a circuit when we have its input(s).

2. Superposition

   Superposition is a powerful concept related to linearity and is applicable to linear circuits with multiple independent inputs. Superposition allows us to examine the impact of each input by itself. Adding the resulting voltage and current (but not power!) derived by considering each input independently allows us to find the total voltage or current (but not power!) due to all inputs.

3. Source transformation

   Source transformation is a wonderful tool that can be used to change the originally given circuit to an equivalent and simpler circuit. This concept is particularly useful when trying to find the Thévenin or Norton equivalent circuits, as we will see later.

4. Thévenin and Norton equivalent circuits

   Finding the Thévenin or Norton equivalent circuits for a given network is a great way to analyze a circuit when we are only interested in its input and output ports. Circuits are usually very complicated and we often only interact with one or two of a circuit's ports: the input and output. Trying to solve the whole network is often unnecessary. The Thévenin and

Norton theorems provide us with a way to find a simple representation of an electrical network with respect to the desired ports. We just have to remember that the procedure is not reversible, meaning that if you only have the Thévenin or Norton equivalent circuit, you cannot recover the original circuit, as many circuits can have the same Thévenin or Norton equivalent representation.

4.2   Linearity

Since we are dealing with linear circuits, it is only logical to take advantage of the mathematical properties of linearity. Linearity has two important properties: a) scalability and b) additivity. In math form, scalability can be described as follows. If we have a function $f(x)$, then

$$f(\alpha x) = \alpha f(x) \qquad \text{for any scalar } \alpha. \tag{4.1}$$

Additivity can be written as

$$f(x + y) = f(x) + f(y). \tag{4.2}$$

Let us look at how these properties translate into our circuit world. We will tackle scalability first. Let us dive into an example right away.

Example 4.2.1

Find the voltage V_x in Figure 4.1 if $V_s = 3$ V.

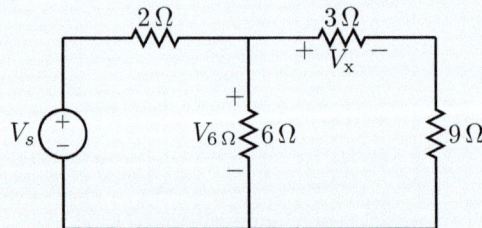

Figure 4.1: Example 4.2.1.

We can use two quick voltage divisions to find V_x. The first one will be to find the voltage across the 6 Ω resistor after we combine it with the series connection of the 3 Ω and 9 Ω resistors. The second one will be to find V_x using the voltage across the 6 Ω resistor.

Here is the first voltage division:

$$V_{6\Omega} = \frac{6 \parallel (3+9)}{[6 \parallel (3+9)] + 2} V_s = \frac{4}{4+2} V_s = 2\,\text{V}.$$

The second voltage division yields:

$$V_x = \frac{3}{3+9} V_{6\Omega} = 0.5\,\text{V}.$$

Let us say that 0.5 V is too low for our desired application and we would rather have 2 V. How much should we increase V_s by?

This is where we could use the scalability property. Instead of resolving the circuit, we can simply observe that

$$V_x(\text{new}) = 2\,\text{V} = 4V_x(\text{old}).$$

Consequently, the input needs to also be scaled by 4×. Hence

$$V_s(\text{new}) = 4V_s(\text{old}) = 4(3) = 12\,\text{V}.$$

The above example clearly demonstrates that if an input to a linear system is scaled by a factor α, the output will also be scaled by the same factor. This is a beautiful property that only applies to linear circuits and helps us in many cases. For instance, let us say that we wanted to find V_x in the previous circuit for twenty different input values of V_s. Instead of solving the circuit twenty times for each individual input, we can solve it for a generic input V_s and simply plug in twenty values for V_s at the end to find the required output.

The next section discusses the second property: additivity or superposition.

4.3 Superposition

Superposition could be quite useful in circuits with multiple independent inputs. By using superposition, we can assess the effect of each input (source) separately and then add these values together to find the total output. While this may or may not be simpler than solving the circuit with all inputs applied simultaneously, it does provide the advantage of understanding the impact of each source on its output separately. This may help us understand what source we need to increase or decrease in order to obtain the desired output.

Let us explain superposition by an example. Specifically, let us take Example 4.2.1 and add an extra source to it. The resulting circuit is shown in Figure 4.2.

Example 4.3.1

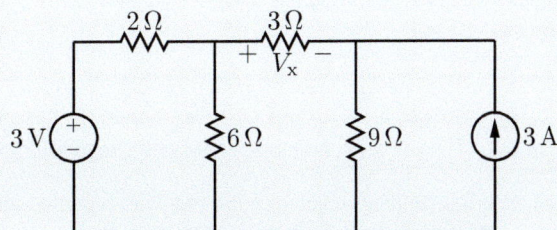

Figure 4.2: Superposition application.

Of course, we already know that we can find the two nodal voltages (nodal analysis) and then find V_x by subtracting them from each other. Alternatively, we can solve the circuit by considering each independent source separately. Figure 4.3 schematically demonstrates the idea. The first circuit has only the voltage source activated (current source has been turned off) with V_{x_1} as the output signal. Only the current source is activated in the second circuit, and the output signal is noted as V_{x_2}.

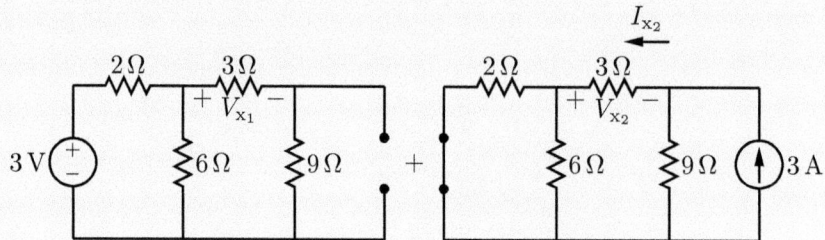

Figure 4.3: Superposition application.

We know from Example 4.2.1 that

$$V_{x_1} = 0.5\,\text{V}.$$

Now we should try to find V_{x_2}. Current division is an easy way to proceed here. First, we will have to find the parallel combination of the $6\,\Omega$ and $2\,\Omega$ resistors and then add the result to the $3\,\Omega$ resistor before we use current division. Second, note that the current resulting from the current division will be entering V_{x_2} from its negative terminal, which will make V_{x_2} negative.

Mathematically, we have

$$I_{x_2} = \frac{9}{9 + [3 + (6 \parallel 2)]}(3) = \frac{9}{9 + 3 + 1.5}(3) = 2\,\text{A}.$$

This makes

$$V_{x_2} = -(3)I_{x_2} = -(3)2 = -6\,\text{V}.$$

Now we can use superposition to find V_x:

$$V_x = V_{x_1} + V_{x_2} = 0.5 - 6 = -5.5\,\text{V}.$$

We encourage you to use nodal analysis to verify this answer; it is a piece of cake.

In conclusion, superposition can help us determine the individual contribution of sources on the circuit. The way to perform superposition is

1. Turn off all independent sources except for one. We do not touch dependent sources. Their values will, of course, change since the input to the circuit has changed.

2. Find the desired variable by solving the circuit for it.

3. Repeat the process for all independent sources (voltages or currents).

4. The final result will be the addition of all partial results.

An important note we need to make here is that you can only add partial voltages and currents, and not power. Power is a nonlinear quantity. The total power can be found by multiplying the total voltage with the total current and not by adding the partial powers. For instance, in case of two independent sources, we can write in math form

$$
\begin{aligned}
P_{\text{tot}} &= V_{\text{tot}} I_{\text{tot}} \\
&= (V_{\text{partial},1} + V_{\text{partial},2}) \cdot (I_{\text{partial},1} + I_{\text{partial},2}) \\
&\neq \underbrace{V_{\text{partial},1} \cdot I_{\text{partial},1}}_{P_{\text{partial},1}} + \underbrace{V_{\text{partial},2} \cdot I_{\text{partial},2}}_{P_{\text{partial},2}}
\end{aligned}
$$

We encourage you to perform this calculation for the $3\,\Omega$ resistor in Example 4.3.1.

Superposition has advantages and drawbacks. The main advantage is that it reveals the impact of each source to the final result. If this is not necessarily desired, superposition may still be used if each of the partial circuits is significantly simpler to solve than the original one. The main drawback of superposition is that it is necessary to solve many circuits to obtain the answer. Unless each of the partial circuits is significantly simpler than the given one, this may require a lot more work. As a result, while superposition is a powerful tool, it does not mean that you should use it every time you see a circuit with multiple sources.

4.4   Linearity and Superposition in the Laboratory

Another way to appreciate the power of these tools is to consider their effectiveness when you do not even know the circuit schematic. This is often the case in real life. Let us consider a simple example to understand this.

Example 4.4.1

Let us consider a linear circuit with two inputs: a voltage source V_{in} and a current source I_{in}. The circuit has a single output voltage V_{out}. How do you determine the relationship between input and output quantities? Simple: since the circuit is linear, you only need two measurements. The reason? Since we have a linear circuit, the output *must* be linked to the inputs through a linear equation. In other words, it *must* be true that

$$
V_{\text{out}} = \alpha V_{\text{in}} + \beta I_{\text{in}}.
$$

Notice that α is unitless while β has the units of Ω. Hence, finding α and β is all we need to do. Conducting two measurements is, as a result, sufficient. Specifically, let us say that we conduct the following measurements:

$$V_{\text{out}} = 1\,\text{V when } V_{\text{in}} = 0\,\text{V and } I_{\text{in}} = 1\,\text{A};$$
$$V_{\text{out}} = 2\,\text{V when } V_{\text{in}} = 1\,\text{V and } I_{\text{in}} = 0\,\text{A}.$$

Clearly, $\alpha = 2$ and $\beta = 1\,\Omega$ from these measurements. Notice that while it is convenient to turn off one source at a time to find the needed relationship, it is not necessary. We could have just as easily conducted the following measurements:

$$V_{\text{out}} = 3\,\text{V when } V_{\text{in}} = 1\,\text{V and } I_{\text{in}} = 1\,\text{A};$$
$$V_{\text{out}} = 7\,\text{V when } V_{\text{in}} = 2\,\text{V and } I_{\text{in}} = 3\,\text{A}.$$

The result would, of course, be the same. Once you know α and β, it is clear that you can find V_{out} for any set of desired inputs. Furthermore, you can determine a variety of possible inputs for any desired value of V_{out}.

4.5   Source Transformation

Source transformation is another useful tool to consider when analyzing a circuit. We are already used to being able to simplify a circuit by combining resistors in series and/or in parallel. Source transformation allows us to manipulate source-resistor pairings to create additional possibilities for creating parallel and series combinations of sources and resistors. Figure 4.4 will help us understand some basic equivalences that will reveal the "magic" we will use later.

Figure 4.4 shows a circuit excited by a voltage source V_s in series with a resistor R. It is probably worth mentioning here that real-life voltage sources cannot be modeled by ideal voltage sources as described so far. They often include losses such as, for example, ohmic losses caused by the finite-conductivity metals they are constructed from. The series resistance R, which is also often called internal resistance, models some of these effects. The series combination of the ideal voltage source V_s with the internal resistor R is often referred to as a realistic voltage source.

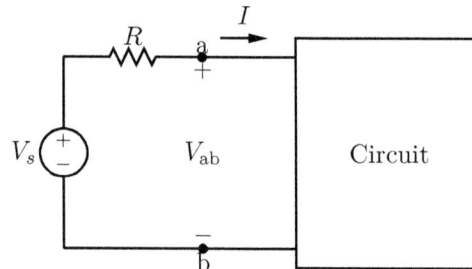

Figure 4.4: A realistic voltage source exciting a circuit via port a-b.

If we would like to find the current I entering the circuit using only what we know, we can write

$$I = \frac{V_s - V_{ab}}{R} = \frac{V_s}{R} - \frac{V_{ab}}{R}.$$ (4.3)

We notice that (4.3) looks exactly like a KCL between two parallel elements, as seen in Figure 4.5. This is where the idea of equivalence comes from. We have two different circuits that give us the same current.

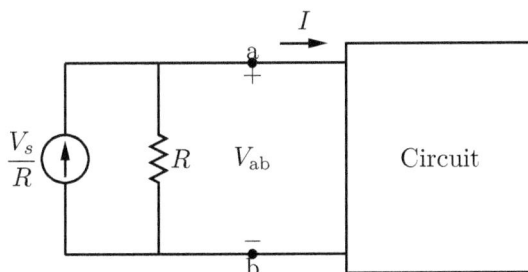

Figure 4.5: Figure 4.4 equivalent circuit.

This is the whole premise behind source transformation theory. As far as port a-b is concerned, the original voltage source V_s with the series resistor R is **equivalent** to a current source $I_s = V_s/R$ in parallel with the same resistor R. Sound familiar? We touched on this subject slightly in Example 3.2.3 with Figure 3.9. Figure 4.6 summarizes this idea.

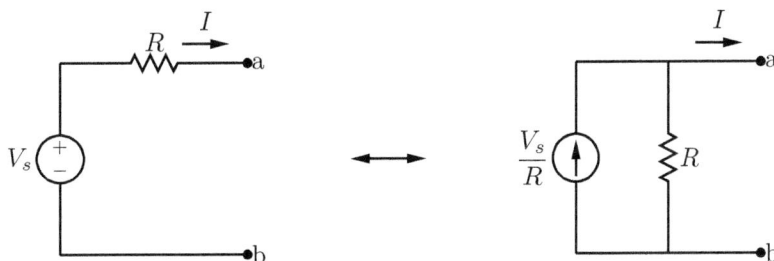

Figure 4.6: Source transformation summary.

Of course, the transformation also works when we start with a current source in parallel with a resistor. This is also a useful model for real-life current sources. Figure 4.7 summarizes the transformation when we start from a current source in parallel with a resistor.

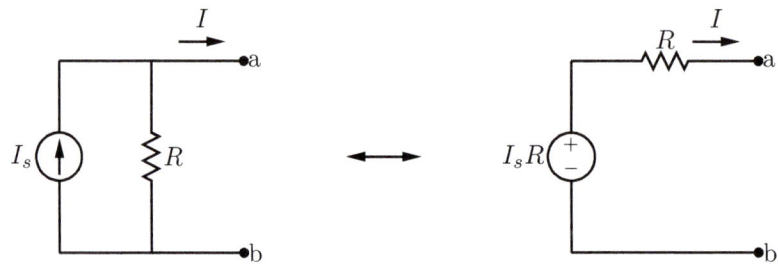

Figure 4.7: Source transformation starting with a current source in parallel with a resistor.

Let us now see an example to demonstrate the power of this technique in analyzing circuits.

Example 4.5.1

Find the voltage V_x in Figure 4.8 using source transformation.

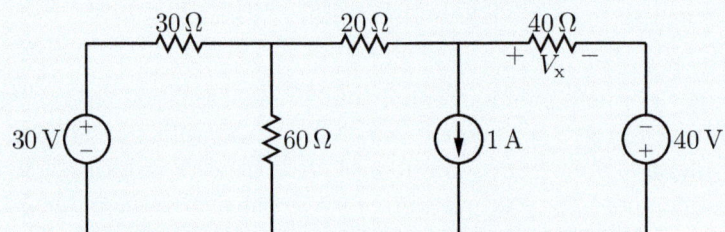

Figure 4.8: Example 4.5.1.

We could solve this problem using nodal analysis (try it if you would like to practice your nodal analysis skills). However, since changing the circuit is not an issue, and we only want to find one variable in the circuit, source transformation can be a better choice. We will start from the far left of the circuit and work our way to the desired variable.

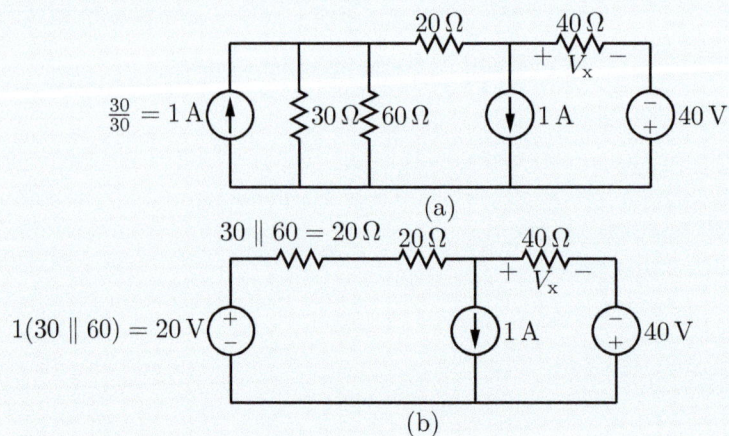

(a)

(b)

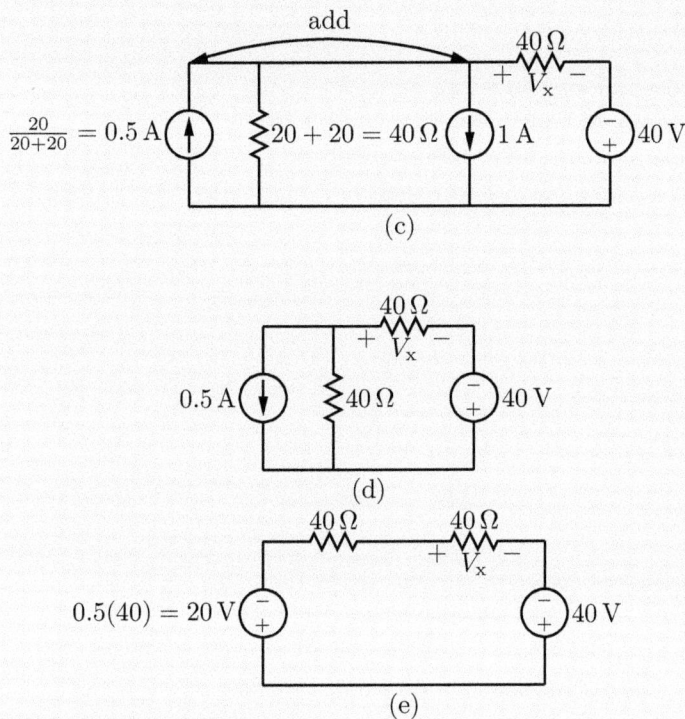

Figure 4.9: The solution to Example 4.5.1.

From the last equivalent circuit, we have

$$V_{\mathrm{x}} = \frac{-20 - (-40)}{2} = 10\,\mathrm{V}.$$

The solution has been significantly simplified using successive source transformations.

4.6 Thévenin and Norton Theorems

The Thévenin and Norton theorems (Thévenin and Norton from now on) are among the most important tools not only for analyzing circuits but also for many disciplines. Not only do they help us simplify circuits, but they also significantly broaden our way of thinking. To help us understand the important concepts behind these theorems, we will divide this section into three different subsections:

1. Understanding Thévenin and Norton by the lab approach.

2. Understanding Thévenin and Norton from a practical perspective.

3. How to find the Thévenin and Norton equivalent circuits.

Although we recommend studying through these sections sequentially, you can also jump directly to the section you are most interested in.

4.6.1 Understanding Thévenin and Norton by the Lab Approach

Consider the following scenario: we are interested in a circuit but it is in a "black box," so we have no way of knowing its components, connections, or schematics (Figure 4.10). All we know about this box is that it contains a linear circuit and it has a single port. This situation occurs in real life quite often: a lot of circuits, particularly older ones, come with no manuals or explanations.

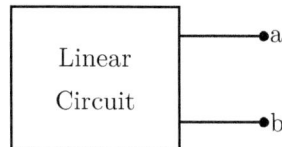

Figure 4.10: Circuit in a box.

Let us think of a way to model what is inside the black box. In other words, we would like to model its behavior by deriving an equivalent circuit without knowing what is inside the box. The only way to interact with this box is through its port a-b. As we will see, it will only take a) a voltmeter/ammeter and b) two measurements to understand its behavior.

First, let us connect our voltmeter to the open circuit terminal as shown in Figure 4.11a and read the resulting voltage. Remember that a good voltmeter has very large internal resistance that emulates an open circuit. To simplify the discussion, we will assume that the voltmeter's resistance is infinite. We will call this the open-circuit voltage V_{oc}. Let us assume, for example, that $V_{oc} = 20$ V. Second, we connect our ammeter to terminal a-b as shown in Figure 4.11b. A good ammeter has very low internal resistance, ideally zero. Hence, we will call the measured current I_{sc}. Let us assume, for example, that the ammeter reading is $I_{sc} = 2A$.

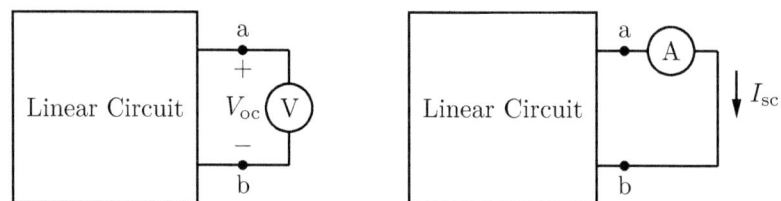

(a) Measuring the open-circuit voltage V_{oc} of the circuit in a box though a voltmeter. (b) Measuring the short-circuit current I_{sc} of the circuit in a box though an ammeter.

Figure 4.11: Taking the circuit measurements.

Based on these readings, we know for sure that the circuit has some sort of power source or a combination of sources (current or voltage) that results in the open-circuit voltage and short-circuit current readings we got. Because of this, we could start guessing what an equivalent circuit might look like. For example, Figure 4.12 shows the linear circuit modeled with a simple voltage source that has the value of the measured open-circuit voltage. Clearly, if we assigned the value $V_{oc} = 20$ V to the model (part (a) of the figure), we would easily satisfy the first measurement (voltmeter). Unfortunately, it would not satisfy the second (ammeter) measurement (part (b) of the figure). In fact, since the ammeter resistance is (theoretically) zero, the measured current would have been (theoretically) infinite. Clearly, this is not the case, so this model is not successful. A similar attempt to model the circuit with an ideal current source with a value of $I_{sc} = 2$ A would also fail, as shown in Figure 4.13.

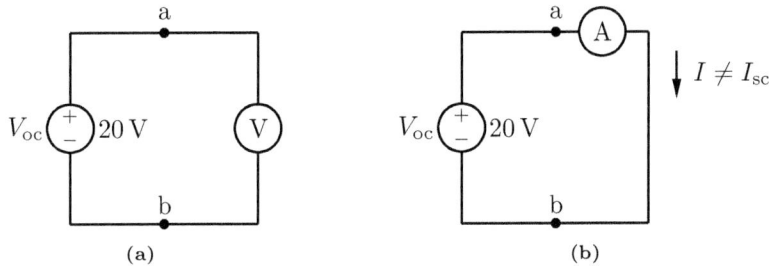

Figure 4.12: Replacing the linear circuit in a box with an ideal voltage source of value V_{oc} does not yield satisfactory results, as it agrees only with the first measurement (voltmeter).

Someone might think that combining both an ideal voltage and an ideal current source in parallel or in series could be successful. Unfortunately, this does not work either. For instance, if we were to add the current source in series with the voltage source, we would have the problem of always getting I_{sc} through *any* load connected to port a-b. Clearly, this is not realistic, as we expect the current to be a function of the connected load. Similarly, if the sources were to be connected in parallel, we would always get V_{oc} across *any* load connected to port a-b. It is clear, therefore, that ideal sources are not sufficient.

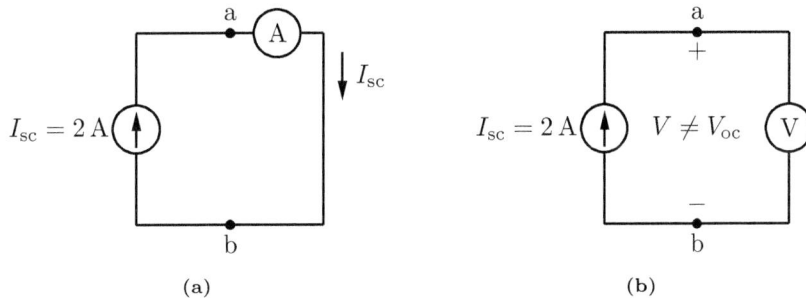

Figure 4.13: Replacing the linear circuit in a box with an ideal current source of value I_{sc} does not yield satisfactory results, as it agrees only with the first measurement (ammeter).

Since we did not succeed with ideal sources, let us try with a model based on realistic sources. Starting with the circuit we had in Figure 4.12, we can add a resistor of value R in series or in parallel with the voltage source. Adding the resistor in parallel will not help us much because we will end up with the same situation of having infinite current through the ammeter as in Figure 4.12(b). On the other hand, adding the resistor in series is really helpful as shown in Figure 4.14. In Figure 4.14(a), we are testing the open-circuit voltage. The value of the resistor has no effect since there is no current flowing through it—that is, there is no voltage drop across the resistor regardless of its value. Consequently, we will still get $V_{oc} = 20$ V for our open-circuit voltage. This is certainly positive, but we need to check the second measurement as well.

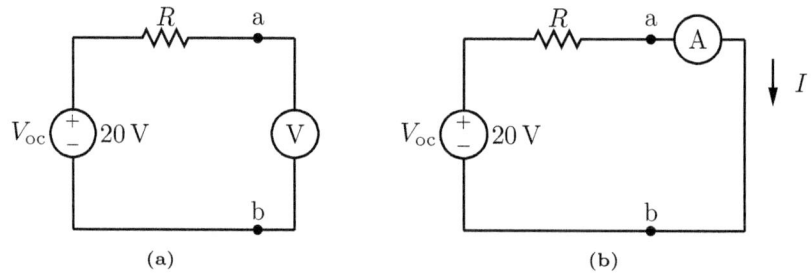

Figure 4.14: Replacing the linear circuit in a box with an ideal voltage source of value V_{oc} in series with a resistor of value R and applying the open- and short-circuit tests.

Since the ammeter has negligible resistance, the value of the current I will indeed depend on the value of the resistor we added:

$$I = \frac{V_{oc}}{R}.$$

In order for this circuit to be equivalent to the circuit in the box, we need the short-circuit current in Figure 4.14(b) to be equal to $I_{sc} = 2$ A. Hence

$$I = I_{sc} = 2 = \frac{V_{oc}}{R},$$
$$R = \frac{V_{oc}}{I_{sc}} = \frac{20}{2},$$
$$R = 10\,\Omega.$$

To summarize, Figure 4.15 shows the unknown linear circuit we started with and its equivalent circuit represented by a voltage source in series with a resistor.

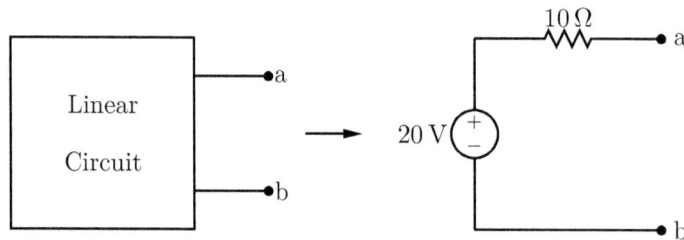

Figure 4.15: The equivalent circuit of Figure 4.10 (the Thévenin equivalent circuit).

What if we were to start with the circuit in Figure 4.13? We will still be facing the choice of adding a resistor in series or in parallel. If we add the resistor in series, we will still have the problem of always getting the current I_{sc} no matter what load is connected to the port. As a result, we will have to go with placing the resistor in parallel as shown in Figure 4.16.

Just as we did with the voltage source, we need to find the value of the resistor. The resistor value will not have an effect on the short-circuit test because it will be shorted. The open circuit test is going to be the determiner

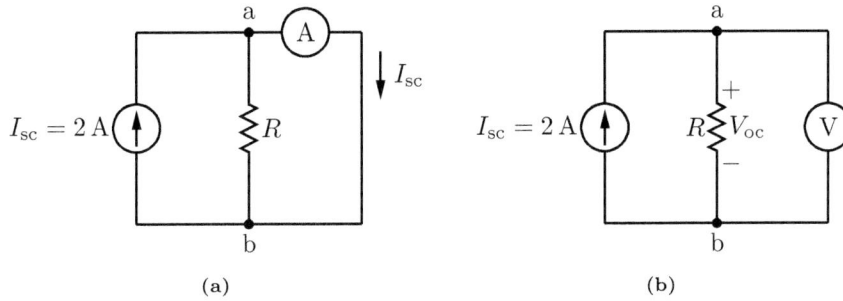

Figure 4.16: Replacing the linear circuit in a box with an ideal current source of value I_{sc} in parallel with a resistor of value R and applying the open- and short-circuit tests.

of the resistor value. Since we need the open-circuit voltage to be $V_{oc} = 20$ V and since

$$V_{oc} = I_{sc}R,$$

we get

$$V_{oc} = I_{sc}R = 20,$$
$$R = \frac{V_{oc}}{I_{sc}} = \frac{20}{2},$$
$$R = 10\,\Omega.$$

The final representation is shown in Figure 4.17. Notice that we arrived at the same value of resistance no matter which path we took. The only difference is the location of the resistor. However, we can solidify our findings further by just combining these results with the source transformation theorem. Source transformation enables us to go back and forth between the two circuit representations of Figure 4.18.

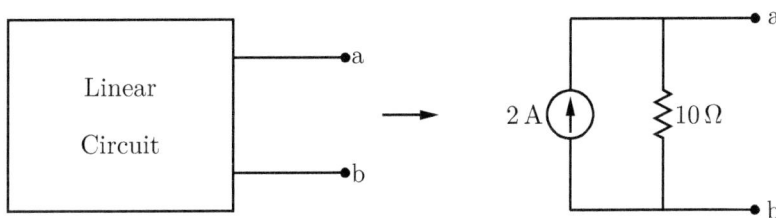

Figure 4.17: The equivalent circuit of Figure 4.10 (the Norton equivalent circuit).

The circuits shown in Figure 4.18 are called the Thévenin and Norton equivalent circuits. As we have seen, both representations are valid to replace the linear circuit in the box. Choosing one or the other is a matter of preference.

We should note here that it is theoretically possible to draw circuit schematics that have no Thévenin or Norton equivalent circuits. These are pathological cases that will receive no further attention in this text; all practically important circuits have both Thévenin and Norton equivalent circuits.

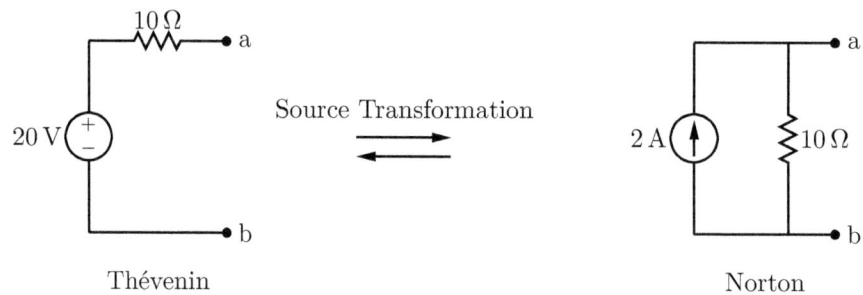

Figure 4.18: Source transformation to switch between the Thévenin and Norton equivalent circuits.

4.6.2   Understanding Thévenin and Norton from a Practical Perspective

Figure 4.19 shows a situation engineers face in many design phases: we have a circuit that does not change but we need to find the effect of a variable load attached to one of its terminals. Let us say that the load R_L could vary from $1\,\Omega$ to $100\,\Omega$ with $1\,\Omega$ increments and we need to measure the current I_L every time the resistor is changed. Remember that we are in the design phase, so there is no physical circuit for us to connect our multimeter to, which means we have to solve the circuit every time we change the load.

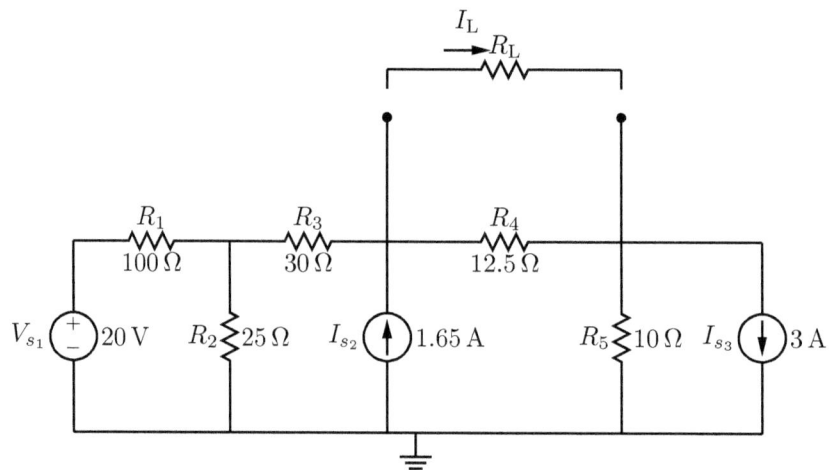

Figure 4.19: A linear circuit with load R_L attached to one of its ports.

Let us estimate the cost of doing this. It seems that nodal analysis would be a (relatively) low-cost approach since we only have three nodal voltages (we would have five meshes if we used mesh analysis). Doing this would mean solving this circuit one hundred times. We would be solving a 3×3 matrix a hundred times!

Compare this approach to the alternative of finding its Thévenin or Norton equivalent circuit at the port we are attaching the load to as shown in Figure 4.20. Finding these circuits would require us to find the equivalent resistance

R_{eq} seen at port a-b and either the open-circuit voltage V_{oc} (Thévenin equivalent) or the short-circuit current I_{sc} (Norton equivalent). In Figure 4.20 we have called the open-circuit voltage V_{Th} and the short-circuit current I_N. This is consistent with the nomenclature found in most books and papers. In other words,

$$V_{Th} = V_{oc},$$

and

$$I_N = I_{sc}.$$

Let us start by finding R_{eq}. We already know the process we need to follow: we need to turn off all independent sources and take a look at the resulting circuit shown in Figure 4.21.

Remember: Shutting down independent sources means replacing a) voltage sources with short circuits and b) current sources with open circuits.

Looking at Figure 4.21, we can quickly find the equivalent resistance:

$$R_{eq} = R_4 \parallel (R_5 + R_3 + (R_1 \parallel R_2)),$$
$$R_1 \parallel R_2 = \left(\frac{R_1 R_2}{R_1 + R_2} \right) = 20\,\Omega,$$
$$R_5 + R_3 + (R_1 \parallel R_2) = 10 + 20 + 20 = 50\,\Omega,$$
$$R_{eq} = R_4 \parallel 50 = 10\,\Omega.$$

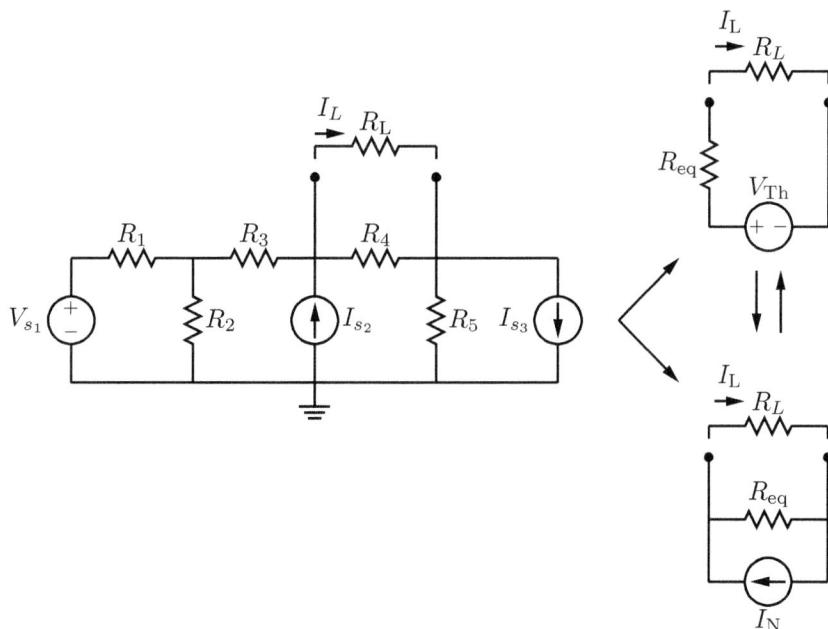

Figure 4.20: Figure 4.19 represented with its Thévenin and Norton equivalent circuits and the load port.

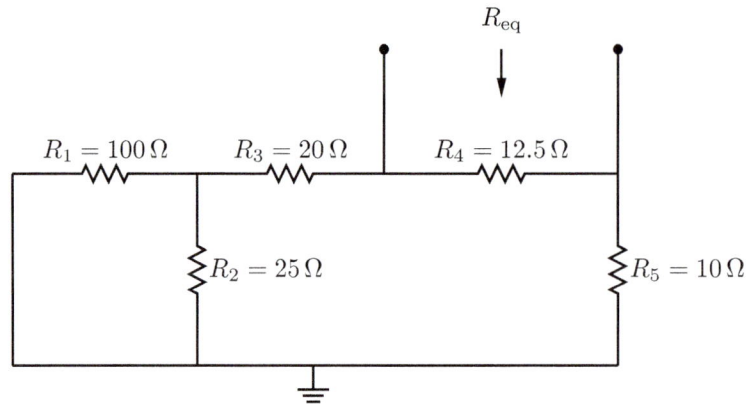

Figure 4.21: All independent sources have been turned off to find R_{eq} at the load terminal.

Note: In simple cases with no dependent sources, we may be able to apply source transformation to find the Thévenin and Norton equivalent circuits. Transforming sources and successively combining sources and resistors in such cases may yield the desired equivalent circuits without applying the formal process, although it is perfectly acceptable to do that.

The next question is which quantity to find: V_{Th} or I_N? Figure 4.22 shows the circuits we need to solve for both cases. The open-circuit (Thévenin) voltage has three unknown nodal voltages. On the other hand, the short-circuit (Norton) path has only two unknowns thanks to R_4 being shorted.

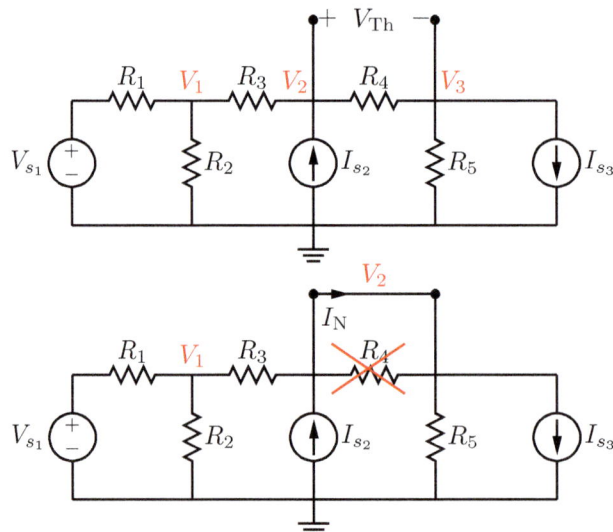

Figure 4.22: Choosing between finding the open-circuit voltage or the short-circuit current.

Since finding the short-circuit current at the load terminal of the circuit in Figure 4.19 will short-circuit the resistor R_4, we can redraw the circuit as seen in Figure 4.23.

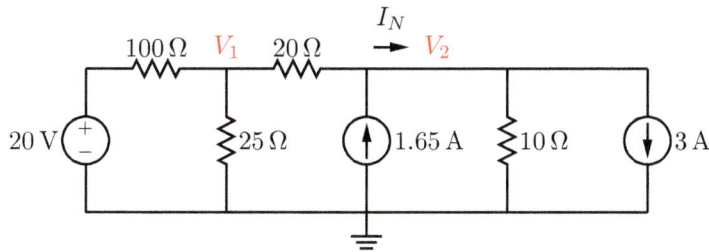

Figure 4.23: The circuit in Figure 4.19 prepared for finding the short-circuit current.

Filling the matrices to solve for the nodal voltages yields

$$\begin{bmatrix} \frac{1}{100} + \frac{1}{25} + \frac{1}{20} & -\frac{1}{20} \\ -\frac{1}{20} & \frac{1}{20} + \frac{1}{10} \end{bmatrix} \begin{bmatrix} V_1 \\ V_2 \end{bmatrix} = \begin{bmatrix} \frac{20}{100} \\ 1.65 - 3 \end{bmatrix}.$$

Inverting the conductance matrix and multiplying by the source vector results in $V_1 = -3$ V and $V_2 = -10$ V.

The short-circuit current I_N is calculated with a KCL:

$$I_N = \frac{V_2}{10} + 3 = 2 \,\text{A}.$$

If we would like to find V_{Th}, there is no need to solve the circuit again. The Thévenin voltage can be calculated using

$$V_{Th} = R_{eq} I_N = (10)(2) = 20 \,\text{V}.$$

Now we can see that it is going to be easy to let the load resistor R_L vary in any desired range and calculate the resulting load current. The load current is calculated using the Norton equivalent circuit using current division:

$$I_L = \frac{R_{eq}}{R_{eq} + R_L} I_N.$$

Alternatively, if we choose to use the Thévenin model, we get

$$I_L = \frac{V_{Th}}{R_{eq} + R_L}.$$

A simple script can be written to vary the load resistance and to calculate the corresponding current.

4.6.3 How to Find the Thévenin and Norton Equivalent Circuits

Now that we can appreciate the usefulness of the Thévenin and Norton equivalent circuits, let us summarize the formal way for finding them. To find the Thévenin or Norton equivalent circuits for any linear circuit, we need to find two of the following three quantities:

- The equivalent resistance at the desired port.

- The open-circuit voltage (V_{Th}) across the desired port.

- The short-circuit current (I_{N}) through the desired port.

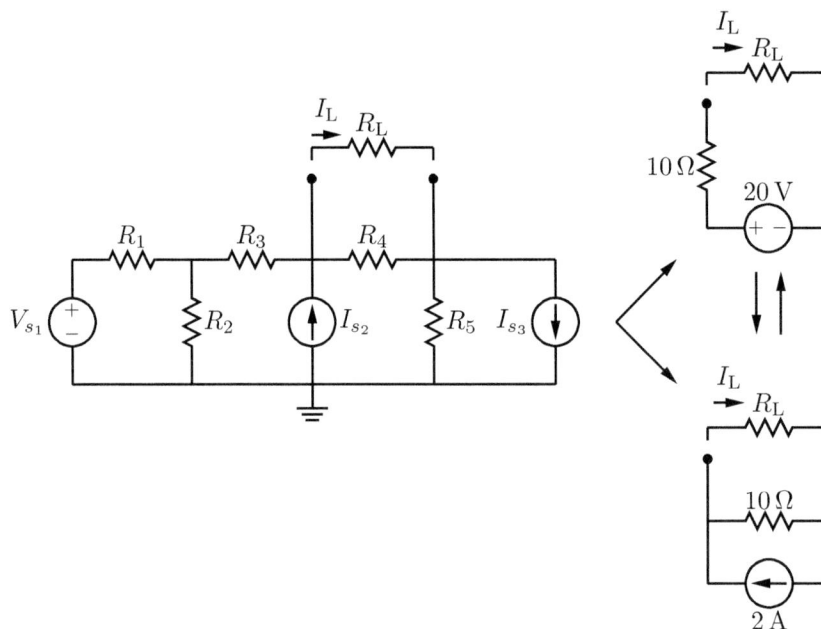

Figure 4.24: The Thévenin and Norton equivalent circuits with respect to port a-b.

It does not matter what we are asked to find; we should always evaluate the circuit and find out the two easiest values to find because

- If you have R_{eq} and V_{Th}, then

$$I_{\text{N}} = \frac{V_{\text{Th}}}{R_{\text{eq}}}.$$

- If you have R_{eq} and I_{N}, then

$$V_{\text{Th}} = I_{\text{N}} R_{\text{eq}}.$$

- If you have I_{N} and V_{Th}, then

$$R_{\text{eq}} = \frac{V_{\text{Th}}}{I_{\text{N}}}.$$

In the previous section, we showed through an example how to find all of these. Here is a summary of the process:

- Equivalent resistance (R_{eq}): Turn off all independent sources (dependent sources remain unchanged) and calculate the resulting resistance at the desired port. Notice that you may have to apply the $i - v$ test if resistors cannot be combined through series and parallel connections, or if the circuit includes dependent sources. In several books and papers, R_{eq} is also called the Thévenin equivalent resistance R_{Th} or Norton equivalent resistance R_{N}. In other words,

$$R_{\mathrm{eq}} = R_{\mathrm{Th}} = R_{\mathrm{N}}.$$

- Open-circuit (Thévenin) voltage (V_{Th}): Leave the desired port open-circuited (i.e., no load connected) and find the voltage across it.

- Short-circuit (Norton) current (I_{N}): Short-circuit the desired port (i.e., connect a short circuit across the port) and find the current through it.

We can now work though a few examples to further demonstrate the process.

Example 4.6.1

For the circuit shown in Figure 4.25, find the Thévenin and Norton equivalent circuits at port a-b.

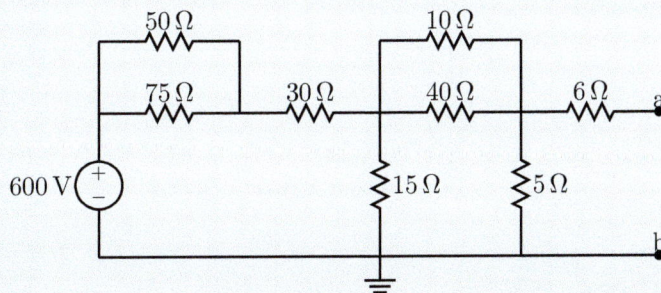

Figure 4.25: Example 4.6.1.

Since the circuit contains no dependent sources, source transformation may be an attractive option. We will try this by starting from the far left and making our way to port a-b.

The 600 V source is connected in series with the parallel combination of the 75 Ω and 50 Ω resistors and the 30 Ω resistor, as seen in Figure 4.26.

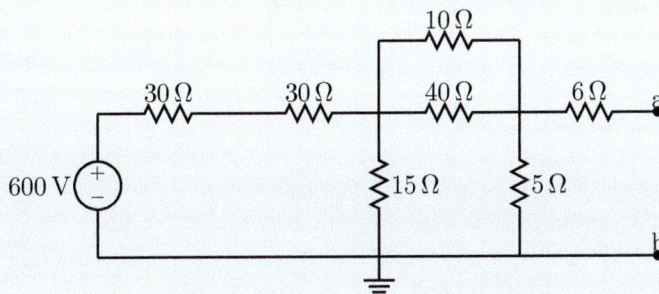

Figure 4.26: First step in solving Example 4.6.1.

Source transformation of the 600 V source and the 60 Ω resistor is shown in Figure 4.27.

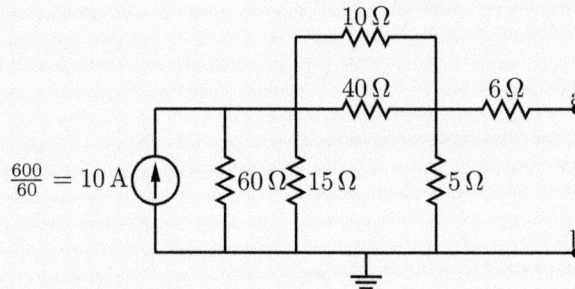

Figure 4.27: Second step in solving Example 4.6.1.

Now we can do another source transformation between the 10 A current source and the parallel combination of the 60 Ω and 15 Ω resistors, as seen in Figure 4.28. In this same step we can also combine the 40 Ω and 10 Ω parallel resistors.

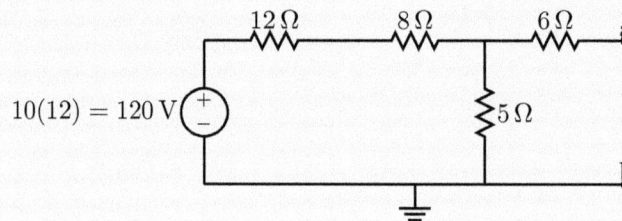

Figure 4.28: Third step in solving Example 4.6.1.

We can perform a source transformation again between the 120 V source and the series combination of the 12 Ω and 8 Ω resistors, as seen in Figure 4.29.

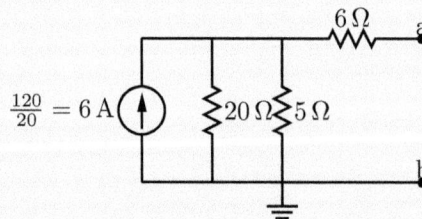

Figure 4.29: Fourth step in solving Example 4.6.1.

One more source transformation will take us to the Thévenin equivalent of the original circuit. We will accomplish this by using source transformation between the 6 A current and the parallel combination of the 20 Ω and 5 Ω resistors, as seen in Figure 4.30.

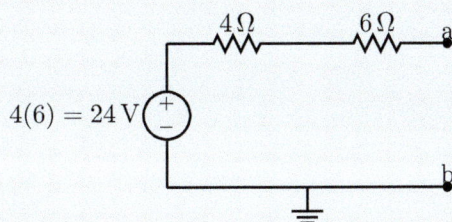

Figure 4.30: Final step in solving Example 4.6.1.

The Norton equivalent can be deduced using the circuit in Figure 4.30. Since $R_{eq} = 10\,\Omega$ and $V_{Th} = 24$ V,

$$I_N = \frac{V_{Th}}{R_{eq}} = \frac{24}{10} = 2.4\,\text{A}.$$

Let us now try an example with a dependent source. We will find V_{Th}, I_N, and R_{eq} separately just so we can cover all methods. We will break this up into three separate examples for ease.

Example 4.6.2

For the circuit shown in Figure 4.31, find the Thévenin voltage V_{Th} at port a-b using the open-circuit voltage method.

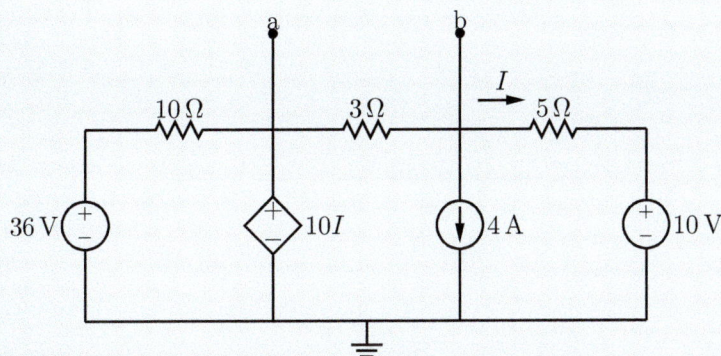

Figure 4.31: Example 4.6.2.

To find V_{Th}, we will have to find the open-circuit voltage V_{ab}. Since we have only two unknown nodes, we can apply nodal analysis to solve this circuit. While it may be tempting to apply source transformation, the circuit will not be simplified. In fact, since the current I is a controlling variable for the $10I$ controlled voltage source, performing a source transformation of the 5 Ω resistor and the 10 V source would be really problematic. Controlling variables should not be touched in a circuit.

The nodal equation for node a is $V_a = 10I$. So we only need an additional equation for the nodal voltage V_b and an equation for I. The current I can be found as

$$I = \left(\frac{V_b - 10}{5}\right),$$

$$V_a = 10I = 2V_b - 20,$$

$$V_a - 2V_b = -20\,\text{V}. \tag{4.4}$$

The node b equation is

$$\left(-\frac{1}{3}\right)V_a + \left(\frac{1}{3} + \frac{1}{5}\right)V_b = -4 + \frac{10}{5}. \tag{4.5}$$

We have two equations, (4.4) and (4.5). Solving for the nodal voltages, we will get $V_a = 110$ V and $V_b = 65$ V. Hence,

$$V_{\text{Th}} = V_{ab} = V_a - V_b = 110 - 65 = 45\,\text{V}.$$

Example 4.6.3

For the circuit shown in Figure 4.32, find the Norton current I_N at port a-b using the short-circuit current method.

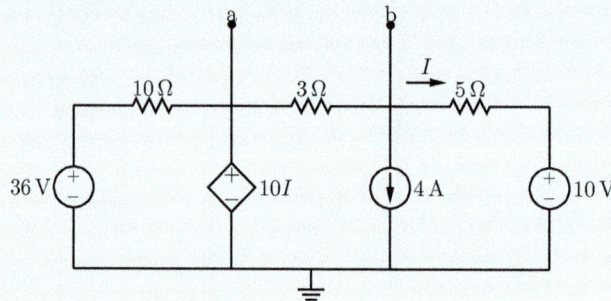

Figure 4.32: Example 4.6.3.

The circuit will change here since we have to short the circuit at port a-b. We have redrawn the circuit in Figure 4.33 with the short circuit between nodes a and b.

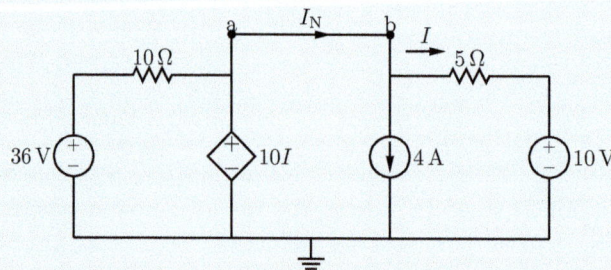

Figure 4.33: Circuit with shorted port a-b to find I_N.

The Norton current can be found using KCL:

$$I_N = I + 4.$$

After shorting port a-b, we end up with only one node, and that node has a voltage

$$V_a = V_b = 10I.$$

The equation for I can be written as

$$I = \frac{V_b - 10}{5} = \frac{10I - 10}{5},$$

$$5I = 10I - 10,$$

$$I = 2\,\text{A}.$$

This makes the Norton current I_N

$$I_N = I + 4 = 2 + 4 = 6\,\text{A}.$$

We can now directly calculate R_{eq}. From Example 4.6.2 and 4.6.3,

$$R_{eq} = \frac{V_{Th}}{I_N} = \frac{45}{6} = 7.5\,\Omega.$$

On the other hand, let us practice finding R_{eq} with the $i - v$ method as well.

Example 4.6.4

For the circuit shown in Figure 4.34, find the equivalent resistance at port a-b using the $i - v$ test (using a 1 A test source).

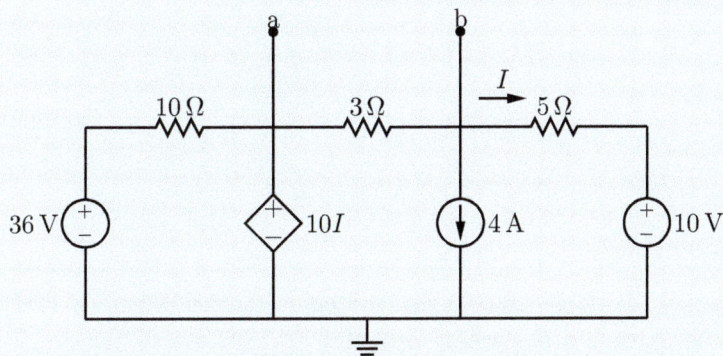

Figure 4.34: Example 4.6.4.

We turn off all independent sources and place the 1 A test source at the a-b port.

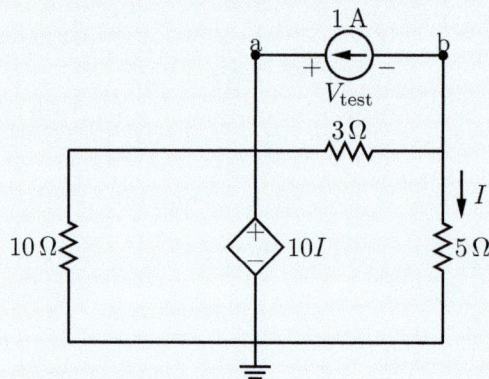

Figure 4.35: Turning off all independent sources and adding the test source.

We have two nodal voltages V_a and V_b, and the test voltage is the difference between these two nodal voltages. The node a equation is

$$V_a = 10I.$$

The current I can be calculated using

$$I = \frac{V_b}{5}.$$

Rewriting node a's equation, we get

$$V_a = 10\frac{V_b}{5} = 2V_b,$$
$$V_a - 2V_b = 0. \tag{4.6}$$

The node b equation is

$$\left(-\frac{1}{3}\right)V_a + \left(\frac{1}{5} + \frac{1}{3}\right)V_b = -1. \tag{4.7}$$

Solving (4.6) and (4.7), we get $V_a = 15$ V and $V_b = 7.5$ V. Then we can find R_{eq}:

$$R_{eq} = \frac{V_{test}}{1} = \frac{V_a - V_b}{1} = 7.5\,\Omega.$$

Of course, Example 4.6.4 yielded the same result as when we used V_{Th} and I_N. Let us try to find R_{eq} again but with a 1 V test source instead.

Example 4.6.5

For the circuit shown in Figure 4.36, find the equivalent resistance at port a-b using the $i - v$ method (using a 1 V test source).

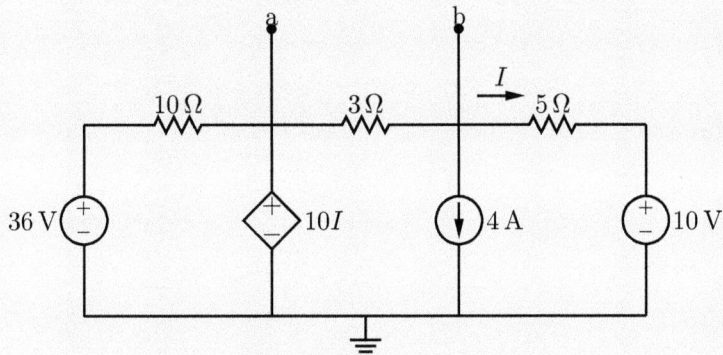

Figure 4.36: Example 4.6.5.

We turn off all independent sources and add the test source at port a-b. For this case, we will need to find I_{test}. Using KCL, we find

$$I_{\text{test}} = \frac{V_a - V_b}{3} - I,$$

$$I_{\text{test}} = \frac{V_a - V_b}{3} - \frac{V_b}{5},$$

$$I_{\text{test}} = \left(\frac{1}{3}\right) V_a - \left(\frac{1}{3} + \frac{1}{5}\right) V_b. \qquad (4.8)$$

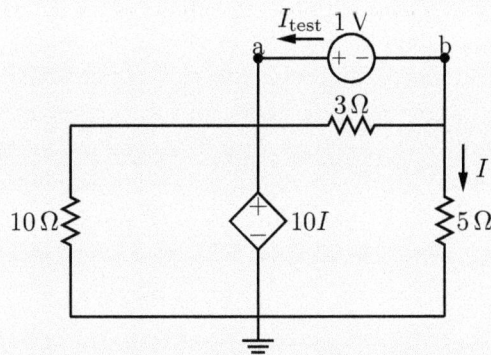

Figure 4.37: Turning off all independent sources and adding the test source.

Because of the test voltage source,

$$V_a - V_b = 1\,\text{V},$$

$$10I - V_b = 1,$$

$$10\frac{V_b}{5} - V_b = 1,$$

$$V_b = 1\,\text{V}.$$

We can now find V_a:

$$V_a = V_{\text{test}} + V_b = 2\,\text{V}.$$

Plugging the values of V_a and V_b into (4.8), we get

$$
\begin{aligned}
I_\text{test} &= \left(\frac{1}{3}\right) V_\mathrm{a} - \left(\frac{1}{3} + \frac{1}{5}\right) V_b, \\
&= \left(\frac{1}{3}\right)(2) - \left(\frac{1}{3} + \frac{1}{5}\right)(1), \\
&= \frac{2}{15}\,\mathrm{A}.
\end{aligned}
$$

Hence

$$
R_\text{eq} = \frac{1}{I_\text{test}} = \frac{15}{2} = 7.5\,\Omega.
$$

4.7   Thévenin and Norton Applications

Replacing an entire (linear) network or system with a source/resistor model is the most important application of the Thévenin and Norton theorems in electrical engineering. Systems interact with the world through ports. In order to study these interactions, we have to model these systems through their ports. Source modeling is the best example to examine.

In the beginning of this book, we introduced the ideal independent source model. In this model, a source can provide constant voltage or current and any power level. Real-life power supplies are limited in what they can supply us with. We can model some of their imperfections by adding series or parallel elements to the ideal source. Since we have only studied resistive circuits so far, we will only model losses using resistors. In future sections we will be able to add more elements to our circuit models. For example, in theory, a car's lead acid battery will be modeled by a 12 V voltage independent source. In real life, the lead acid battery will be modeled by a 12 V voltage independent source in series with a resistor, as seen in Figure 4.38, to account for losses.

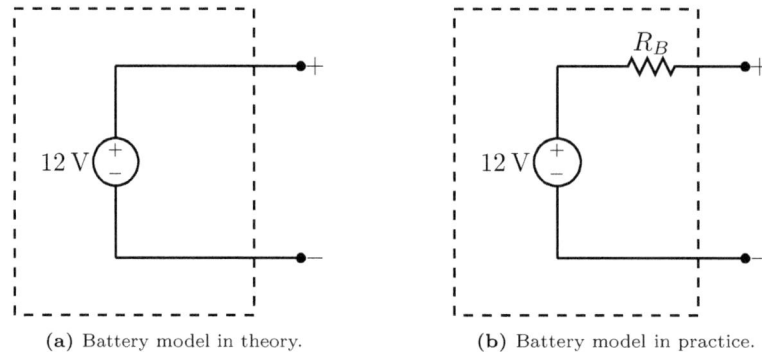

(a) Battery model in theory.   (b) Battery model in practice.

Figure 4.38: The model of a lead acid battery (a) in theory and (b) in practice.

The same can be said about a cell phone battery. The battery actually has internal resistance that limits its performance. A battery's internal resistance tends to increase as the battery ages. Furthermore, battery leakage and capacity also tend to become worse over time. Let us look at a real-life example that happened to us.

Application 4.7.1

On a cold December day we attempted to start our car. After failing a couple of times, we decided to check if the battery was dead.

Figure 4.39: Measuring a car battery.

We measured the voltage across the battery and the battery was fully charged. If the battery was fully charged, why was it not able to start the car? Now, remember that the voltage measured was the open-circuit voltage. Since that was fine, the problem had to be with the current supplied by the battery to the load (starter). Since the battery was old, it was taken to the shop for further testing. As expected, the battery was delivering only 60% of its rated current, which was not enough to start the engine. How did that happen? Over time, the internal resistance of the battery had increased, which in turn decreased the cold cranking current.

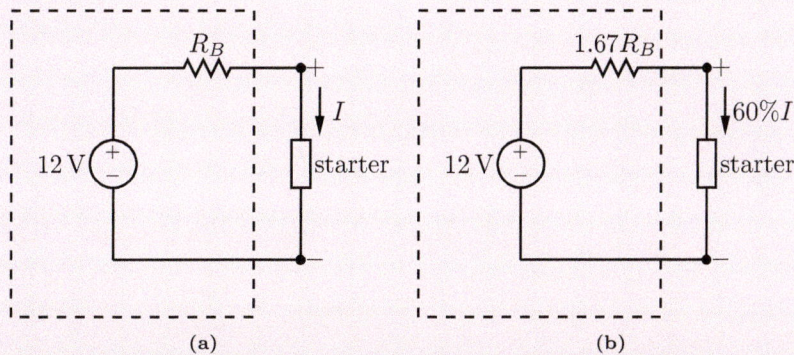

Figure 4.40: New and old lead acid battery models.

4.8 Exercises

Section 4.2

4.1. Find the output voltage V_o for the circuit in Figure 4.41. If we wanted the output voltage to be $V_o = 12\,\text{V}$, find the new value for the source.

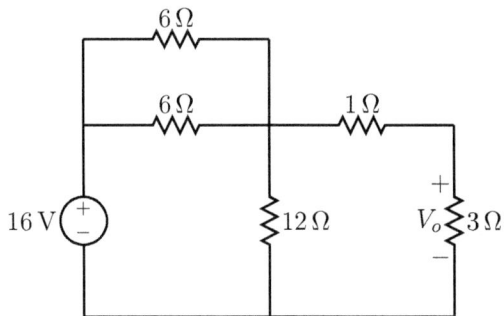

Figure 4.41

4.2. Find the output voltage V_o for the circuit in Figure 4.42 for the current of source values shown in Table 4.2.

I_s	V_o
$3\,\text{A}$	
$-6\,\text{A}$	
$1.5\,\text{A}$	

Table 4.2

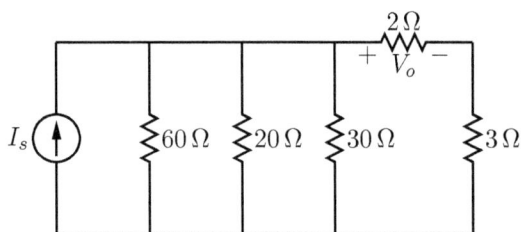

Figure 4.42

4.3. For the circuit shown in Figure 4.43, the output voltage is $V_o = 5\,\text{V}$. Find the value of the source voltage V_s. If the value of the voltage source were changed to $V_s = 15\,\text{V}$, find V_o.

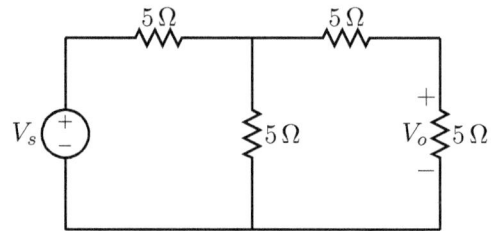

Figure 4.43

Section 4.3

4.4. For the circuit in Figure 4.44, find V_o and I_o using superposition.

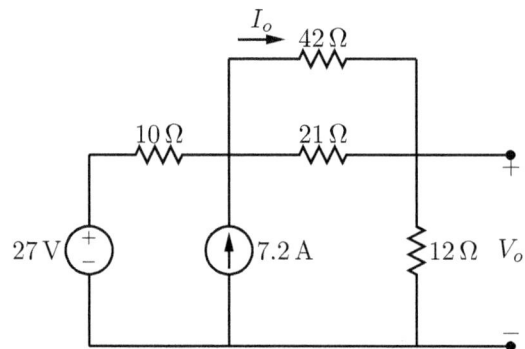

Figure 4.44

4.5. For the circuit in Figure 4.45, find I_o. If a 2 A current source were added to terminals a and b, find the new value for I_o. (Hint: Use superposition.)

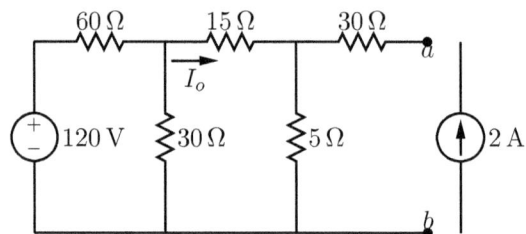

Figure 4.45

4.6. For the circuit in Figure 4.46, find V_o using superposition. If the 15 V voltage source were replaced with a 30 V source, find the new value for V_o. (Hint: Use linearity.)

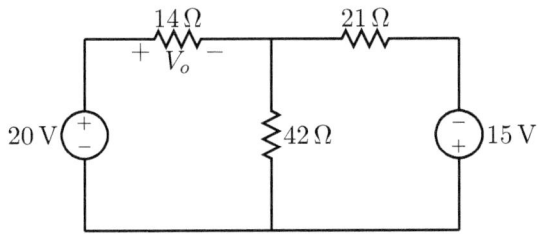

Figure 4.46

4.7. The 30 V voltage source in Figure 4.47 was turned off by accident. Find the output voltage V_o. If the 30 V voltage source were turned on again, find the new value of the output voltage V_o.

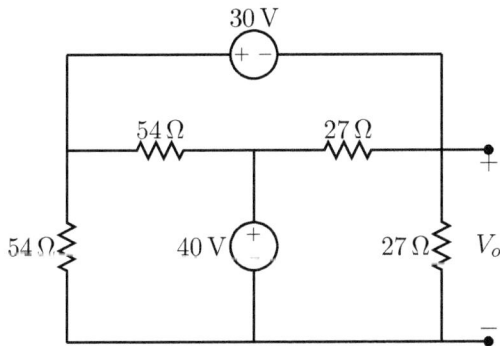

Figure 4.47

4.8. Using superposition, find the value of the output current I_o in Figure 4.48. We would like to study the effect of doubling and halving the value of the 10 A current source. Find the output current I_o for each value in Table 4.3. (Hint: Use linearity.)

I_s	I_o
10 A	
20 A	
5 A	

Table 4.3

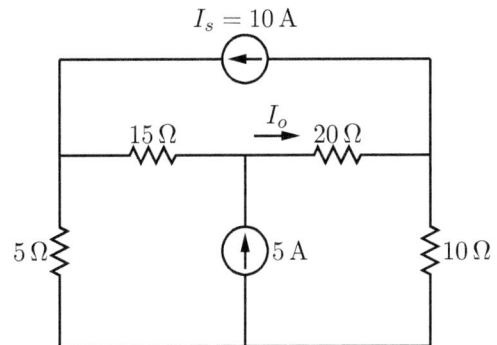

Figure 4.48

Section 4.5

4.9. Using source transformation, find V_o in Figure 4.49.

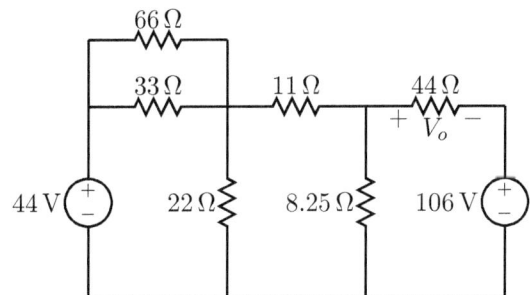

Figure 4.49

4.10. Using source transformation, find V_o and I_o in Figure 4.50.

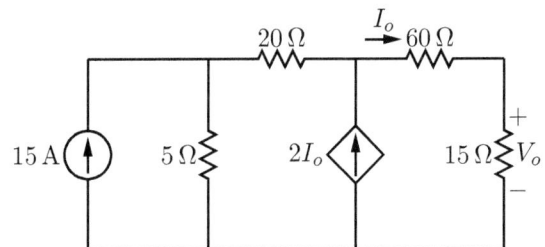

Figure 4.50

4.11. Using source transformation, find I_o in Figure 4.51.

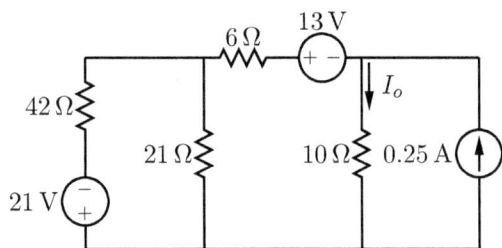

Figure 4.51

4.12. Using source transformation, find the power delivered by the 62.6 V voltage source in Figure 4.52.

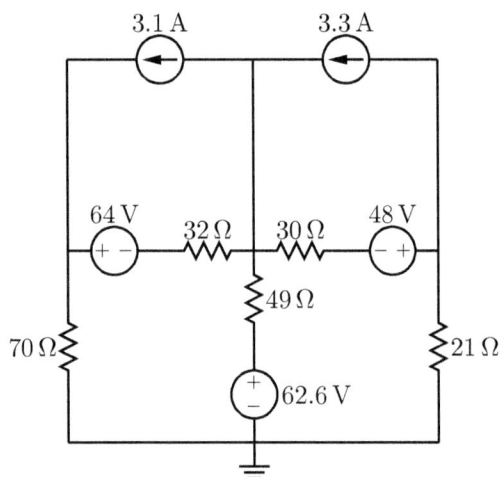

Figure 4.52

Section 4.6

4.13. Find the Thévenin equivalent at terminals a and b in Figure 4.53. (Hint: Use source transformation.)

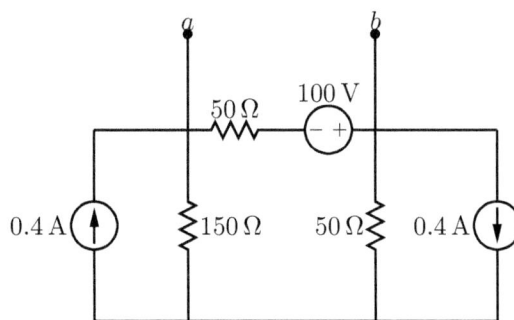

Figure 4.53

4.14. Find the Norton equivalent at terminals a and b in Figure 4.54. (Hint: Use source transformation.)

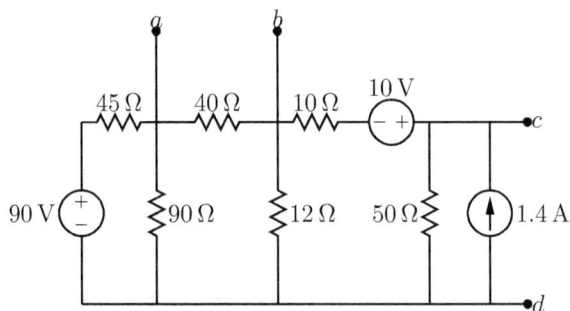

Figure 4.54

4.15. Find the Norton and Thévenin equivalents at terminals c and d in Figure 4.55. (Hint: Use source transformation)

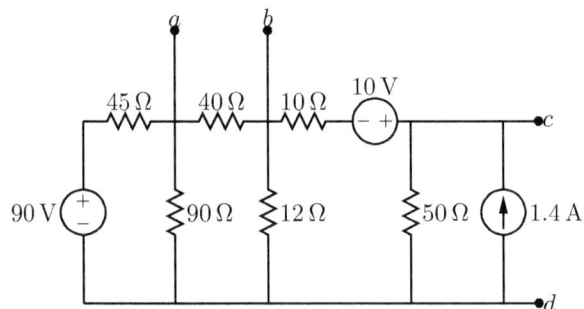

Figure 4.55

4.16. A source is represented by its Thévenin equivalent in Figure 4.56a. Find the current I leaving this source when it is attached to the circuit in Figure 4.56b at terminals a and b. (Hint: Find the Thévenin equivalent of the circuit.)

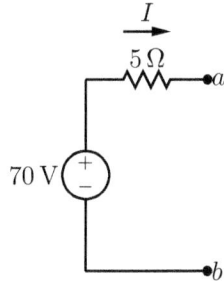

(a)

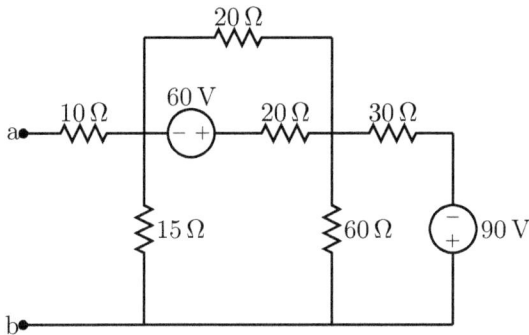

(b)

Figure 4.56

4.17. Find the Thévenin equivalent at terminals a and b in Figure 4.57.

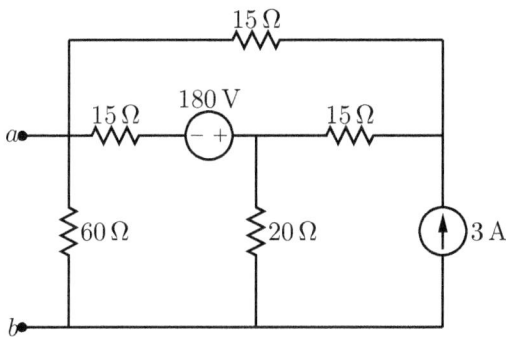

Figure 4.57

4.18. Find the Thévenin and Norton equivalents at terminals a and b in Figure 4.58.

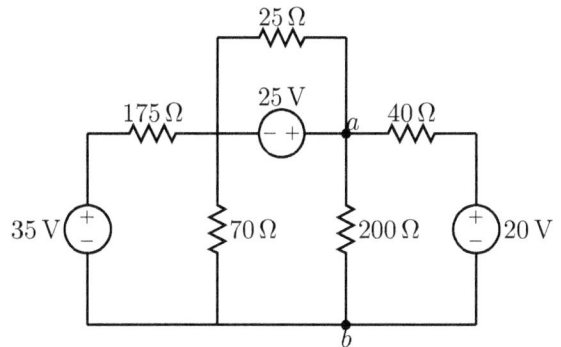

Figure 4.58

4.19. Find the Thévenin and Norton equivalents at terminals a and b in Figure 4.59.

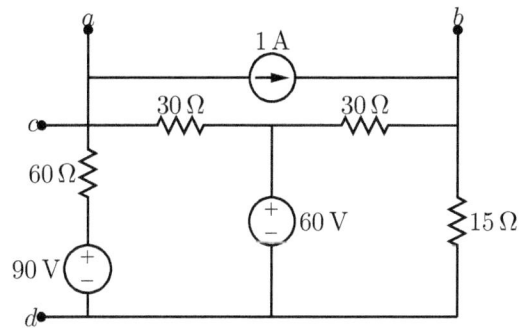

Figure 4.59

4.20. Find the Thévenin equivalent at terminals c and d in Figure 4.60.

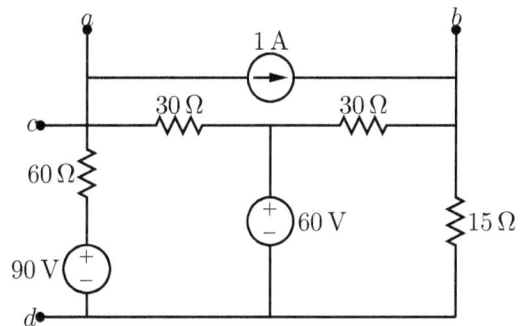

Figure 4.60

4.21. Find the Thévenin equivalent at terminals a and b in Figure 4.61.

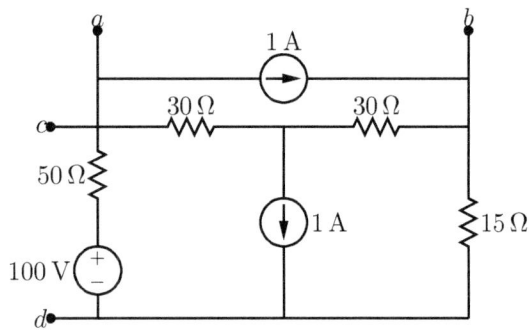

Figure 4.61

4.22. Find the Thévenin equivalent at terminals c and d in Figure 4.62.

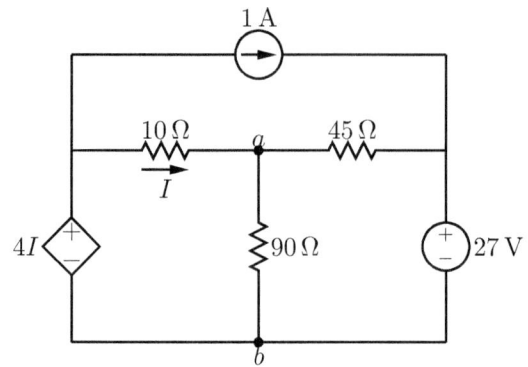

Figure 4.62

4.23. Find the Thévenin equivalent at terminals a and b in Figure 4.63.

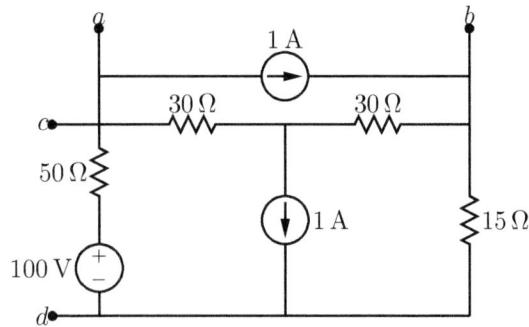

Figure 4.63

4.24. Find the Thévenin equivalent at terminals a and b in Figure 4.64.

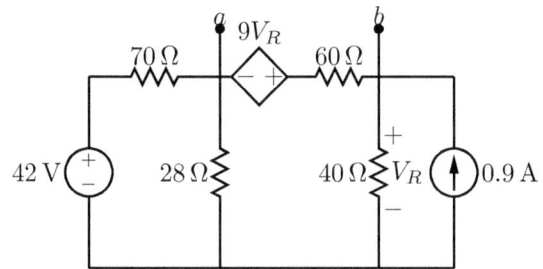

Figure 4.64

4.25. Find the Thévenin equivalent at terminals a and b in Figure 4.65.

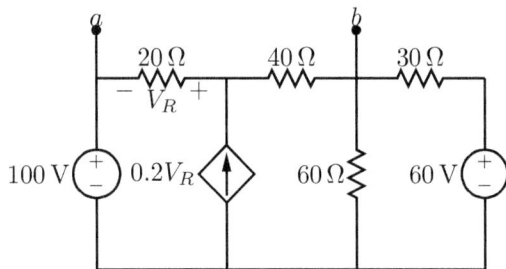

Figure 4.65

4.26. Find the Thévenin and Norton equivalents at terminals a and b in Figure 4.66.

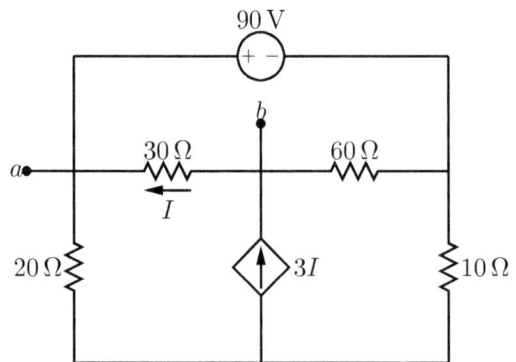

Figure 4.66

Chapter 5

Capacitors and Inductors

5.1 Introduction

Resistors are not the only basic building blocks of a circuit. By adding capacitors and inductors in our discussion, we can model a lot of real-life circuits, such as radios, motors, generators, etc. We are still in the realm of linear passive elements, but contrary to resistors, capacitors and inductors are energy storage elements. A capacitor stores energy in its electric field, while an inductor stores energy in its magnetic field. In this chapter, we will learn the basics about capacitors and inductors as well as how they interact with other devices in a circuit, and explore their dual nature. We will also compare them to resistors and emphasize their differences and similarities.

5.2 Capacitors

Capacitors are passive elements that store electric field energy. They are widely used in various shapes and sizes in many applications. Capacitors in real-life applications typically range in magnitude from picofarads (pF) to hundreds of farads (F). You can find them in your everyday electronics such as your phone and radio, as well as in electric cars or aircrafts, or you can find them hanging on transmission line poles as shunt capacitor banks. The application areas for capacitors are nearly endless.

The parallel plate image is perhaps the simplest and most common concept picture of a capacitor. It consists of two parallel metal sheets with a dielectric medium (insulator) in between them. When a voltage difference is applied to the plates, negative charges accumulate at one plate while positive charges accumulate on the opposite side as seen in Figure 5.1.

While we use the parallel plate structure to discuss capacitors, basic properties of the derived $i - v$ characteristic and related conditions are true for all capacitors regardless of their structures. We remember from basic physics that the opposite charges on the parallel plates create an electric field in between them where electric energy is stored. The amount of charge q deposited on the

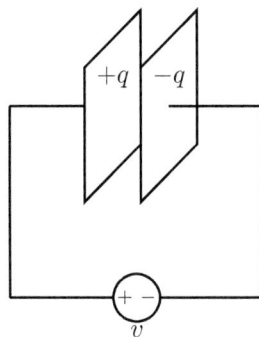

Figure 5.1: Parallel plate capacitor.

plates is proportional to the applied voltage V. The constant of proportionality is the capacitance C of the structure—that is,

$$q = Cv. \tag{5.1}$$

The capacitance C depends solely on the geometry of the structure and its material properties. It does not depend on the applied voltage. For a parallel plate capacitor, we can increase its capacitance by increasing the surface area of the plates, or by decreasing the distance between the plates, or by changing the material between the plates. From (5.1), the unit for capacitance is coulombs per volt, which is called farads.

$$1\,\mathrm{F} = \frac{1\,\mathrm{C}}{1\,\mathrm{V}}. \tag{5.2}$$

The capacitance of the parallel plate capacitor shown in Figure 5.2 can be approximated as

$$C = \frac{\epsilon A}{d} \tag{5.3}$$

where A is the area of the plate in meters squared, d is the distance between the plates in meters, and ϵ is the permittivity of the dielectric between the two plates in farads per meter. This expression is derived from electrostatics for infinite plates and is approximately true when the plate dimensions are much larger than the distance d. While this expression ignores edge effects (fringing fields), in practice it is within 20–30% of the true capacitance value.

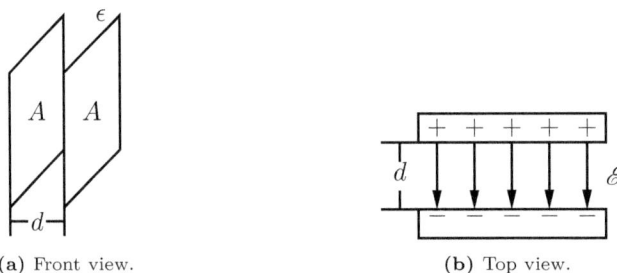

(a) Front view.   (b) Top view.

Figure 5.2: Front and top views of a parallel plate capacitor.

An ideal capacitor holds a uniform charge distribution indefinitely and has an electric field $\mathscr{E}$ in straight lines as seen in Figure 5.2. Real-life capacitors are leaky (i.e., they lose charge over time) and the electric field fringes at its edges. Most of the time the energy loss is modeled by a resistor in parallel to an ideal capacitor.

A capacitor is represented in circuit diagrams as two parallel plates as seen in Figure 5.3.

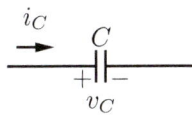

(a) Capacitor circuit representation (passive sign convention).

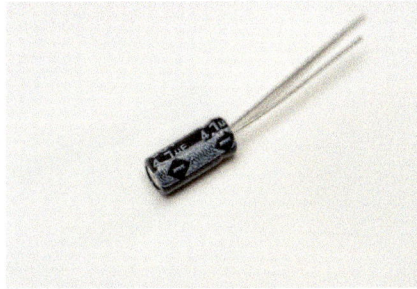

(b) An actual 4.7 μF capacitor.

Figure 5.3: Capacitor as a circuit element.

Application 5.2.1

Nature's parallel plate capacitor

Rain clouds and the ground make nature's biggest parallel plate capacitor. The water droplets in the clouds are negatively charged and the ground is positively charged. Air sandwiched in between acts as the dielectric.

Figure 5.4: Nature's parallel plate capacitor.

As you will learn in future classes, every capacitor has a breakdown voltage. This is the voltage difference between the capacitor's plates that results in dielectric breakdown and may cause extensive damage. In nature, when the cloud-ground voltage difference exceeds the air breakdown field, the two "plates" are shorted and current flows through. Discharging current into the ground causes lightning.

Application 5.2.2

Capacitors as power factor correctors

If you live in a place that has aerial distribution transmission lines, you will likely see transformers that step down the voltage before transmitting it to buildings. Every once in a while, especially in front of industrial places such as hospitals, malls, and offices, you will find a capacitor bank connected in parallel with the transformer.

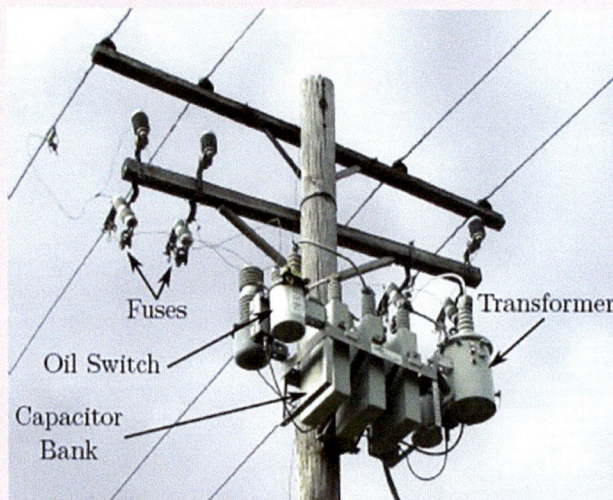

Figure 5.5: Power factor correction using capacitors.

These capacitor banks act as power factor correctors offsetting the effects of the highly inductive loads in the area. Power factor correction is not discussed in detail here. It leads, though, to more efficient power transfer.

5.2.1   Finding the Current Through a Capacitor

Starting from (5.1), we can find the current–voltage relationship of the capacitor—that is, the equivalent governing equation to Ohm's law for resistors. From the current definition, we know that

$$i_C(t) = \frac{dq}{dt}. \tag{5.4}$$

Taking the derivative of both sides of (5.1) and assuming constant capacitance yields

$$\frac{dq}{dt} = C\frac{dv_C(t)}{dt}. \tag{5.5}$$

Using (5.4) in (5.5), we find the current through the capacitor:

$$\boxed{i_C(t) = C\frac{dv_C(t)}{dt}.} \tag{5.6}$$

This is the basic current–voltage (or $i - v$) characteristic of the capacitor. Similarly to Ohm's law for resistors, we can use this equation to find the voltage or current of a capacitor if we know one of the two. Let us look at a couple of examples.

Example 5.2.1

The voltage across the 2 F capacitor in Figure 5.6 is $v_C(t) = t^3 \cos(\pi t)$ V. Find $i_C(t)$. Then find $i_C(0)$, $i_C(0.5)$, and $i_C(1)$.

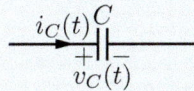

Figure 5.6: Example 5.2.1.

We have

$$i_C(t) = C\frac{dv_C(t)}{dt} = 6t^2 \cos(\pi t) - 2\pi t^3 \sin(\pi t) \text{ A}.$$

Also,

$$i_C(0) = 0 \text{ A}.$$

$$i_C(0.5) = 6t^2 \cos(\pi t) - 2\pi t^3 \sin(\pi t) = -2\pi(0.5^3)(1) = -0.25\pi \text{ A}.$$

$$i_C(1) = 6t^2 \cos(\pi t) - 2\pi t^3 \sin(\pi t) = 6(1)(-1) = -6 \text{ A}.$$

Example 5.2.2

Figure 5.7 shows a plot of the voltage across a 2 F capacitor with respect to time. Find the current $i_C(t)$ through the capacitor.

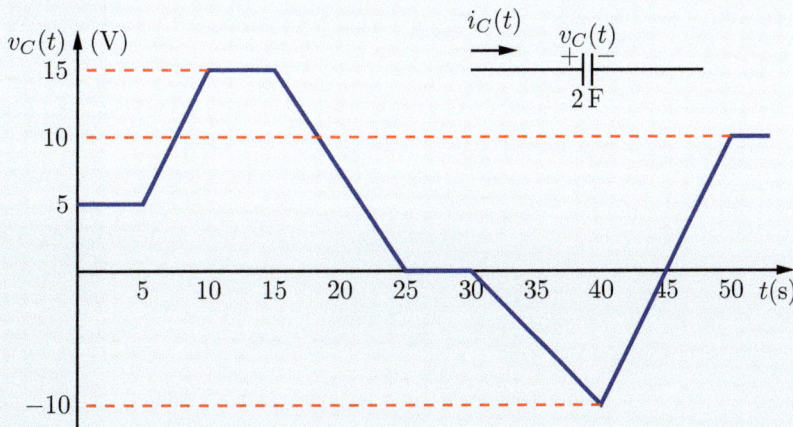

Figure 5.7: Voltage plot for Example 5.2.2.

We start with (5.6). Since the voltage derivative is discontinuous, we will have to do the calculations in the appropriate time intervals. For convenience, we repeat (5.6) here.

$$i_C(t) = C\frac{dv_C(t)}{dt}.$$

For $t < 5\,\mathrm{s}$, the voltage is constant, so

$$i_C(t) = 2\frac{d5}{dt} = 0\,\mathrm{A}.$$

For the interval $5\,\mathrm{s} < t < 10\,\mathrm{s}$, the voltage is increasing linearly and we can find it by using the equation of a line $y = mx + b$.

$$v_C(t) = 2t - 5,$$
$$i_C(t) = 2\frac{d(2t - 5)}{dt} = 4\,\mathrm{A}.$$

For the interval $10\,\mathrm{s} < t < 15\,\mathrm{s}$, the voltage is constant again:

$$i_C(t) = 2\frac{d15}{dt} = 0\,\mathrm{A}.$$

For the interval $15\,\mathrm{s} < t < 25\,\mathrm{s}$, the voltage is decreasing linearly and we can find it by using the equation of a line $y = mx + b$.

$$v_C(t) = -1.5t + 37.5$$
$$i_C(t) = 2\frac{d(-1.5t + 37.5)}{dt} = -3\,\mathrm{A}.$$

For the interval $25\,\mathrm{s} < t < 30\,\mathrm{s}$, the voltage is constant again:

$$i_C(t) = 0\,\mathrm{A}.$$

For the interval $30\,\mathrm{s} < t < 40\,\mathrm{s}$, the voltage is decreasing linearly and we can find it by using the equation of a line $y = mx + b$.

$$v_C(t) = -t + 30$$
$$i_C(t) = 2\frac{d(-t + 30)}{dt} = -2\,\mathrm{A}.$$

For the interval $40\,\mathrm{s} < t < 50\,\mathrm{s}$, the voltage is increasing linearly and we can find it by using the equation of a line $y = mx + b$.

$$v_C(t) = 2t - 90$$
$$i_C(t) = 2\frac{d(2t - 90)}{dt} = 4\,\mathrm{A}.$$

To summarize,

$$
i_C(t) = \begin{cases}
0 & t < 5\,\text{s}, \\
4 & 5\,\text{s} < t < 10\,\text{s}, \\
0 & 10\,\text{s} < t < 15\,\text{s}, \\
-3 & 15\,\text{s} < t < 25\,\text{s}, \\
0 & 25\,\text{s} < t < 30\,\text{s}, \\
-2 & 30\,\text{s} < t < 40\,\text{s}, \\
4 & 40\,\text{s} < t < 50\,\text{s}, \\
0 & t > 50\,\text{s}.
\end{cases} \quad \text{A}
$$

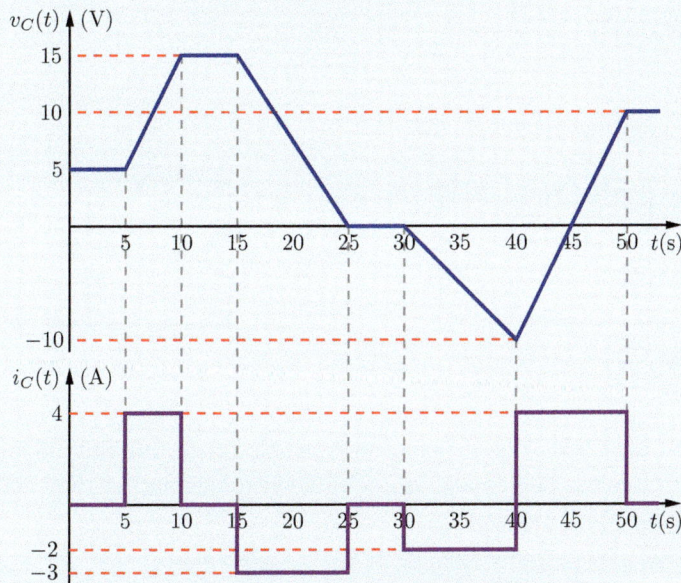

Figure 5.8: Voltage and current plots for Example 5.2.2.

Let us examine (5.6) more carefully to further understand a capacitor's behavior; (5.6) is repeated here for convenience:

$$
i_C(t) = C \frac{\mathrm{d}v_C(t)}{\mathrm{d}t}.
$$

The first thing we notice is that we have a derivative—more specifically, the first time derivative of the voltage across the capacitor. This means that the voltage of a capacitor must be **continuous** and cannot change abruptly. Why? Because a sudden jump in its voltage would result in infinite current, which is impossible. In math form we can write that for *every* time t, $v_C(t^-) = v_C(t^+)$.

The voltage of a capacitor is always continuous: $v_C(t^-) = v_C(t^+)$.

Note that just because the voltage cannot change abruptly, the current does not necessarily behave in the same manner. Current through the capacitor can change instantaneously to compensate for the changes happening to the circuit. We will see more about this in the next chapter.

This derivative also means that the capacitor behaves as an **open circuit** in DC circuits. This is because in DC circuits

$$i_C(t) = C\frac{\mathrm{d}v_C(t)}{\mathrm{d}t} = C \cdot 0 = 0\,\mathrm{A}. \tag{5.7}$$

Zero current indeed indicates an open circuit. Similarly, a capacitor in steady state (i.e., no changes) acts as an open circuit. Keep in mind that although the capacitor acts as an open circuit, it can still have a voltage drop across it. Take, for example, the circuit in Figure 5.9. The capacitor has zero current through it, but the voltage across the capacitor is

$$v_C = \frac{R_2}{R_1 + R_2}V_s. \tag{5.8}$$

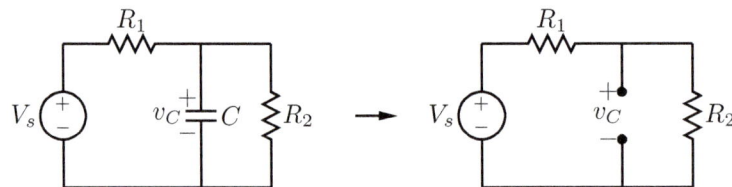

Figure 5.9: Capacitor in steady state or constant voltage conditions.

We will need this information later when we try to find initial conditions in first-order circuits. This is also important when calculating the energy stored in a capacitor.

Example 5.2.3

The circuit in Figure 5.10 has been in the given state for a long time. Find the voltage across the capacitor.

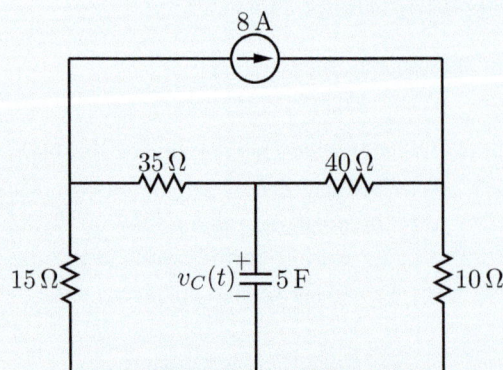

Figure 5.10: Example 5.2.3.

Since the circuit has not changed in a long time, the capacitor is under steady-state conditions and behaves as an open circuit as seen in Figure 5.11.

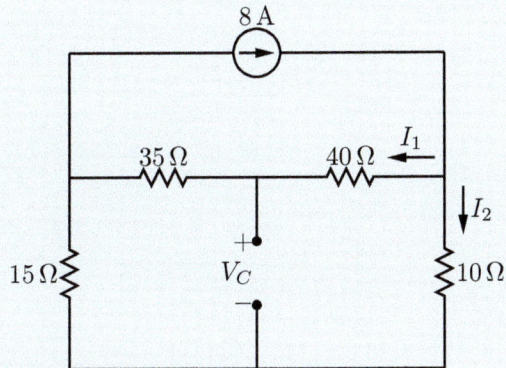

Figure 5.11: Example 5.2.3 in steady state.

We have

$$V_C = -40I_1 + 10I_2.$$

Using current division,

$$I_1 = \frac{(10 + 15)}{(10 + 15) + (40 + 35)}(8) = 2\,\text{A}.$$

Using KCL,

$$I_2 = 8 - I_1 = 6\,\text{A}.$$

Hence

$$V_C = -40(2) + 10(6) = -20\,\text{V}.$$

Example 5.2.4

The circuit in Figure 5.12 has been in the given state for a long time. Find the voltage across both capacitors.

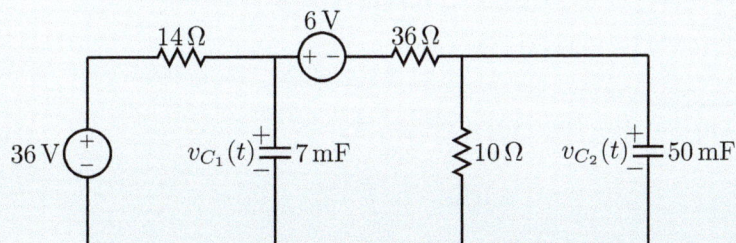

Figure 5.12: Example 5.2.4.

Since the circuit has not changed in a long time, the capacitors are in steady-state conditions and both behave as open circuits as seen in Figure 5.13.

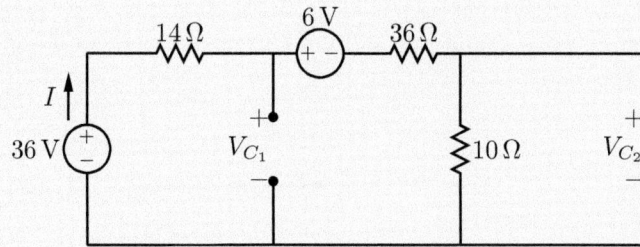

Figure 5.13: Example 5.2.4 in steady state.

We can find the total current I as

$$I = \frac{36 - 6}{14 + 36 + 10} = 0.5\,\text{A}.$$

Now we can find the voltages across the capacitors:

$$V_{C_1} = 36 - 14I = 29\,\text{V},$$

and

$$V_{C_2} = 10I = 5\,\text{V}.$$

5.2.2   Finding the Voltage Across a Capacitor

Let us now understand how to find the voltage or current of a capacitor. If we know the voltage, finding the current is straightforward through (5.6). So we will focus on finding the voltage if we know the current. Let us start by rearranging (5.6) to isolate the derivatives:

$$dv_C = \frac{1}{C} i_C(t)\, dt. \tag{5.9}$$

Integrating both sides of (5.9) and changing the integration variables (from t to x) to avoid confusion, we get

$$\int_{-\infty}^{v_C(t)} dv_C = \frac{1}{C} \int_{-\infty}^{t} i_C(x)\, dx,$$

$$v_C(t) - v_C(-\infty) = \frac{1}{C} \int_{-\infty}^{t} i_C(x)\, dx. \tag{5.10}$$

Assuming that the capacitor voltage at $-\infty$ is zero ($v_C(-\infty) = 0$), (5.10) becomes

$$v_C(t) = \frac{1}{C} \int_{-\infty}^{t} i_C(x)\, dx. \tag{5.11}$$

While integrating from $-\infty$ may be hard to understand at first, you may think about it this way: Let us assume a typical circuit where everything is constant until some time t_o. For example, at t_0 a switch may be flipped or a source may be turned on or off. Up until t_0, everything will be constant (steady-state or DC conditions), so it makes sense to split the integral as follows:

$$v_C(t) = \underbrace{\frac{1}{C} \int_{-\infty}^{t_0} i_C(x) \, dx}_{a} + \underbrace{\frac{1}{C} \int_{t_0}^{t} i_C(x) \, dx}_{b}. \qquad (5.12)$$

We see that we end up with two parts. Part a represents the voltage across the capacitor at the time t_0 and we can use $v_C(t_0)$ to represent that voltage. This is often called the **initial condition** for a circuit since it provides the initial capacitor voltage before a change takes place. Substituting part a in (5.12) with $v_C(t_0)$ and rearranging, we get

$$\boxed{v_C(t) = \frac{1}{C} \int_{t_0}^{t} i_C(x) \, dx + v_C(t_0).} \qquad (5.13)$$

Example 5.2.5

A 2 mF capacitor was initially charged to 10 V. Find the voltage across the capacitor for $t \geq 0$ if the current through the capacitor is $i_C(t) = 4\sin(2t)$ mA. Also, find the voltage at $t = \frac{\pi}{4}$, $t = \frac{\pi}{2}$, and $t = \pi$.

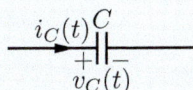

Figure 5.14: Example 5.2.5.

We use (5.13) to find the voltage across the capacitor as

$$v_C(t) = \frac{1}{2\,\text{mF}} \int_0^t 4\sin(2t)\,\text{mA} \, dx + v(0).$$

$$v_C(t) = \int_0^t 2\sin(2x) \, dx + 10.$$

$$v_C(t) = -2\frac{\cos(2x)}{2}\Big|_0^t + 10 = 11 - \cos(2t) \text{ V}.$$

$$v_C\left(\frac{\pi}{4}\right) = 11 - \cos\left(2\frac{\pi}{4}\right) = 11 \text{ V}.$$

$$v_C\left(\frac{\pi}{2}\right) = 11 - \cos\left(2\frac{\pi}{2}\right) = 11 - (-1) = 12 \text{ V}.$$

$$v_C(\pi) = 11 - \cos(2\pi) = 11 - 1 = 10 \text{ V}.$$

Example 5.2.6

The current through a 5 F is shown in Figure 5.15. Find the voltage across the capacitor for all t. The capacitor was initially uncharged.

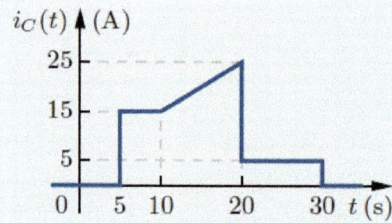

Figure 5.15: Example 5.2.6.

$$
i_C(t) = \begin{cases}
0 & 0\,\mathrm{s} < t < 5\,\mathrm{s}, \\
15 & 5\,\mathrm{s} < t < 10\,\mathrm{s}, \\
t+5 & 10\,\mathrm{s} < t < 20\,\mathrm{s}, \quad \mathrm{A} \\
5 & 20\,\mathrm{s} < t < 30\,\mathrm{s}, \\
0 & t > 30\,\mathrm{s}.
\end{cases}
$$

We will use (5.13) to find the voltage across the capacitor. For $5\,\mathrm{s} \le t \le 10\,\mathrm{s}$,

$$
v_C(t) = \frac{1}{5}\int_5^t 15\ \mathrm{d}x + v_C(0) = 3t\Big|_5^t + 0 = 3t - 15\ \mathrm{V}.
$$

We will need the voltage at $t = 10\,\mathrm{s}$ for the next step.

$$
v_C(10\,\mathrm{s}) = 3(10) - 15 = 15\ \mathrm{V}
$$

For $10\,\mathrm{s} \le t \le 20\,\mathrm{s}$,

$$
v_C(t) = \frac{1}{5}\int_{10}^t (x+5)\ \mathrm{d}x + v_C(10) = \frac{1}{5}\Big(\frac{x^2}{2} + 5x\Big)\Big|_{10}^t + 15 = \frac{t^2}{10} + t - 5\ \mathrm{V}.
$$

Then,

$$
v_C(20\,\mathrm{s}) = \frac{(20)^2}{10} + 20 - 5 = 55\ \mathrm{V}.
$$

For $20\,\mathrm{s} \le t \le 30\,\mathrm{s}$,

$$
v_C(t) = \frac{1}{5}\int_{20}^t (5)\ \mathrm{d}x + v_C(20) = \frac{1}{5}(5t)\Big|_{20}^t + 55 = t + 35\ \mathrm{V}.
$$

The final voltage value will be ($t \ge 30\,\mathrm{s}$).

$$
v_C(30\,\mathrm{s}) = 30 + 35 = 65\ \mathrm{V}.
$$

5.2.3   Energy Storage in a Capacitor

Now that we have an equation for the voltage and current, we can start thinking about the energy stored in a charged capacitor. We have already learned that the instantaneous power can always be calculated as

$$
p_C(t) = v_C(t)i_C(t). \tag{5.14}
$$

Using (5.6) in the power equation, we get

$$p_C(t) = v_C(t) C \frac{\mathrm{d}v_C(t)}{\mathrm{d}t}. \tag{5.15}$$

To find the energy stored in the capacitor, we need to integrate the power over the period during which the capacitor is charging:

$$w_C(t) = \int_{-\infty}^{t} p_C(x) \, \mathrm{d}x = C \int_{-\infty}^{t} v_C(x) \frac{\mathrm{d}v_C(x)}{\mathrm{d}x} \, \mathrm{d}x = \left. \frac{C}{2} v_C^2(x) \right|_{-\infty}^{t}$$

$$= \frac{1}{2} C(v_C^2(t) - v_C^2(-\infty)). \tag{5.16}$$

Assuming the capacitor is not charged at $t = -\infty$, $(v_C(t = -\infty) = 0)$, we can write

$$\boxed{w_C(t) = \frac{1}{2} C v_C^2(t).} \tag{5.17}$$

For a constant voltage (DC), we can rewrite (5.17) as

$$W_C = \frac{1}{2} C V_C^2. \tag{5.18}$$

Let us look at an example to solidify this idea.

Example 5.2.7

For the circuit in Figure 5.16, find

1. The energy stored in the capacitor at $t = \infty$.

2. The energy stored in the capacitor from $t = 0$ to $t = \infty$.

The switch next to the 5 Ω resistor has been closed for a long time and is opened at $t = 0$. The switch next to the 2 Ω resistor has been open for a long time and is closed at $t = 0$.

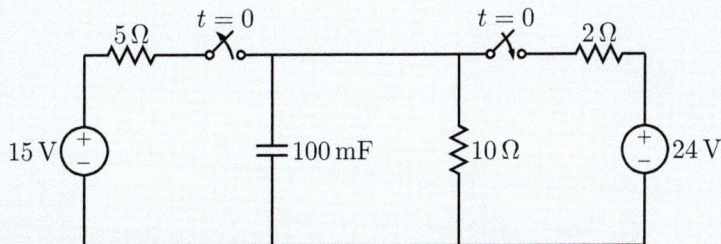

Figure 5.16: Example 5.2.7.

To find the energy stored in the capacitor at $t = \infty$, we need to look at the final state of the circuit at $t = \infty$. The circuit at $t = \infty$ is shown in Figure 5.17

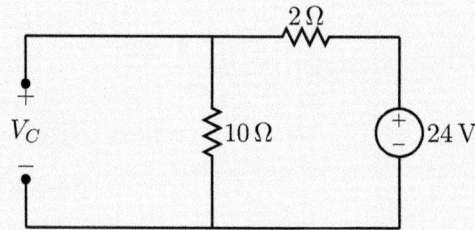

Figure 5.17: Circuit drawn at $t = \infty$ for Example 5.2.7.

The voltage across the capacitor can be obtained by using voltage division and is

$$V_C = \frac{10}{10 + 2}(24) = 20\,\text{V}.$$

The energy stored in the capacitor is

$$W_C = \frac{1}{2}CV_C^2 = \frac{1}{2}(100\text{m})(20^2) = 20\,\text{J}.$$

Now let us find the energy stored in the capacitor from $t = 0$ to $t = \infty$. We know that the capacitor had a charge before the switches changed states since it has a nonzero voltage for $t < 0$. Consequently, we need to find the energy stored in the capacitor at $t = 0$ and subtract it from the energy at $t = \infty$ since $v_C(0^-) = v_c(0^+)$, $w_C(0^-) = w_C(0^+)$. Hence, we will find the energy stored at $t = 0$ from the circuit before any changes happened—that is, at $t = 0^-$. The circuit at $t = 0^-$ is shown in Figure 5.18.

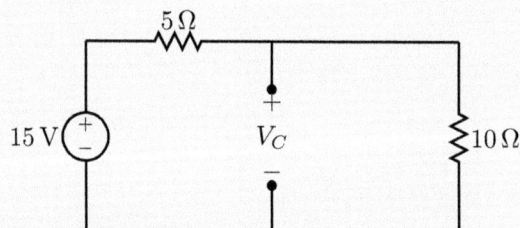

Figure 5.18: Circuit shown at $t = 0^-$ for Example 5.2.7.

The voltage across the capacitor can be obtained by using voltage division:

$$V_C = \frac{10}{10 + 5}(15) = 10\,\text{V}.$$

Consequently, the energy stored in the capacitor at $t = 0^-$ is

$$w_C(t = 0^-) = \frac{1}{2}Cv_C^2(0^-) = \frac{1}{2}(100\text{m})(10^2) = 5\,\text{J}.$$

Hence the energy stored in the capacitor from $t = 0$ to $t = \infty$ is

$$w_C(t > 0) = w_C(t = \infty) - w_C(t = 0^-) = 20 - 5 = 15\,\text{J}.$$

You may wonder why we had to find the energy $w_C(t = 0)$ from the circuit state at $t = 0^-$ and not from the circuit state at $t = 0^+$. The reason is that it was easy to find the capacitor voltage at $t = 0^-$ since the circuit was unchanged from $t = -\infty$ to $t = 0^-$. Hence the capacitor behaved as an open circuit during this entire time.

Let us go back to Example 5.2.4 and try to calculate the energy stored in each capacitor and the total energy stored in the circuit.

Example 5.2.8

We would like to find the energy stored in each capacitor in Figure 5.19 and the total energy stored in the circuit.

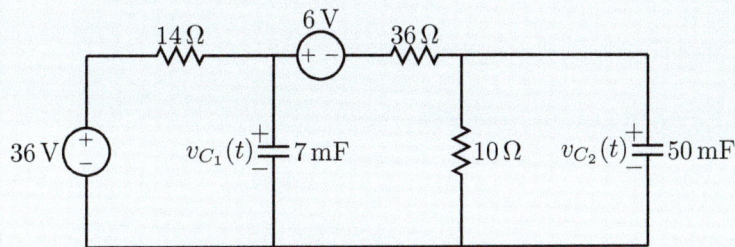

Figure 5.19: Example 5.2.8.

From Example 5.2.4, we know that the voltages across the capacitors are

$$V_{C_1} = 36 - 14I = 29\,\text{V},$$

and

$$V_{C_2} = 10I = 5\,\text{V}.$$

The energy stored in each capacitor is found as

$$W_{C_1} = \frac{1}{2}C_1 V_{C_1}^2 = \frac{1}{2}(7)(29^2) = 2.944\,\text{J},$$

and

$$W_{C_2} = \frac{1}{2}C_2 V_{C_2}^2 = \frac{1}{2}(50)(5^2) = 0.625\,\text{J}.$$

The total energy stored in the circuit is

$$W_{C_1} + W_{C_2} = 3.569\,\text{J}.$$

5.2.4   Capacitor Connections

Just like resistors, capacitors can be connected in multiple ways: we may connect them in series, parallel, delta, or wye formations. We will examine here the series and parallel connections.

Capacitors in Series

Let us start by taking a KVL around the circuit shown in Figure 5.20:

$$v_s(t) = v_{C_1} + v_{C_2} + \cdots + v_{C_N}. \tag{5.19}$$

We recall from (5.13) that

$$v_C(t) = \frac{1}{C} \int_{t_0}^{t} i_C(x) \, \mathrm{d}x + v_C(t_0). \tag{5.20}$$

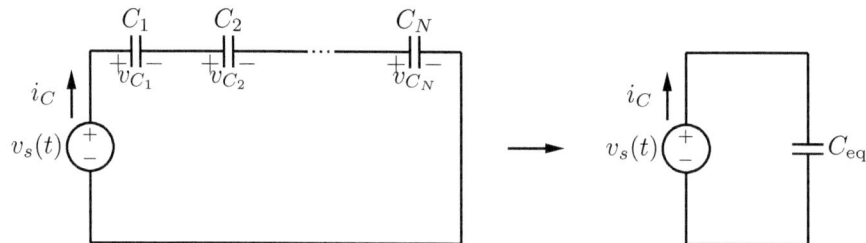

Figure 5.20: Capacitors connected in series.

Since the capacitors are connected in series, the current is the same through all of them. Using (5.20) in (5.19),

$$v_s(t) = \frac{1}{C_1} \int_{t_0}^{t} i_C(x) \, \mathrm{d}x + v_{C_1}(t_0) + \frac{1}{C_2} \int_{t_0}^{t} i_C(x) \, \mathrm{d}x + v_{C_2}(t_0) + \ldots$$
$$+ \frac{1}{C_N} \int_{t_0}^{t} i_C(x) \, \mathrm{d}x + v_{C_N}(t_0). \tag{5.21}$$

This can be rewritten as

$$v_s(t) = \left(\frac{1}{C_1} + \frac{1}{C_2} + \ldots + \frac{1}{C_N} \right) \int_{t_0}^{t} i_C(x) \, \mathrm{d}x + v_{C_1}(t_0) + v_{C_2}(t_0) + \ldots$$
$$+ v_{C_N}(t_0). \tag{5.22}$$

We can rewrite (5.22) as

$$v_s(t) = \left(\frac{1}{C_{\mathrm{eq}}} \right) \int_{t_0}^{t} i_C(x) \, \mathrm{d}x + v_{C_1}(t_0) + v_{C_2}(t_0) + \ldots + v_{C_N}(t_0), \tag{5.23}$$

where

$$\boxed{\frac{1}{C_{\mathrm{eq}}} = \sum_{i=1}^{N} \frac{1}{C_i}.} \tag{5.24}$$

Consequently, we proved that:

1. The equivalent capacitance of a group of capacitors connected in series is calculated by $\frac{1}{C_{\mathrm{eq}}} = \left(\frac{1}{C_1} + \frac{1}{C_2} + \ldots + \frac{1}{C_N} \right)$. Thus capacitors connected in series behave as resistors connected in parallel.

2. The initial voltage across C_{eq} is given by $v_{C\text{eq}}(t_0) = v_{C_1}(t_0) + v_{C_2}(t_0) + \ldots + v_{C_N}(t_0)$.

Capacitors in Parallel

Figure 5.21 shows a group of capacitors connected in parallel.

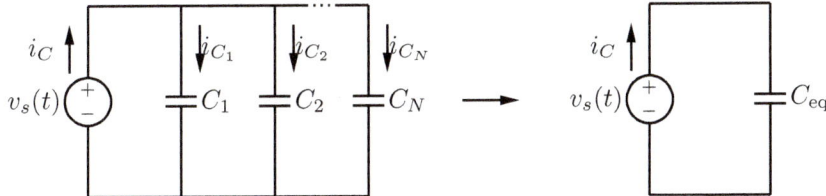

Figure 5.21: Capacitors connected in parallel.

We can start with adding the currents using KCL:

$$i_C = i_{C_1} + i_{C_2} + \ldots + i_{C_N}. \tag{5.25}$$

We remember that

$$i_C(t) = C \frac{dv_C(t)}{dt}, \tag{5.26}$$

and since the voltage is the same across all capacitors, we have

$$i_C = C_1 \frac{dv(t)}{dt} + C_2 \frac{dv(t)}{dt} + \ldots + C_N \frac{dv(t)}{dt}$$

$$i_C = (C_1 + C_2 + \ldots + C_N) \frac{dv(t)}{dt} = C_{\text{eq}} \frac{dv(t)}{dt}. \tag{5.27}$$

Just as expected, capacitors connected in parallel are summed to find the equivalent capacitance:

$$\boxed{C_{\text{eq}} = \sum_{i=1}^{n} C_n.} \tag{5.28}$$

Consequently, we proved that:

1. The equivalent capacitance of a group of capacitors connected in parallel is calculated as $C_{\text{eq}} = C_1 + C_2 + \ldots + C_N$. Thus capacitors connected in parallel behave as resistors connected in series.

2. The initial voltage across C_{eq} is the same as the initial voltage across each of the capacitors (capacitors connected in parallel share the same voltage).

Example 5.2.9

In this example we will find the equivalent capacitance at port a-b in Figure 5.22.

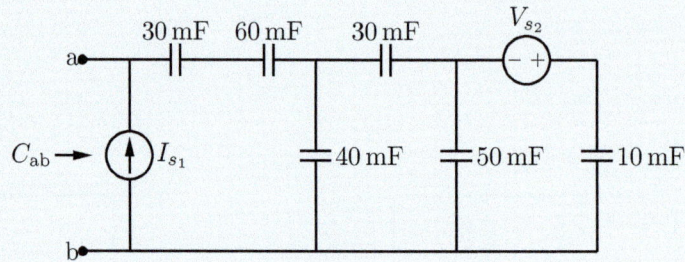

Figure 5.22: Example 5.2.9.

We remember from our resistive circuits chapter that we need to turn off all independent sources to find the equivalent resistance. We can do the same when trying to find the capacitance, as seen in Figure 5.23.

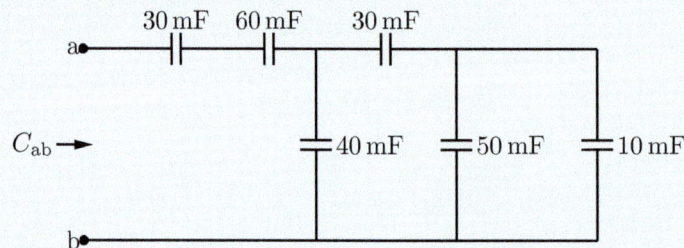

Figure 5.23: Circuit after turning off all independent sources for Example 5.2.9.

We will start from the far right and work our way to port a-b. The first two capacitors are in parallel, so they get summed up. We then have the sum in series with the 30 mF capacitor. Hence

$$C_{eq_1} = \frac{30(50 + 10)}{30 + 50 + 10} = 20\,\text{mF}.$$

Now C_{eq_1} is in parallel with the 40 mF capacitor, so they just get summed up as

$$C_{eq_2} = C_{eq_1} + 40 = 20 + 40 = 60\,\text{mF}.$$

Now the rest of the capacitors are in series, so we have

$$C_{ab} = \frac{1}{\frac{1}{30} + \frac{1}{60} + \frac{1}{C_{eq_2}}} = \frac{1}{\frac{1}{30} + \frac{1}{60} + \frac{1}{60}} = 15\,\text{mF}.$$

Example 5.2.10

In this example, we will find the equivalent capacitance at port c-d in Figure 5.24.

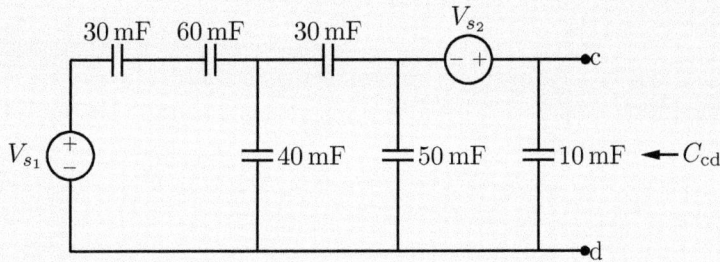

Figure 5.24: Example 5.2.10.

We start again by turning off all independent sources, as seen in Figure 5.25.

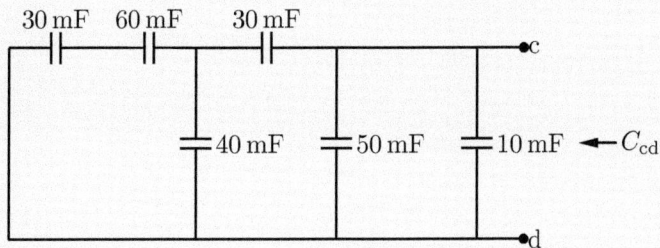

Figure 5.25: Circuit after turning off all dependent sources for Example 5.2.10.

We will start from the far left and work our way to port c-d. The first two capacitors are in series, so

$$C_{\text{eq}_1} = \frac{30(60)}{30 + 60} = 20\,\text{mF}.$$

Now C_{eq_1} is in parallel with the 40 mF capacitor. Their sum will be in series with the 30 mF capacitor, so all together become

$$C_{\text{eq}_2} = \frac{(20 + 40)(30)}{20 + 40 + 30} = 20\,\text{mF}.$$

C_{eq_2} is now in parallel with the 10 mF and 50 mF capacitors. Hence

$$C_{\text{cd}} = C_{\text{eq}_2} + 50 + 10 = 80\,\text{mF}.$$

5.3 Inductors

Inductors are passive elements that store magnetic energy. Inductance is measured in henries (H) and is widely used in everyday life. Some of the most common uses for inductors include transformers, radio-frequency identification (RFID) circuits, and motor starters. For instance, transformers are used in our everyday electronics to step voltages up or down. You can also see high-voltage transformers hanging from transmission line poles. Inductors vary widely in shape and size depending on the application, frequency of operation, and power-handling needs.

5.3.1   Self-Inductance

To better understand inductors and inductance, let us start with an example. We will not get into detailed derivations here; we will just present a few basic ideas. Consider a loop with a current I. We remember from Ampere's law that if we have a current through a wire, we will have a magnetic field $\vec{B}$. In this example, the $\vec{B}$ field resulting from the circulating current can be found using the Biot-Savart law. The magnetic field penetrates the surface area S of the loop, resulting in a magnetic flux Φ through the loop defined by

$$\Phi = \int_S \vec{B} \cdot \vec{dS}. \tag{5.29}$$

The magnetic field is measured in teslas (T) and dS is in square meters. The magnetic flux is measured in weber (Wb).

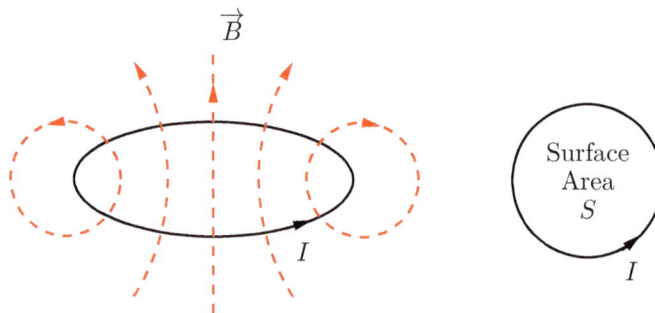

Figure 5.26: Magnetic field due to current in a closed loop.

The self-inductance, or simply inductance, of the current loop in Figure 5.26 is defined as

$$L = \frac{\Phi}{I}. \tag{5.30}$$

If we had N turns of the loop, the inductance would be given by

$$L = \frac{N\Phi}{I} = \frac{\lambda}{I} \tag{5.31}$$

where λ is the flux linkage defined by

$$\lambda = N\Phi. \tag{5.32}$$

Before we go any further, let us examine (5.31) a bit more. From our discussions on resistors and capacitors, we recall that neither element is dependent on voltage or current; they only depend on device geometry and material. This is true for inductors as well. While in the inductance-defining equation we see the current I in the denominator, it is important to remember that the associated flux is proportional to the magnetic field and the magnetic field is proportional to the current. Consequently, the flux-over-current ratio is constant. Consequently, inductance depends only on device geometry and material.

Let us look at an example to further consolidate this fact. Specifically, we will look at a solenoid as a good inductance calculation example.

Figure 5.27 shows a solenoid with N turns and length l. Every turn (loop) has a surface area S. We assume that all space is filled with air. We also assume that a current I flows through this solenoid. We will assume here that the length l is very large, so we can ignore the end effects. Let us also define the number of turns per unit length n as

$$n = \frac{N}{l}. \tag{5.33}$$

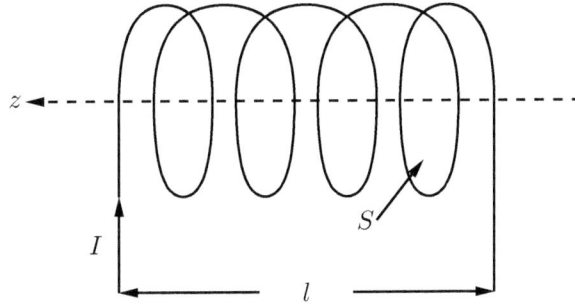

Figure 5.27: Schematic of a solenoid.

Without going through the proof, the magnetic field $\overrightarrow{B}$ is found to be

$$\overrightarrow{B} = \mu_o n I \overrightarrow{a_z} \tag{5.34}$$

where μ_o is the permeability of air (or free space). If the volume around the solenoid were filled with a magnetic material with permeability μ, we would have used this permeability in the previous equation. The magnetic flux through each turn is

$$\Phi = \int_S \overrightarrow{B} \cdot \overrightarrow{dS} = B_z S = \mu_o n S I. \tag{5.35}$$

Since we have N turns, the inductance can be found using (5.31) as

$$L = \frac{N\Phi}{I} = \frac{N\mu_o n S I}{I} = N\mu_o n S. \tag{5.36}$$

We can use (5.33) in (5.36) to further find

$$L = N\mu_o n S = N\mu_o \frac{N}{l} S = \mu_o \frac{N^2}{l} S. \tag{5.37}$$

As discussed before, the inductance depends only on the geometry of the inductor (number of turns, length, and surface area) and material (air in this case). We also notice that the inductance depends on the square of the number of turns (N^2). As a result, increasing the number of turns can increase the inductance dramatically.

5.3.2 The Induced Voltage or Electromotive Force (emf)

Faraday's law tells us that a current is inducted in a closed loop if we have a varying magnetic field through the loop even in the absence of any explicit circuit source. Faraday experimentally discovered this law by using two magnetically coupled loops as shown in Figure 5.28. The primary loop in this figure includes an inductor connected to a battery through a switch. The secondary loop is not excited through a source but is connected to a galvanometer, which is a very sensitive ammeter. Faraday discovered that when the switch was closed or opened a reading in the galvanometer would be recorded indicating current flow in the secondary loop. However, this current was nonzero only during the switch closing or opening times; the current was zero during steady-state conditions for both switch positions. He observed similar effects when he moved one loop relative to the other. Faraday concluded that it was the magnetic field **change** that would induce a current in the closed loop.

To explain the existence of this current, we define an electromotive force (emf) that is essentially an induced voltage (V_{emf}) around the closed loop. It is this induced voltage that generates the current observed in the loop. If the loop is open, there will be no current, of course, but there will still be the induced emf. In fact, this can be easily measured with a voltmeter in volts. While we are talking about a voltage, we call it emf. The reason for this rather unusual name is because we try to differentiate it from the electrostatic voltages we are familiar with that obey KVL—that is, they are zero around any closed path. The emf is not zero despite the fact that it is defined in a closed path. It has, however, exactly the same impact and generates a current in the loop in exactly the same way a voltage source would have if it had been connected in the loop.

Although we make it sound simple here, it in fact took Faraday over ten years to conduct conclusive experiments. While the historical account of these experiments is not the primary focus here, the equation that was later developed to capture this effect, known as Faraday's law, is important and is listed below.

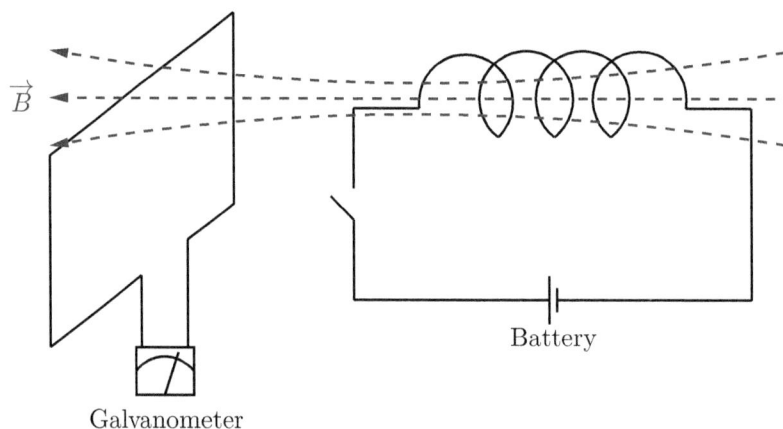

Figure 5.28: Faraday's induction experiment.

$$\oint_c \vec{E} \cdot \vec{\mathrm{d}l} = -\frac{\mathrm{d}}{\mathrm{d}t} \int_S \vec{B} \cdot \vec{\mathrm{d}S}. \tag{5.38}$$

The left-hand side of this equation is the induced emf, V_{emf}, and is measured in volts (V)—that is,

$$\oint_c \vec{E} \cdot \vec{\mathrm{d}l} = V_{\text{emf}}. \tag{5.39}$$

The right-hand side of (5.38) can be written more simply as

$$\Phi = \int_S \vec{B} \cdot \vec{dS}. \tag{5.40}$$

If we substitute that into Faraday's law, we get

$$\oint_c \vec{E} \cdot \vec{dl} = -\frac{d\Phi}{dt}. \tag{5.41}$$

Consequently, a simpler version of Faraday's law is

$$V_{\text{emf}} = -\frac{d\Phi}{dt}. \tag{5.42}$$

This form more clearly shows that a change in magnetic flux results in an induced voltage. The negative sign, known as Lenz's law, indicates that the induced voltage generates a current whose magnetic field opposes the change caused by the external magnetic field. While we will not discuss this further here, Lenz's law is a direct consequence of conservation of energy.

Let us look at an example that demonstrates these principles in a relatively simple setting.

Example 5.3.1

The circuit in Figure 5.29 is a 2×1 m^2 loop on the $x-y$ plane inside a uniform field $\vec{B} = -10\,t\,\vec{a_z}$ (T). Find V_1 and V_2.

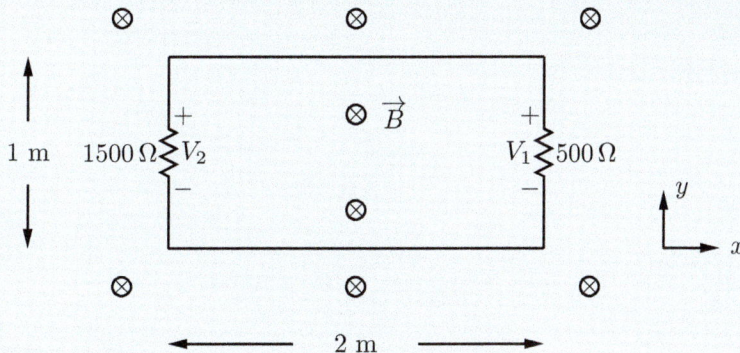

Figure 5.29: Example 5.3.1.

Looking at the circuit, we might think that since we have no source and since the resistors are in parallel, we can write $v_1 = v_2 = 0$. This would be the electrostatic solution, but it is not the case for time-varying fields. If we calculate the flux through the loop, we get

$$\Phi = \int_S \vec{B} \cdot \vec{dS} = -10\,t(2)(1) = -20\,t\,(\text{Wb}).$$

Since the resulting flux is changing with time, we are going to have an induced emf voltage:

$$V_{\text{emf}} = -\frac{\text{d}\Phi}{\text{d}t} = 20\,\text{V}.$$

We can redraw the circuit with the induced voltage as seen in Figure 5.30. How did we determine the polarity of the voltage source and the current direction?

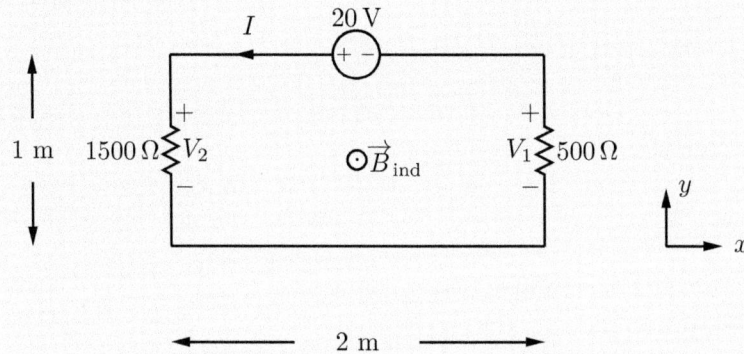

Figure 5.30: Redrawing Figure 5.29 of Example 5.3.1.

We used Lenz's law to determine them. We know that the induced voltage has to cause a current whose magnetic field opposes the change caused by the external flux. Since the original B field was increasing in the $-z$ direction, the induced B field should be in the $+z$, which by the right-hand rule will create a current going counterclockwise. Knowing the direction of the current sets the polarity of the voltage source: the generated current leaves the source's positive terminal. We can now use voltage division to find the voltages:

$$V_1 = -\frac{500}{1500 + 500}(20) = -5\,\text{V}$$

and

$$V_2 = \frac{1500}{1500 + 500}(20) = 15\,\text{V}.$$

Based on this background, we are now ready to derive the $i-v$ characteristic of an inductor. Let us rewrite (5.42) for the induced voltage across an inductor without the negative sign.

$$v(t) = \frac{\text{d}\Phi}{\text{d}t}. \tag{5.43}$$

We will take care of the negative sign later by correctly assigning the voltage polarity across the inductor with respect to the current flow. From (5.30) we have

$$\Phi = Li(t). \tag{5.44}$$

Taking the derivative of (5.44) with respect to time yields

$$\frac{\mathrm{d}\Phi}{\mathrm{d}t} = L\frac{\mathrm{d}i(t)}{\mathrm{d}t}. \tag{5.45}$$

This leads us to write

$$v(t) = \frac{\mathrm{d}\Phi}{\mathrm{d}t} = L\frac{\mathrm{d}i(t)}{\mathrm{d}t}. \tag{5.46}$$

Hence we can write the voltage across an inductor as

$$\boxed{v_L(t) = L\frac{\mathrm{d}i_L(t)}{\mathrm{d}t}.} \tag{5.47}$$

This is the basic current–voltage (or $i - v$) characteristic of the inductor. Similarly to Ohm's law for resistors, we can use this equation to find the voltage or current of an inductor if we know one of the two. We will see examples of this later.

Figure 5.31 shows the circuit element used to portray the inductor along with the voltage polarity and current directions. Notice that the indicated passive sign convention is consistent with Lenz's law. One way to convince yourself about this is to think about the following. Consider the situation of $i(t)$ being imposed by an external source and confirm that the induced voltage $v(t)$ has the correct polarity. For example, consider the case where the imposed $i(t)$ is increasing with respect to time. Then $v(t)$, with the indicated passive convention polarity, should be positive, since it would tend to induce an opposite current to oppose the imposed change. From a conservation of energy point of view, in the presence of an increasing $i(t)$, the inductor would build up a potential barrier by storing magnetic field energy. In other words, the inductor would get charged exactly as you would expect. The externally imposing current would need to overcome this potential barrier to continue flowing. The higher the rate of change of the current, the higher the voltage barrier (i.e., stored energy) across the inductor would become. Similarly, if $i_L(t)$ is decreasing versus time, $v_L(t)$ should be negative since the inductor would try to induce a current in the same direction as $i_L(t)$. In this case, the inductor is being discharged, leading to a reduction in the stored magnetic field energy and potential barrier across it.

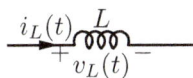

(a) Inductor circuit representation (passive sign convention).

(b) An actual 1 mH inductor.

Figure 5.31: Inductor as a circuit element.

This discussion should perhaps now make the passive sign convention even clearer, not just for the inductor but also for the resistor and capacitor. For example, let us look at the resistor again. If an external current $i_L(t)$ is imposed, a positive voltage barrier $v_L(t)$ (in the passive sign convention) is developed across it. This makes it harder for the current to keep flowing. Why should this be the case? Because the resistor takes the energy of the imposed $i_L(t)$ and turns it into heat. Heat does not depend on the rate of change of $i_L(t)$ but only on its magnitude. Therefore, we had concluded that $v_L(t)$ should be proportional to $i_L(t)$ and thus we arrived at Ohm's law. Similar arguments can be stated for the capacitor. The capacitor, of course, does not consume energy but stores it in the form of electric field energy—that is, in the form of stored charge. Hence, as the current is increasing, the capacitor is being charged, leading to a positive voltage drop (when considered in the passive sign convention), as it should be. The resulting potential barrier is positive since energy is being taken away from the imposed current and is stored in the capacitor in the form of electric field energy.

Example 5.3.2

The current through the 2 H inductor shown in Figure 5.32 is

$$i_L(t) = (5t+3)^2 \, \text{A}, \, t \text{ is in s.}$$

Find the inductor voltage $v_L(t)$.

Figure 5.32: Example 5.3.2.

We have

$$i_L(t) = 25t^2 + 30t + 9.$$

Using (5.47), we get

$$v_L(t) = L\frac{\mathrm{d}i_L(t)}{\mathrm{d}t} = 2\frac{\mathrm{d}(25t^2 + 30t + 9)}{\mathrm{d}t} = 100t + 60 \, \text{V}.$$

Example 5.3.3

The current through a 20 H inductor is plotted in Figure 5.33. Find the voltage across the inductor $v_L(t)$.

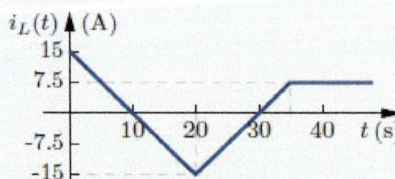

Figure 5.33: Example 5.3.3.

We need to find the current first.

$$i_L(t) = \begin{cases} 15 & t \leq 0, \\ -\frac{3}{2}t + 15 & 0 \leq t \leq 20, \\ \frac{3}{2}t - 45 & 20 \leq t \leq 35, \\ 7.5 & t \geq 35. \end{cases} \quad \text{A, } t \text{ is in s.}$$

Now we can use (5.47).

$$v_L(t) = \begin{cases} 0 & t < 0, \\ -30 & 0 < t < 20, \\ 30 & 20 < t < 35, \\ 0 & t > 35. \end{cases} \quad \text{V, } t \text{ is in s.}$$

Let us examine (5.47) more carefully to further understand an inductor's behavior. The equation is repeated below for convenience:

$$v_L(t) = L\frac{\mathrm{d}i_L(t)}{\mathrm{d}t}. \tag{5.48}$$

The first thing we notice is that we have a derivative—more specifically, the derivative of the current through the inductor. This means that the current of an inductor must be **continuous** and cannot change abruptly. Why? Because a sudden jump in its current would result in an infinite voltage drop, which is impossible to achieve. In math form, we can write that for **every** time t, $i_L(t^-) = i_L(t^+)$.

The current through an inductor is always continuous: $i_L(t^-) = i_L(t^+)$.

Note that just because the current cannot change abruptly, it does not mean that the voltage behaves in the same manner. Voltage across the inductor can change instantaneously to compensate for the changes happening to the circuit.

Let us examine what happens to the inductor when the current is not changing or under DC conditions. Figure 5.34 shows a simple circuit containing an inductor.

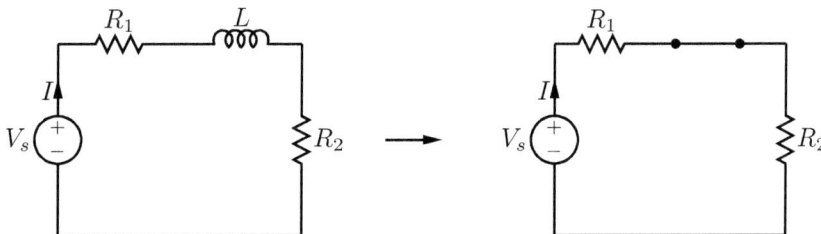

Figure 5.34: The inductor under DC conditions.

The voltage across the inductor is

$$v_L(t) = L\frac{\mathrm{d}i_L(t)}{\mathrm{d}t},$$

and since the current is constant, the derivative of a constant is zero. This means that the voltages across the inductor is zero, or

$$v_L = 0.$$

A zero voltage across an element can be represented by a short circuit as seen on the right-hand circuit of Figure 5.34. So the inductor in DC or steady-state conditions behaves as a short circuit. Now, when the voltage across the inductor is zero, it does not mean that the current through it is zero. For example, in this case the current through the inductor can be calculated by using Ohm's law as

$$I = \frac{V_s}{R_1 + R_2}.$$

The same thing happens to the inductor in Figure 5.35. The inductor becomes a short circuit and will end up shorting R_2. The current through the inductor will be equal to the current I since R_2 is shorted. Hence,

$$I = \frac{V_s}{R_1}.$$

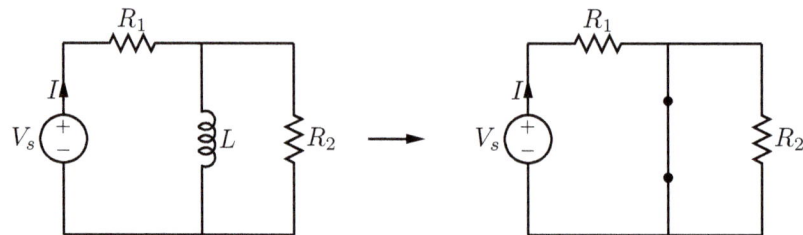

Figure 5.35: The inductor in DC conditions.

Example 5.3.4

For the circuit in Figure 5.36, find the currents through the inductors under DC conditions.

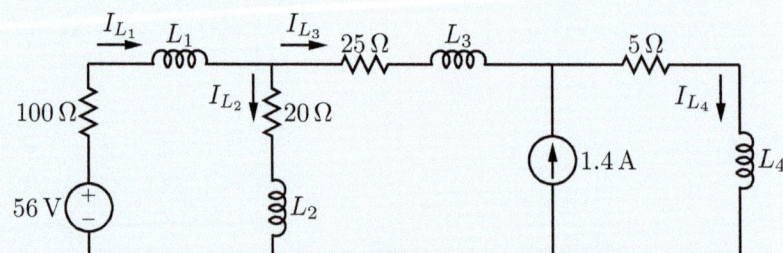

Figure 5.36: Example 5.3.4.

Under DC conditions all inductors behave as short circuits. We can use nodal analysis to first find the nodal voltages and then find every current needed. The DC-state circuit is shown in Figure 5.37.

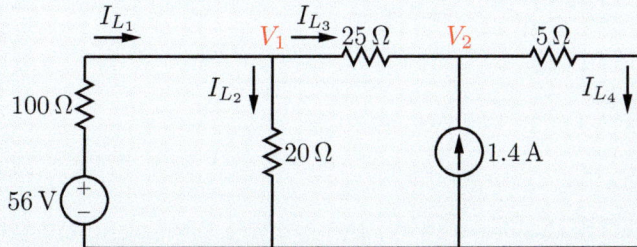

Figure 5.37: Example 5.3.4 redrawn under DC conditions.

The nodal analysis matrix is

$$\begin{bmatrix} \dfrac{1}{100} + \dfrac{1}{20} + \dfrac{1}{25} & -\dfrac{1}{25} \\[2ex] -\dfrac{1}{25} & \dfrac{1}{5} + \dfrac{1}{25} \end{bmatrix} \begin{bmatrix} V_1 \\[2ex] V_2 \end{bmatrix} = \begin{bmatrix} \dfrac{56}{100} \\[2ex] 1.4 \end{bmatrix}, \text{ or}$$

$$\begin{bmatrix} V_1 \\[2ex] V_2 \end{bmatrix} = \begin{bmatrix} 8.5 \\[2ex] 7.25 \end{bmatrix} \text{V}$$

Now we can find the currents needed. Indeed,

$$I_{L_1} = \frac{56 - 8.5}{100} = 0.475 \,\text{A}.$$

$$I_{L_2} = \frac{8.5}{20} = 0.425 \,\text{A}.$$

$$I_{L_3} = \frac{8.5 - 7.25}{25} = 0.05 \,\text{A}.$$

$$I_{L_4} = \frac{7.25}{5} = 1.45 \,\text{A}.$$

5.3.3 Finding the Current through an Inductor

Now that we understand the $i - v$ characteristic of an inductor, let us try to investigate how to find the current through the inductor in terms of the voltage across it. The circuit in Figure 5.38 shows a simple circuit with a voltage source and an inductor.

To find the current in terms of the voltage, let us start by rearranging (5.47) to isolate the derivatives:

$$\mathrm{d}i_L = \frac{1}{L} v_L(t) \,\mathrm{d}t. \tag{5.49}$$

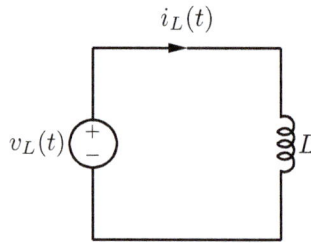

Figure 5.38: Simple inductor circuit.

Integrating both sides of (5.49) and changing the integration variables (from t to x) to avoid confusion, we get

$$\int_{-\infty}^{i_L(t)} \mathrm{d}i_L = \frac{1}{L} \int_{-\infty}^{t} v_L(x)\,\mathrm{d}x$$

$$i_L(t) - i_L(-\infty) = \frac{1}{L} \int_{-\infty}^{t} v_L(x)\,\mathrm{d}x. \tag{5.50}$$

Assuming that the inductor current at $-\infty$ is zero (i.e., $i_L(-\infty) = 0$), (5.50) becomes

$$i_L(t) = \frac{1}{L} \int_{-\infty}^{t} v_L(x)\,\mathrm{d}x. \tag{5.51}$$

Just like the capacitor case, we can assume that at some point in time t_0 a switch may be flipped or a source may be turned on or off. Up until t_0, everything is constant (steady-state or DC conditions). Therefore, it makes sense to split the integral as follows:

$$i_L(t) = \underbrace{\frac{1}{L} \int_{-\infty}^{t_0} v_L(x)\,\mathrm{d}x}_{a} + \underbrace{\frac{1}{L} \int_{t_0}^{t} v_L(x)\,\mathrm{d}x}_{b}. \tag{5.52}$$

Again, just like the capacitor case, we end up with two parts. Part a represents the current through the inductor at time t_0 and we can use $i_L(t_0)$ to indicate this current. This is often called the **initial condition** for a circuit since it provides the initial inductor current before a change takes place. Substituting part a in (5.52) with $i_L(t_0)$ and rearranging, we get

$$\boxed{i_L(t) = \frac{1}{L} \int_{t_0}^{t} v_L(x)\,\mathrm{d}x + i_L(t_0).} \tag{5.53}$$

Example 5.3.5

The voltage across the 2 H inductor shown in Figure 5.39 is $v_L(t) = e^{-5t} + 5\,\mathrm{V}$. If the inductor was charged with 0.9 A at $t = 0\,\mathrm{s}$, find the current $i_L(t)$ through the inductor.

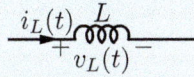

Figure 5.39: Voltage plot for Example 5.3.5.

Using (5.53), we get

$$i_L(t) = \frac{1}{2} \int_0^t e^{-5x} + 5 \, dx + 0.9 = -0.1e^{-5t} + 2.5t + 1 \, \text{A}.$$

Example 5.3.6

Figure 5.40 shows a plot of the voltage across a 5 H inductor with respect to time. Find the current $i_L(t)$ through the inductor assuming the inductor initial current to be -12 A.

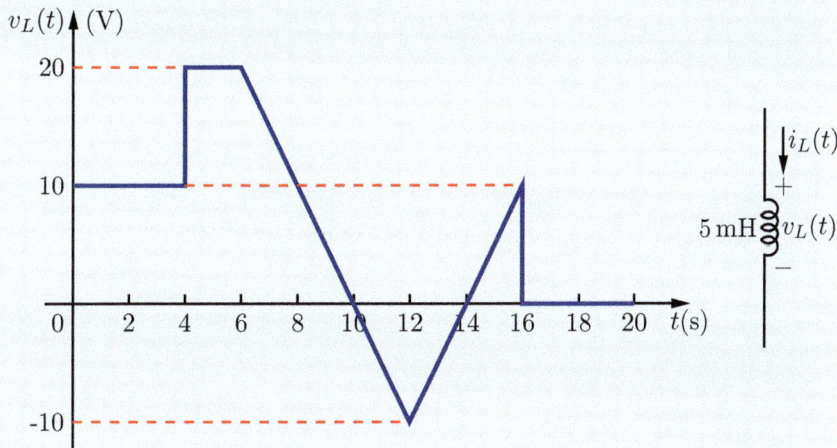

Figure 5.40: Voltage plot for Example 5.3.6.

We know that we will start with (5.53) and we know that we will have to do it in intervals. For convenience, we repeat this equation below as

$$i_L(t) = \frac{1}{L} \int_{t_0}^t v_L(x) \, dx + i_L(t_0).$$

For $0 \, \text{s} \le t \le 4$ s, the voltage is constant; hence

$$i_L(t) = \frac{1}{5} \int_0^t 10 \, dx + i_L(t_0) = 2t - 12 \, \text{A}.$$

For all subsequent time intervals, we will have to calculate the initial current. For convenience, we should find it after each interval. For $t = 4$ s, we have

$$i_L(4) = 2t - 12 = 2(4) - 12 = -4 \, \text{A}.$$

For the interval $4\,\mathrm{s} \leq t \leq 6\,\mathrm{s}$, the voltage is also constant:

$$i_L(t) = \frac{1}{5} \int_4^t 20\,\mathrm{d}x + i_L(4) = \frac{20}{5}(t-4) + i_L(4) = 4t - 20\,\mathrm{A}.$$

The initial current for $t = 6\,\mathrm{s}$ is

$$i_L(6) = 4t - 20 = 4(6) - 20 = 4\,\mathrm{A}.$$

For the interval $6 \leq t \leq 12$ s, the voltage is decreasing linearly and we can find it by using the equation of a line $y = mx + b$. Hence,

$$v_L(t) = -5t + 50,$$

$$i_L(t) = \frac{1}{5} \int_6^t (-5x + 50)\,\mathrm{d}x + i_L(6) = -\frac{x^2}{2} + 10x \bigg|_6^t + i_L(6)$$

$$= -0.5t^2 + 10t - 38\,\mathrm{A}.$$

The initial current for $t = 12\,\mathrm{s}$ is

$$i_L(12) = -0.5t^2 + 10t - 38 = -0.5(12)^2 + 10(12) - 38 = 10\,\mathrm{A}.$$

For the interval $12\,\mathrm{s} \leq t \leq 16\,\mathrm{s}$, the voltage is increasing linearly and we can find it by using the equation of a line as before.

$$v_L(t) = 5t - 70,$$

$$i_L(t) = \frac{1}{5} \int_{12}^t (5x - 70)\,\mathrm{d}x + i_L(12) = \frac{x^2}{2} - 14x \bigg|_{12}^t + i_L(12)$$

$$= 0.5t^2 - 14t + 106\,\mathrm{A}.$$

The initial current for $t = 16\,\mathrm{s}$ is

$$i_L(16) = 0.5t^2 - 14t + 106 = 0.5(16)^2 - 14(16) + 106 = 10\,\mathrm{A}.$$

For the interval $t \geq 16\,\mathrm{s}$, the voltage is zero. Hence,

$$i_L(t) = \frac{1}{5} \int_{16}^t 0\,\mathrm{d}x + i_L(16) = 0 + i_L(16) = 10\,\mathrm{A}.$$

To summarize,

$$i_L(t) = \begin{cases} -12 & t \leq 0, \\ 2t - 12 & 0 \leq t \leq 4, \\ 4t - 20 & 4 \leq t \leq 6, \\ -0.5t^2 + 10t - 38 & 6 \leq t \leq 12, \\ 0.5t^2 - 14t + 106 & 12 \leq t \leq 16, \\ 10 & t \geq 16. \end{cases} \quad \mathrm{A},\ t \text{ is in s.}$$

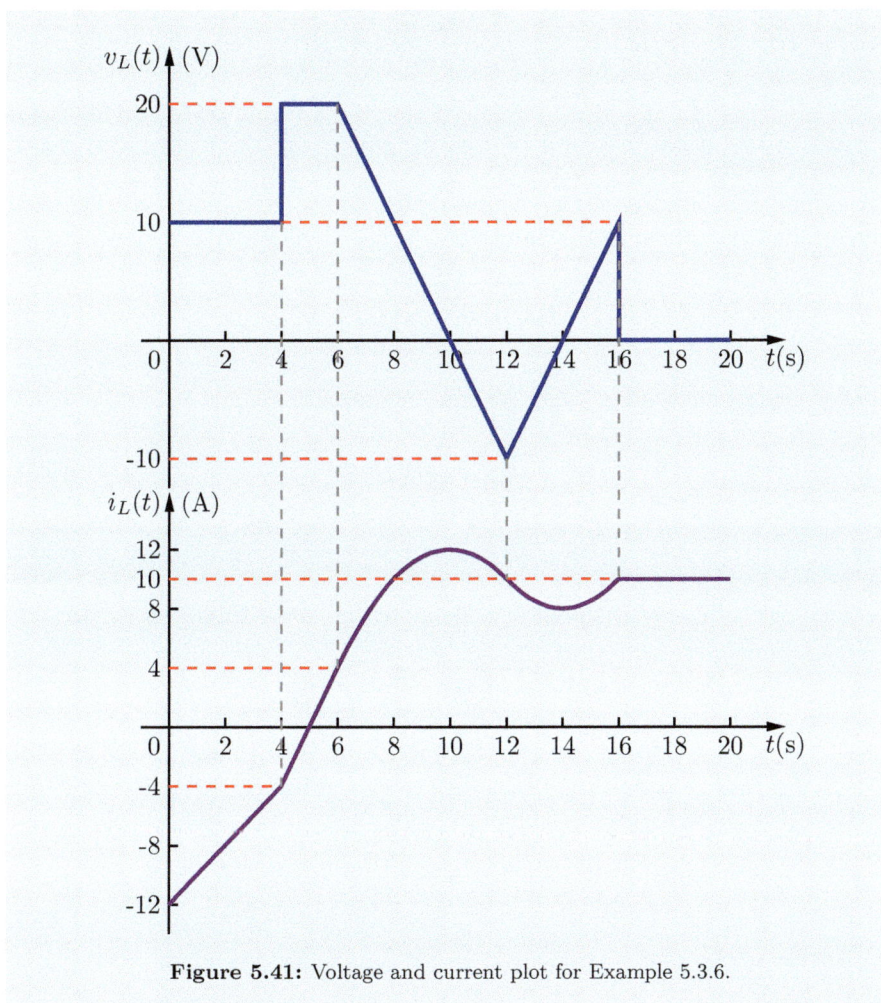

Figure 5.41: Voltage and current plot for Example 5.3.6.

5.3.4 Energy Storage

Let us now focus on the energy stored in an inductor. We remember our most important equation for the instantaneous power:

$$p_L(t) = v_L(t)i_L(t). \tag{5.54}$$

Using (5.47) for the voltage, we get

$$p_L(t) = L\frac{\mathrm{d}i_L(t)}{\mathrm{d}t}i_L(t). \tag{5.55}$$

To find the energy stored in the inductor, we need to integrate the power over the period in which the inductor is charging. If the inductor starts charging at t_0, then

$$w_L(t) = \int_{-\infty}^{t} p_L(x)\,dx = L\int_{-\infty}^{t}\frac{di_L(x)}{dx}i_L(x)\,dx$$

$$= L\int_{-\infty}^{t} i_L(x)\,dx = \frac{L}{2}i_L(x)^2\Big|_{-\infty}^{t} = \frac{L}{2}(i_L^2(t) - i_L^2(-\infty)) \qquad (5.56)$$

We assume the inductor starts with zero charge, so $i_L(-\infty) = 0$. This will make the total energy stored in the inductor

$$\boxed{w_L(t) = \frac{1}{2}Li_L^2(t).} \qquad (5.57)$$

For a constant voltage (DC), we can rewrite (5.57) as

$$W_L = \frac{1}{2}LI_L^2. \qquad (5.58)$$

Example 5.3.7

Find the total energy stored in both inductors in the following circuit.

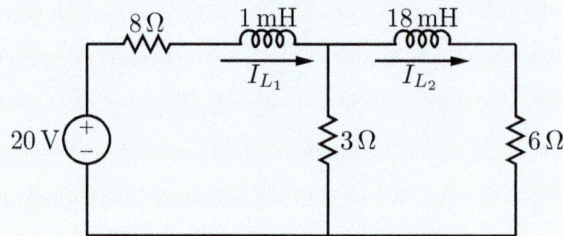

Figure 5.42: Example 5.3.7

Since the circuit is in DC conditions, the inductors behave as short circuits. We can redraw the circuit with the inductors as short circuits as seen in Figure 5.43. The goal is to find the total energy stored. Hence we need the currents through each inductor.

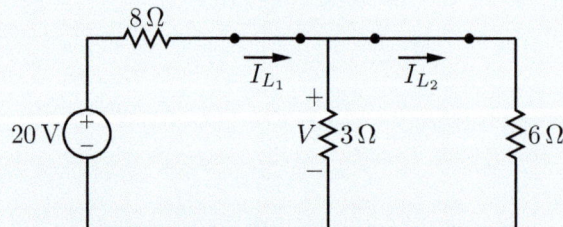

Figure 5.43: Redrawing the circuit in Figure 5.42 for DC conditions.

Voltage division can be used to find the voltage across the $8\,\Omega$ resistor and the parallel combination of the $3\,\Omega$ and $6\,\Omega$ resistors. Indeed,

$$6 \parallel 3 = 2\,\Omega,$$

$$V_{8\Omega} = \frac{8}{8+2}(20) = 16\,\text{V}.$$

Using KVL, we get

$$V = 20 - V_{8\Omega} = 20 - 16 = 4\,\text{V}.$$

Now we can use Ohm's law to find the currents:

$$I_{L_1} = \frac{V_{8\Omega}}{8} = \frac{16}{8} = 2\,\text{A},$$

and

$$I_{L_2} = \frac{V}{6} = \frac{4}{6} = \frac{2}{3}\,\text{A}.$$

The energy stored in each inductor is

$$W_{L_1} = \frac{1}{2}L_1 I_{L_1}^2 = 0.5(1\text{m})(2)^2 = 2\,\text{mJ},$$

and

$$W_{L_2} = \frac{1}{2}L_2 I_{L_2}^2 = 0.5(18\text{m})\left(\frac{2}{3}\right)^2 = 4\,\text{mJ}.$$

How about trying a DC example with an added capacitor? Let us work on one such case below.

Example 5.3.8

Find the total energy stored in the inductor and in the capacitor in the following circuit.

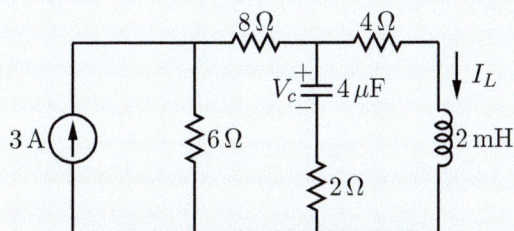

Figure 5.44: Example 5.3.8.

In order to find the total energy, we need to find the voltage across the capacitor and the current through the inductor. The capacitor behaves as an open circuit and the inductor behaves as a short circuit. The modified circuit is shown in Figure 5.45.

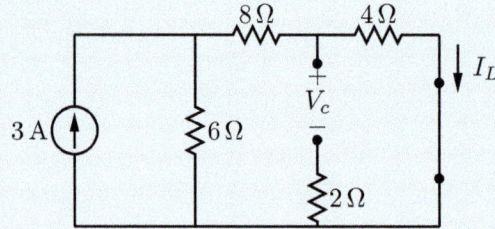

Figure 5.45: Redrawing the circuit in Figure 5.44 for DC conditions.

We can use current division to find the inductor current:

$$I_L = \frac{6}{8 + 4 + 6}(3) = 1\,\text{A}.$$

The capacitor voltage is the same as the voltage across the 4 Ω resistor. Hence

$$V_c = 4(I_L) = 4(1) = 4\text{V}.$$

The energy stored in the capacitor is

$$W_c = \frac{1}{2}CV^2 = 0.5(4\mu)(4^2) = 32\,\mu\text{J},$$

and the energy stored in the inductor is

$$W_L = \frac{1}{2}LI_L^2 = 0.5(2\text{m})(1^2) = 1\,\text{mJ}.$$

5.3.5   Inductor Connections

In this section we examine the two most common ways to connect inductors together: in series and in parallel. From the capacitor section we found out that, with respect to connections, capacitors are the dual of resistors. Here we will see that inductors behave like resistors.

Inductors in Series

We start with inductors connected in series. The circuit in Figure 5.46 has N inductors connected in series. Since we have a series connection, the current through all inductors is the same.

A KVL around the circuit yields

$$v_s(t) = v_{L_1} + v_{L_2} + \ldots + v_{L_N}. \tag{5.59}$$

We remember that

$$v_L(t) = L\frac{di_L(t)}{dt}. \tag{5.60}$$

Plugging (5.60) into (5.59) yields

$$v_s(t) = L_1\frac{di_L(t)}{dt} + L_2\frac{di_L(t)}{dt} + \ldots + L_N\frac{di_L(t)}{dt}. \tag{5.61}$$

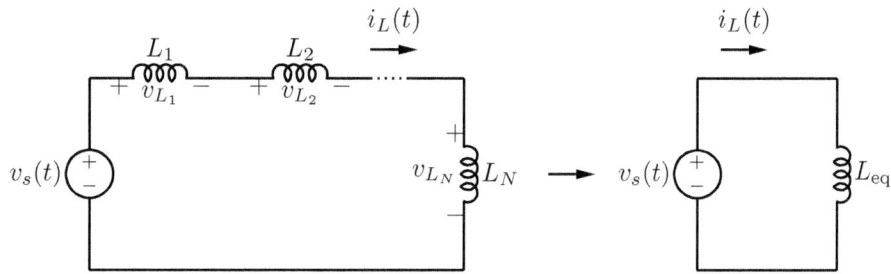

Figure 5.46: Inductors connected in series.

Since the current is the same through all inductors, the current can be taken as a common factor and (5.61) can be rewritten as

$$v_s(t) = (L_1 + L_2 + \ldots + L_N)\frac{\mathrm{d}i_L(t)}{\mathrm{d}t} = L_{\text{eq}}\frac{\mathrm{d}i_L(t)}{\mathrm{d}t}. \tag{5.62}$$

We just found out that inductors are similar to resistors when connected in series, where

$$\boxed{L_{\text{eq}} = \sum_{n=1}^{N} L_n.} \tag{5.63}$$

Inductors in Parallel

The circuit in Figure 5.47 has N inductors connected in parallel. Since we have a parallel connection, the voltage across all inductors is the same.

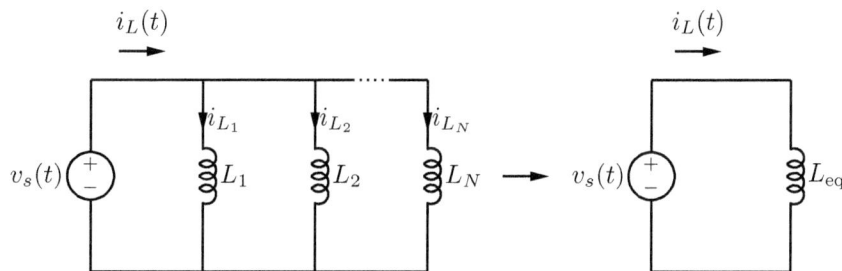

Figure 5.47: Inductors connected in parallel.

We start with KCL to add all currents:

$$i(t) = i_{L_1} + i_{L_2} + \ldots + i_{L_N}. \tag{5.64}$$

Since we know the voltage, we can find the current using

$$i_L(t) = \frac{1}{L}\int_{t_0}^{t} v_L(x)\,\mathrm{d}x + i_L(t_0). \tag{5.65}$$

We can now plug (5.65) into the KCL equation to get

$$i_L(t) = \frac{1}{L_1} \int_{t_0}^{t} v_s(x)\,dx + i_{L_1}(t_0) + \frac{1}{L_2} \int_{t_0}^{t} v_s(x)\,dx + i_{L_2}(t_0) + \ldots$$

$$+ \frac{1}{L_N} \int_{t_0}^{t} v_s(x)\,dx + i_{L_N}(t_0). \tag{5.66}$$

The voltage across a parallel connection is the same; hence

$$i_L(t) = \left(\frac{1}{L_1} + \frac{1}{L_2} + \ldots + \frac{1}{L_N} \right) \int_{t_0}^{t} v_s(x)\,dx + i_{L_1}(t_0) + i_{L_2}(t_0) + \ldots + i_{L_N}(t_0)$$

$$= \frac{1}{L_{eq}} \int_{t_0}^{t} v_s(x)\,dx + i_{L_{eq}}(t_0). \tag{5.67}$$

The equivalent inductance of a parallel connection is

$$\boxed{\frac{1}{L_{eq}} = \sum_{n=1}^{N} \frac{1}{L_n}.} \tag{5.68}$$

Notice also that the initial conditions of inductors connected in parallel are added together.

Example 5.3.9

Find the equivalent inductance L_{eq} as seen from port a-b.

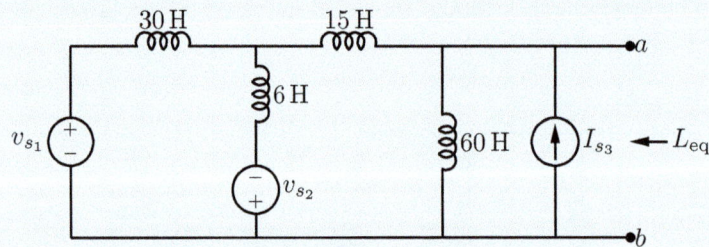

Figure 5.48: Example 5.3.9.

We first need to turn off all independent sources and then find the equivalent inductance as seen in Figure 5.49

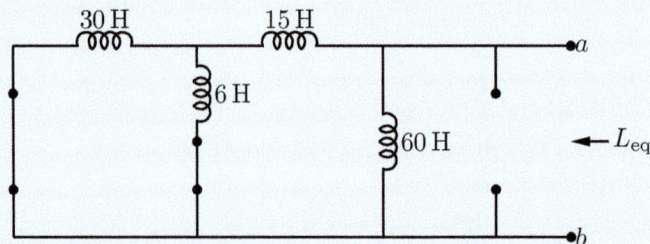

Figure 5.49: Redrawing the circuit in Figure 5.48 with all independent sources off.

Now we can proceed to finding the equivalent inductance as

$$L_{eq} = ((30 \parallel 6) + 15) \parallel 60 = 15\,\text{H}.$$

5.4   (Super) Capacitors versus Batteries

Before we compare batteries to super capacitors, we need to explore the difference between a capacitor and a super capacitor.

The capacitors we have studied so far are made with a dielectric in between two metal plates. Super capacitors, on the other hand, use interactions between ions in an electrolyte, which allows for higher capacitance. A trade-off of this is a limitation on the maximum voltage that can applied on the super capacitor. This is why super capacitors are considered low-voltage devices.

The most widely used batteries are lithium ion batteries. Batteries use an electrochemical reaction to charge and discharge. Although efficient in energy storage, the chemicals in the battery can degrade due to charging and discharging cycles. This will render the battery useless after a certain number of cycles. Batteries are considered energy-dense devices. Why? Because of their ability to store large amounts of energy in a small volume. An example of this is your cell phone battery. You expect your phone to last you at least the whole day, which means you want it to store a lot of energy and slowly dispense it during the day. Of course, the drawback of this is that the phone battery will take a couple of hours to recharge.

Super capacitors can sustain millions of charging and discharging cycles due to the way they electrostatically store energy. This also allows them to charge and discharge faster than a battery, which makes them ideal for quick power bursts. Super capacitors and capacitors in general are considered power-dense devices. This is because of their ability to discharge high amounts of power for a small volume. An example is the flash in your camera: the capacitor is small enough to fit inside the device and also has the ability to discharge fast and to recharge again before you take the next picture. Another example is regenerative breaking, in which rapid charging and discharging is more important than the amount of energy stored.

Clearly, we can see here that there is no winner, as batteries and super capacitors serve different functions and so far one can't replace the other—instead, they complete each other. Which one is used depends on the application: if long-term energy storage is your goal, then batteries are the appropriate choice; if rapid charging and discharging are more important, then a super capacitor is the right choice. Keep in mind that super capacitors are more expensive and can be bigger in volume for the same energy rating.

5.5   A Comparison of Passive Circuit Elements

In this section we present a summary table that compares all passive circuit elements we have studied so far.

Element	Resistor	Capacitor	Inductor
Symbol			
Common equation	$R = \rho \dfrac{l}{A}$	$C = \dfrac{\epsilon A}{d}$	$L = \dfrac{N\Phi}{I}$ (note here that the current will be canceled)
Current	$i_R(t) = \dfrac{v(t)}{R}$	$i_C(t) = C\dfrac{dv_C(t)}{dt}$	$i_L(t) = \dfrac{1}{L}\int_{t_0}^{t} v_L(x)\,dx + i_L(t_0)$
Voltage	$v(t) = Ri(t)$	$v_C(t) = \dfrac{1}{C}\int_{t_0}^{t} i_C(x)\,dx + v_C(t_0)$	$v_L(t) = L\dfrac{di_L(t)}{dt}$
Energy	Dissipates it	Stores it in electric field	Stores it in magnetic field
Energy	$w(t) = \dfrac{1}{R}\int_{0}^{t} v_R^2(x)\,dx$ or $w(t) = R\int_{0}^{t} i_R^2(x)\,dx$	$w(t) = \dfrac{1}{2}Cv_C(t)^2$	$w(t) = \dfrac{1}{2}Li_L(t)^2$
At DC	Stays the same	Open circuit	Short circuit
Voltage can change abruptly	Yes	No	Yes
Current can change abruptly	Yes	Yes	No
In series	$R_{eq} = R_1 + R_2$	$\frac{1}{C_{eq}} = \frac{1}{C_1} + \frac{1}{C_2}$	$L_{eq} = L_1 + L_2$
In parallel	$\frac{1}{R_{eq}} = \frac{1}{R_1} + \frac{1}{R_2}$	$C_{eq} = C_1 + C_2$	$\frac{1}{L_{eq}} = \frac{1}{L_1} + \frac{1}{L_2}$

5.6 Exercises

Section 5.2.1

5.1. The voltage across the 3 F capacitor in Figure 5.50 is $v_C(t) = 3t^3 + 2t + 5$ V. Find $i_C(t)$, then find the current at time $t = 1$ s.

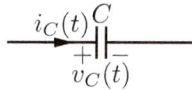

Figure 5.50

5.2. The voltage across the 0.5 F capacitor in Figure 5.51 is $v_C(t) = (7t - 6)^2$ V. Find $i_C(t)$.

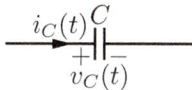

Figure 5.51

5.3. The voltage across the 1 F capacitor in Figure 5.52 is $v_C(t) = t\sin(\pi t)$ V. Find $i_C(t)$. Also find the current at times $t = 0, 0.5, 1$.

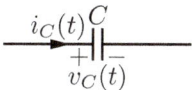

Figure 5.52

5.4. The voltage across the 2 F capacitor in Figure 5.53 is $v_C(t) = t^3 e^{4t}$ V. Find $i_C(t)$.

Figure 5.53

5.5. The voltage across a 5 F capacitor is shown in Figure 5.54. Find the current $i_C(t)$ for all t.

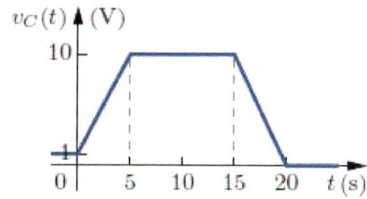

Figure 5.54

5.6. The voltage across a 10 F capacitor is shown in Figure 5.55. Find the current $i_C(t)$ for all t.

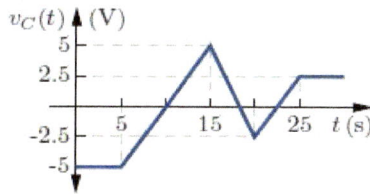

Figure 5.55

5.7. The circuit in Figure 5.56 has been in this state for a long time. Find the voltage V_C.

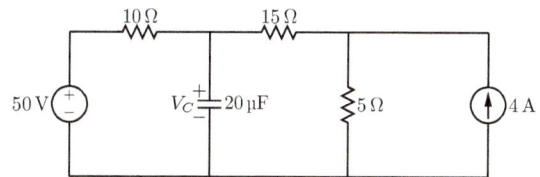

Figure 5.56

5.8. The circuit in Figure 5.57 has been in this state for a long time. Find the voltage V_C.

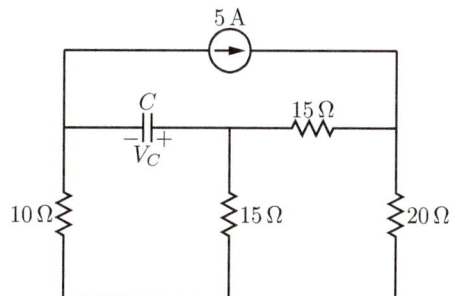

Figure 5.57

5.9. The circuit in Figure 5.58 has been in this state for a long time. Find the voltage V_C.

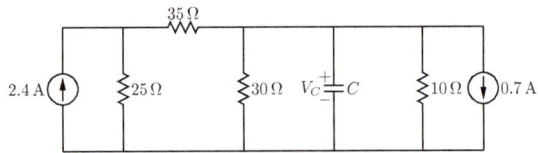

Figure 5.58

Section 5.2.2

5.10. The 2 mF capacitor in Figure 5.59 has been initially charged to $V_C(0) = -4\,\text{V}$. Find the voltage across the capacitor for all t if the current $i_C(t) = 6t^2 - 8t + 4\,\text{mA}$.

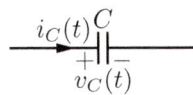

Figure 5.59

5.11. At $t = 1\,\text{s}$, the $1/3\,\text{F}$ capacitor in Figure 5.60 was charged to $12\,\text{V}$. Find the voltage across the capacitor for all t if the current $i_C(t) = (4t - 2)^2\,\text{A}$.

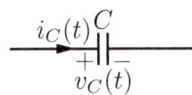

Figure 5.60

5.12. The current through the $1/5\,\text{mF}$ capacitor in Figure 5.61 is $i_C(t) = 10\cos(5t)\,\text{mA}$ for $t > \pi/5\,\text{s}$. If the capacitor was initially charged to $5\,\text{V}$, find the voltage across the capacitor. Also, find $v_C(\frac{\pi}{10})$ and $v_C(\frac{2\pi}{5})$.

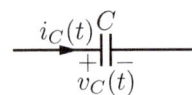

Figure 5.61

5.13. The current through a $1\,\text{F}$ capacitor is shown in Figure 5.62. Find the voltage across the capacitor for all t. The capacitor was initially charged to $5\,\text{V}$.

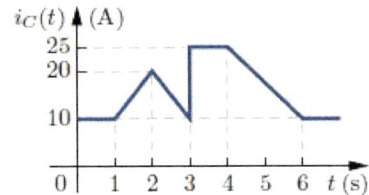

Figure 5.62

5.14. The current through a $4\,\text{F}$ capacitor is shown in Figure 5.63. Find the voltage across the capacitor for all t. The capacitor was initially charged to $2\,\text{V}$.

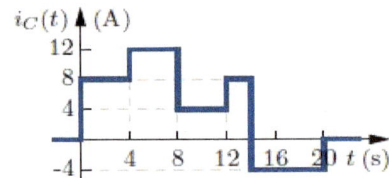

Figure 5.63

Section 5.2.3

5.15. The circuit in Figure 5.64 has been in this state for a long time. Find the total energy stored in the circuit.

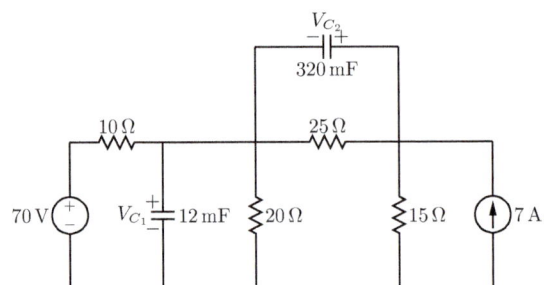

Figure 5.64

5.16. The circuit in Figure 5.65 has been in this state for a long time. The energy stored in each capacitor is $W_{C_1} = 225\,\text{J}$ and $W_{C_2} = 450\,\text{J}$. Find the values of the capacitors.

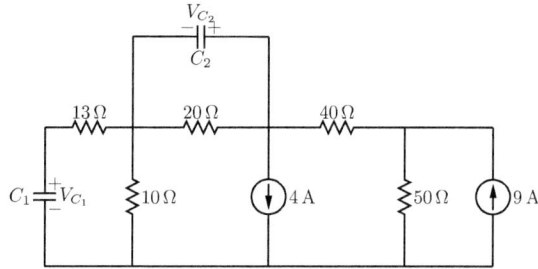

Figure 5.65

5.17. For the circuit in Figure 5.66, find the energy stored in the capacitor from $t = 1$ to $t = \infty$.

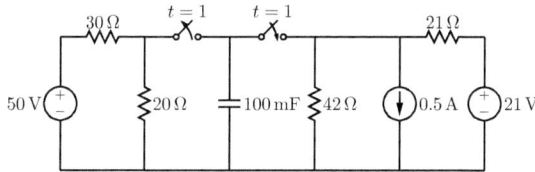

Figure 5.66

Section 5.2.4

5.18. For the circuit in Figure 5.67, find the equivalent capacitance at a-b.

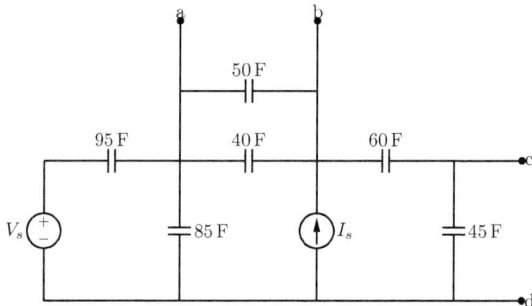

Figure 5.67

5.19. For the circuit in Figure 5.68, find the equivalent capacitance at c-d.

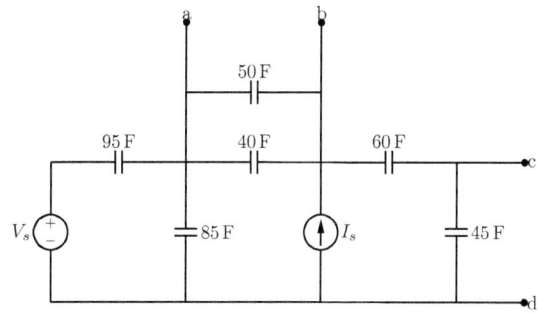

Figure 5.68

5.20. For the circuit in Figure 5.69, find the equivalent capacitance at a-b.

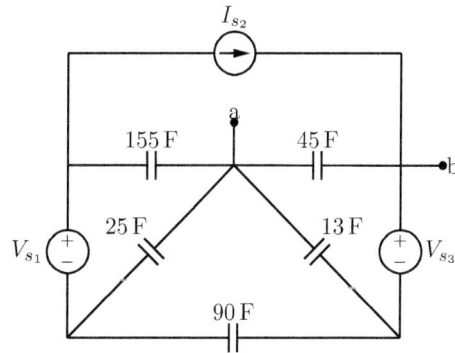

Figure 5.69

5.21. For the circuit in Figure 5.70, find the equivalent capacitance at a-b.

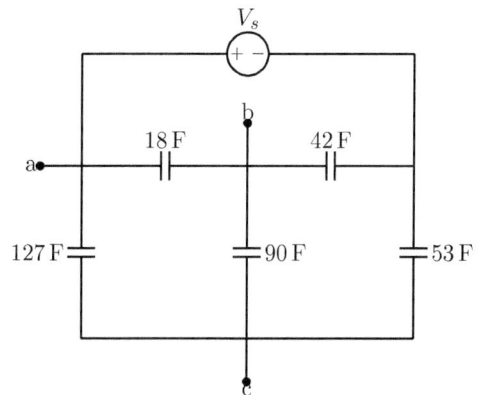

Figure 5.70

5.22. For the circuit in Figure 5.71, find the equivalent capacitance at b-c.

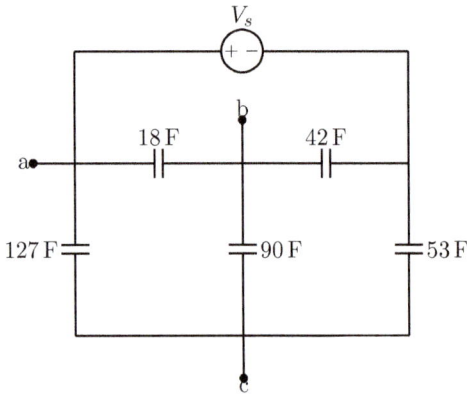

Figure 5.71

5.23. For the circuit in Figure 5.72, find the equivalent capacitance at a-b.

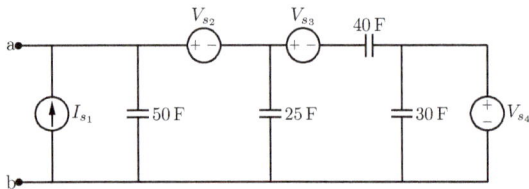

Figure 5.72

Section 5.3.2

5.24. The current through the 5 H inductor shown in Figure 5.73 is

$$i_L(t) = e^{20t} + \sin(3t) \text{ A}.$$

Find the voltage $v_L(t)$.

Figure 5.73

5.25. The current through the $\frac{1}{3}$ H inductor shown in Figure 5.74 is

$$i_L(t) = \cos(\pi t^3) \text{ A}.$$

Find the voltage $v_L(t)$.

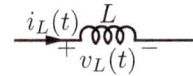

Figure 5.74

5.26. The current through a 40 H inductor is plotted in Figure 5.75. Find the voltage across the inductor $v_L(t)$.

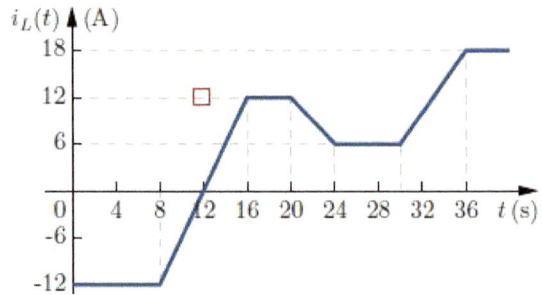

Figure 5.75

5.27. For the circuit in Figure 5.76, find the currents through the inductors under DC conditions.

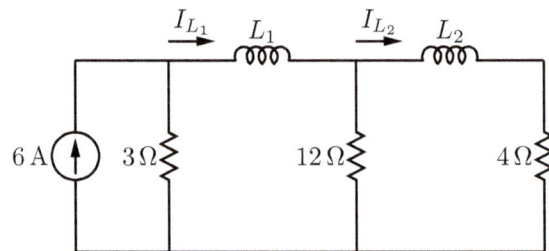

Figure 5.76

5.28. For the circuit in Figure 5.77, find the currents through the inductors under DC conditions.

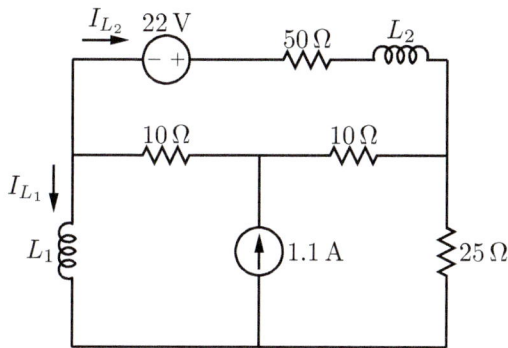

Figure 5.77

5.29. For the circuit in Figure 5.78, find the currents through the inductors under DC conditions.

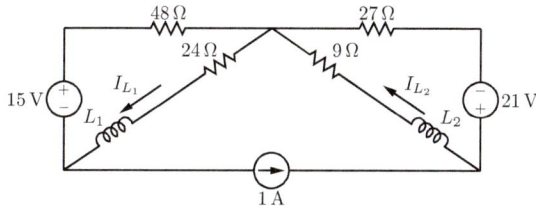

Figure 5.78

Section 5.3.3

5.30. The voltage across the 1 H inductor shown in Figure 5.79 is $v_L(t) = 1 - e^{-2t}$ V. The inductor was charged with 2 A at $t = 0$ s. Find the current $i_L(t)$.

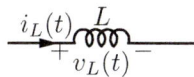

Figure 5.79

5.31. The voltage across the 2 H inductor shown in Figure 5.80 is $v_L(t) = (6t - 1)^2$ V. The inductor was charged with -3 A at $t = 1$ s. Find the current $i_L(t)$.

Figure 5.80

5.32. The voltage across a 50 H inductor is plotted in Figure 5.81. The inductor was charged with -1 A at $t = 0$ s. Find the current $i_L(t)$.

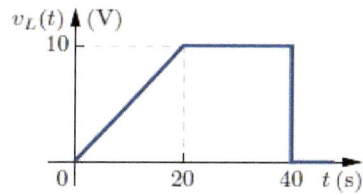

Figure 5.81

5.33. The voltage across a 5 H inductor is plotted in Figure 5.82. The inductor was charged with 10 A at $t = 0$ s. Find the current $i_L(t)$.

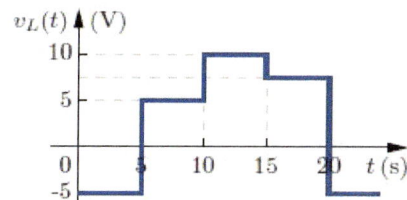

Figure 5.82

Section 5.3.4

5.34. Find the energy stored in the 20 mH inductor in Figure 5.83.

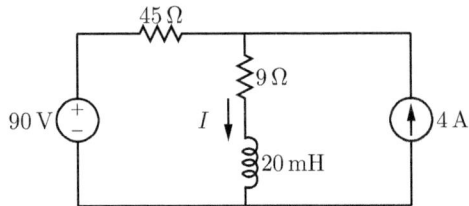

Figure 5.83

5.35. Find the total energy stored in the circuit shown in Figure 5.84 under DC conditions.

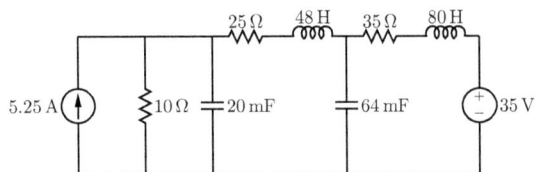

Figure 5.84

5.36. Find the total energy stored in the circuit shown in Figure 5.85.

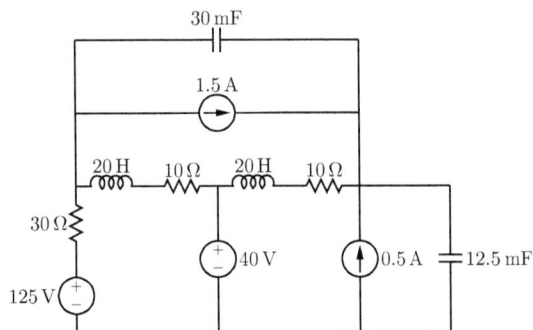

Figure 5.85

Section 5.3.5

5.37. Find the equivalent inductance at port a-b in the circuit shown in Figure 5.86.

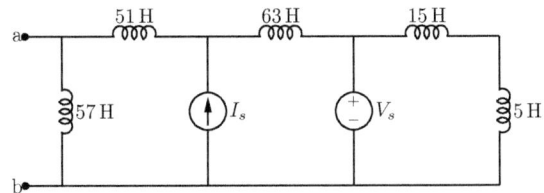

Figure 5.86

5.38. Find the equivalent inductance at port a-b in the circuit shown in Figure 5.87.

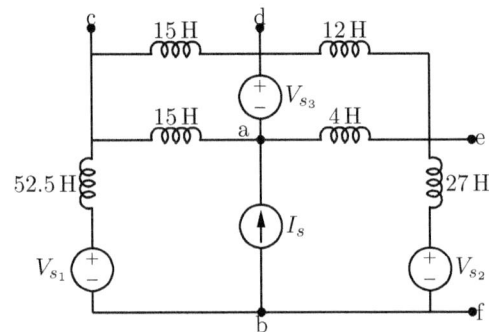

Figure 5.87

5.39. Find the equivalent inductance at port c-d in the circuit shown in Figure 5.87.

5.40. Find the equivalent inductance at port e-f in the circuit shown in Figure 5.87.

5.41. Find the equivalent inductance at port a-b in the circuit shown in Figure 5.88.

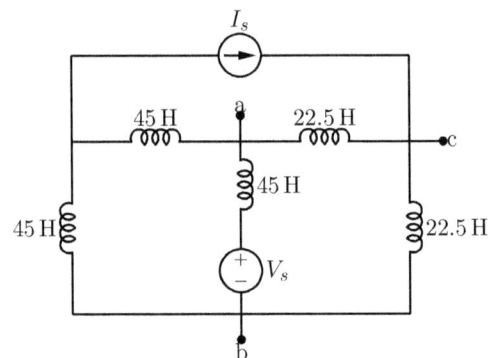

Figure 5.88

5.42. Find the equivalent inductance at port c-b in the circuit shown in Figure 5.88.

Chapter 6

First-Order Circuits

Introduction

In the previous chapter, we studied capacitors and inductors. We know that they are passive elements that store energy and can get charged and discharged. In charging or discharging such elements, we often find ourselves dealing with RC or RL circuits. Let's briefly discuss a couple of characteristic cases. First, let us remember that ideal energy-storage elements can hold their charges or currents indefinitely, but in real life they slowly leak their energy. A nonideal capacitor (inductor) can often be modeled as an ideal capacitor (inductor) with a shunt (series) resistor. In such cases we deal with RC or RL circuits as models for characterizing discharging operations. Second, consider a (super) capacitor powering a load (e.g., a wireless sensor). This is again a case where discharging takes place. Of course, we can think of similar cases during charging cycles. For example, writing a memory cell in a digital circuit resembles charging an RC circuit. Starting a simple motor can also be understood by an RL circuit.

In many charging and discharging applications, several key questions arise. For instance, we would like to know how long it takes for the capacitor (inductor) to charge or discharge. In computer chips, this often determines the speed of a digital circuit. Another important idea is to understand how the voltage or current of an energy-storage element varies versus time. Knowing this can help us avoid dangerous situations and/or select the appropriate elements that can handle the potentially high voltage or current spikes that may occur. To help us understand these concepts, we will tackle circuits with either a capacitor or an inductor. Circuits that simultaneously include both inductors and capacitors are not discussed here.

An RC or RL circuit is described by a differential equation that includes up to the first derivative of the unknown voltage or current. Hence, we will call them first-order circuits. Circuits that include both a capacitor and an inductor include differential equations with second-order derivatives. These are called second-order circuits and are not discussed here.

6.1   The Complete Response of an RC Circuit

In this section we start with RC circuits. We will jump directly into an example circuit that will include the complete response. The advantage of this approach is that we will be able to see all important concepts without increasing the complexity.

Figure 6.1 shows a circuit that includes two switches, sw_1 and sw_2. The first switch, sw_1, is assumed to have been closed for a long time and is opened at $t = t_0$. On the other hand, the second switch, sw_2, has been in its open state for a long time and is closed at $t = t_0$.

> The best way to understand **first-order circuits** is to look at the circuit throughout the entire operating window, starting from $t = -\infty$ and going through $t = \infty$.

Clearly, you need to focus closely on the moments that changes occur (e.g., sources are turned on/off or switches are turned on/off). Let us apply this methodology here for our basic RC circuit of Figure 6.1.

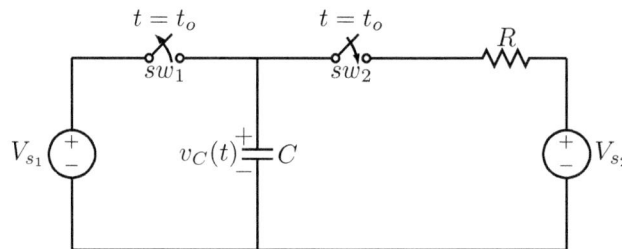

Figure 6.1: The basic RC circuit.

1. From $t = -\infty$ to $t = t_0$, the capacitor is only connected to the source V_{s_1}. Clearly this means that $v_C(t) = V_{s_1}$ for $-\infty < t \leq t_0$. Since no changes exist during this time interval, the capacitor behaves as an open circuit. The equivalent circuit during the $-\infty < t \leq t_0$ time interval is shown in Figure 6.2a. The voltage of the capacitor at t_0 is the most important result of our analysis here since it will become the **initial condition** of the capacitor for the rest of the time.

2. At $t = t_0$ two changes occur: both switches are flipped. Since changes occur, we cannot assume steady-state conditions anymore for the entire time interval $t_0 \leq t < \infty$. Thus the capacitor cannot be modeled anymore as an open circuit and we need to consider its transient behavior. The circuit during this time interval is shown in Figure 6.2b. During this stage the capacitor is connected to the second source V_{s_2} through the resistor R. While we will soon present the mathematical analysis during this transient behavior, we can first intuitively understand what happens. If $V_{s_2} > v_C(t_0)$, current will start flowing from V_{s_2} toward the capacitor, transferring charges so that an equilibrium state is eventually established.

We can expect that after sufficient time has passed (i.e., at $t = \infty$), the capacitor will be charged to the V_{s_2} voltage—that is, $v_C(\infty) = V_{s_2}$. On the other hand, if $V_{s_2} < v_C(t_0)$, current will start flowing from the capacitor to the source until the two voltages become equal again. Hence, in this case the final capacitor voltage will still be $v_C(\infty) = V_{s_2}$. You may wonder here: OK, this makes sense, but what is the role of the resistor then? The resistor (along with the capacitor value) dictates how long these operations take. In other words, you can imagine that if the resistance is very high, the charge flow will face a high barrier, leading to a slow charge or discharge of the capacitor. We will see this more clearly when we go through the mathematics.

3. At $t = \infty$, equilibrium will be reached as we discussed. This means that we can consider the circuit operating under steady-state conditions at $t = \infty$. The equivalent circuit for this time is shown in Figure 6.2c.

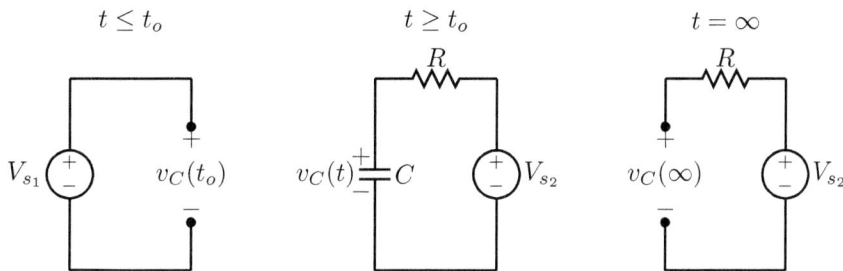

(a) Steady state before switching. **(b)** Transient between switching. **(c)** Steady state after switching.

Figure 6.2: The different stages of our basic RC circuit shown in Figure 6.1.

So far we have a good intuitive understanding of how the voltage across the capacitor behaves versus time. We lack, however, the quantitative understanding of the problem. Specifically, we cannot yet answer two critical questions:

1. How long does it take for the capacitor to go back to steady-state conditions after a switching event? Just saying "after a very long time (i.e., at $t = \infty$)" is not very useful from a practical perspective.

2. What does the voltage look like during the transient (charging or discharging) period? In other words, although we may be able to easily guess the starting and ending voltage values, are we missing the behavior in the middle?

To answer these questions, we will have to look at the transient circuit in Figure 6.2b. Figure 6.3 shows the transient circuit redrawn.

Where do we start? It is best to start by focusing on the voltage across the capacitor because it is *guaranteed* to be continuous as a function of time (remember that $v_C(t^-) = v_C(t^+)$), including, of course, between switching intervals. Since we are looking for the capacitor voltage, let us start with applying KVL around the loop. Applying KVL yields

$$V_{s_2} = Ri(t) + v_C(t). \tag{6.1}$$

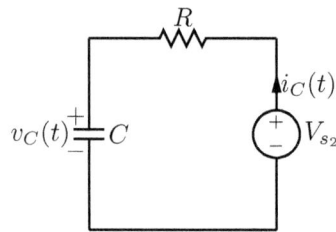

Figure 6.3: Transient circuit of Figure 6.1.

We remember from the capacitor section in chapter 5 that the current through a capacitor can be written as

$$i_C(t) = C\frac{\mathrm{d}v_C(t)}{\mathrm{d}t}. \tag{6.2}$$

If we plug our current equation (6.2) into the KVL equation (6.1), we will obtain a first-order differential equation:

$$V_{s_2} = RC\frac{\mathrm{d}v_C(t)}{\mathrm{d}t} + v_C(t). \tag{6.3}$$

The unknown quantity in a differential equation is a function, not a variable. The unknown function is the capacitor voltage $v_C(t)$ that appears under the derivative. Solving the differential equation requires an integration to remove the derivative. To prepare for this, we can rearrange (6.3) by separating the derivative from the other quantities. This yields

$$v_C(t) - V_{s_2} = -RC\frac{\mathrm{d}v_C(t)}{\mathrm{d}t}. \tag{6.4}$$

Now we can separate the derivatives—that is,

$$\frac{\mathrm{d}v_C(t)}{v_C(t) - V_{s_2}} = -\frac{\mathrm{d}t}{RC}. \tag{6.5}$$

We can integrate the left-hand side of (6.5) with respect to $v_C(t)$ where the voltage across the capacitor will go from the initial value $v_C(t_0)$ to the general value $v_C(t)$. Notice that we are going to $v_C(t)$ instead of ∞ because we want the general solution for the capacitor voltage. Similarly, the right-hand side of (6.5) can be integrated with respect to time from t_0 to a general t. Therefore,

$$\int_{v_C(t_0)}^{v_C(t)} \frac{\mathrm{d}v_C(t)}{v_C(t) - V_{s_2}} = -\int_{t_0}^{t} \frac{\mathrm{d}t}{RC}. \tag{6.6}$$

We remember from math that

$$\int_{a}^{b} \frac{1}{x}\,\mathrm{d}x = \ln(x)\Big|_{a}^{b} = \ln(b) - \ln(a). \tag{6.7}$$

Integrating (6.6) by using (6.7) yields

$$\ln(v_C(t) - V_{s_2})\Big|_{v_C(t_0)}^{v_C(t)} = -\frac{t}{RC}\Big|_{t_0}^{t}. \tag{6.8}$$

Solving and substituting, we get

$$\ln(v_C(t) - V_{s_2}) - \ln(v_C(t_0) - V_{s_2}) = -\frac{1}{RC}(t - t_0). \tag{6.9}$$

Another useful property to remember here is

$$\ln(x) - \ln(y) = \ln\left(\frac{x}{y}\right). \tag{6.10}$$

Using this property in (6.10) results in

$$\ln\frac{(v_C(t) - V_{s_2})}{(v_C(t_0) - V_{s_2})} = -\frac{1}{RC}(t - t_0). \tag{6.11}$$

To get rid of the natural logarithm on the left-hand side of the equation, we can take the exponential of both sides since

$$e^{\ln(x)} = x. \tag{6.12}$$

Applying (6.12) to (6.11), we get

$$\frac{v_C(t) - V_{s_2}}{v_C(t_0) - V_{s_2}} = e^{-\frac{1}{RC}(t-t_0)}. \tag{6.13}$$

Rearranging (6.13) yields

$$v_C(t) = V_{s_2} + (v_C(t_0) - V_{s_2})\, e^{-\frac{1}{RC}(t-t_0)}. \tag{6.14}$$

We also know that at $v_C(t_0) = V_{s_1}$. Hence,

$$v_C(t) = V_{s_2} + (V_{s_1} - V_{s_2})\, e^{-\frac{1}{RC}(t-t_0)}. \tag{6.15}$$

A more intuitive way to write this equation is the following:

$$\boxed{v_C(t) = \text{final value} + (\text{initial value} - \text{final value})\, e^{-\frac{1}{RC}(t-t_0)}.} \tag{6.16}$$

This is an important equation since it gives the general solution for an RC circuit. You should memorize it and learn how to apply it in RC circuits. Sometimes, you may see this equation written as

$$v_C(t) = v_C(\infty) + (v_C(t_0) - v_C(\infty))\, e^{-\frac{1}{RC}(t-t_0)}. \tag{6.17}$$

Either way you write the equation, the important thing to remember is how to find the initial and final values of the circuit. Specifically,

- The initial value is found by drawing the circuit right before the switching event(s)—that is, at $t = t_0^-$. If the circuit was in steady-state conditions before the switching event(s), replace the capacitor with an open circuit.

- The final value is found by drawing the circuit at $t = \infty$ and replacing the capacitor with an open circuit.

Let us conduct a quick check to see if our equation is valid. Since we know the initial and final values for the capacitor voltage, let us make sure that we can recover these unknown values from our equation. At $t = t_0$, we have

$$
\begin{aligned}
v_C(t_0) &= V_{s_2} + (V_{s_1} - V_{s_2}) \, e^{-\frac{1}{RC}(t_0 - t_0)} \\
&= V_{s_2} + (V_{s_1} - V_{s_2}) \, (1) \\
&= V_{s_1}.
\end{aligned}
\tag{6.18}
$$

This is indeed the correct value. Now let us check at $t = \infty$. In this case,

$$
\begin{aligned}
v_C(\infty) &= V_{s_2} + (V_{s_1} - V_{s_2}) \, e^{-\frac{1}{RC}(\infty - t_0)} \\
&= V_{s_2} + (V_{s_1} - V_{s_2}) \, (0) \\
&= V_{s_2}.
\end{aligned}
\tag{6.19}
$$

This is also correct.

We can also rewrite (6.15) as follows:

$$
v_C(t) = \underbrace{V_{s_2}}_{\text{steady state}} + \underbrace{(V_{s_1} - V_{s_2}) \, e^{-\frac{1}{RC}(t - t_0)}}_{\text{transient state}}.
\tag{6.20}
$$

or

$$
\boxed{v_C(t) = \underbrace{\text{final value}}_{\text{steady state}} + \underbrace{(\text{initial value} - \text{final value}) \, e^{-\frac{1}{RC}(t - t_0)}}_{\text{transient state}}.}
\tag{6.21}
$$

Once we know the capacitor voltage, we can easily find its current by applying (6.2). Note though that the current may not be continuous.

As we discussed, the voltage across the capacitor will start at an initial value V_{s_1} and transition to the new steady-state value V_{s_2}. The rate of that transition depends on the value RC, which is defined as the *time constant* τ.

$$
\tau = RC.
\tag{6.22}
$$

Then we can rewrite (6.20) in terms of the time constant τ:

$$
v_C(t) = V_{s_2} + (V_{s_1} - V_{s_2}) \, e^{-\frac{(t - t_0)}{\tau}}.
\tag{6.23}
$$

We can see now that if we want to minimize the duration of the transient's effects (most of the time this is desirable to speed up the circuit), we will have to minimize the time constant τ. This concept is better visualized with time-domain plots.

Figure 6.4 displays the voltage response for three different time constants, $\tau_1 < \tau_2 < \tau_3$, and for $V_{s_2} > V_{s_1}$ (i.e., the capacitor is being charged). We can see that the smaller the time constant τ, the faster the capacitor achieves its final charged value (steady state). We can replot Figure 6.4 but this time with

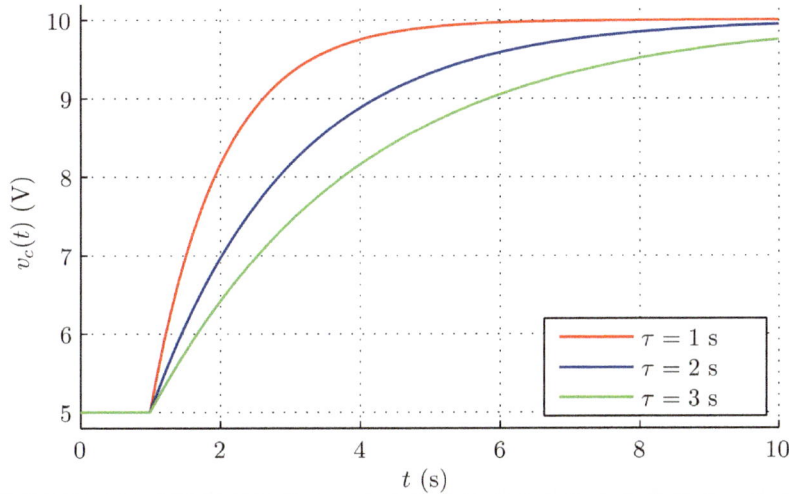

Figure 6.4: Charging capacitor with $V_{s_1} = 5\,\text{V}$ and $V_{s_2} = 10\,\text{V}$ for different values of the time constant τ.

$V_{s_2} < V_{s_1}$. With the final value being smaller than the starting value, we can say that the capacitor is being discharged as seen in Figure 6.5.

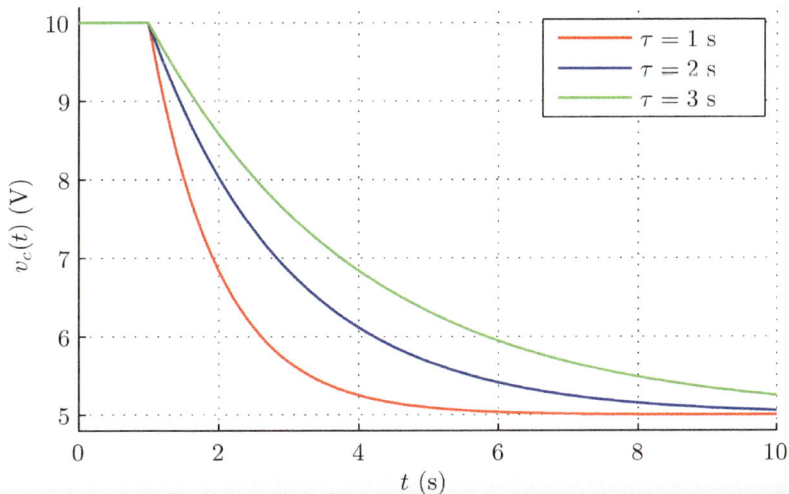

Figure 6.5: Discharging capacitor with $V_{s_1} = 10\,\text{V}$ and $V_{s_2} = 5\,\text{V}$ for different values of the time constant τ.

It is also interesting to note that it theoretically takes infinite time for the capacitor to reach its final value. However, in practice we do not need to wait this long. To better understand this, let us consider our **charging** cycle with $V_{s_1} = 0\,\text{V}$ and $t_0 = 0$. It is easy to see from (6.16) that it takes

- One time constant (RC) for the voltage to reach $1 - \frac{1}{e} \simeq 63.2\%$ of its final value.

- Three time constants $(3RC)$ for the voltage to reach 95% of its final value.

- Five time constants $(5RC)$ for the voltage to reach 99.3% of its final value.

Similarly, let us consider our **discharging** cycle with $V_{s_2} = 0\,\text{V}$ and $t_0 = 0$. It is easy to see from (6.16) that it takes

- One time constant (RC) for the voltage to decay to $\frac{1}{e} \simeq 36.8\%$ of its initial value.

- Three time constants $(3RC)$ for the voltage to decay to 5% of its initial value.

- Five time constants $(5RC)$ for the voltage to decay to 0.7% of its initial value.

We were able to divide (6.23) into a steady-state part and a transient part. We can also rearrange the equation according to the following voltages:

$$v_C(t) = \underbrace{V_{s_1} e^{-\frac{(t-t_0)}{\tau}}}_{\text{natural response}} + \underbrace{V_{s_2}\left(1 - e^{-\frac{(t-t_0)}{\tau}}\right)}_{\text{forced response}}. \tag{6.24}$$

This emphasizes what we call a **natural** and a **forced** response. The natural response is due to the energy initially stored in the capacitor. It is the response we would get if there were no driving voltage applied to the capacitor after t_0. To better understand this, let us draw the circuit of Figure 6.1 without the voltage source V_{s_2} (driving voltage source after switching). This circuit is shown in Figure 6.6. Clearly, the capacitor voltage for $t \geq t_0$ is now given by

$$v_C(t) = V_{s_1} e^{-\frac{(t-t_0)}{\tau}}. \tag{6.25}$$

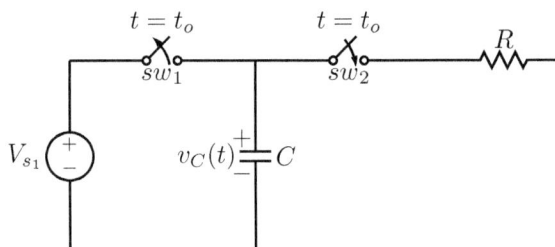

Figure 6.6: Natural response RC circuit.

The forced response, just like the name suggests, is the result of having an independent source in the circuit after switching. It is the response we would get if there were no initial charge in the capacitor.

Equation (6.23) is the most general solution we can get for a first-order circuit with a capacitor. The only problem we might face is that the equation was derived for a circuit with one capacitor and one resistor. What if we had multiple capacitors and/or multiple resistors? We have learned many techniques that help us simplify our circuits, so why not use that now? If we manage to simplify our circuit to one resistor and one capacitor (e.g., by using Thévenin's or Norton's theorem), we will be able to find the voltage across the capacitor easily. We will only need to know the initial condition of the capacitor (note that combining capacitors will result in the combination of their initial conditions no matter how they are connected), the steady-state capacitor voltage, and the time constant. We can write an even more generalized form of the capacitor voltage as seen in (6.26) and (6.27).

$$v_{C_{eq}}(t) = v_{C_{eq}}(\infty) + \left(v_{C_{eq}}(t_0) - v_{C_{eq}}(\infty)\right) e^{-\frac{(t-t_0)}{\tau}} \qquad (6.26)$$

where

$$\tau = R_{eq}C_{eq}. \qquad (6.27)$$

An example is the best way to demonstrate the use of (6.26) and (6.27).

Example 6.1.1

From Figure 6.7, find $i(t)$ for $t > 0$.

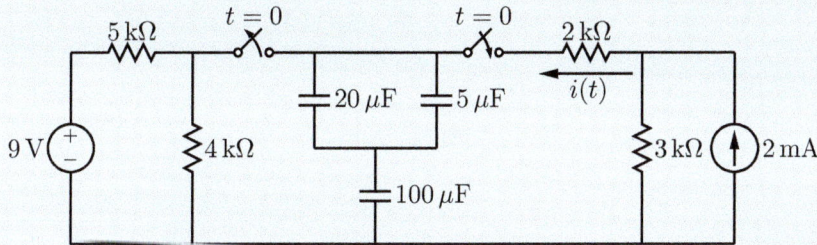

Figure 6.7: Example 6.1.1.

Since we only know how to solve an RC circuit, we need to simplify the given circuit to a simple RC circuit. First, let us combine the three capacitors into one equivalent capacitance as

$$C_{eq} = \frac{(20+5)100}{(20+5)+100} = 20\,\mu F.$$

We can redraw the circuit with the equivalent capacitance as seen in Figure 6.8.

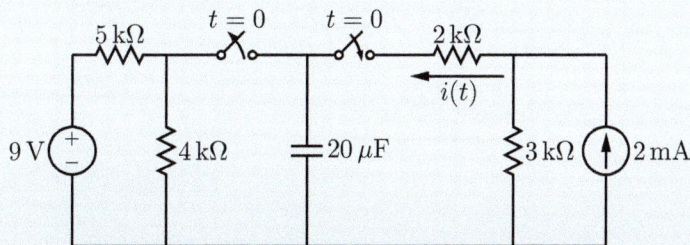

Figure 6.8: The circuit from Figure 6.7 with combined capacitors.

Now we can find the equivalent capacitor's initial condition. At $t < 0$, we can redraw the circuit as seen in Figure 6.9.

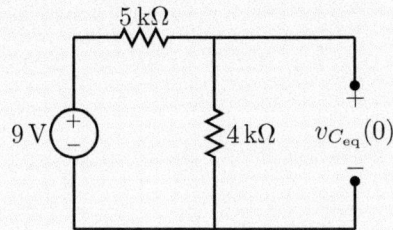

Figure 6.9: The circuit from Figure 6.7 before switching.

We can use voltage division to find the voltage across the capacitor as

$$v_{C_{eq}}(0) = \frac{4}{4+5}(9) = 4\,\text{V}.$$

The circuit after switching can be seen in Figure 6.10.

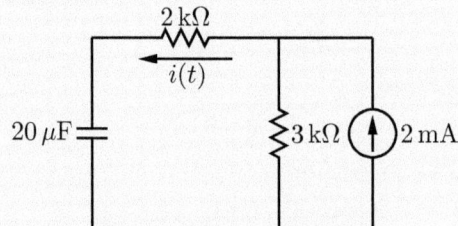

Figure 6.10: The circuit from Figure 6.7 after switching.

We can use our tools to simplify our circuit. For example, we can use source transformation to change the current source in parallel with a resistor into a voltage source in series with the same resistor, as seen in Figure 6.11.

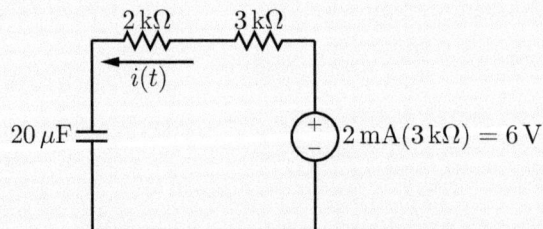

Figure 6.11: The circuit from Figure 6.10 after source transformation.

The resulting circuit enables us to find the equivalent resistance to find the time constant τ.

$$R_{eq} = 2\,\text{k}\Omega + 3\,\text{k}\Omega = 5\,\text{k}\Omega.$$

The resulting time constant τ is

$$\tau = R_{eq}C_{eq} = 5\,k\Omega(20\,\mu F) = 0.1\,s.$$

We also need to find the capacitor's final voltage. The capacitor acts as an open circuit, as seen in Figure 6.12, and the voltage across it will be

$$v_{C_{eq}}(\infty) = 6\,V.$$

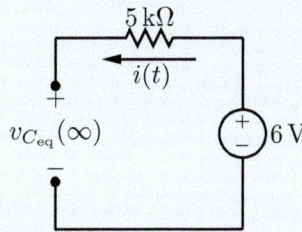

Figure 6.12: The circuit from Figure 6.11 at steady state.

Now we have everything we need to find the capacitor's voltage with respect to time $v_C(t)$. Indeed, by using

$$v_{C_{eq}}(t) = v_{C_{eq}}(\infty) + \left(v_{C_{eq}}(t_0) - v_{C_{eq}}(\infty)\right) e^{-\frac{(t-t_0)}{\tau}},$$

we obtained

$$v_{C_{eq}}(t) = 6 + (4-6)\,e^{-\frac{(t-0)}{0.1}}.$$

Simplifying this equation yields

$$v_{C_{eq}}(t) = 6 - 2e^{-10t}\,V.$$

The current $i(t)$ for $t > 0$ can now be found using

$$i(t) = C\frac{dv_{C_{eq}}(t)}{dt} = 20\,\mu F(-10)(-2)e^{-10t} = 0.4e^{-10t}\,mA.$$

Notice that $i(0^+) = 0.4$ mA, while as you can see from the original circuit, $i(0^-) = 0$ A. As expected, the current is discontinuous at $t = 0$.

Example 6.1.2

Find the voltage $v_C(t)$ for $t > t_0$ for the circuit shown in Figure 6.13.

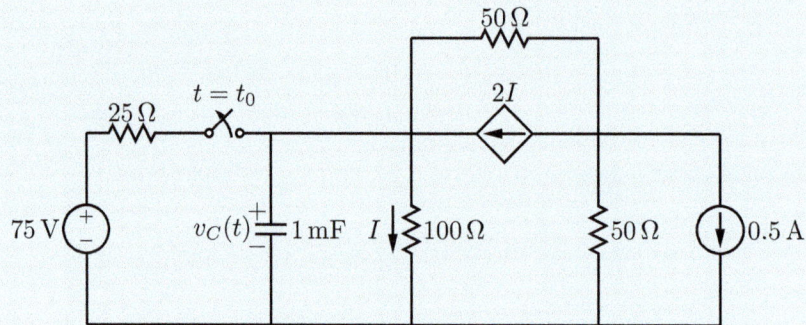

Figure 6.13: Example 6.1.2

We will have to solve this circuit in three different stages. First, we need to find the initial value of the voltage across the capacitor. Second, we need to find the final value of the voltage across the capacitor. Third, we have to find the value of the equivalent resistance seen by the capacitor. Since we have a dependent source, we will have to use our $i - v$ test for this part. So let us start with finding the capacitor's initial value.

The circuit just before switching $(t < t_0)$ is shown in Figure 6.14.

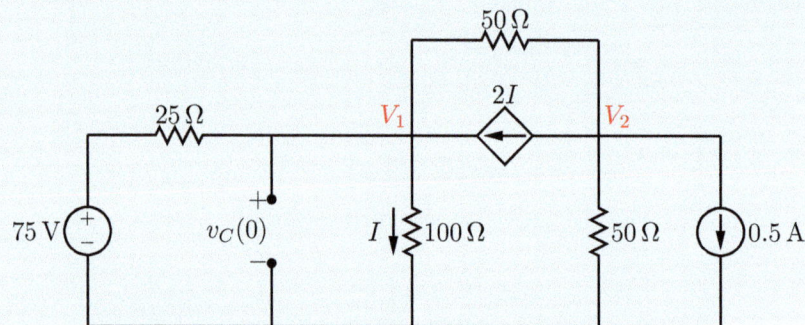

Figure 6.14: The circuit in Figure 6.13 just before switching.

The matrix form of the nodal equations is

$$
\begin{bmatrix} \dfrac{1}{25} + \dfrac{1}{100} + \dfrac{1}{50} - \dfrac{2}{100} & -\dfrac{1}{50} \\[2mm] -\dfrac{1}{50} + \dfrac{2}{100} & \dfrac{1}{50} + \dfrac{1}{50} \end{bmatrix} \begin{bmatrix} V_1 \\[2mm] V_2 \end{bmatrix} = \begin{bmatrix} \dfrac{75}{25} \\[2mm] -0.5 \end{bmatrix}.
$$

Solving for the nodal voltages, we get

$$
\begin{bmatrix} V_1 \\ V_2 \end{bmatrix} = \begin{bmatrix} 55 \\ -12.5 \end{bmatrix} \text{ V}.
$$

Hence,

$$
v_C(0) = 55 \text{ V}.
$$

Now let us try to find the final capacitor voltage. The circuit at $t = \infty$ is shown in Figure 6.15.

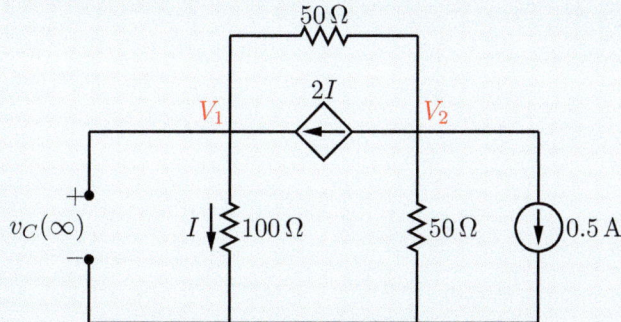

Figure 6.15: The circuit in Figure 6.13 a long time after switching.

We will use nodal analysis again to find the capacitor's voltage. The final matrix is

$$\begin{bmatrix} \dfrac{1}{100} + \dfrac{1}{50} - \dfrac{2}{100} & -\dfrac{1}{50} \\[2ex] -\dfrac{1}{50} + \dfrac{2}{100} & \dfrac{1}{50} + \dfrac{1}{50} \end{bmatrix} \begin{bmatrix} V_1 \\[2ex] V_2 \end{bmatrix} = \begin{bmatrix} 0 \\[2ex] -0.5 \end{bmatrix}.$$

Solving for the nodal voltages, we get

$$\begin{bmatrix} V_1 \\ V_2 \end{bmatrix} = \begin{bmatrix} -25 \\ -12.5 \end{bmatrix} \text{V}.$$

Hence,

$$v_C(\infty) = -25\,\text{V}.$$

Now it is time to use the $i - v$ test to find the equivalent resistance seen by the capacitor. We need to turn off all independent sources in the circuit **after** the switching, then add a test source at the terminal we want. We chose here to use a 1 A test source as seen in Figure 6.16.

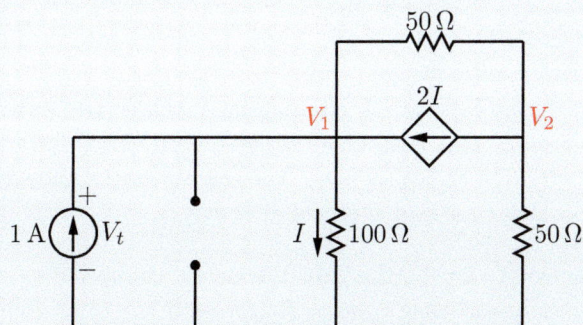

Figure 6.16: Using the $i - v$ test to find the equivalent resistance of the circuit after switching.

We will use nodal analysis again to find V_t. The final matrix is

$$\begin{bmatrix} \dfrac{1}{100} + \dfrac{1}{50} - \dfrac{2}{100} & -\dfrac{1}{50} \\[3mm] -\dfrac{1}{50} + \dfrac{2}{100} & \dfrac{1}{50} + \dfrac{1}{50} \end{bmatrix} \begin{bmatrix} V_1 \\[2mm] V_2 \end{bmatrix} = \begin{bmatrix} 1 \\[2mm] 0 \end{bmatrix}.$$

Solving for the nodal voltages, we get

$$\begin{bmatrix} V_1 \\ V_2 \end{bmatrix} = \begin{bmatrix} 100 \\ 0 \end{bmatrix} \text{V}.$$

Hence,

$$V_t = V_1 = 100\,\text{V}.$$

This yields

$$\underline{R_\text{eq} = 100\,\Omega.}$$

We can now find the time constant τ as

$$\tau = R_\text{eq} C = 100\,\Omega(1\,\text{mF}) = 0.1\,\text{s}.$$

Putting everything together in the general equation (6.26), we get

$$v_{C_\text{eq}}(t) = v_{C_\text{eq}}(\infty) + \left(v_{C_\text{eq}}(t_0) - v_{C_\text{eq}}(\infty) \right) e^{-\frac{(t-t_0)}{\tau}}$$

$$v_C(t) = -25 + (55 - (-25)) e^{-\frac{(t-t_0)}{0.1}}$$

$$\underline{\underline{v_C(t) = -25 + 80e^{-10(t-t_0)}\,\text{V}.}}$$

6.2   The Complete Response of an RL Circuit

Following our RC discussion, in this section we will solve the RL circuit. We will find out how the inductor charges and discharges just like we did with the capacitor. We will start with a general circuit and then separate the natural and forced responses. As you will see, the discussion is analogous to what we did before. It is a good idea to use this opportunity to consolidate your understanding of the first-order circuit concepts.

Figure 6.17 shows a basic RL circuit that includes two switches, sw_1 and sw_2. The first switch, sw_1, is assumed to have been closed for a long time and is opened at $t = t_0$. On the other hand, the second switch, sw_2, has been in its open state for a long time and is closed at $t = t_0$.

1. From $t = -\infty$ to $t = t_0$, the inductor is only connected to the source I_{s_1}. Clearly this means that $i_L(t) = I_{s_1}$ for $-\infty < t \le t_0$. Since no changes exist during this time interval, the inductor behaves as a short circuit. The equivalent circuit during the $-\infty < t \le t_0$ time interval is shown in Figure 6.18a. The current through the inductor at t_0 is the most important

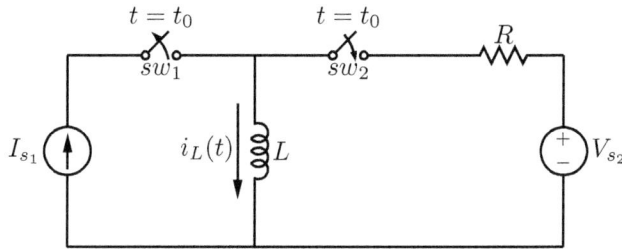

Figure 6.17: The basic RL circuit.

result of our analysis here since it will become the *initial condition* of the inductor for the rest of the time.

2. At $t = t_0$, two changes occur: both switches are flipped. Since changes occur, we cannot assume steady-state conditions anymore for the entire time interval $t_0 \leq t < \infty$. Thus the inductor cannot be modeled anymore as a short circuit and we need to consider its transient behavior. The circuit during this time interval is shown in Figure 6.18b. During this stage, the inductor is connected to the second source V_{s_2} through the resistor R. While we will soon present the mathematical analysis during this transient behavior, we can first intuitively understand what happens. If $(V_{s_2}/R) > i_L(t_0)$, the current through the inductor will begin to increase until an equilibrium state is eventually established. We can expect that after sufficient time has passed (i.e., at $t = \infty$), the inductor will be charged to the (V_{s_2}/R) current—that is, $i_L(\infty) = (V_{s_2}/R)$. On the other hand, if $(V_{s_2}/R) < i_L(t_0)$, the current through the inductor will start decreasing until the current through the inductor becomes (V_{s_2}/R). Hence, in this case the final inductor current will still be $i_L(\infty) = (V_{s_2}/R)$. The resistor along with the inductor value dictate how long these operations will take.

3. At $t = \infty$, equilibrium will be reached, as we discussed. This means that we can consider the circuit to be operating under steady-state conditions at $t = \infty$. The equivalent circuit for this time is shown in Figure 6.18c.

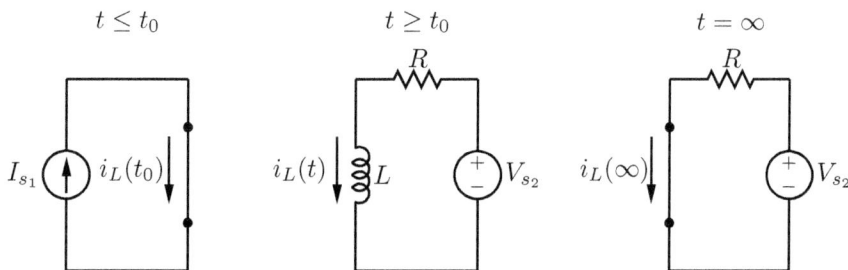

(a) Steady state before switching. (b) Transient between switching. (c) Steady state after switching.

Figure 6.18: The different stages of the basic RL circuit shown in Figure 6.17.

Just like we did in the basic RC circuit analysis, we will answer the following critical questions:

1. How long does it take for the inductor to go back to steady state after the switching events?

2. How does the current look during the transient (charging or discharging) period?

To answer these questions, we will have to look at the transient circuit in Figure 6.18b. Figure 6.19 shows the transient circuit redrawn.

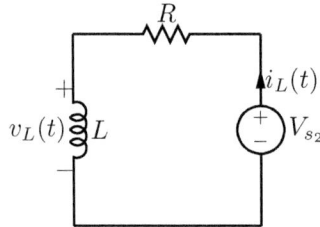

Figure 6.19: Transient circuit of Figure 6.17.

We will focus here on the current through the inductor because it is *guaranteed* to be continuous as a function of time (remember that $i_L(t^-) = i_L(t^+)$), including, of course, between switching intervals. There are a couple of things to remember before we start our analysis: just before the switches are activated $(t < t_0)$, the inductor behaves as a short circuit and the initial current is

$$i_L(t_0) = I_{s_1}. \tag{6.28}$$

Once steady state is reached, the inductor acts as a short circuit and the steady-state inductor current can be calculated as

$$i_L(\infty) = \frac{V_{s_2}}{R}. \tag{6.29}$$

Since we are looking for the inductor current, let us start with applying KVL around the loop in Figure 6.19. Applying KVL yields

$$V_{s_2} = Ri_L(t) + v_L(t). \tag{6.30}$$

We remember that the voltage across the inductor is

$$v_L(t) = L\frac{\mathrm{d}i_L(t)}{\mathrm{d}t}. \tag{6.31}$$

Substituting back into the KVL in (6.30) yields a first-order differential equation:

$$V_{s_2} = Ri_L(t) + L\frac{\mathrm{d}i_L(t)}{\mathrm{d}t}. \tag{6.32}$$

Dividing both sides of (6.32) by R yields

$$\frac{V_{s_2}}{R} = i_L(t) + \frac{L}{R}\frac{\mathrm{d}i_L(t)}{\mathrm{d}t}. \tag{6.33}$$

Using (6.29) in (6.33) gives us

$$i_L(\infty) = i_L(t) + \frac{L}{R}\frac{di_L(t)}{dt}. \tag{6.34}$$

Here we have again a differential equation. As we did in the RC section, we can rearrange (6.34) by separating the derivative from the other quantities—that is,

$$-\frac{L}{R}\frac{di_L(t)}{dt} = i_L(t) - i_L(\infty). \tag{6.35}$$

Now we can separate the derivatives as

$$\frac{di_L(t)}{i_L(t) - i_L(\infty)} = -\frac{R}{L}\,dt. \tag{6.36}$$

We can integrate the left-hand side of (6.36) with respect to $i_L(t)$ where the current through the inductor will go from the initial value $i_L(t_0)$ to the general value $i_L(t)$. Notice that we are going to $i_L(t)$ instead of ∞ because we want the general solution for the inductor current. Similarly, the right-hand side of (6.36) can be integrated with respect to time from t_0 to a general t.

$$\int_{i_L(t_0)}^{i_L(t)} \frac{di_L(t)}{i_L(t) - i_L(\infty)} = -\frac{R}{L}\int_{t_0}^{t} dt. \tag{6.37}$$

This looks exactly like what we did for the capacitor. We remember that

$$\int_a^b \frac{1}{x}\,dx = \ln(x)\Big|_a^b = \ln(b) - \ln(a), \tag{6.38}$$

and that

$$\ln(x) - \ln(y) = \ln\left(\frac{x}{y}\right). \tag{6.39}$$

Applying (6.38) and (6.39) to (6.37) gives

$$\ln\left(\frac{i_L(t) - i_L(\infty)}{i_L(t_0) - i_L(\infty)}\right) = -\frac{R}{L}(t - t_0). \tag{6.40}$$

We know what comes next—we need to get rid of the natural logarithm.

$$\frac{i_L(t) - i_L(\infty)}{i_L(t_0) - i_L(\infty)} = e^{-\frac{R}{L}(t-t_0)}. \tag{6.41}$$

Rearranging (6.41) yields

$$i_L(t) = i_L(\infty) + (i_L(t_0) - i_L(\infty))e^{-\frac{R}{L}(t-t_0)}. \tag{6.42}$$

A more intuitive way to write this equation is the following:

$$\boxed{i_L(t) = \text{final value} + (\text{initial value} - \text{final value})\,e^{-\frac{R}{L}(t-t_0)}.} \tag{6.43}$$

This is an important equation since it gives the general solution for an RL circuit. You should memorize it and learn how to apply it in RL circuits. Sometimes, you may see this equation written as

$$i_L(t) = i_L(\infty) + (i_L(t_0) - i_L(\infty))\, e^{-\frac{R}{L}(t-t_0)}. \qquad (6.44)$$

Either way you write the equation, the important thing to remember is how to find the initial and final values of the circuit. Specifically,

- The initial value is found by drawing the circuit right before the switching event(s)—that is, at $t = t_0^-$. If the circuit was in steady-state conditions before the switching event(s), replace the inductor with a short circuit.

- The final value is found by drawing the circuit at $t = \infty$ and replacing the inductor with a short circuit.

Let us conduct a quick check to see if our equation is valid. Since we know the initial and final values for the inductor current, let us make sure that we can recover these values from our equation. At $t = t_0$,

$$
\begin{aligned}
i_L(t_0) &= \frac{V_{s_2}}{R} + \left(I_{s_1} - \frac{V_{s_2}}{R}\right) e^{-\frac{R}{L}(t_0 - t_0)} \\
&= \frac{V_{s_2}}{R} + \left(I_{s_1} - \frac{V_{s_2}}{R}\right)(1) \\
&= I_{s_1}. \qquad (6.45)
\end{aligned}
$$

This is indeed the correct value. Now let us check at $t = \infty$:

$$
\begin{aligned}
i_L(\infty) &= \frac{V_{s_2}}{R} + \left(I_{s_1} - \frac{V_{s_2}}{R}\right) e^{-\frac{R}{L}(\infty - t_0)} \\
&= \frac{V_{s_2}}{R} + \left(I_{s_1} - \frac{V_{s_2}}{R}\right)(0) \\
&= \frac{V_{s_2}}{R}. \qquad (6.46)
\end{aligned}
$$

This is also correct.

Just as in the capacitor's case, we have a time constant that will determine the rate of charge and discharge of the inductor in this circuit. We will define the time constant τ as

$$\tau = \frac{L}{R}. \qquad (6.47)$$

Rewriting the inductor current equation in terms of the time constant yields

$$i_L(t) = \underbrace{i_L(\infty)}_{\text{steady state}} + \underbrace{(i_L(t_0) - i_L(\infty))e^{-\frac{(t-t_0)}{\tau}}}_{\text{transient}} \qquad (6.48)$$

or

$$i_L(t) = \underbrace{\text{final value}}_{\text{steady state}} + \underbrace{(\text{initial value} - \text{final value})\, e^{-\frac{(t-t_0)}{\tau}}}_{\text{transient state}}. \qquad (6.49)$$

We can also write (6.48) in terms of the natural response and the forced response as

$$i_L(t) = \underbrace{i_L(t_0)e^{-\frac{(t-t_0)}{\tau}}}_{\text{natural response}} + \underbrace{i_L(\infty)\left(1 - e^{-\frac{(t-t_0)}{\tau}}\right)}_{\text{forced response}}. \tag{6.50}$$

We can generalize (6.48) for multiple inductors and resistors by finding the equivalent inductance L_{eq} and resistance R_{eq}.

$$i_{L_{\text{eq}}}(t) = i_{L_{\text{eq}}}(\infty) + (i_{L_{\text{eq}}}(t_0) - i_{L_{\text{eq}}}(\infty))e^{-\frac{(t-t_0)}{\tau}} \tag{6.51}$$

and

$$\tau = \frac{L_{\text{eq}}}{R_{\text{eq}}}. \tag{6.52}$$

Once we know the inductor current, we can easily find the voltage across it by applying (6.31). Note though that the voltage may not be continuous.

Example 6.2.1

From Figure 6.20, find $i(t)$ for $t > 1$ s.

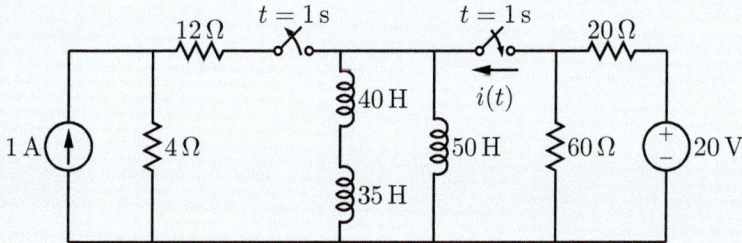

Figure 6.20: Example 6.2.1.

Since the desired current for $t > 1$ s is going to be the equivalent inductor's current $i_{L_{\text{eq}}}$, the plan is going to be the following:

1. Find L_{eq}.

2. Find the initial current (before switching) through the equivalent inductance $i_{L_{\text{eq}}}(1)$.

3. Simplify the post-switching circuit to find R_{eq}.

4. Calculate τ using (6.52).

5. Find the steady state value of the equivalent inductor current $i_{L_{\text{eq}}}(\infty)$.

6. Fill in (6.51) to find the total current.

Let us start with finding the equivalent inductance as

$$L_{\text{eq}} = 50 \parallel (40 + 35) = 30 \, \text{H}.$$

The circuit after finding the equivalent inductance is shown in Figure 6.21.

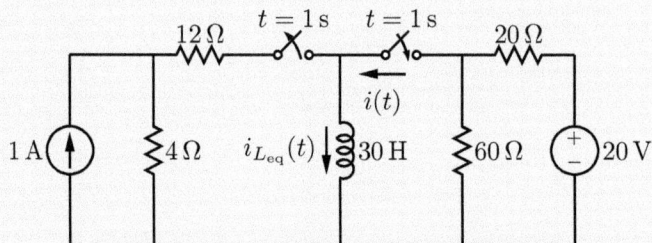

Figure 6.21: The circuit in Figure 6.20 after finding the equivalent inductance.

The initial current $i_{L_{eq}}(1)$ can be found using current division as shown in Figure 6.22.

$$i_{L_{eq}}(1) = \frac{4}{4+12}(1) = 0.25 \, \text{A}.$$

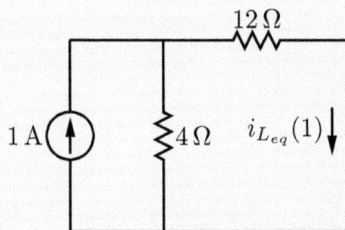

Figure 6.22: The circuit in Figure 6.20 before switching.

We now focus on the circuit after the switching events (transient state). In Figure 6.23, we did a source transformation to enable us to find the total resistance. We can now find R_{eq}, which will enable us to find τ, and we can also find the steady-state current through the inductor.

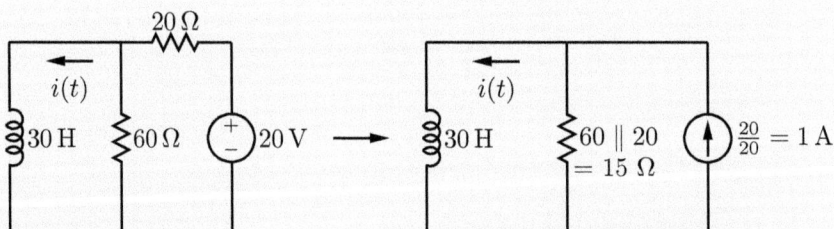

Figure 6.23: Applying source transformation to the circuit in Figure 6.20 after switching.

We have

$$R_{eq} = 60 \parallel 20 = \frac{60(20)}{60+20} = 15 \, \Omega.$$

Now we can find τ as

$$\tau = \frac{L_{eq}}{R_{eq}} = \frac{30}{15} = 2 \, \text{s}.$$

The steady-state current can easily be found from the last-stage circuit in Figure 6.23. At steady state, the inductor becomes a short circuit that will result in shorting the resistor, and all the source current will end up going through the inductor. Hence,

$$i_{L_{eq}}(\infty) = 1\,\text{A}.$$

We can now plug everything we have into the general inductor current equation found in (6.51) to get

$$i(t) = 1 + (0.25 - 1)e^{-\frac{t-1}{2}}.$$

Cleaning up, we get

$$i(t) = 1 - 0.75\,e^{-0.5(t-1)}\,\text{A}.$$

Example 6.2.2

From Figure 6.24, find $i_L(t)$ for $t \geq t_0$.

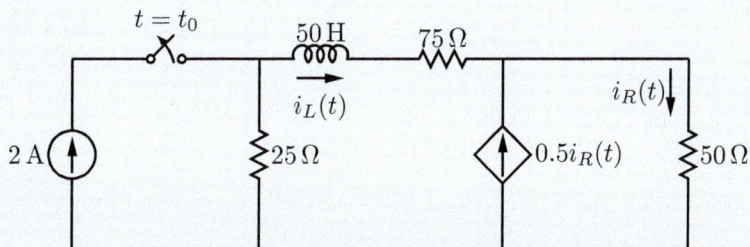

Figure 6.24: Example 6.2.2.

We will start by finding the inductor's initial current $i_L(t_0)$. We do so by shorting the inductor, as seen in Figure 6.25.

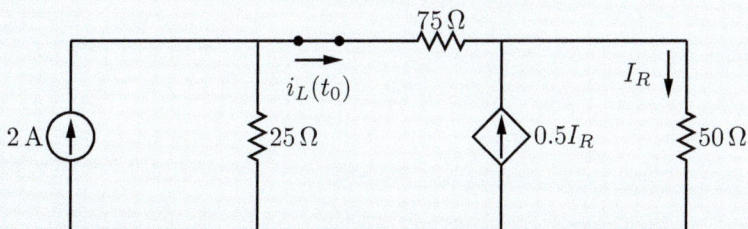

Figure 6.25: The circuit in Example 6.2.2 just before switching.

We can use source transformation on the independent source to make the circuit easier or we can solve for two nodal equations. We choose here to use source transformation for practice, as seen in Figure 6.26.

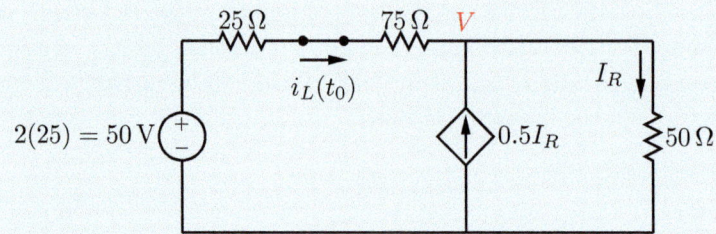

Figure 6.26: The circuit in Example 6.2.2 just before switching.

We have one nodal equation to write,

$$\left(\frac{1}{100} + \frac{1}{50}\right) V = 0.5 I_R + \frac{50}{100}.$$

The current I_R can be written as a function of the nodal voltage:

$$I_R = \frac{V}{50}.$$

Substituting and recognizing, we get

$$\left(\frac{1}{100} + \frac{1}{50} - \frac{0.5}{50}\right) V = \frac{50}{100}.$$

Then

$$V = 25 \, \text{V}.$$

Now we can find the inductor's initial current as

$$i_L(t_0) = \frac{50 - 25}{100} = 0.25 \, \text{A}.$$

The final step here is to try to find the equivalent resistance seen by the inductor. Since we have a controlled source, we will need to use our $i - v$ test. Applying a 1 V test voltage at the terminals of the inductor yields Figure 6.27.

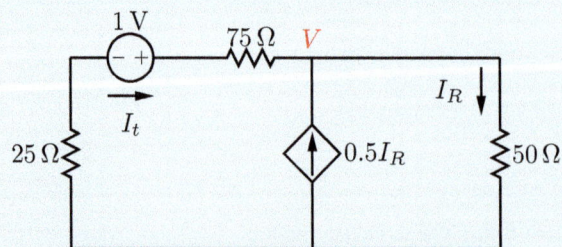

Figure 6.27: Using the $i - v$ test to find the equivalent resistance.

We have one nodal equation to write:

$$\left(\frac{1}{100} + \frac{1}{50}\right) V = 0.5 I_R + \frac{1}{100}.$$

The current I_R can be written as a function of the nodal voltage,

$$I_R = \frac{V}{50}.$$

Substituting and recognizing,

$$\left(\frac{1}{100} + \frac{1}{50} - \frac{0.5}{50}\right)V = \frac{1}{100}.$$

Then

$$V = 0.5\,\text{V}.$$

The test current can now be calculated as

$$I_t = \frac{1 - 0.5}{100} = \frac{1}{200}\,\text{A}.$$

The equivalent resistance seen by the inductor is

$$R_\text{eq} = \frac{1}{I_t} = 200\,\Omega.$$

Since the final value of the current through the inductor is zero, we can put everything we have together and find the inductor current for all $t \geq t_0$ as

$$i_L(t) = i_L(\infty) + (i_L(t_0) - i_L(\infty))e^{-\frac{(t-t_0)}{\tau}} = 0.25e^{-\frac{(t-t_0)}{\tau}}$$

and

$$\tau = \frac{L_\text{eq}}{R_\text{eq}} = \frac{50}{200} = 0.25\,\text{s}.$$

Hence,

$$i_L(t) = 0.25e^{-4(t-t_0)}.$$

6.3 Modeling Switching Mathematically

Switching in circuits can sometimes be represented mathematically instead of drawing an actual switch. Let us look at the simple circuit in Figure 6.28. The

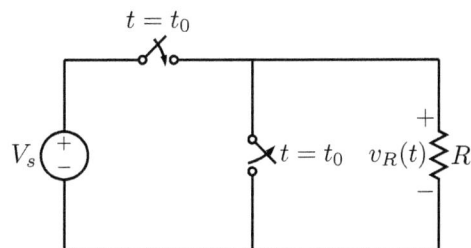

Figure 6.28: Simple example for switching in a circuit.

circuit in Figure 6.28 tells us that the voltage is zero across the resistor until the switch is closed at time t_0. After the switch is closed, the voltage across the resistor is V_s. We can write this by separating our expressions in time intervals as follows:

$$v_R(t) = \begin{cases} 0 & t < t_0, \\ V_s & t > t_0. \end{cases} \tag{6.53}$$

Another way to represent the switching is by using the unity function or unit step function $u(t)$. The unit step function is zero before time t, and it jumps to a value of 1 (hence the unity) at time t. Figure 6.29 shows the plot of the unity function. From Figure 6.29, we can write the unity function as

$$u(t) = \begin{cases} 0 & t < 0, \\ 1 & t > 0. \end{cases} \tag{6.54}$$

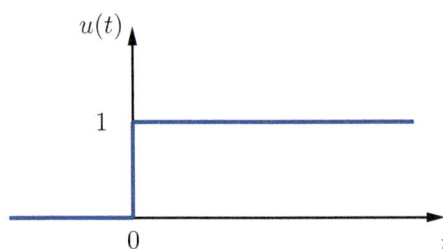

Figure 6.29: The unity step function.

What if we do not want to start the switching at time 0? We can shift the unity step function in time as seen in Figure 6.30. The step function in the figure has been shifted by t_0, where time t_0 is a time value grater than zero. From Figure 6.30 we can write the unity function as

$$u(t - t_0) = \begin{cases} 0 & t < t_0, \\ 1 & t > t_0. \end{cases} \tag{6.55}$$

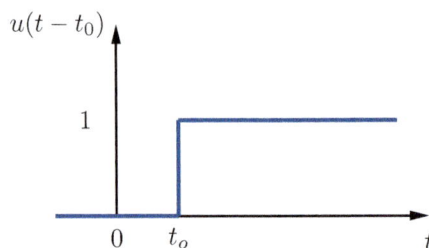

Figure 6.30: The unity step function with a shift in time.

Going back to the circuit in Figure 6.28, we can use (6.55) to write the voltage across the resistor as

$$v_R(t) = V_s u(t - t_0). \tag{6.56}$$

The circuit representation can be seen in Figure 6.31.

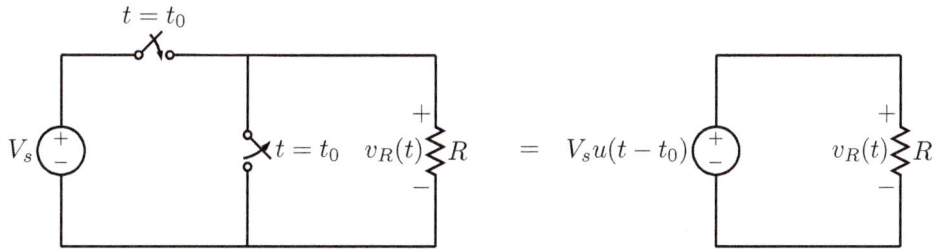

Figure 6.31: Switching using the unity step function.

So far we have modeled the switch turning on. What if we want to model turning the switch off, as seen in Figure 6.32? The circuit in Figure 6.32 tells

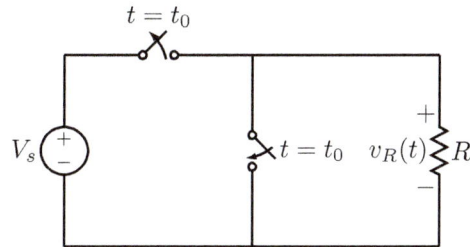

Figure 6.32: Simple example for switching in a circuit.

us that the voltage is V_s across the resistor until the switch is opened at time t_0. After the switch is opened, the voltage across the resistor is zero. We can also write this by separating our expressions in time intervals as follows:

$$v_R(t) = \begin{cases} V_s & t < t_0, \\ 0 & t > t_0. \end{cases} \qquad (6.57)$$

The unity function to model turning off a switch at $t = 0$ is $u(-t)$, as seen in Figure 6.33.

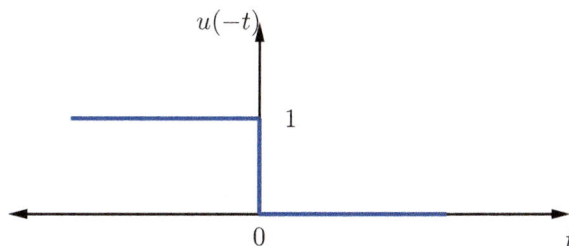

Figure 6.33: The mirror image of the unity step function.

To model turning off a switch at $t = t_0$, the unity function will be $u(t_0 - t)$, as seen in Figure 6.34.

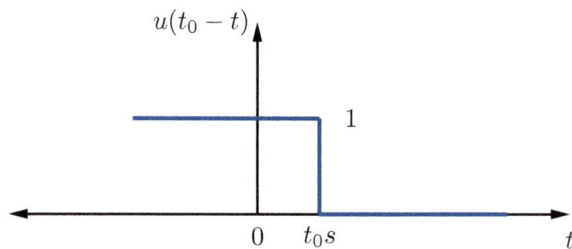

Figure 6.34: The unity step function with a shift in time.

Going back to the circuit in Figure 6.32, we can use the function $u(t_0 - t)$ to write the voltage across the resistor as

$$v_R(t) = V_s u(t_0 - t). \tag{6.58}$$

The circuit representation can be seen in Figure 6.35.

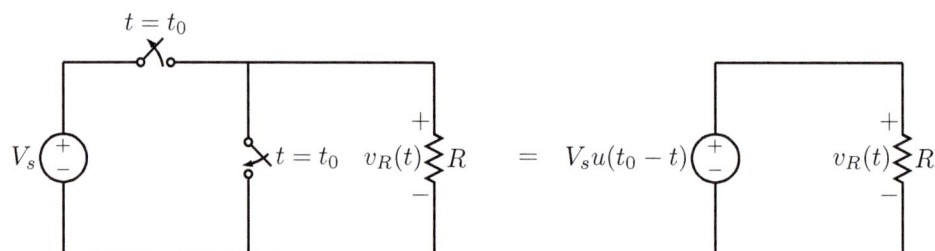

Figure 6.35: Switching using the unity step function.

Example 6.3.1

The voltage change caused by switching events across the resistor in Figure 6.36 is modeled by adding two in-series voltage sources as a function of $u(t)$. Find the voltage $v_R(t)$ for all t.

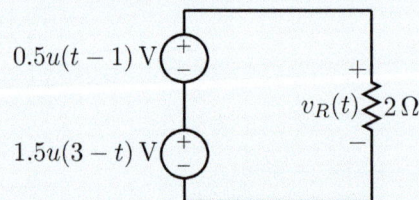

Figure 6.36: Example 6.3.1.

The voltage across the resistor is

$$v_R(t) = 0.5u(t - 1) + 1.5u(3 - t)\,\text{V}.$$

The easiest way to find the voltage across the resistor is to plot the voltages with respect to time and add them in the plot.

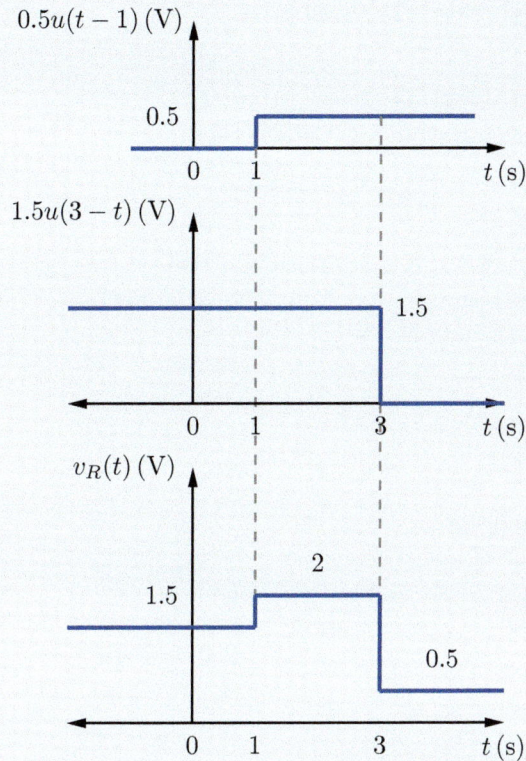

Figure 6.37: Voltage plot for Example 6.3.1.

We can write the voltage across the resistor from the plot as

$$v_R(t) = \begin{cases} 1.5 & t < 1\,\text{s}, \\ 2 & 1\,\text{s} < t < 3\,\text{s}, \\ 0.5 & t > 3\,\text{s}. \end{cases} \quad (\text{V})$$

Example 6.3.2

The current change caused by switching events through the resistor in Figure 6.38 is modeled by adding two parallel current sources as a function of $u(t)$. Find the voltage across the resistor $v_R(t)$ for all t.

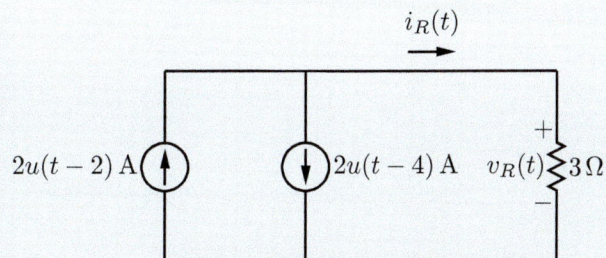

Figure 6.38: Example 6.3.2.

The voltage across the resistor is

$$v_R(t) = 3i_R(t)\,\text{V}.$$

The current through the resistor is

$$i_R(t) = 2[u(t-2) - u(t-4)]\,\text{A}.$$

As in Example 6.3.1, it is easier to plot the currents and subtract them graphically, as seen in Figure 6.39. We can write the current through the resistor from the plot as

$$i_R(t) = \begin{cases} 0 & t < 2\,\text{s}, \\ 2 & 2\,\text{s} < t < 4\,\text{s}, \quad \text{(A)} \\ 0 & t > 4\,\text{s}. \end{cases}$$

The voltage will be

$$v_R(t) = Ri_R(t) = 3i_R(t) = \begin{cases} 0 & t < 2\,\text{s}, \\ 6 & 2\,\text{s} < t < 4\,\text{s}, \quad \text{(V)} \\ 0 & t > 4\,\text{s}. \end{cases}$$

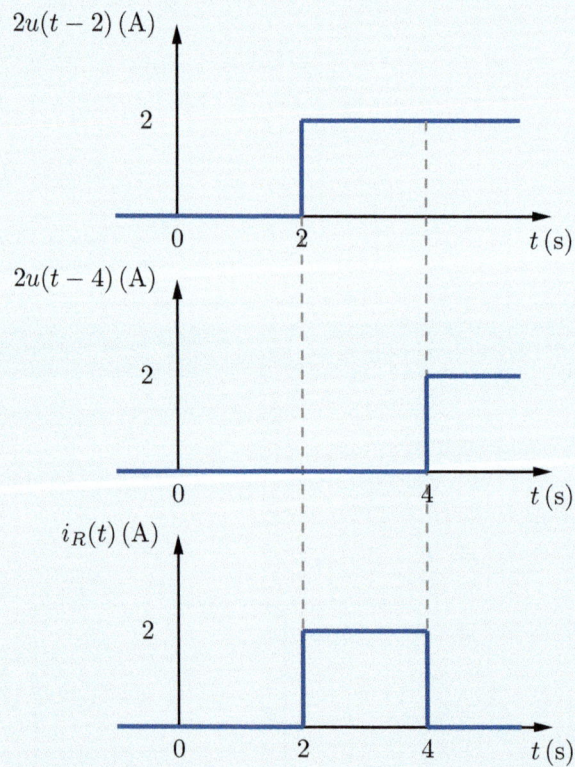

Figure 6.39: Current plot for Example 6.3.2.

Example 6.3.3

Find the current $i_L(t)$ and the voltage $v_C(t)$ in the circuit of Figure 6.40 for $t \geq t_0$.

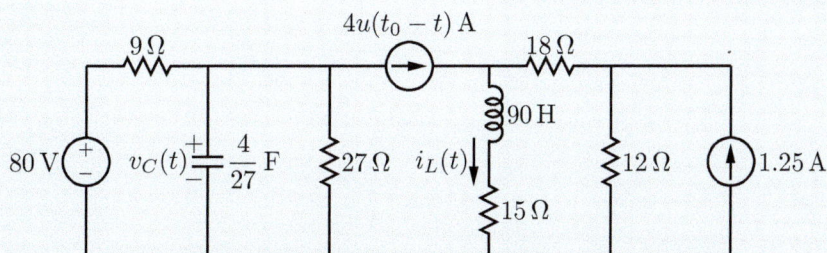

Figure 6.40: Example 6.3.3.

We need to find the capacitor's initial voltage and the inductor's initial current first. We see that the current source bridging our RC and RL circuits is only there for $t \leq t_0$, so we can redraw the circuit with our capacitor open and inductor shorted to find the initial conditions. The circuit is shown in Figure 6.41.

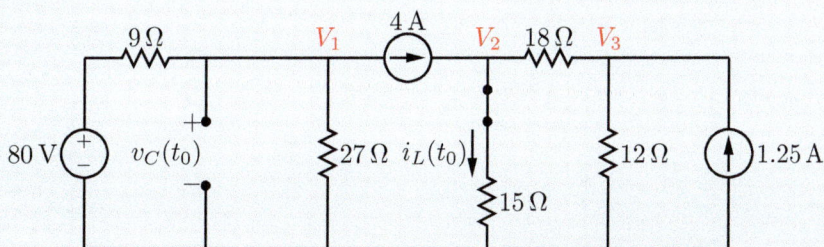

Figure 6.41: The circuit just before switching.

We can use nodal analysis to find the node voltages first. We notice that if we use a source transformation on the 1.25 A source, we can reduce the number of nodal equations. The reduced circuit is shown in Figure 6.42.

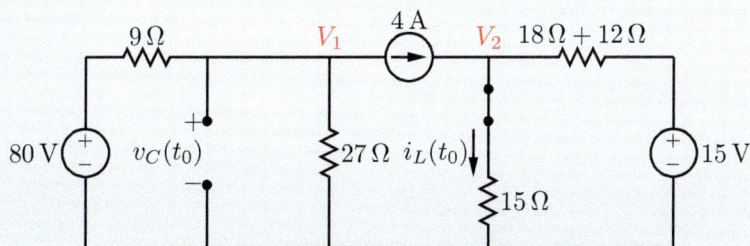

Figure 6.42: The reduced circuit just before switching.

The final matrix for the nodal equations is

$$
\begin{bmatrix} \dfrac{1}{9} + \dfrac{1}{27} & 0 \\[2ex] 0 & \dfrac{1}{15} + \dfrac{1}{30} \end{bmatrix}
\begin{bmatrix} V_1 \\[2ex] V_2 \end{bmatrix} =
\begin{bmatrix} \dfrac{80}{9} - 4 \\[2ex] \dfrac{15}{30} + 4 \end{bmatrix}.
$$

Solving for the nodal voltages, we get

$$
\begin{bmatrix} V_1 \\ V_2 \end{bmatrix} = \begin{bmatrix} 33 \\ 45 \end{bmatrix} \text{ V}.
$$

The capacitor's initial voltage is

$$
v_C(t_0) = V_1 = \underline{33\,\text{V}.}
$$

The inductor's initial current is

$$
i_L(t_0) = \frac{V_2}{15} = \underline{3\,\text{A}.}
$$

After the 4 A source is turned off, we will have two separate circuits as seen in Figure 6.43.

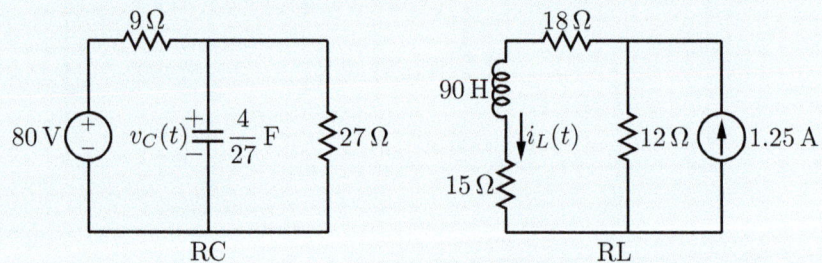

Figure 6.43: The two separate circuits after switching.

We can now find the capacitor's final voltage and the inductor's final current. We will do that by using the same circuits in Figure 6.43 but with opening the capacitor and shorting the inductor as seen in Figure 6.44.

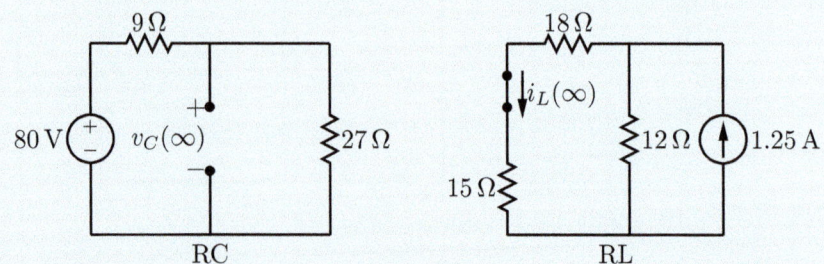

Figure 6.44: The two separate circuits at $t = \infty$.

We can find the capacitor's final voltage using voltage division:

$$v_C(\infty) = \frac{27}{27 + 9}80 = \underline{60\,\text{V}}.$$

We can also find the inductor's final current using current division:

$$i_L(\infty) = \frac{12}{12 + 18 + 15}1.25 = \underline{\frac{1}{3}\,\text{A}}.$$

The final step is to find the equivalent resistances seen by the capacitor and the inductor. We do that by turning off all independent sources and then finding the equivalent resistances. The circuits are shown in Figure 6.45.

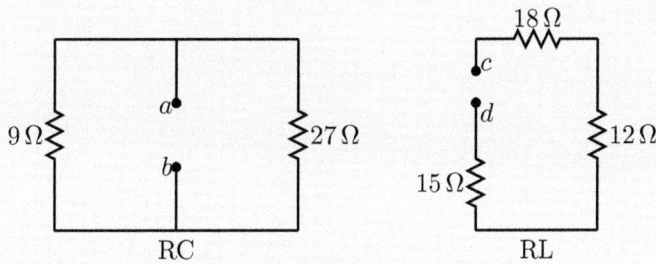

Figure 6.45: Finding the equivalent resistance for the two separate circuits.

The resistance seen by the capacitor is

$$R_{ab} = 9 \parallel 27 = \underline{6.75\,\Omega}.$$

This makes the time constant for the RC circuit

$$\tau_{RC} = R_{ab}C = 6.75\frac{4}{27} = \underline{1\,\text{s}}.$$

The resistance seen by the inductor is

$$R_{cd} = 12 + 18 + 15 = \underline{45\,\Omega}.$$

This makes the time constant for the RL circuit

$$\tau_{RL} = \frac{L}{R_{cd}} = \frac{90}{45} = \underline{2\,\text{s}}.$$

We can now find our capacitor's voltage and inductor's current using the general equation

$$x(t) = x(\infty) + (x(t_0) - x(\infty))\,e^{-\frac{(t - t_0)}{\tau}}.$$

For the capacitor, we get

$$v_C(t) = 60 + (33 - 60)e^{-\frac{(t-t_0)}{1}},$$

$$v_C(t) = 60 - 27e^{-t}\,\text{V}.$$

For the inductor, we get

$$i_L(t) = \frac{1}{3} + (3 - \frac{1}{3})e^{-\frac{(t-t_0)}{2}},$$

$$i_L(t) = \frac{1}{3} + \frac{8}{3}e^{-0.5t}\,\text{A}.$$

6.4 Exercises

Section 6.1

6.1. For the circuit in Figure 6.46, find the voltage $v(t)$.

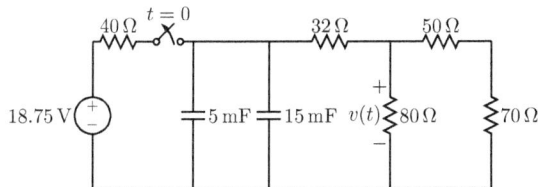

Figure 6.46

6.2. For the circuit in Figure 6.47, find $v(t)$ and $i(t)$.

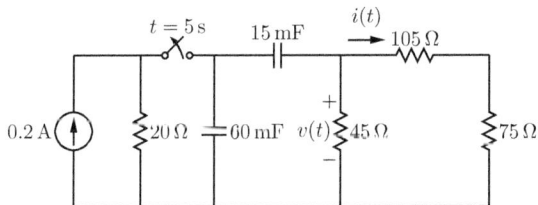

Figure 6.47

6.3. For the circuit in Figure 6.48, $v(t) = 15e^{-2(t-2)}$ V. Find:

a) The switching time t_0.

b) The value of the capacitor C in millifarads.

c) The current $i(t)$.

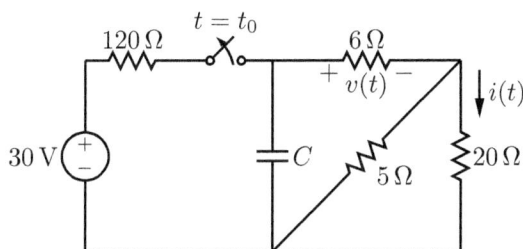

Figure 6.48

6.4. For the circuit in Figure 6.49, find $v_C(t)$.

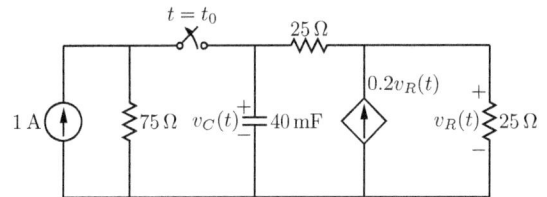

Figure 6.49

6.5. For the circuit in Figure 6.50, find $v(t)$. The initial conditions of each capacitor is found in Table 6.1.

C	$V(t_o)$
12 mF	15 V
30 mF	−5 V
42 mF	10 V

Table 6.1

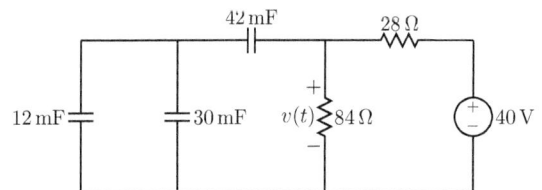

Figure 6.50

6.6. For the circuit in Figure 6.51, find $v_C(t)$.

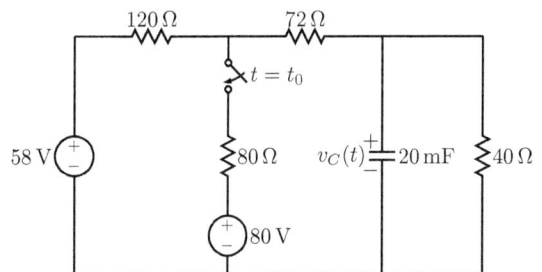

Figure 6.51

6.7. For the circuit in Figure 6.52:

a) Find the voltage $v(t)$.

b) If a 40 mF capacitor were connected to the circuit between points a and b, find the new voltage $v(t)$.

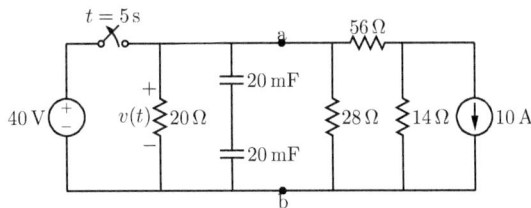

Figure 6.52

6.8. For the circuit in Figure 6.53, the capacitor C was initially uncharged at time zero. If $\tau = 0.6\,\text{s}$, find the value of the capacitor and the voltage $v_C(t)$.

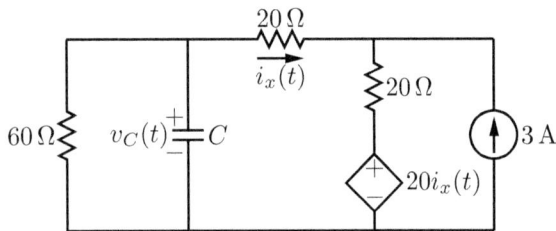

Figure 6.53

Section 6.2

6.9. For the circuit in Figure 6.54, find the current $i(t)$.

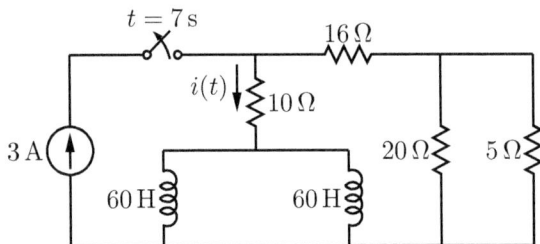

Figure 6.54

6.10. For the circuit in Figure 6.55, the initial current through the inductor is $i_L(t_o) = 5\,\text{A}$. Find the current $i_L(t)$.

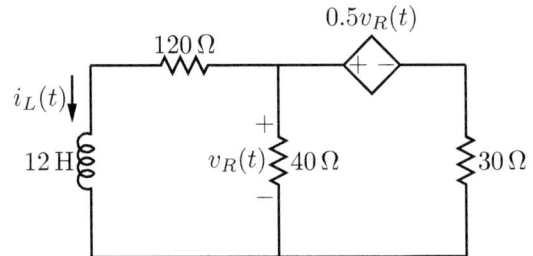

Figure 6.55

6.11. For the circuit in Figure 6.56, find the voltage $v_R(t)$.

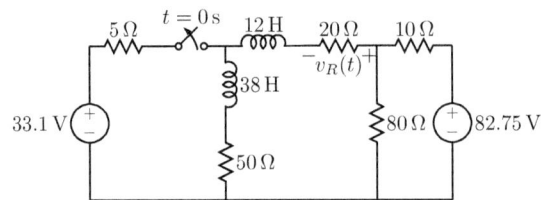

Figure 6.56

Section 6.3

6.12. For the circuit in Figure 6.57, the voltage $v_s(t) = 70u(1 - t)\text{V}$. Find the voltage $v_R(t)$.

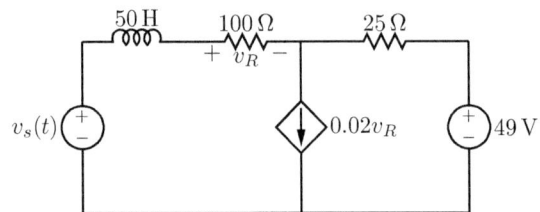

Figure 6.57

6.13. For the circuit in Figure 6.58, $v_{s_1}(t) = 110u(3 - t)\text{V}$, and $v_{s_2}(t) = 99u(t - 3))\text{V}$. Find the voltage $v_C(t)$.

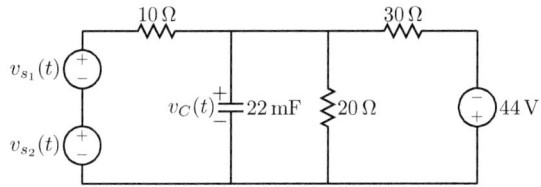

Figure 6.58

6.14. For the circuit in Figure 6.59, $i_s(t) = 5u(4 - t)$A, and $v_s(t) = 80u(t - 4)$V. Find the current $i_L(t)$.

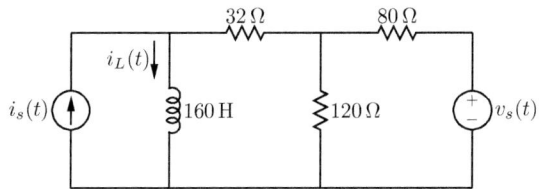

Figure 6.59

Chapter 7

AC Circuits

Introduction

The circuits we have studied so far have mostly included direct current (DC) sources. Cell phone batteries, automobile batteries, and solar cells are examples of DC sources. Historically, DC power was invented first. Furthermore, DC circuits are easier to analyze and design. While DC circuits are quite common, the reality is that AC circuits are far more common. Any plug in our house, school, or office is an AC source. The power lines you see outside are distribution networks to send AC power to our buildings. Although there are many different waveforms of interest, the term "AC sources" primarily refers to sources that produce sinusoidal signals. There are many reasons for this. The vast majority of applications utilize sinusoidal signals: power distribution networks, WiFi routers, cell phones, wireless chargers, microwave ovens, GPS systems, wireless sensors, and weather radars are just a few examples of systems based on sinusoidal signals. Sinusoids are easy to generate, efficient to transmit, and easy to analyze. Another wonderful property of sinusoids is that any other (practical) signal can be written as a sum or integral of sinusoids (Fourier theory). This means that any (practical) signal can be analyzed as a superposition of sinusoids. Before we jump into AC circuit analysis, we will review sinusoidal signals and our main analysis tools: phasors and impedances.

7.1 Sinusoids in General

We will begin by reviewing the basics of sinusoids. Figure 7.1 shows a time plot of two sine waves:

$$v_1(t) = V_m \sin(\omega t) \,(\text{V}), \tag{7.1}$$
$$v_2(t) = V_m \sin(\omega t + \phi) \,(\text{V}). \tag{7.2}$$

The quantities we need to recognize here are:

1. The peak value V_m. We will refer to V_m as the magnitude of the sine wave.

2. The time period T. The time period is the time it takes for the signal to repeat itself.

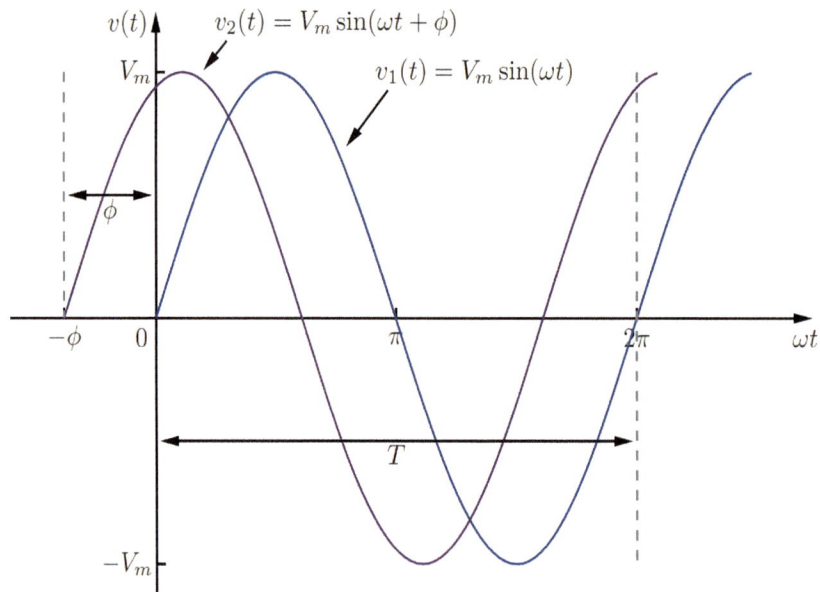

Figure 7.1: Two sinusoids.

3. The frequency f. The frequency can be found by using the time period T:

$$f = \frac{1}{T} \text{ (Hz).}$$

4. The radial frequency ω:

$$\omega = 2\pi f \text{ (rad/s).}$$

5. The phase shift ϕ. We notice that the signal $v_2(t)$ is shifted from $v_1(t)$ by the angle ϕ. A positive shift ϕ results in the signal $v_2(t)$ reaching its peak value before the signal $v_1(t)$. We often express this by saying that $v_2(t)$ leads $v_1(t)$ by ϕ degrees. If we have a negative shift, $v_2(t)$ will lag $v_1(t)$ by ϕ degrees.

Example 7.1.1

For the sinusoidal function

$$v(t) = 24 \sin\left(\frac{2\pi}{3}t + 70°\right) \text{ (V),}$$

1. The magnitude of the sine wave is

$$V_m = 24 \text{ V.}$$

2. The radial frequency ω of the sine wave is

$$\omega = \frac{2\pi}{3} \text{ rad/s}.$$

3. The frequency f of the sine wave is

$$f = \frac{\omega}{2\pi} = \frac{\frac{2\pi}{3}}{2\pi} = \frac{1}{3} \text{ Hz}.$$

4. The time period T of the sine wave is

$$T = \frac{1}{f} = 3 \text{ s}.$$

5. The phase shift ϕ of the sine wave is

$$\phi = 70°.$$

7.2 Sinusoidal Voltages and Currents in Linear Circuits

In an AC circuit, the driving force is a sinusoid. Figure 7.2 shows a load attached to a sinusoidal voltage source. Note that in some circuits, an AC source is drawn with a wave instead of the usual $\pm$ sign to indicate that the source is sinusoidal. However, the voltage polarity still needs to be explicitly defined when we need to analyze a circuit. The voltage $v(t)$ is expressed as

$$v(t) = V_m \cos(\omega t + \theta) \, (\text{V}). \tag{7.3}$$

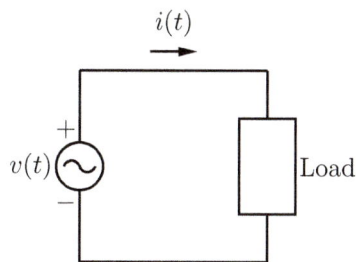

Figure 7.2: A simple AC circuit.

The voltage in (7.3) is a cosine wave with an amplitude V_m V, an angular frequency ω rad/s, and a phase shift of θ. The phase shift θ can be expressed in radians or degrees. From what we have studied so far, the load in Figure 7.2 can be a resistor, an inductor, a capacitor, or a combination of these. No matter what the load is, the current in steady state will *always* be a cosine wave with the same frequency but perhaps with a different amplitude and phase shift.

Why? Let us find the answer by looking into simple cases first. If the load is resistive, the current will be equal to the voltage divided by a constant. If the load is inductive or capacitive, the current will be obtained by either integrating or differentiating the voltage. In all cases, the frequency of the resulting cosine will not change. The current in more complex loads can be viewed as a linear combination of the aforementioned cases. We will discuss this again in more detail later.

> The main result we need to remember is that in linear systems, the output (voltage or current) signal will (in steady state) have the *same* frequency as the input signal. Hence, it is sufficient to compute the resulting magnitude and phase.

This is a very important property that will prove very useful in solving AC circuits.

7.3   Limitations of Conventional Methods in AC Circuits

We could, theoretically, apply the methods we have learned so far to analyze AC circuits. We will see in this section that these methods, albeit correct, are not very convenient when applied directly to AC circuits. Let us see a couple of examples that progressively reveal this.

Consider first the circuit in Figure 7.3. We have two AC voltages in series with $v_1(t) = V_{m_1} \cos(\omega t + \theta_1)$ and $v_2(t) = V_{m_2} \cos(\omega t + \theta_2)$. The first thing we may attempt is to add the voltage sources together since they are in series. Notice that both sources are at the same frequency. Two cosines at the same frequency can be easily combined into one by using the identity in (7.4).

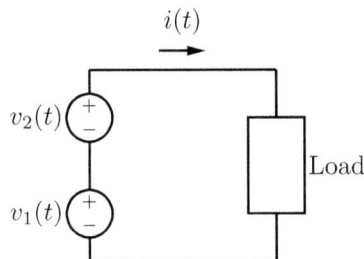

Figure 7.3: A simple AC circuit with two sources.

$$A\cos(\omega t + \alpha) + B\cos(\omega t + \beta) = X\cos(\omega t + \phi)$$

where

$$X = \sqrt{\left([A\cos(\alpha) + B\cos(\beta)]^2 + [A\sin(\alpha) + B\sin(\beta)]^2\right)}$$

and

$$\phi = \tan^{-1}\left[\frac{A\sin(\alpha) + B\sin(\beta)}{A\cos(\alpha) + B\cos(\beta)}\right] \tag{7.4}$$

Let us look at a specific numerical example.

Example 7.3.1

For Figure 7.4, let $v_1(t) = 5\cos(100t + 30°)$ V and $v_2(t) = 2\cos(100t + 60°)$ V. Find the total voltage across the load.

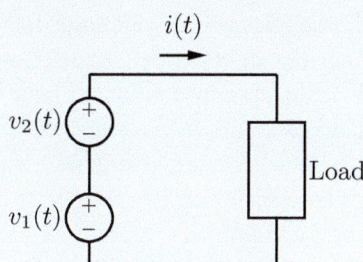

Figure 7.4: An AC circuit for Example 7.3.1.

To find the total voltage across the load, we need to add the voltage sources. Using the identity in (7.4):

$$v_1(t) + v_2(t) = 5\cos(100t + 30°) + 2\cos(100t + 60°) = X\cos(100t + \phi)$$

where

$$X = \sqrt{\left([5\cos(30°) + 2\cos(60°)]^2 + [5\sin(30°) + 2\sin(60°)]^2\right)} = 6.81\,\text{V}$$

and

$$\phi = \tan^{-1}\left[\frac{5\sin(30°) + B\sin(60°)}{5\cos(30°) + 2\cos(60°)}\right] = 38.45°$$

$$v_1(t) + v_2(t) = 6.81\cos(100t + 38.45°)\,\text{V}$$

While this was not too bad, let us look at a more practical circuit, shown in Figure 7.5. In this case, we want to find the nodal voltages $v_1(t)$, $v_2(t)$, and $v_3(t)$. We can obviously apply nodal analysis as we have learned so far. Nodal analysis at node 1 yields

$$\frac{v_1(t) - v_{s_1}}{R_1} + C_1\frac{\mathrm{d}v_1(t)}{\mathrm{d}t} + \frac{1}{L_1}\int_{-\infty}^{t}[v_1(x) - v_2(x)]\,\mathrm{d}x = 0.$$

We can quickly see that this is not the best way to do this. We will need to simultaneously solve three integrodifferential equations. You can imagine how complex things would become if we were to attempt to analyze more realistic

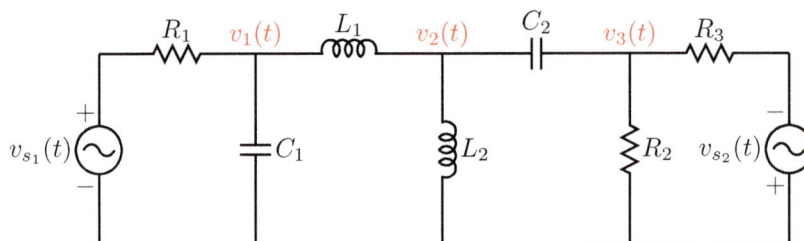

Figure 7.5: A more involved AC circuit.

circuits with more nodes and components. This is why we will introduce a new concept called **phasors**, and we will apply our methods in a new domain called the **phasor** or **frequency domain**. In the frequency domain, integrations and differentiations will be converted to simple multiplications or divisions by the imaginary number i, or what we will in this case call j, which is $\sqrt{-1}$. (The symbol i is reserved for current in electrical engineering.) We will find out that any circuit element will transform into what we will call an **impedance** Z, which will help us treat all elements as resistors. As a result, all integrodifferential equations will be reduced to simple algebraic equations. Before we dive into this magical world where a circuit like the one in Figure 7.5 is really easy to solve, let us review complex numbers, as they are the key to open the door to this world.

7.4   Review of Complex Numbers

As mentioned above, the imaginary unit $\sqrt{-1}$ is represented by the symbol j in electrical engineering. Hence, we write

$$j = \sqrt{-1}. \tag{7.5}$$

Complex numbers have two possible representations (forms): rectangular and polar representations. The rectangular form emphasizes the real and imaginary parts of the complex number. This is written as

$$X = \underbrace{a}_{\text{Real}} + j \underbrace{b}_{\text{Imaginary}} \tag{7.6}$$

For the complex number in (7.6), a is the real part and b is the imaginary part:

$$\Re(X) = a, \tag{7.7}$$

$$\Im(X) = b. \tag{7.8}$$

We can also write the complex number X in (7.6) in polar form as

$$X = r\underline{/\theta} \tag{7.9}$$

where r is the length or magnitude of the complex number X, and θ is the angle. The angle θ can be in radians or degrees:

$$|X| = r, \tag{7.10}$$

$$\text{angle}(X) = \theta. \tag{7.11}$$

Due to Euler's identity, we can write the polar form shown in (7.9) as

$$X = r\underline{/\theta} = re^{j\theta}. \tag{7.12}$$

To avoid the angle ambiguity, we will restrict the angle θ from $-180°$ to $180°$. In other words,

$$-\pi < \theta \le \pi. \tag{7.13}$$

This is, of course, only one of the possible choices. We could have also chosen the angle to be between $0°$ to $360°$. We will, however, use (7.13) for consistency throughout this text. To transfer an angle from radians to degrees, we can use

$$\theta° = \theta(\text{rad})\frac{180°}{\pi}, \tag{7.14}$$

and to transfer an angle from degrees to radians, we can write

$$\theta(\text{rad}) = \theta°\frac{\pi}{180°}. \tag{7.15}$$

Since the complex number X in (7.6) is the same as in (7.9), it is useful to explore the relationship between the variables $a, b, r,$ and θ. The best way to do this is to employ the geometric representation of the complex plane, as seen in Figure 7.6.

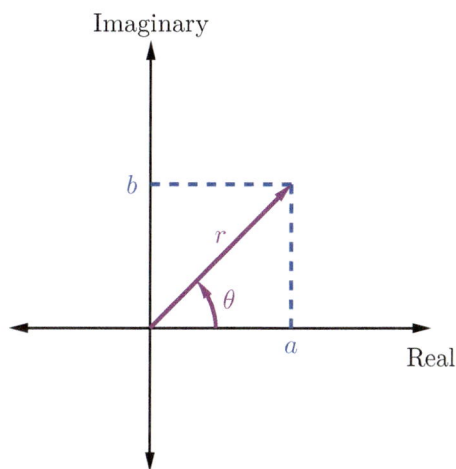

Figure 7.6: The complex plane.

Specifically, we can relate our variables using the Pythagorean theorem as follows:

$$a = |X|\cos(\text{angle}(X)) = r\cos(\theta) \tag{7.16}$$

$$b = |X|\sin(\text{angle}(X)) = r\sin(\theta) \tag{7.17}$$

and

$$r = \sqrt{\Re(X)^2 + \Im(X)^2} = \sqrt{a^2 + b^2} \tag{7.18}$$

$$\theta = \tan^{-1}\left(\frac{\Im(X)}{\Re(X)}\right) = \tan^{-1}\left(\frac{b}{a}\right) \tag{7.19}$$

We have to be *very* careful with (7.19). Using this equation without considering the complex plane geometric representation can easily lead us to the wrong angle calculation. In many programming languages (such as Matlab), you will have to use the function "atan2" to obtain the correct angle. Also, atan2 results in an angle that is in radians, so you will have to use (7.14) to find the angle in degrees if needed. Let us look at an example to demonstrate this.

Example 7.4.1

Find the polar form for the following complex numbers: $X_1 = 1 + j$, $X_2 = 1 - j$, $X_3 = -1 - j$, and $X_4 = -1 + j$.

Figure 7.7 shows the complex numbers drawn on the complex plane.

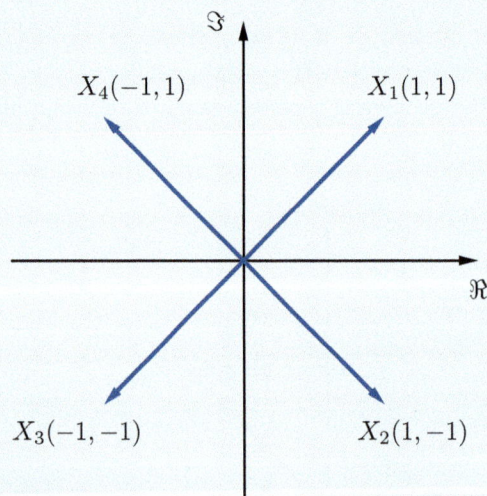

Figure 7.7: Complex numbers drawn on the complex plane for Example 7.4.1.

We can use (7.18) and (7.19) to find the polar forms for the given complex numbers. The first thing we notice is that X_1, X_2, X_3, and X_4 are going to have the same magnitude, so we can quickly use (7.18) to find the magnitudes:

$$|X_1| = |X_2| = |X_3| = |X_4|.$$

$$|X_1| = \sqrt{\Re(X_1)^2 + \Im(X_1)^2} = \sqrt{(1)^2 + (1)^2} = \sqrt{2}.$$

Now we can divert our attention to the angles. For $X_1 = 1 + j$, we have

$$\theta_1 = \tan^{-1}\left(\frac{\Im(X_1)}{\Re(X_1)}\right) = \tan^{-1}\left(\frac{1}{1}\right) = 45°.$$

Consequently,

$$X_1 = \sqrt{2} \underline{/45°}.$$

For $X_2 = 1 - j$, we have

$$\theta_2 = \tan^{-1}\left(\frac{\Im(X_2)}{\Re(X_2)}\right) = \tan^{-1}\left(\frac{-1}{1}\right) = -45°.$$

Hence,

$$X_2 = \sqrt{2} \underline{/-45°}.$$

For $X_3 = -1 - j$, we could write (careful—this is wrong!)

$$\theta_3 = \tan^{-1}\left(\frac{\Im(X_3)}{\Re(X_3)}\right) = \tan^{-1}\left(\frac{-1}{-1}\right) = 45° \text{ (wrong!)}.$$

This would incorrectly yield

$$X_3 = \sqrt{2} \underline{/45°} \text{ (wrong!)}.$$

We have a problem here since we have incorrectly found that $X_1 = X_3$. Look back at Figure 7.7 for the right solution. This is what we meant when we noted that applying (7.19) needs to be done with special care. So what do we do here? The safe way to proceed is to draw or mentally recall the geometric representation on the complex plane. From the figure, we can see that X_3 is 180° away from X_1. Thus,

$$\theta_3 = \theta_1 + 180° = 45° + 180° = 225°.$$

Alternatively, we could also write

$$\theta_3 = 225° - 360° = -135°.$$

Based on (7.13), we will choose $\theta_3 = -135°$. Hence,

$$X_3 = \sqrt{2} \underline{/-135°}.$$

We will face the same problem for $X_4 = -1 + j$. A careless calculation would yield

$$\theta_4 = \tan^{-1}\left(\frac{\Im(X_4)}{\Re(X_4)}\right) = \tan^{-1}\left(\frac{1}{-1}\right) = -45° \text{ (wrong!)}.$$

This would incorrectly result in

$$X_4 = \sqrt{2} \,\underline{/-45^\circ} \text{ (wrong!)}.$$

Again, we will have to consider the complex plane representation to find the correct angle:

$$\theta_4 = \theta_1 + 90^\circ = 45^\circ + 90^\circ = 135^\circ.$$

Therefore,

$$X_4 = \sqrt{2} \,\underline{/135^\circ}.$$

Figure 7.8 shows all complex numbers with the angles correctly indicated.

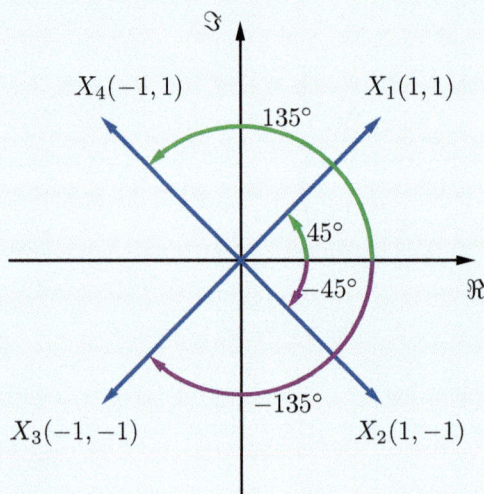

Figure 7.8: Complex numbers drawn on the complex plane with their respective angles for Example 7.4.1.

You may be asking yourself, "Why do we need two ways for writing a complex number? Can't we just stick to one form?" There are two main reasons that both forms are important in AC circuit analysis. The first reason is that, as we will soon discover, complex numbers represent real physical and measurable quantities in electrical and electronic systems. For instance, many times we find ourselves thinking in terms of designing a system to control the phase or magnitude of a complex number that represents an important physical quantity (e.g., power). As a result, we are often concerned about one particular part of a complex number. Sometimes, this may be its real part, while in other cases its magnitude may be of primary importance. Consequently, from a practical point of view, it is very important to be able to understand both forms and easily translate one to another.

The second reason is that some mathematical operations naturally favor one form versus the other. For instance, it is easier to add two numbers when you have them in their rectangular forms (real/imaginary parts), while multiplying and dividing two numbers can be more easily done when they are expressed in their polar representations. Of course, scientific calculators can do these operations regardless of the form you have the numbers in. This tempts some students to think that learning how to perform mathematical calculations by hand is unimportant, since in real life they will often have access to a computer or scientific calculator. However, this way of thinking misses one important aspect of real-life engineering: basic-principles calculations, or back-of-the-envelope calculations. Real-life systems are complicated and are designed by employing powerful computing engines. It is often necessary, though, to double-check the result provided by a "black box" computer against a back-of-the-envelope calculation. Although such calculations are based on significantly simplified assumptions, they are nevertheless easily verifiable and interpretable by humans. This is a critical skill, particularly for engineers in leadership positions whose decisions affect a large number of people with lasting effects.

Based on the above, let us discuss some aspects of the two different forms.

1. Addition and subtraction are easily performed between complex numbers in rectangular form.

    Example: if
    $$X = a + jb, \ Y = c + jd,$$

    then
    $$X - Y = (a - c) + j(b - d).$$

2. Multiplication and division are easier in polar form.

    Example: if
    $$X = r_1 \underline{/\theta_1}, \ Y = r_2 \underline{/\theta_2},$$

    then
    $$XY = r_1 r_2 \underline{/\theta_1 + \theta_2}$$

    and
    $$\frac{X}{Y} = \frac{r_1}{r_2} \underline{/\theta_1 - \theta_2}.$$

3. Taking the reciprocal is easier in polar form.

    Example: if
    $$X = r \underline{/\theta},$$

    then
    $$\frac{1}{X} = \frac{1}{r} \underline{/-\theta}.$$

4. Squaring or taking the square root of a complex number is easier in polar form.

    Example: if
    $$X = r \underline{/\theta},$$

    then
    $$\sqrt{X} = \sqrt{r} \underline{/\frac{\theta}{2}}.$$

    In general,
    $$X^{\frac{1}{n}} = r^{\frac{1}{n}} \underline{/\frac{\theta}{n}} \tag{7.20}$$

and

$$X^2 = r^2\underline{/2\theta}.$$

In general,

$$X^n = r^n\underline{/n\theta}. \tag{7.21}$$

5. The complex conjugate can be easily expressed in both forms.
   If $X = a + jb = r\underline{/\theta}$, then

$$X^* = a - jb = r\underline{/-\theta}. \tag{7.22}$$

6. Some other very helpful facts are summarized below.

$$j = 1\underline{/90°} \tag{7.23}$$

and

$$\frac{1}{j} = -j = 1\underline{/-90°}. \tag{7.24}$$

Example: if

$$X = r\underline{/\theta},$$

then

$$jX = r\underline{/\theta + 90°}$$

and

$$\frac{X}{j} = -jX = r\underline{/\theta - 90°}.$$

Example 7.4.2

For the complex numbers:

$$X = 100 + j54, \qquad Y = 34\underline{/128°}.$$

Find the following:

1. $X + Y$

   Since we are trying to find their sum, it is best to have both complex numbers in rectangular form. Hence,

   $$X + Y = (100 + j54) + (-20.93 + j26.79) = 79.07 + j80.79.$$

2. $\dfrac{X}{Y}$

   For division, it is better to have both numbers in polar form:

   $$\frac{X}{Y} = \frac{113.65\underline{/28.37°}}{34\underline{/128°}} = \frac{113.65}{34}\underline{/28.37° - 128°} = 3.34\underline{/-99.63°}.$$

3. X^* and Y^*

Conjugation can be done in both forms, so we can use the original expressions to find

$$X^* = 100 - j54, \qquad Y^* = 34\underline{/-128^\circ}.$$

4. jX and jY

Multiplication by j is easily done in the polar form, but it can also be done in the rectangular form. Specifically,

$$jX = -54 + j100, \qquad jY = 34\underline{/-142^\circ}.$$

7.5 Phasors

A phasor representation is a method or a way to treat sinusoids as complex numbers or vectors. We represent a sinusoidal signal as a complex number, which will make operations of sinusoids much easier. The only requirement of this method is that **all sinusoids involved have to oscillate at the same frequency**. If we had multiple frequencies in our circuit, we would have to use our powerful method of superposition. We will demonstrate this later in this chapter.

Let us see now how a physical quantity is represented by a complex number. We will use a cosine wave as our reference. Although this may seem like an arbitrary choice in the beginning, it is the norm in literature. Besides, sine waves are simply cosine waves shifted by 90°; they can be easily understood once we are done with cosines. Let us consider the signal

$$x(t) = A\cos(\omega t + \theta^\circ) \tag{7.25}$$

with a magnitude A, angular frequency ω, and phase (shift) θ. The phasor of this cosine signal is defined as

$$\tilde{X} \triangleq A\underline{/\theta^\circ}. \tag{7.26}$$

Notice here that we added the tilde ($\sim$) symbol to indicate that this variable $\tilde{X}$ is not only a complex number but also a phasor representation of a cosine signal. Notice that two of three critical characteristics of $x(t)$ are preserved in its phasor form: the magnitude A and phase θ. What about the frequency ω? As we discussed earlier in this chapter, the output of a linear system in steady state is guaranteed to be a sinusoid at the same frequency as the input sinusoid. Consequently, there is no reason to preserve this information in our calculations in the phasor domain. Since only the magnitude and phase may be affected, we have only preserved these quantities.

In order to write a sine signal in its phasor form, we need to write it in terms of a cosine wave, which simply requires a 90° phase shift. As a result, for the signal

$$y(t) = B\sin(\omega t + \beta^\circ), \tag{7.27}$$

we can write

$$y(t) = B\sin(\omega t + \beta^\circ) = B\cos(\omega t + \beta^\circ - 90^\circ). \tag{7.28}$$

Now we can transfer the signal $y(t)$ to the phasor domain:

$$\tilde{Y}(\omega) = B\underline{/\beta^\circ - 90^\circ}. \tag{7.29}$$

As we learned in the previous section, a positive phase shift by 90° can be represented as a multiplication by j (7.23) and a negative phase shift by 90° can be represented as a multiplication by $-j$ (7.24). As a result, we can rewrite (7.29) as

$$\tilde{Y} = -jB\underline{/\beta^\circ}. \tag{7.30}$$

7.5.1 Leading, Lagging, or Being in Phase

Let us consider the signal we have been working with so far,

$$x(t) = A\cos(\omega t + \theta^\circ), \tag{7.31}$$

and let us introduce a few more signals to examine:

$$y(t) = 0.5A\cos(\omega t + \theta^\circ),$$
$$z(t) = A\cos(\omega t + \theta^\circ + \phi^\circ),$$
$$w(t) = 1.5A\cos(\omega t + \theta^\circ - \phi^\circ),$$
$$u(t) = A\cos(\omega t + \theta^\circ - 90^\circ),$$
$$v(t) = 2A\cos(\omega t + \theta^\circ \pm 180^\circ).$$

All signals have the same frequency but vary in magnitude and/or phase. We can say that:

1. Signal $y(t)$ is in phase with signal $x(t)$ even though they differ in magnitude.

2. Signal $z(t)$ leads $x(t)$ by ϕ°.

3. Signal $w(t)$ lags $x(t)$ by ϕ°.

4. Signal $u(t)$ lags $x(t)$ by 90°.

5. Signal $v(t)$ is the negative of $x(t)$.

This is just an introduction to the terminology of leading and lagging. It emphasizes the fact that absolute phases do not matter much but phase differences do. We will see it again later and we will understand why it is important to know the relative phase among signals.

Example 7.5.1

The voltage and current for an element are

$$v(t) = 200\cos(5\pi t + 20^\circ)\,\text{V},$$

$$i(t) = 2\cos(5\pi t - 40°)\,\text{A}.$$

Find the angle between the voltage and current. Does the current lead or lag the voltage?

The angle between the voltage and current is

$$\theta_{vi} = \theta_v - \theta_i = 20 - (-40) = 60°.$$

Thus, the current lags the voltage by 60°.

The Theory Behind Phasors

As in every new concept in life, the phasor definition in (7.26) may seem a little strange in the beginning. To make it a bit easier to accept, let us review its mathematical basis. It starts with the famous Euler's identity which states that

$$e^{\pm j\theta} = \cos(\theta) \pm j\sin(\theta). \tag{7.32}$$

We remember from our complex numbers discussion that for a complex number

$$X = a + jb, \tag{7.33}$$

we can write

$$\Re(X) = a, \tag{7.34}$$
$$\Im(X) = b. \tag{7.35}$$

Applying (7.34) and (7.35) to (7.32), we get

$$\Re(e^{\pm j\theta}) = \cos(\theta), \tag{7.36}$$
$$\Im(e^{\pm j\theta}) = \pm\sin(\theta). \tag{7.37}$$

Since we said that in Electrical Engineering we use cosines as the basis for our phasor domain, we will focus on (7.36). Also, since (7.36) is not affected by the $\pm$ sign, we can rewrite it as

$$\Re(e^{j\theta}) = \cos(\theta). \tag{7.38}$$

We can apply (7.38) to our general voltage $v(t) = V_m\cos(\omega t + \theta)$ representation to get

$$v(t) = V_m\cos(\omega t + \theta) = V_m\,\Re(e^{j(\omega t+\theta)}) = \Re(V_m e^{j(\omega t+\theta)}). \tag{7.39}$$

Remember also that

$$e^{x+y} = e^x e^y. \tag{7.40}$$

Applying this identity to (7.39) yields

$$v(t) = \Re(V_m e^{j\omega t} e^{j\theta}) = \Re(V_m e^{j\theta} e^{j\omega t}). \tag{7.41}$$

Here is where we will define our phasor (also remember (7.12)) as

$$\tilde{V} = V_m e^{j\theta} = V_m \underline{/\theta}.$$
(7.42)

Using (7.42) to rewrite (7.41) we have

$$v(t) = \Re(\tilde{V} e^{j\omega t}).$$
(7.43)

What does (7.43) tell us? It tells us that a cosine signal in the time domain could be represented by a phasor $\tilde{V}$ that rotates in the frequency domain at a speed of ω rad/s. That is why we say that the frequency domain is a different domain. In the time domain, the cosine is a wave that alternates between two peaks $\pm V_m$ (and moves linearly in the positive time direction) as time passes. In the phasor domain, that wave is now a vector with a magnitude equal to the peak value of the wave and an angle that represents the phase shift of the wave. This vector rotates instead of moving linearly.

It is important to emphasize again that in our analysis, we drop the term $e^{j\omega t}$ because we know that all signals have the same frequency when we go to the phasor domain (assuming a single frequency excitation). Consequently, **a phasor is never a function of time**.

7.6   Kirchhoff's Laws in the Phasor Domain

We can quickly prove that Kirchhoff's laws remain valid in the phasor domain just as they are in the time domain.

7.6.1   Kirchhoff's Current Law (KCL)

In the time domain, we write KCL as

$$\sum_{n=1}^{N} i_n = 0.$$
(7.44)

Let us consider sinusoidal currents at the same frequency leaving the node in Figure 7.9 and write them as

$$i_1(t) = I_{m_1} \cos(\omega t + \theta_1^\circ),$$
$$i_2(t) = I_{m_2} \cos(\omega t + \theta_2^\circ),$$
$$\vdots$$
$$i_N(t) = I_{m_N} \cos(\omega t + \theta_N^\circ).$$
(7.45)

So KCL applied to the currents in (7.45) will result in

$$\sum_{n=1}^{N} I_{m_n} \cos(\omega t + \theta_n^\circ)$$
$$= I_{m_1}\cos(\omega t + \theta_1^\circ) + I_{m_2}\cos(\omega t + \theta_2^\circ) + \ldots + I_{m_N}\cos(\omega t + \theta_N^\circ) = 0. \quad (7.46)$$

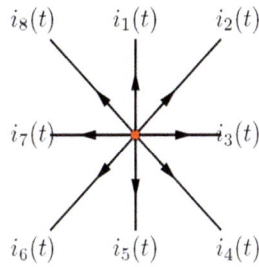

Figure 7.9: KCL in the time domain.

By applying Euler's identity, we can express this equation with the currents as

$$\Re\left\{\sum_{n=1}^{N} I_{m_n}\underline{/\theta_n^\circ}\right\} = \Re\left\{I_{m_1}\underline{/\theta_1^\circ} + I_{m_2}\underline{/\theta_2^\circ} + \ldots + I_{m_N}\underline{/\theta_N^\circ}\right\} = 0. \qquad (7.47)$$

Since $I_{m_n}\underline{/\theta_n^\circ} = \tilde{I}_n$, we can rewrite (7.47) in terms of the phasor currents as

$$\boxed{\sum_{n=1}^{N} \tilde{I}_n = \tilde{I}_1 + \tilde{I}_2 + \ldots + \tilde{I}_N = 0.} \qquad (7.48)$$

Notice that we dropped the $\Re\{\,\}$ in the last equation. As we will see in the next example, we will take the real part of the frequency domain representation at the end of our calculations when we express our final answer in the time domain.

In conclusion, KCL is applied in the phasor domain in the same manner as in the time domain: the sum of currents leaving a node will be zero. Figure 7.10 shows KCL applied to both the time and frequency domains.

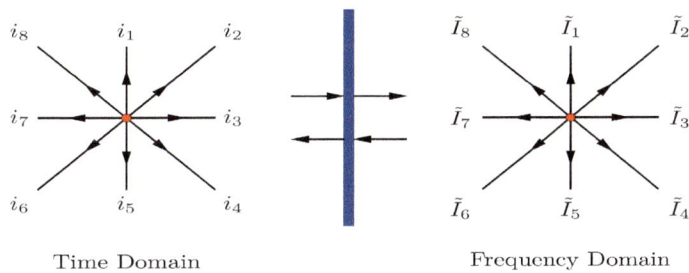

Figure 7.10: KCL in the time and phasor domains.

Example 7.6.1

For the circuit in Figure 7.11, the currents are given as

$$i_s(t) = 25\cos(\omega\pi t + 30°)\,\text{A},$$
$$i_1(t) = 22\cos(\omega\pi t + 65°)\,\text{A},$$
$$i_2(t) = 4\cos(\omega\pi t - 15°)\,\text{A},$$
$$i_3(t) = 12\cos(\omega\pi t - 35°)\,\text{A}.$$

Find the output currents $i_{o_1}(t)$, $i_{o_2}(t)$, and $i_{o3}(t)$.

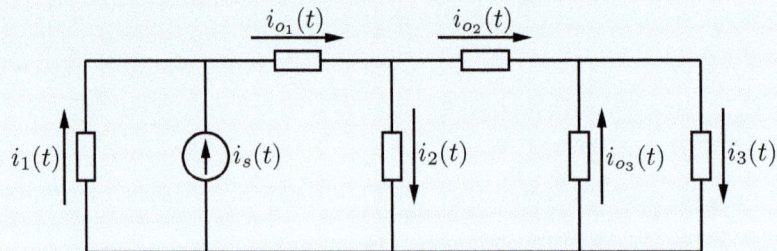

Figure 7.11: Example 7.6.1.

Since all currents have the same frequency, we can transfer the currents to the phasor domain and then find the output currents using KCL in the phasor domain. Hence,

$$i_s(t) = 25\cos(\omega\pi t + 30°) \iff \tilde{I}_s = 25\underline{/30°}.$$
$$i_1(t) = 22\cos(\omega\pi t + 65°) \iff \tilde{I}_1 = 22\underline{/65°}.$$
$$i_2(t) = 4\cos(\omega\pi t - 15°) \iff \tilde{I}_2 = 4\underline{/-15°}.$$
$$i_3(t) = 12\cos(\omega\pi t - 35°) \iff \tilde{I}_3 = 12\underline{/-35°}.$$

Using KCL, we have

$$\tilde{I}_{o_1} = \tilde{I}_1 + \tilde{I}_s = 44.83\underline{/46.35°}\,\text{A}.$$
$$\tilde{I}_{o_2} = \tilde{I}_{o_1} - \tilde{I}_2 = 43.06\underline{/51.02°}\,\text{A}.$$
$$\tilde{I}_{o_3} = \tilde{I}_3 - \tilde{I}_{o_2} = 43.89\underline{/-113.2°}\,\text{A}.$$

Now we can transfer the phasor currents to the time domain as

$$i_{o_1}(t) = 44.83\cos(\omega\pi t + 46.35°)\,\text{A}.$$
$$i_{o_2}(t) = 43.06\cos(\omega\pi t + 51.02°)\,\text{A}.$$
$$i_{o_3}(t) = 43.89\cos(\omega\pi t - 113.2°)\,\text{A}.$$

7.6.2 Kirchhoff's Voltage Law (KVL)

KVL is valid for any closed loop and is written as

$$\sum_{n=1}^{N} v_n(t) = 0. \tag{7.49}$$

Figure 7.12 shows a simple AC circuit with a sinusoidal source and multiple elements in a closed loop. As we have already emphasized before KVL, all voltages are at the same frequency.

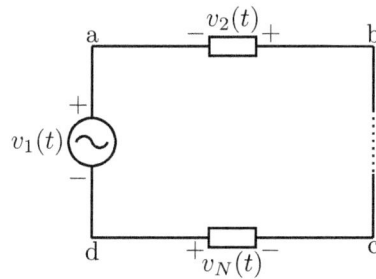

Figure 7.12: KVL in the time domain.

We can write the voltages in Figure 7.12 in general as

$$v_1(t) = V_{m_1} \cos(\omega t + \theta_1^\circ),$$
$$v_2(t) = V_{m_2} \cos(\omega t + \theta_2^\circ),$$
$$\vdots$$
$$v_N(t) = V_{m_N} \cos(\omega t + \theta_N^\circ). \tag{7.50}$$

Applying KVL results in

$$\sum_{n=1}^{N} V_{m_n} \cos(\omega t + \theta_n^\circ)$$
$$= V_{m_1} \cos(\omega t + \theta_1^\circ) + V_{m_2} \cos(\omega t + \theta_2^\circ) + \ldots + V_{m_N} \cos(\omega t + \theta_N^\circ) = 0. \tag{7.51}$$

By applying Euler's identity again, we can write the voltages in (7.51) as (we have dropped the $\Re\{\,\}$):

$$\sum_{n=1}^{N} V_{m_n}\underline{/\theta_n^\circ} = V_{m_1}\underline{/\theta_1^\circ} + V_{m_2}\underline{/\theta_2^\circ} + \ldots + V_{m_N}\underline{/\theta_N^\circ} = 0. \tag{7.52}$$

Since $V_{m_n}\underline{/\theta_n^\circ} = \tilde{V}_n$, we can rewrite (7.52) in terms of the phasor voltages as

$$\boxed{\sum_{n=1}^{N} \tilde{V}_n = 0.} \tag{7.53}$$

Just like KCL, KVL is applied in the phasor domain in the same manner as in the time domain: the sum of voltages around a closed loop is zero. Figure 7.13 shows KVL applied to both the time and phasor domains.

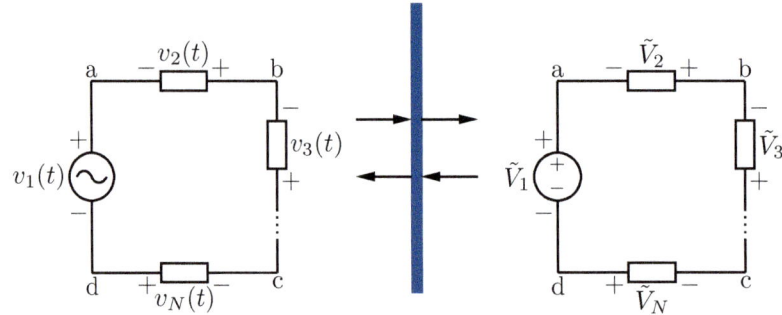

Figure 7.13: KVL in the time and phasor domains.

Example 7.6.2

For the circuit in Figure 7.14, the voltages are given as

$$v_1(t) = 13\cos(\omega t + 50°)\,\text{V},$$
$$v_2(t) = 42\cos(\omega t - 20°)\,\text{V},$$
$$v_3(t) = 20\cos(\omega t + 153°)\,\text{V}.$$

Find the output voltages $v_{o_1}(t)$ and $v_{o_2}(t)$.

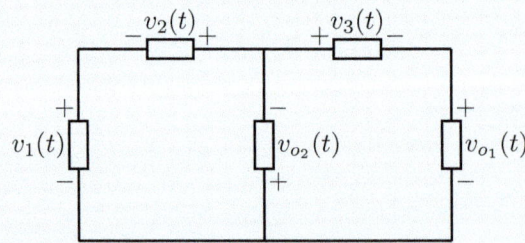

Figure 7.14: Example 7.6.2.

Since all voltages have the same frequency, we can transfer the voltages to the phasor domain and then find the output voltages using KVL in the phasor domain. Hence,

$$v_1(t) = 13\cos(\omega t + 50°) \iff \tilde{V}_1 = 13\underline{/50°}.$$
$$v_2(t) = 42\cos(\omega t - 20°) \iff \tilde{V}_2 = 42\underline{/-20°}.$$
$$v_3(t) = 20\cos(\omega t + 153°) \iff \tilde{V}_3 = 20\underline{/153°}.$$

Using KVL, we have

$$\tilde{V}_{o_1} = -\tilde{V}_3 + \tilde{V}_2 + \tilde{V}_1 = 67.01\underline{/-11.61°}\,\text{V},$$
$$\tilde{V}_{o_2} = -\tilde{V}_1 - \tilde{V}_2 = 48.03\underline{/174.74°}\,\text{V}.$$

Now we can transfer the phasor voltages to the time domain as

$$v_{o_1}(t) = 67.01\cos(\omega t - 11.61°)\,\text{V},$$
$$v_{o_2}(t) = 48.03\cos(\omega t + 174.74°)\,\text{V}.$$

7.7 The Concept of Impedance

Based on the previous sections, we know that we can take our sinusoidal circuit to the phasor domain. We also learned how we can transform our sources. In this section, we focus on the rest of the linear circuit elements that we have studied. Before we discuss each element separately, it is worth discussing our strategy first. The key idea is to find the $i - v$ characteristic for each element in the frequency (phasor) domain based on its corresponding $i - v$ characteristic in the time domain. For instance, what is the frequency domain representation of the $i_c(t) = C\frac{dv_c}{dt}$ equation for capacitors? As we discuss in the next sections, the answer is that all passive elements we have discussed so far (i.e., resistors, inductors, and capacitors) get transformed into elements that obey the same general equation (assuming passive sign convention),

$$\tilde{V}(\omega) = Z(\omega)\tilde{I}(\omega) \quad \text{or} \quad Z(\omega) = \frac{\tilde{V}(\omega)}{\tilde{I}(\omega)}, \tag{7.54}$$

where $Z(\omega)$ is the impedance of the element. The impedance of an element is the effective resistance that this element presents to AC current flow. In other words, we can say that, for a given AC voltage difference, any element opposes or impedes the flow of current through it. We can formally state that

> **Impedance** is a complex number (not a phasor) that is equal to the ratio of the phasor voltage over the phasor current of an element. It is defined only for AC signals and is measured in Ω. In general, it is a function of frequency and is a measure of the difficulty that AC current faces when flowing through a device.

Looking at (7.54), we can see three important properties.

- In the frequency domain, all elements obey Ohm's law, with the impedance (generalized resistance) playing the same role as resistance does for DC signals.

- Impedance is, in general, a function of frequency. Consequently, the behavior of an entire circuit or system is a function of frequency. The response of a system as a function of frequency is often an important question in circuit design.

- The impedance of an element combines both resistance and reactance effects. Consequently, some elements or groups of elements may impede current flow primarily due to high resistance or high reactance, while others may have comparable resistances and reactances.

When we studied resistors, we got introduced to the conductance concept. We learned that conductance G is the inverse of resistance R and is measured in S. Similarly, we can define **admittance** as the inverse of impedance, or

$$Y = \frac{1}{Z}.$$

(7.55)

Just like conductance, admittance is measured in S. Note that the admittance is a complex number and not a phasor. Let us now focus on finding the frequency-domain $i - v$ characteristics and the associated impedances and admittances of resistors, inductors, and capacitors.

7.7.1   The Resistor

A resistor is the "friendliest" element we have since it always obeys Ohm's law no matter what the source is. Figure 7.15 shows a simple circuit in the time domain driven by a sinusoidal source. Remember that the exciting AC source is of the form

$$v(t) = V_m \cos(\omega t + \theta^\circ).$$

(7.56)

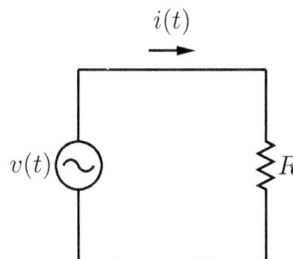

Figure 7.15: An AC circuit with a resistor.

We start the circuit transformation by finding the current in the time domain. Using Ohm's law yields

$$i(t) = \frac{v(t)}{R} = \frac{1}{R} V_m \cos(\omega t + \theta^\circ).$$

(7.57)

We can now transfer both the current and voltage into the frequency (phasor) domain by using (7.26) as

$$\tilde{V}(\omega) = V_m \underline{/\theta^\circ}, \tag{7.58}$$

$$\tilde{I}(\omega) = \frac{1}{R} V_m \underline{/\theta^\circ} = \frac{\tilde{V}(\omega)}{R} = \frac{\tilde{V}(\omega)}{Z}. \tag{7.59}$$

Equation (7.59) explicitly states that Ohm's law also applies in the frequency (phasor) domain without any changes. Hence, looking back at (7.54), we find that the impedance of a resistor is **equal** to its resistance, or

$$\boxed{Z = R.} \tag{7.60}$$

Additionally, the admittance of a resistor is equal to its conductance. Hence,

$$\boxed{Y = \frac{1}{Z} = \frac{1}{R} = G.} \tag{7.61}$$

Figure 7.16 shows both the time and frequency domains of a resistor side by side. The blue boundary is there to represent a portal or barrier that reminds you that once you step into the frequency domain, you do not have a time variable; and once you come out of the phasor domain, you do not have phasors anymore.

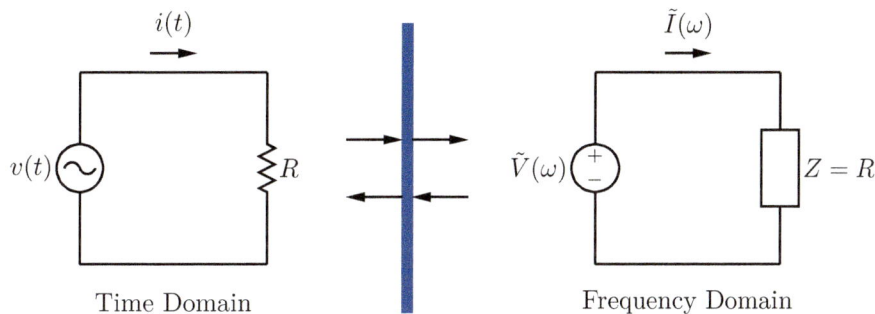

Figure 7.16: Resistor representation in the time and frequency domains.

Looking at (7.59) again, we can see that the current differs from the voltage by the constant R. Consequently, they both have the same phase θ°. In other words, the voltage drop across a resistor is always **in phase** with the current through it. Figure 7.17 shows a resistor's $i - v$ plot in both the time and frequency domains.

7.7.2   The Capacitor

The time-domain $i - v$ characteristic of a capacitor is given by

$$i(t) = C \frac{dv(t)}{dt}. \tag{7.62}$$

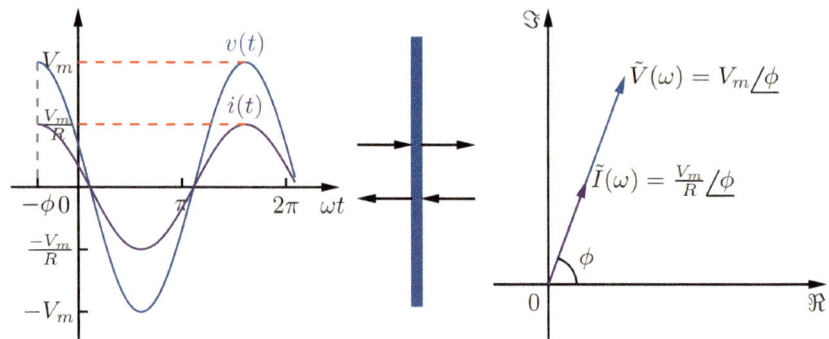

Figure 7.17: Resistor's $i - v$ plot in both the time and frequency domains.

Consequently, for the AC circuit shown in Figure 7.18, the time-domain current is given by

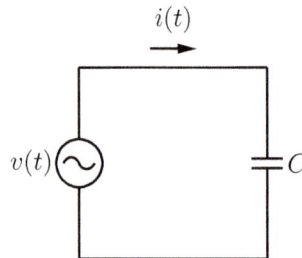

Figure 7.18: An AC circuit with a capacitor.

$$i(t) = C\frac{dv(t)}{dt} = C\frac{d}{dt}(V_m \cos(\omega t + \theta°)). \tag{7.63}$$

The derivative of the cosine is

$$\frac{d}{dt}(V_m \cos(\omega t + \theta°)) = -\omega V_m \sin(\omega t + \theta°). \tag{7.64}$$

Plugging (7.64) into (7.63) yields

$$i(t) = C\frac{dv(t)}{dt} = -\omega C V_m \sin(\omega t + \theta°). \tag{7.65}$$

We know that from (7.28), we can write a sine in terms of cosine as

$$i(t) = C\frac{dv(t)}{dt} = -\omega C V_m \cos(\omega t + \theta° - 90°). \tag{7.66}$$

Now we are ready to move into the phasor domain. We can use (7.29) to transform (7.66) into the phasor domain as

$$\tilde{I}(\omega) = -\omega C V_m \underline{/\theta° - 90°}. \tag{7.67}$$

We learned from (7.30) that the $-90°$ phase shift can be replaced by a multiplication by $-j$ as

$$\tilde{I}(\omega) = -(-j)\omega C V_m \underline{/\theta°} = \underbrace{j\omega C}_{a} \underbrace{V_m \underline{/\theta°}}_{b}. \tag{7.68}$$

We remember that the voltage source in the circuit in Figure 7.18 is a cosine and its phasor domain representation is

$$\tilde{V}(\omega) = V_m \underline{/\theta°}. \tag{7.69}$$

We recognize part b of (7.68) as the voltage. Thus, we can rewrite (7.68) as

$$\tilde{I}(\omega) = j\omega C \, \tilde{V}(\omega). \tag{7.70}$$

If we compare (7.70) to (7.54), we can rewrite (7.70) as

$$\tilde{V}(\omega) = \frac{\tilde{I}(\omega)}{j\omega C} = \tilde{I}(\omega)Z \tag{7.71}$$

where the impedance Z of a capacitor is

$$\boxed{Z = \frac{1}{j\omega C} = \frac{-j}{\omega C}.} \tag{7.72}$$

Figure 7.19 shows both the time and frequency domains of a capacitor side by side. Once again, the blue bar is there to represent a portal or barrier to remind you that once you step into the frequency domain, you do not have a time variable, and once you come out of the phasor domain, you do not have phasors anymore.

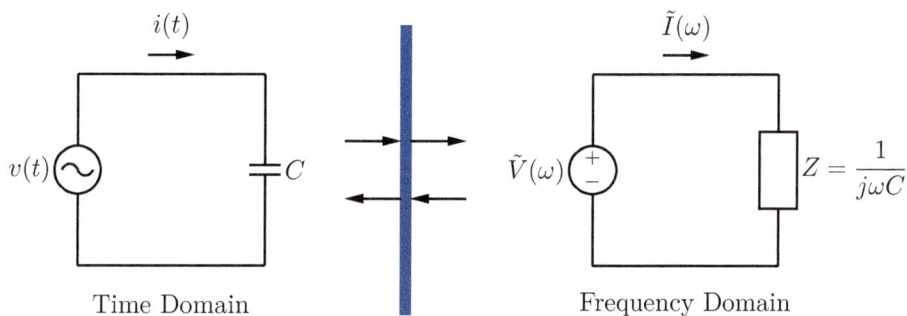

Figure 7.19: Capacitor representation in the time and frequency domains.

We can see from (7.70) that the current **leads** the voltage by $90°$ (or the voltage **lags** the current by $90°$). The $90°$ is explainable since the derivative of the cosine voltage is a sine wave, and that introduces the $90°$ shift. What about the lag? We remember from our capacitor section in chapter 5 that the voltage across a capacitor cannot change instantaneously but the current can. Therefore, the voltage will always be delayed with respect to the current— hence the leading of the current. Figure 7.20 shows the capacitor's $i - v$ plot in both the time and frequency domains. The admittance for the capacitor can be written as

$$Y = \frac{1}{Z} = j\omega C. \tag{7.73}$$

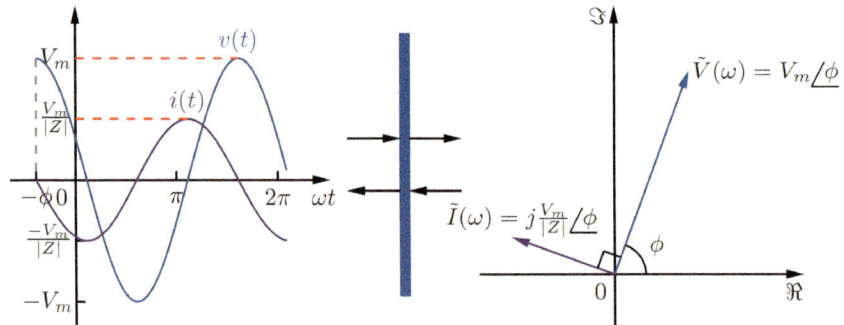

Figure 7.20: Capacitor's $i - v$ plot in both the time and frequency domains.

7.7.3   The Inductor

It is easier to start the derivation for the inductor's impedance when excited with a current source, as shown in Figure 7.21.

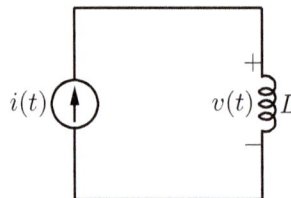

Figure 7.21: An AC circuit with an inductor.

The voltage drop across the inductor in the time domain is given by

$$v(t) = L\frac{di(t)}{dt}. \tag{7.74}$$

From Figure 7.21, the voltage is

$$v(t) = L\frac{di(t)}{dt} = L\frac{d}{dt}(I_m \cos(\omega t + \theta°)). \tag{7.75}$$

The derivative of the cosine is

$$\frac{d}{dt}(I_m \cos(\omega t + \theta°)) = -\omega I_m \sin(\omega t + \theta°). \tag{7.76}$$

Plugging (7.76) into (7.75) yields

$$v(t) = L\frac{di(t)}{dt} = -\omega L I_m \sin(\omega t + \theta°). \tag{7.77}$$

We know from (7.28) that we can write a sine in terms of a cosine as

$$v(t) = L\frac{di(t)}{dt} = -\omega L I_m \cos(\omega t + \theta° - 90°). \tag{7.78}$$

Now we are ready to move into the phasor domain. We can use (7.29) to transform (7.78) into the phasor domain as

$$\tilde{V}(\omega) = -\omega L I_m \underline{/\theta° - 90°}. \tag{7.79}$$

We learned from (7.30) that the $-90°$ phase shift can be replaced by a multiplication by $-j$ as

$$\tilde{V}(\omega) = -(-j)\omega L I_m \underline{/\theta°} = \underbrace{j\omega L}_{a} \underbrace{I_m \underline{/\theta°}}_{b}. \tag{7.80}$$

We remember that the current in the circuit in Figure 7.21 was a cosine and its phasor domain representation is

$$\tilde{I}(\omega) = I_m \underline{/\theta°}. \tag{7.81}$$

We recognize part b of (7.80) as the current, so we can rewrite (7.80) as

$$\tilde{V}(\omega) = j\omega L \tilde{I}(\omega). \tag{7.82}$$

Comparing (7.82) to (7.54), we can rewrite (7.82) as

$$\tilde{V}(\omega) = \tilde{I}(\omega)Z, \tag{7.83}$$

where the impedance Z of an inductor is

$$\boxed{Z = j\omega L.} \tag{7.84}$$

Figure 7.22 shows both the time and frequency domains of an inductor side by side. Once again, the blue bar is there to help you remember that the representations shown are valid in two different domains.

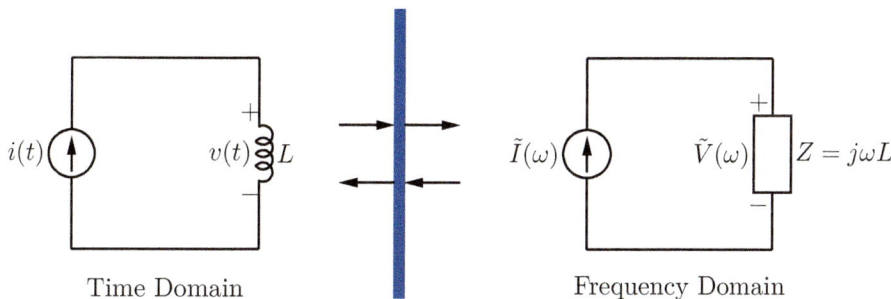

Figure 7.22: Inductor representation in the time and frequency domains.

We can see from (7.82) that the voltage **leads** the current by 90° (or the current **lags** the voltage by 90°). The 90° is explainable since the derivative of the cosine current is a sine wave and that introduces the 90° shift. The current lag is explained by remembering that the current through the inductor cannot change instantaneously but the voltage can. Therefore, the current will always be behind the voltage or will lag it. Figure 7.20 shows the inductor's $i - v$ plot in both the time and frequency domains.

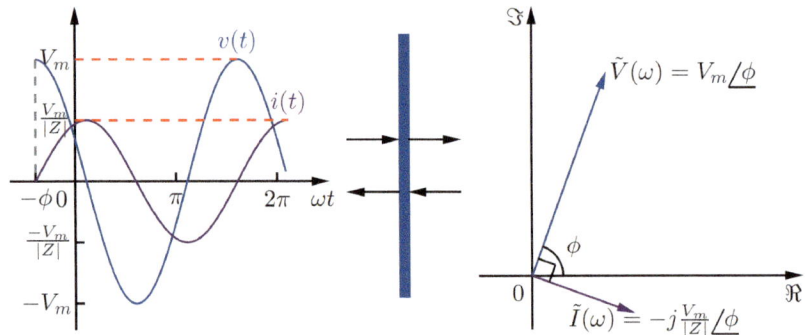

Figure 7.23: Inductor's $i - v$ plot in both the time and frequency domains.

The admittance of an inductor can be written as

$$Y = \frac{1}{Z} = \frac{1}{j\omega L}. \tag{7.85}$$

The General Load

For a general AC load, such as the one shown in Figure 7.24, we will have

$$v(t) = V_m \cos(\omega t + \theta_v)\,(\text{V}), \tag{7.86}$$

and

$$i(t) = I_m \cos(\omega t + \theta_i)\,(\text{A}). \tag{7.87}$$

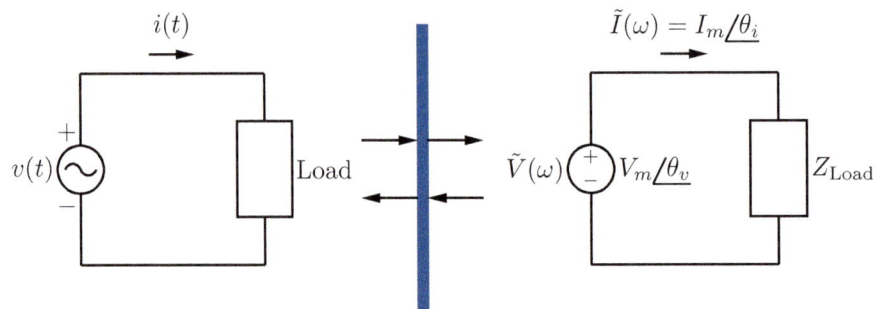

Figure 7.24: An AC circuit with a general load in the time and phasor domains.

The voltage and current in the phasor domain are

$$\tilde{V} = V_m\underline{/\theta_v}, \tag{7.88}$$

$$\tilde{I} = I_m\underline{/\theta_i}. \tag{7.89}$$

We can find the load impedance by using Ohm's law as

$$Z_{\text{Load}} = \frac{V_m\underline{/\theta_v}}{I_m\underline{/\theta_i}} = \frac{V_m}{I_m}\underline{/\theta_v - \theta_i}. \tag{7.90}$$

We can see from (7.90) that

> the angle of the impedance is equal to the phase difference between the voltage and current.

In general, since the impedance is a complex number, we can write it in rectangular form as

$$Z_{\text{Load}} = R + jX. \tag{7.91}$$

The real part of the impedance is the **resistance**, and the imaginary part is the **reactance**.

Let us now revisit the circuit in Figure 7.5. We can use what we learned to transfer all its elements into impedances in the phasor domain. Figure 7.25 shows the circuit in the time domain and its transformation to the phasor domain.

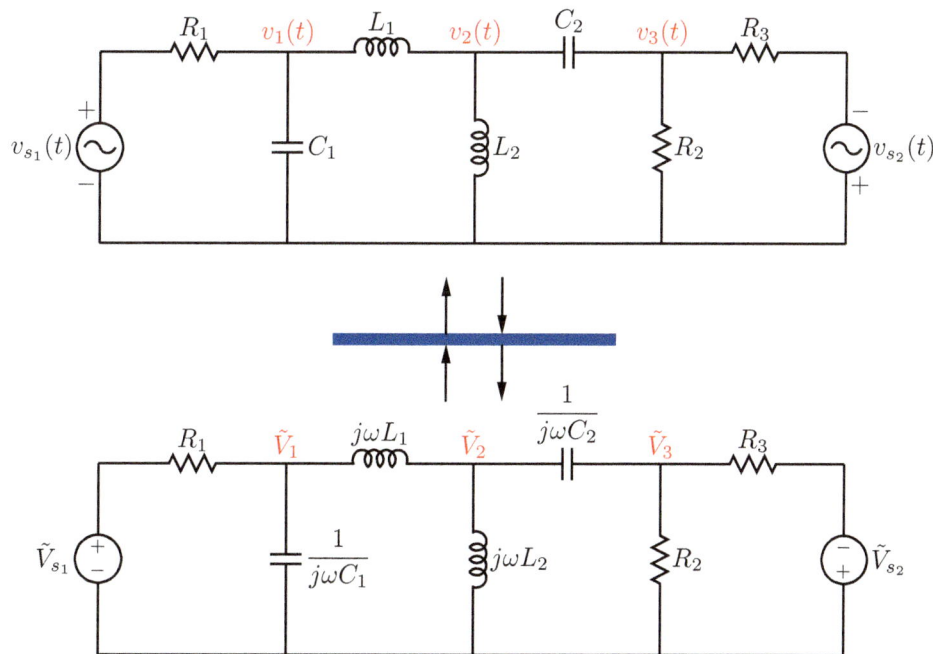

Figure 7.25: The time and phasor domains of Figure 7.5.

Now we can write the nodal equations for the circuit. Starting with node 1, we have

$$\frac{\tilde{V}_1 - \tilde{V}_{s_1}}{R_1} + j\omega C_1 \tilde{V}_1 + \frac{\tilde{V}_1 - \tilde{V}_2}{j\omega L_1} = 0. \tag{7.92}$$

The same can be done for the rest of the nodes. We will be revisiting nodal and mesh analyses for the phasor domain soon. The point here is to show that we will end up with algebraic equations instead of integrodifferential equations. Yes, the equations may contain complex numbers, but they are much easier to solve than the time-domain ones.

7.7.4   Impedances and Admittances: Series and Parallel Connections

Now that we learned how to transfer all our elements to the frequency domain, it is time to explore the connection of these elements in the frequency domain. We will first take a look at impedances and admittances connected in series and then in parallel.

Impedances and Admittances Connected in Series

Impedances are treated in the same way as resistances. Consequently, impedances connected in series are added together. Let us quickly prove it. Figure 7.26 shows N impedances connected in series. We will use KVL and Ohm's law to find the

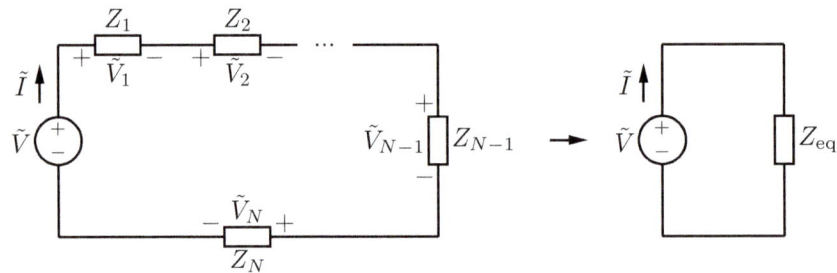

Figure 7.26: N impedances connected in series.

common current $\tilde{I}$. Starting with a KVL around the loop, we get

$$-\tilde{V} + \tilde{V}_1 + \tilde{V}_2 + \cdots + \tilde{V}_{N-1} + \tilde{V}_N = 0,$$
$$\tilde{V} = \tilde{V}_1 + \tilde{V}_2 + \cdots + \tilde{V}_{N-1} + \tilde{V}_N. \tag{7.93}$$

We can now use Ohm's law to get

$$\tilde{V} = \tilde{I}Z_1 + \tilde{I}Z_2 + \cdots + \tilde{I}Z_{N-1} + \tilde{I}Z_N. \tag{7.94}$$

Since the current $\tilde{I}$ is the same through every impedance, we can rewrite (7.94) as

$$\tilde{V} = \tilde{I}(Z_1 + Z_2 + \cdots + Z_{N-1} + Z_N) \tag{7.95}$$

and from that we can calculate the current $\tilde{I}$ as

$$\tilde{I} = \frac{\tilde{V}}{(Z_1 + Z_2 + \cdots + Z_{N-1} + Z_N)}. \tag{7.96}$$

Just as expected, impedances in series are added together to yield the equivalent impedance

$$\boxed{Z_{\text{eq}} = \sum_{n=1}^{N} Z_n.} \tag{7.97}$$

Equation (7.96) can now be rewritten as

$$Z_{\text{eq}} = \frac{\tilde{V}}{\tilde{I}}. \tag{7.98}$$

We can also rewrite (7.96) in terms of admittance:

$$\tilde{I} = \frac{\tilde{V}}{\left(\frac{1}{Y_1} + \frac{1}{Y_2} + \cdots + \frac{1}{Y_{N-1}} + \frac{1}{Y_N}\right)} = \frac{\tilde{V}}{\sum_{n=1}^{N} \frac{1}{Y_n}}. \tag{7.99}$$

To find the equivalent admittance, we write

$$\frac{\tilde{I}}{\tilde{V}} = \frac{1}{\sum_{n=1}^{N} \frac{1}{Y_n}} = Y_{\text{eq}}, \tag{7.100}$$

$$\boxed{Y_{\text{eq}} = \frac{1}{\sum_{n=1}^{N} \frac{1}{Y_n}}.} \tag{7.101}$$

Impedances and Admittances Connected in Parallel

Figure 7.27 shows N impedances connected in parallel. Let us start by finding the current $\tilde{I}$.

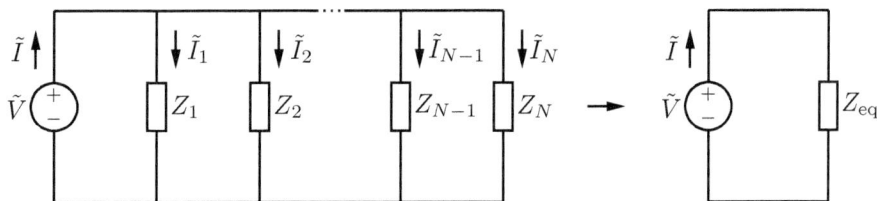

Figure 7.27: Impedances connected in parallel.

Using KCL to sum the currents, we get

$$-\tilde{I} + \tilde{I}_1 + \tilde{I}_2 + \cdots + \tilde{I}_{N-1} + \tilde{I}_N = 0$$
$$\tilde{I} = \tilde{I}_1 + \tilde{I}_2 + \cdots + \tilde{I}_{N-1} + \tilde{I}_N. \tag{7.102}$$

Now we can use Ohm's law, since we know the voltage across the impedances.

$$\tilde{I} = \frac{\tilde{V}}{Z_1} + \frac{\tilde{V}}{Z_2} + \cdots + \frac{\tilde{V}}{Z_{N-1}} + \frac{\tilde{V}}{Z_N},$$

$$\tilde{I} = \tilde{V}\left(\frac{1}{Z_1} + \frac{1}{Z_2} + \cdots + \frac{1}{Z_{N-1}} + \frac{1}{Z_N}\right). \tag{7.103}$$

Rearranging this yields

$$\frac{\tilde{I}}{\tilde{V}} = \frac{1}{Z_{\text{eq}}} = \sum_{n=1}^{N}\frac{1}{Z_n} \tag{7.104}$$

or

$$Z_{\text{eq}} = \frac{1}{\sum_{n=1}^{N}\frac{1}{Z_n}}. \tag{7.105}$$

If we replace the impedances in (7.103) with admittances, we get

$$\tilde{I} = \tilde{V}\left(Y_1 + Y_2 + \cdots + Y_{N-1} + Y_N\right). \tag{7.106}$$

This result shows that admittances connected in parallel are summed together, or

$$Y_{\text{eq}} = \sum_{n=1}^{N} Y_n. \tag{7.107}$$

Note that in Figures 7.26 and 7.27 we used a general rectangle to represent an impedance element. This is because we want to emphasize that the analysis method is the same regardless of the nature of the element. As you acquire more experience, you may stop using rectangles to replace the elements, choosing instead to directly apply the phasor method in the original circuit. However, in the beginning, we advise you to do so. This will help reinforce the idea that the time domain is different than the phasor domain.

Example 7.7.1

Find the equivalent impedance for the circuit in Figure 7.28 at port a-b. For this example, $v_s(t) = 15.3\cos(100t - 15°)\,\text{V}$. Also find the current $i_s(t)$.

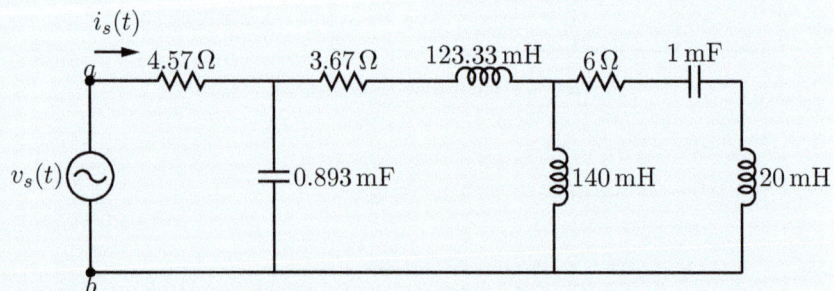

Figure 7.28: Example 7.7.1.

We will first need to transfer the circuit to the phasor domain using the radial frequency $\omega = 100\,\mathrm{rad/s}$.

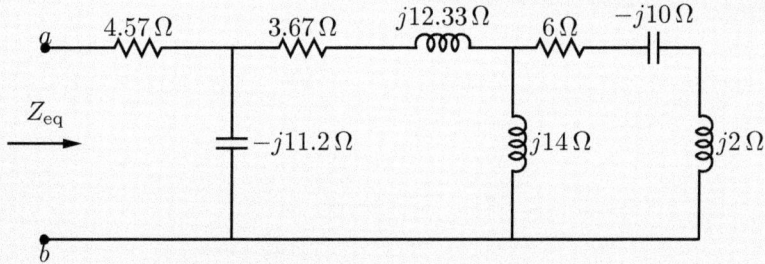

Figure 7.29: Figure 7.28 in phasor domain.

We will start from the far right of the circuit in Figure 7.29 and make our way to the left. It will be better to write the polar and rectangular forms of the impedances we find to make our solution easier.

$$Z_1 = 6 - j10 + j2 = 6 - j8\,\Omega = 10\underline{/-53.13^\circ}\,\Omega.$$

Z_1 is in parallel to the 140 mH inductor, and this is where we notice that having both forms of the complex number helps.

$$Z_2 = \frac{j14Z_1}{Z_1 + j14} = \frac{14\underline{/90^\circ}\,10\underline{/-53.13^\circ}}{6 - j8 + j14}$$

$$= \frac{140\underline{/36.87^\circ}}{6 + j6} = \frac{140\underline{/36.87^\circ}}{8.49\underline{/45^\circ}} = 16.5\underline{/-8.13^\circ}\,\Omega = 16.33 - j2.33\,\Omega.$$

Z_2 is added to the next inductor and resistor to yield

$$Z_3 = Z_2 + 3.67 + j12.33 = 16.33 - j2.33 + 3.67 + j12.33$$

$$= 20 + j10\,\Omega = 22.36\underline{/26.57^\circ}\,\Omega.$$

Z_3 is in parallel to the next capacitor. Hence,

$$Z_4 = \frac{-j11.2\,Z_3}{Z_3 - j11.24} = \frac{(11.2\underline{/-90^\circ})\,(22.36\underline{/26.57^\circ})}{20 + j10 - j11.2}$$

$$= \frac{250.43\underline{/-63.43^\circ}}{20 - j1.2} = \frac{250.43\underline{/-63.43^\circ}}{20.04\underline{/-3.43^\circ}}$$

$$= 12.5\underline{/-60^\circ}\,\Omega = 6.25 - j10.82\,\Omega.$$

Now we can find the total equivalent impedance as

$$Z_{\mathrm{eq}} = 4.57 + 6.25 - j10.82 = 10.82 - j10.82\,\Omega = 15.3\underline{/-45^\circ}\,\Omega.$$

Notice that the equivalent impedance can be written as a resistor in series with a capacitor. The value of that capacitor can be found using

$$C_{\text{eq}} = \frac{1}{\omega|\Im(Z_{\text{eq}})|} = \frac{1}{100(10.82)} = 0.924\,\text{mF}.$$

To find the current $i_s(t)$, we need to first find the current phasor:

$$\tilde{I}_s = \frac{\tilde{V}_s}{Z_{\text{eq}}} = \frac{15.3\underline{/-15^\circ}}{15.3\underline{/-45^\circ}} = 1\underline{/30^\circ}\,\text{A}.$$

The current leads the voltage, which is a verification that the equivalent impedance is capacitive.

The current $i_s(t)$ will therefore be

$$i_s(t) = 1\cos(100t + 30^\circ)\,\text{A}.$$

7.8 Exercises

Section 7.1

7.1. The voltage across an element is

$$v(t) = 50\cos(3\pi t + 20°)\,\text{V}.$$

Find the signal's:

a) magnitude V_m,

b) radial (angular) frequency ω,

c) frequency f,

d) time period T,

e) phase shift ϕ.

7.2. The current through an element is

$$i(t) = 2\cos(20\pi t - 30°)\,\text{A}.$$

Find the singnal's:

a) magnitude I_m,

b) radial (angular) frequency ω,

c) frequency f,

d) time period T,

e) phase shift ϕ.

7.3. The voltage across an element is

$$v(t) = 100\sin(\pi t + 10°)\,\text{V}.$$

Find the signal's:

a) magnitude V_m,

b) radial (angular) frequency ω,

c) frequency f,

d) time period T,

e) phase shift ϕ.

7.4. The current through an element is

$$i(t) = 12\sin(300\pi t - 60°)\,\text{A}.$$

Express the current in cosine form, then find the signal's:

a) magnitude I_m,

b) radial (angular) frequency ω,

c) frequency f,

d) time period T,

e) phase shift ϕ.

Section 7.4

7.5. Find the polar form for the following complex numbers:

a) $100 + j54$.   b) $83 - j77$.

7.6. Find the polar form for the following complex numbers:

a) $64 + j50$.   b) $47 - j100$.

7.7. Find the polar form for the following complex numbers:

a) $14 + j69$.   b) $78 - j41$.

7.8. Find the rectangular form for the following complex numbers:

a) $34\underline{/128°}$.   b) $56\underline{/-168°}$.

7.9. Find the rectangular form for the following complex numbers:

a) $90\underline{/204°}$.   b) $10\underline{/-76°}$.

7.10. Find the rectangular form for the following complex numbers:

a) $43\underline{/50°}$.   b) $87\underline{/-300°}$.

7.11. For the complex numbers

$$X = 83 - j77, \qquad Y = 56\underline{/-168°},$$

find the following:

a) $X + Y$

b) $\dfrac{X}{Y}$

c) X^* and Y^*

d) jX and jY

7.12. For the complex numbers

$$X = 64 + j50, \qquad Y = 90\underline{/-156°},$$

find the following:

  a) $X + Y$

  b) $\dfrac{X}{Y}$

  c) X^* and Y^*

  d) jX and jY

7.13. For the complex numbers

$$X = 47 - j100, \qquad Y = 10\underline{/-76°},$$

find the following:

  a) $X + Y$

  b) $\dfrac{X}{Y}$

  c) X^* and Y^*

  d) jX and jY

7.14. For the complex numbers

$$X = 14 + j69, \qquad Y = 43\underline{/50°},$$

find the following:

  a) $X + Y$

  b) $\dfrac{X}{Y}$

  c) X^* and Y^*

  d) jX and jY

Section 7.5

7.15. Find the phasor form of the following signals:

$$x(t) = -16\cos(79t + 127°).$$
$$y(t) = -16\sin(79t + 127°).$$

7.16. Find the phasor form of the following signals:

$$x(t) = 46\cos(18t - 22°).$$
$$y(t) = 46\sin(18t - 22°).$$

7.17. Find the phasor form of the following signals:

$$x(t) = 95\cos(97t - 92°).$$
$$y(t) = 95\sin(97t - 92°).$$

7.18. Find the phasor form of the following signals:

$$x(t) = 58\cos(6\pi t - 100°).$$
$$y(t) = -63\sin(6\pi t + 90°).$$

7.19. Find the phasor form of the following signals:

$$x(t) = 10\cos(2\pi t - 180°).$$
$$y(t) = -80\sin(2\pi t + 115°).$$

Section 7.5.1

7.20. The voltage and current for an element are:

$$v(t) = 50\cos(2\pi t + 30°)\,\text{V},$$
$$i(t) = 10\cos(2\pi t - 60°)\,\text{A}.$$

Find the angle between the voltage and current θ_{vi}. Does the current lead or lag the voltage?

7.21. The voltage and current for an element are:

$$v(t) = 140\cos(100\pi t - 10°)\,\text{V},$$
$$i(t) = 20\cos(100\pi t + 80°)\,\text{A}.$$

Find the angle between the voltage and current θ_{vi}. Does the current lead or lag the voltage?

7.22. The voltage and current for an element are:

$$v(t) = 250\cos(120\pi t)\,\text{V},$$
$$i(t) = 10\cos(120\pi t - 30°)\,\text{A}.$$

Find the angle between the voltage and current θ_{vi}. Does the current lead or lag the voltage?

Section 7.6.1

7.23. For the circuit in Figure 7.30, the source currents are given as:

$$i_{s_1}(t) = 21\cos(\omega\pi t + 30°)\,\text{A}.$$
$$i_{s_2}(t) = 15\sin(\omega\pi t + 20°)\,\text{A}.$$

Use KCL to find $i_o(t)$.

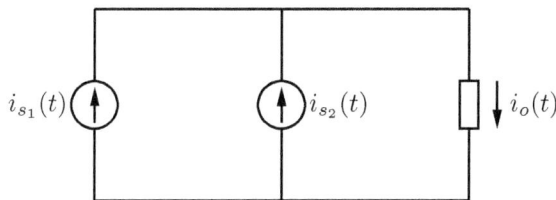

Figure 7.30

7.24. For the circuit in Figure 7.31, the source currents are given as:

$$\tilde{I}_1 = 6 + j4\,\text{A}.$$
$$\tilde{I}_2 = 7 + j4\,\text{A}.$$

Use KCL to find $\tilde{I}_s$.

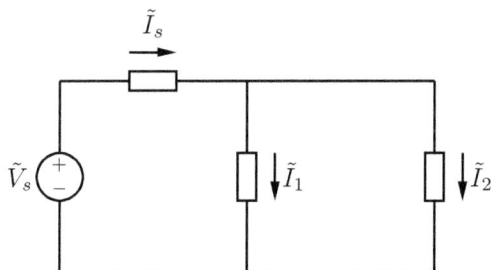

Figure 7.31

7.25. For the circuit in Figure 7.32, the source currents are given as:

$$\tilde{I}_s = 17 + j10\,\text{A}.$$
$$\tilde{I}_1 = 10 - j5\,\text{A}.$$
$$\tilde{I}_4 = 37 + j6\,\text{A}.$$

Use KCL to find $\tilde{I}_2$ and $\tilde{I}_3$.

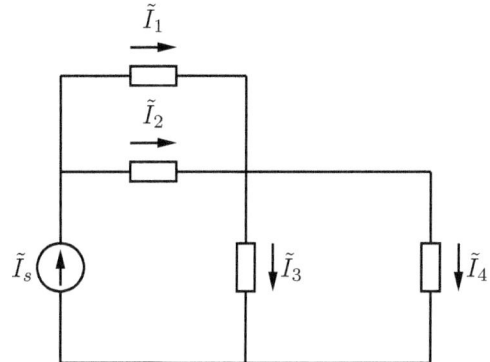

Figure 7.32

7.26. For the circuit in Figure 7.33,

a) $\tilde{I}_1 = \tilde{I}_2 = \dfrac{\tilde{I}_s}{3}$,

b) $\tilde{I}_4 = 26 - j19\,\text{A}$,

c) $\tilde{I}_5 = 19 + j2\,\text{A}$.

Use KCL to find $\tilde{I}_s$, $\tilde{I}_1$, $\tilde{I}_2$, and $\tilde{I}_3$.

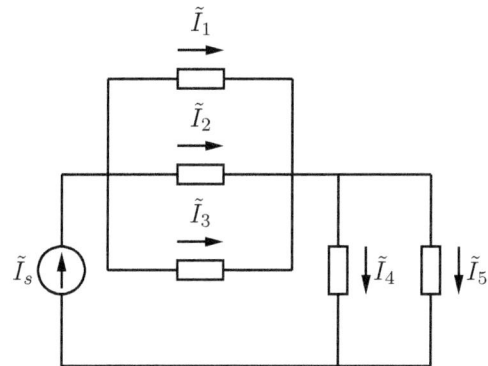

Figure 7.33

7.27. For the circuit in Figure 7.34, use KCL to find $\tilde{I}_1$ and $\tilde{I}_2$.

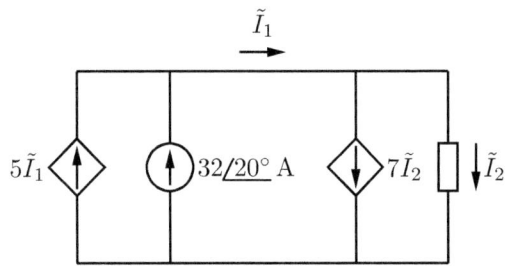

Figure 7.34

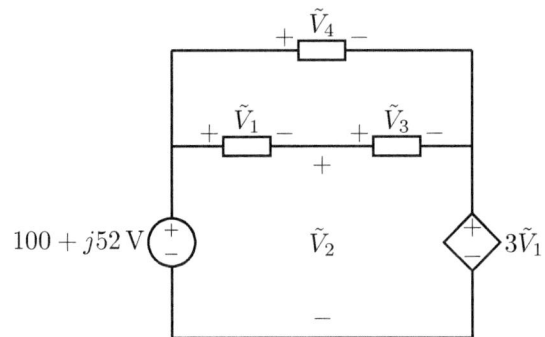

Figure 7.37

Section 7.6.2

7.28. For the circuit in Figure 7.35,

$$v_s(t) = 85\cos(\omega\pi t + 45°)\,\text{V}.$$
$$v_1(t) = 28\cos(\omega\pi t + 10°)\,\text{V}.$$

Using KVL, find the output voltage $v_o(t)$.

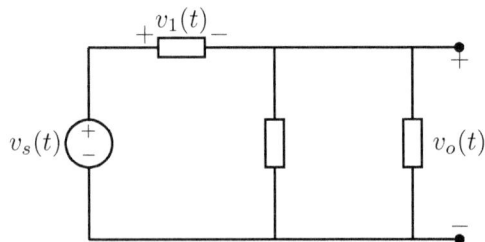

Figure 7.35

Section 7.7

7.31. The circuit in Figure 7.38 has a radial frequency of $\omega = 10\,\text{rad/s}$. Find the equivalent impedance seen by the voltage source in the phasor domain, then find the corresponding elements in the time domain.

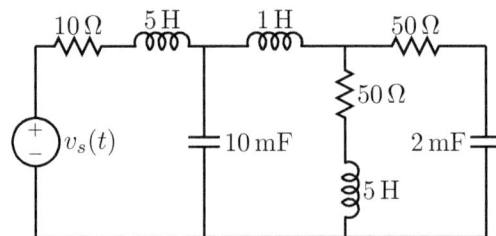

Figure 7.38

7.29. For the circuit in Figure 7.36, use KVL to find $\tilde{V}_1$ and $\tilde{V}_2$.

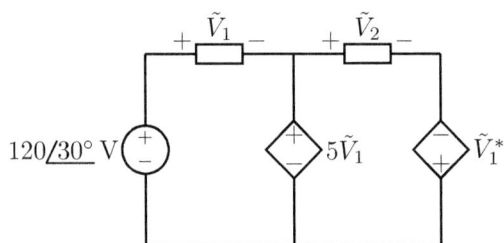

Figure 7.36

7.32. The circuit in Figure 7.39 has a radial frequency of $\omega = 50\,\text{rad/s}$. Find the equivalent impedance seen by the voltage source in the phasor domain, then find the elements in the time domain.

7.30. For the circuit in Figure 7.37, use KVL to find $\tilde{V}_2$, $\tilde{V}_3$, and $\tilde{V}_4$ if $\tilde{V}_1 = 50 - j8\,\text{V}$.

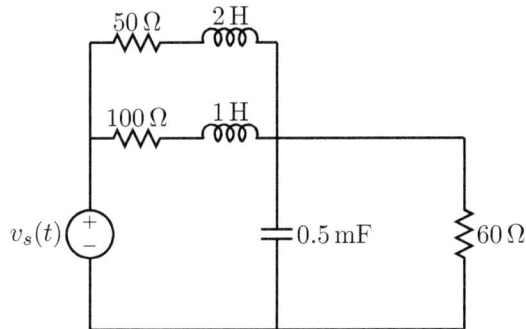

Figure 7.39

7.33. For the circuit in Figure 7.40, find the equivalent impedance at terminal a-b. Does the equivalent impedance contain a capacitor or inductor?

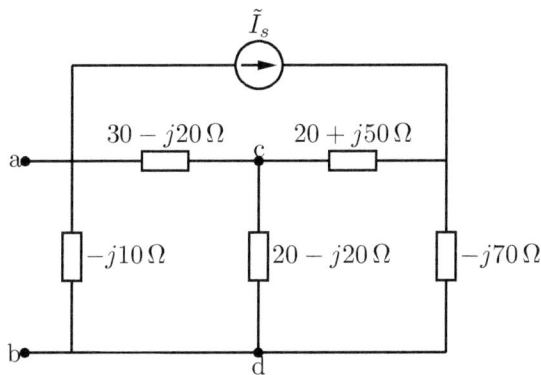

Figure 7.40

7.34. For the circuit in Figure 7.41, find the equivalent impedance at terminal c-d.

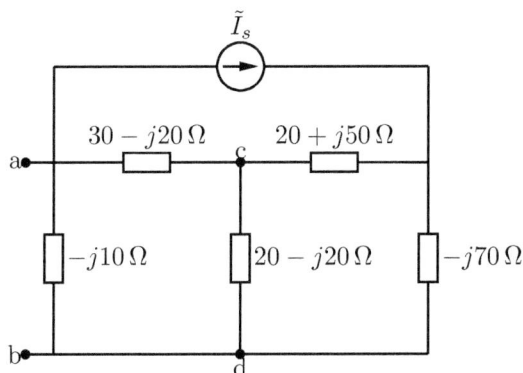

Figure 7.41

7.35. For the circuit in Figure 7.42, find the equivalent impedance at terminal a-b if $Z_1 = 100 - j100\,\Omega$, $Z_2 = 100 + j100\,\Omega$, and $Z_3 = 50 - j100\,\Omega$.

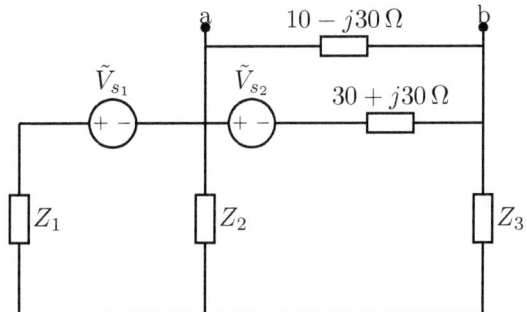

Figure 7.42

7.36. For the circuit in Figure 7.43, the load impedance is $Z_{\text{load}} = 100 - j50\,\Omega$. We would like to find the mystery element that will make the equivalent impedance seen by the source be $Z_{\text{eq}} = 100\,\Omega$. State if the element is a capacitor, a resistor, an inductor, or a combination of any of these.

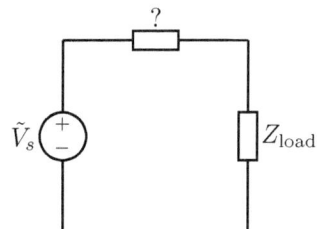

Figure 7.43

7.37. For the circuit in Figure 7.44, the load impedance is $Z_{\text{load}} = 50\,\Omega$. We would like to find the mystery element that will make the equivalent impedance seen by the source be $Z_{\text{eq}} = 50 - 30\,\Omega$. State if the element is a capacitor, a resistor, an inductor, or a combination of any of these.

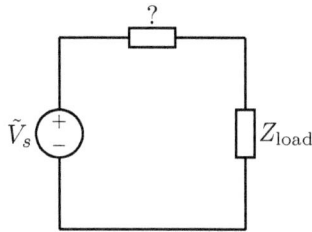

Figure 7.44

7.38. For the circuit in Figure 7.45, the load impedance is $Z_{\text{load}} = 20 - j30\,\Omega$. We would like to find the mystery element that will make the equivalent impedance seen by the source be $Z_{\text{eq}} = 32.5\,\Omega$. State if the element is a capacitor, a resistor, an inductor, or a combination of any of these.

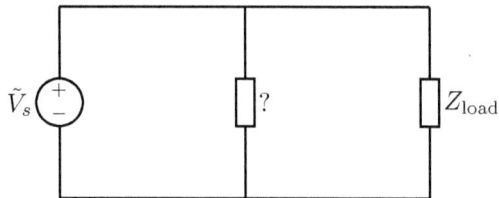

Figure 7.45

7.39. For the circuit in Figure 7.46, the load impedance is $Z_{\text{load}} = 50 + j50\,\Omega$. We would like to find the mystery element that will make the equivalent impedance seen by the source be $Z_{\text{eq}} = 50\,\Omega$. State if the element is a capacitor, a resistor, an inductor, or a combination of any of these.

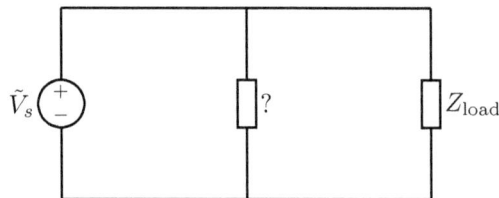

Figure 7.46

7.40. For the circuit in Figure 7.47, find the equivalent impedance seen by the independent voltage source.

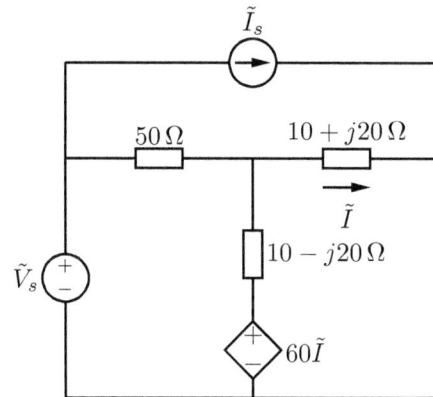

Figure 7.47

7.41. For the circuit in Figure 7.48, find the equivalent impedance at the terminal a-b.

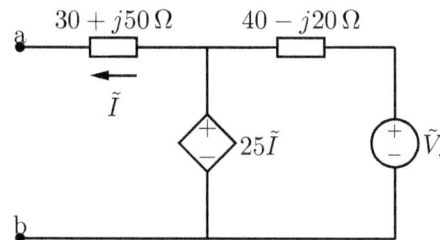

Figure 7.48

Chapter 8

Sinusoidal Steady-State Analysis

Introduction

We managed so far to find the friendliest way to look at circuits driven by sinusoidal sources. We are able to transfer the circuits to the phasor domain where resistors, capacitors, and inductors become impedances that are treated as (complex) resistors. Now it is time to look at the analysis techniques we learned in chapter 3 and to see how to apply them in analyzing our AC circuits. We will revisit, for instance, nodal and mesh analyses, linearity, superposition, and source transformation. We will also see the application of the Thévenin and Norton theorems in AC circuits. All techniques work the same way, as we have learned in DC circuits. Furthermore, we will examine some important power transfer concepts in AC circuits.

Here are some things to keep in mind before we start:

1. The circuits are always driven by sinusoidal sources only.

2. A mixture of sources at different frequencies is allowed. However, in this case, we will need to use superposition to deal with different frequencies.

> Remember: We can only add signals of different frequencies in the time domain. Phasors that represent signals at different frequencies can never be added together.

3. For this chapter, we are in the steady-state zone. This means that both the circuit itself as well as its sources do not change versus time. We assume in other words that all transient phenomena due to switching events, such as the ones we considered in first order derivatives, have died out.

8.1 Nodal and Mesh Analyses

Since Kirchhoff's laws are applicable in the phasor domain, we can use nodal and mesh analyses in the phasor domain, as they are direct applications of

Kirchhoff's laws. For both methods, we will start with transferring the circuit we have into the phasor domain; then we can perform our analysis the same way we learned for DC circuits.

8.1.1   Nodal Analysis

Nodal analysis is based on applying Kirchhoff's current law (KCL) at every essential node. Let us start with a symbolic example with more than one complication for a challenge. The circuit in Figure 8.1 is already in the phasor domain. Notice again that we use a rectangular element to draw an impedance to emphasize that the nature of the element is not important in the way we analyze the circuit. We would like to write the nodal equations for the circuit. Dummy variables in blue are added for the branches with floating voltage sources to make the solution simpler.

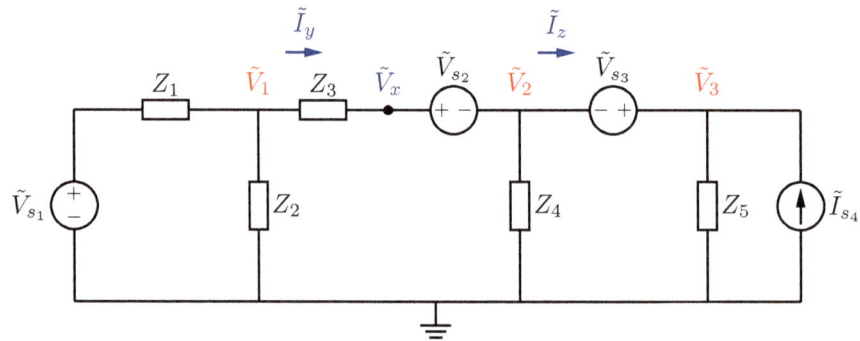

Figure 8.1: General circuit in the phasor domain.

We start with a KCL at node 1:

$$\frac{\tilde{V}_1 - \tilde{V}_{s_1}}{Z_1} + \frac{\tilde{V}_1}{Z_2} + \tilde{I}_y = 0. \tag{8.1}$$

We now need to find an equation for the dummy variable $\tilde{I}_y$. This is

$$\tilde{I}_y = \frac{\tilde{V}_1 - \tilde{V}_x}{Z_3}. \tag{8.2}$$

We also need an equation for $\tilde{V}_x$. We have

$$\tilde{V}_x - \tilde{V}_2 = \tilde{V}_{s_2},$$
$$\tilde{V}_x = \tilde{V}_{s_2} + \tilde{V}_2. \tag{8.3}$$

Plugging (8.3) into (8.2) yields

$$\tilde{I}_y = \frac{\tilde{V}_1}{Z_3} - \frac{\tilde{V}_{s_2}}{Z_3} - \frac{\tilde{V}_2}{Z_3}. \tag{8.4}$$

Plugging (8.4) into (8.1) yields

$$\frac{\tilde{V}_1 - \tilde{V}_{s_1}}{Z_1} + \frac{\tilde{V}_1}{Z_2} + \frac{\tilde{V}_1}{Z_3} - \frac{\tilde{V}_{s_2}}{Z_3} - \frac{\tilde{V}_2}{Z_3} = 0. \tag{8.5}$$

Regrouping and rearranging (8.5) to get the final form of the node 1 equation results in

$$\left(\frac{1}{Z_1} + \frac{1}{Z_2} + \frac{1}{Z_3}\right)\tilde{V}_1 - \left(\frac{1}{Z_3}\right)\tilde{V}_2 = \frac{\tilde{V}_{s_1}}{Z_1} + \frac{\tilde{V}_{s_2}}{Z_3}. \tag{8.6}$$

Now we can move on to apply KCL at node 2. This equation is

$$\frac{\tilde{V}_2}{Z_4} - \tilde{I}_y + \tilde{I}_z = 0. \tag{8.7}$$

We already have an equation for $\tilde{I}_y$, so plugging (8.4) into (8.7) and rearranging yields the final form of the node 2 equation:

$$\frac{\tilde{V}_2}{Z_4} - \frac{\tilde{V}_1}{Z_3} + \frac{\tilde{V}_{s_2}}{Z_3} + \frac{\tilde{V}_2}{Z_3} + \tilde{I}_z = 0$$

$$-\left(\frac{1}{Z_3}\right)\tilde{V}_1 + \left(\frac{1}{Z_3} + \frac{1}{Z_4}\right)\tilde{V}_2 = -\frac{\tilde{V}_{s_2}}{Z_3} - \tilde{I}_z. \tag{8.8}$$

KCL at node 3 gives

$$-\tilde{I}_z + \frac{\tilde{V}_3}{Z_5} - \tilde{I}_{s_4} = 0,$$

$$\frac{\tilde{V}_3}{Z_5} = \tilde{I}_{s_4} + \tilde{I}_z. \tag{8.9}$$

We notice that if we sum (8.8) and (8.9) together, we will be able to get rid of the remaining dummy variable $\tilde{I}_z$:

$$-\left(\frac{1}{Z_3}\right)\tilde{V}_1 + \left(\frac{1}{Z_3} + \frac{1}{Z_4}\right)\tilde{V}_2 + \left(\frac{1}{Z_5}\right)\tilde{V}_3 = \tilde{I}_{s_4} - \frac{\tilde{V}_{s_2}}{Z_3}. \tag{8.10}$$

We need an extra equation, and that equation can be obtained using the relationship between nodes 2 and 3:

$$\tilde{V}_3 - \tilde{V}_2 = \tilde{V}_{s_3}. \tag{8.11}$$

We can now solve the system of (8.6), (8.10), and (8.11) to find the nodal voltages. In matrix form, we get

$$\begin{bmatrix} \dfrac{1}{Z_1} + \dfrac{1}{Z_2} + \dfrac{1}{Z_3} & -\dfrac{1}{Z_3} & 0 \\[2ex] -\dfrac{1}{Z_3} & \dfrac{1}{Z_3} + \dfrac{1}{Z_4} & \dfrac{1}{Z_5} \\[2ex] 0 & -1 & 1 \end{bmatrix} \begin{bmatrix} \tilde{V}_1 \\[2ex] \tilde{V}_2 \\[2ex] \tilde{V}_3 \end{bmatrix} = \begin{bmatrix} \dfrac{\tilde{V}_{s_1}}{Z_1} + \dfrac{\tilde{V}_{s_2}}{Z_3} \\[2ex] \tilde{I}_{s_4} - \dfrac{\tilde{V}_{s_2}}{Z_3} \\[2ex] \tilde{V}_{s_3} \end{bmatrix}$$

We observe here that nodal analysis is applied exactly the same way as if we had a resistive circuit. Of course there are some differences in the final matrix: a) our equations contain impedances which are, in general, complex numbers, and b) the voltages are phasors. After we find the nodal voltages in the phasor domain, we can transfer the voltages back to the time domain as we have seen before.

Let us now look at a numerical example to solidify this idea.

Example 8.1.1

For the circuit shown in Figure 8.2, $v_{s_1}(t) = 50\cos(50t + 36.87°)\,\mathrm{V}$ and $i_{s_2}(t) = 2\sin(50t)\,\mathrm{A}$. Find the nodal voltages $v_1(t)$ and $v_2(t)$ and the current $i(t)$.

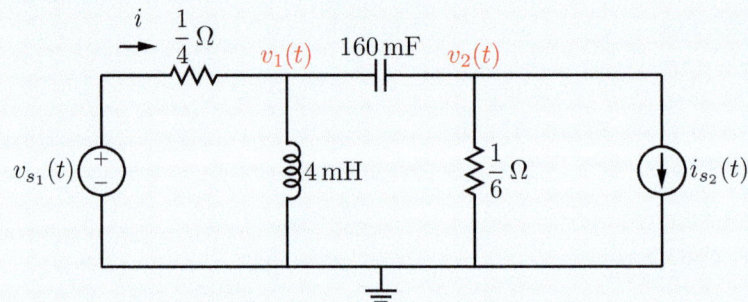

Figure 8.2: The circuit for Example 8.1.1 shown in the time domain.

The first thing we need to do is to transfer the circuit in Figure 8.2 into the phasor domain. We notice that both sources are at the same frequency. Hence we can solve the circuit by considering both sources simultaneously. To find the phasor of the current source, we need to express it as a cosine by using the trigonometric identity

$$y(t) = B\sin(\omega t + \beta°) = B\cos(\omega t + \beta° - 90°).$$

Thus, we can write the current source as

$$i(t) = 2\sin(50t) = 2\cos(50t - 90°)\,\mathrm{A}.$$

Now we can transfer the circuit to the phasor domain, as seen in Figure 8.3.

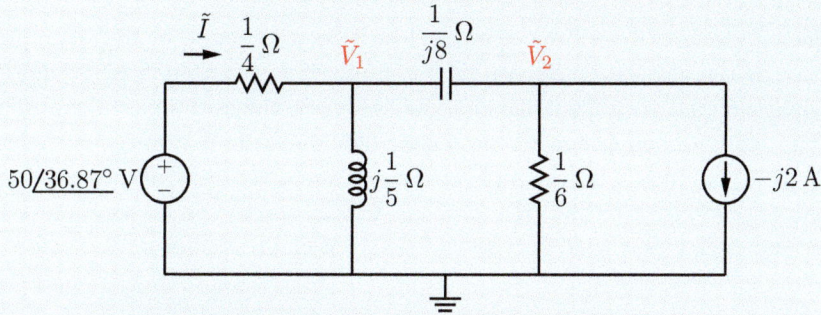

Figure 8.3: The circuit in Figure 8.2 in the phasor domain.

The node 1 equation, after rearranging the terms, is

$$\left(\frac{1}{1/4} + \frac{1}{j/5} + \frac{1}{1/j8}\right)\tilde{V}_1 - \left(\frac{1}{1/j8}\right)\tilde{V}_2 = \frac{50\underline{/36.87°}}{1/4},$$

$$(4 - j5 + j8)\,\tilde{V}_1 - (j8)\,\tilde{V}_2 = 200\underline{/36.87°},$$

$$(4 + j3)\,\tilde{V}_1 - (j8)\,\tilde{V}_2 = 200\underline{/36.87°}.$$

Similarly, the node 2 equation becomes

$$-\left(\frac{1}{1/j8}\right)\tilde{V}_1 + \left(\frac{1}{1/j8} + \frac{1}{1/6}\right)\tilde{V}_2 = j2,$$

$$-(j8)\,\tilde{V}_1 + (6 + j8)\,\tilde{V}_2 = j2$$

In matrix form, we have

$$\begin{bmatrix} 4 + j3 & -j8 \\ -j8 & 6 + j8 \end{bmatrix} \begin{bmatrix} \tilde{V}_1 \\ \tilde{V}_2 \end{bmatrix} = \begin{bmatrix} 200\underline{/36.87°} \\ j2 \end{bmatrix}.$$

Solving the system of equations, we get

$$\begin{bmatrix} \tilde{V}_1 \\ \tilde{V}_2 \end{bmatrix} = \begin{bmatrix} 15 + j19.53 \\ 0.4 + j19.82 \end{bmatrix} = \begin{bmatrix} 24.63\underline{/52.5°} \\ 19.82\underline{/88.84°} \end{bmatrix} \text{V}.$$

The needed current is given (in the phasor domain) by

$$\tilde{I} = \frac{\tilde{V}_{s_1} - \tilde{V}_1}{1/4} = 4(40 + j30 - 15 - j19.53)$$

$$= 100 + j41.88\,\text{A} = 108.4\underline{/22.72°}\,\text{A}.$$

Now we can transfer everything into the time domain:

$$v_1(t) = 24.63\cos(50t + 52.5°)\,\text{V},$$

$$v_2(t) = 19.82\cos(50t + 88.84°)\,\text{V},$$

$$i(t) = 108.4\cos(50t + 22.72°)\,\text{A}.$$

8.1.2   Mesh Analysis

Mesh analysis utilizes KVL around closed loops (meshes) to find the currents through these loops. We have to remember that mesh analysis is limited to planar circuits. Just like nodal analysis, we follow the next steps: a) transfer our sinusoidally driven circuit to the phasor domain, b) find all mesh currents and any other unknowns in the phasor domain, and c) transfer all needed currents and voltages back to the time domain. Let us start with an example using a specific circuit that is already in the phasor domain as shown in Figure 8.4.

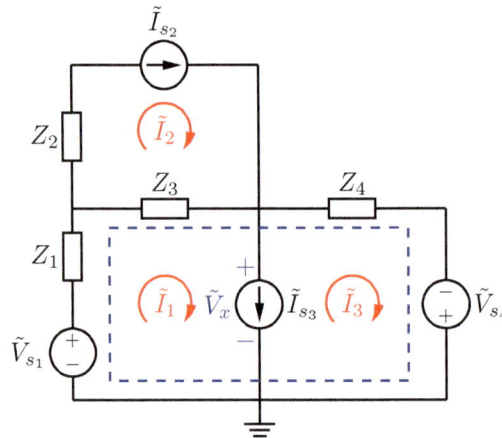

Figure 8.4: A circuit in the phasor domain with all mesh currents shown.

We can identify three mesh currents in Figure 8.4. As we have learned before, we always choose to work with the simple loops shown in red. In this case, we will end up with the current source $\tilde{I}_{s_3}$ being in both loop 1 and loop 3. One way to deal with it is to write equations for all three simple meshes and use the dummy variable $\tilde{V}_x$ to express the voltage drop across the $\tilde{I}_{s_3}$ source. Alternatively, we can choose to write one of the equations for the bigger loop shown in blue for our third equation. Choosing one of these methods is a matter of personal preference. Obviously, both techniques yield the same result. Here we choose to work with the blue loop.

Let us now write the mesh equations. The equation for loop 2 is

$$\tilde{I}_2 = \tilde{I}_{s_2}. \tag{8.12}$$

KVL around the blue loop yields

$$-\tilde{V}_{s_1} + Z_1\tilde{I}_1 + Z_3(\tilde{I}_1 - \tilde{I}_2) + Z_4\tilde{I}_3 - \tilde{V}_{s_4} = 0. \tag{8.13}$$

Rearranging (8.13) yields

$$(Z_1 + Z_3)\tilde{I}_1 - (Z_3)\tilde{I}_2 + (Z_4)\tilde{I}_3 = \tilde{V}_{s_1} + \tilde{V}_{s_4}. \tag{8.14}$$

We get the third equation by relating mesh currents $\tilde{I}_1$ and $\tilde{I}_3$ to each other:

$$\tilde{I}_1 - \tilde{I}_3 = \tilde{I}_{s_3}. \tag{8.15}$$

In matrix form, the mesh equations are written as

$$\begin{bmatrix} Z_1 + Z_3 & -Z_3 & Z_4 \\ 0 & 1 & 0 \\ 1 & 0 & -1 \end{bmatrix} \begin{bmatrix} \tilde{I}_1 \\ \tilde{I}_2 \\ \tilde{I}_3 \end{bmatrix} = \begin{bmatrix} \tilde{V}_{s_1} + \tilde{V}_{s_4} \\ \tilde{I}_{s_2} \\ \tilde{I}_{s_3}. \end{bmatrix}$$

Let us now work with a specific numerical example.

Example 8.1.2

For the circuit shown in Figure 8.5, $v_{s_1}(t) = 3\cos(100t)\,\text{V}$, $i_{s_2}(t) = 2\cos(100t + 30°)\,\text{A}$, $i_{s_3}(t) = \cos(100t + 60°)\,\text{A}$, and $v_{s_4}(t) = 4\cos(100t + 90°)$. Find the voltage drop across the inductor $v_L(t)$ and the voltage drop across the capacitor $v_C(t)$.

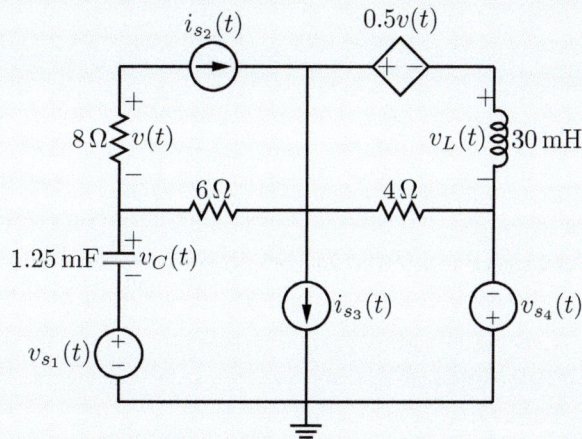

Figure 8.5: The circuit for Example 8.1.2 shown in the time domain.

The first thing we need to do is transfer the circuit in Figure 8.5 into the phasor domain. The transformation is shown in Figure 8.6. We will have to find the mesh currents first, then move on to find the voltages.

We will choose the blue loop path again because we will deal with fewer unknowns. KVL around it yields

$$-3\underline{/0°} - j8\tilde{I}_1 + 6(\tilde{I}_1 - \tilde{I}_2) + 4(\tilde{I}_3 - \tilde{I}_4) - 4\underline{/90°} = 0.$$

Rearranging gives

$$(6 - j8)\tilde{I}_1 - 6\tilde{I}_2 + 4\tilde{I}_3 - 4\tilde{I}_4 = 3\underline{/0°} + 4\underline{/90°} = 3 + j4.$$

Now we need the equation that relates $\tilde{I}_1$ and $\tilde{I}_3$ to the source current $\tilde{I}_{s_3}$. This is

$$\tilde{I}_1 - \tilde{I}_3 = \tilde{I}_{s_3} = 1\underline{/60°}\,\text{A}.$$

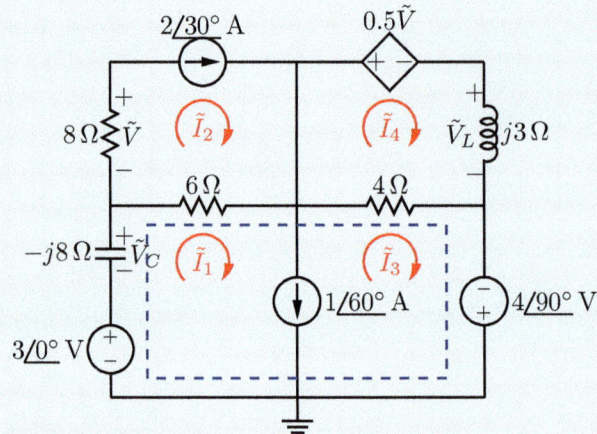

Figure 8.6: The circuit in Figure 8.5 in the phasor domain.

The mesh 2 equation is

$$\tilde{I}_2 = 2\underline{/30°} \text{ A}.$$

The mesh 4 equation is

$$0.5\tilde{V} + j3\tilde{I}_4 + 4(\tilde{I}_4 - \tilde{I}_3) = 0.$$

We can rearrange the mesh 4 equation as

$$-4\tilde{I}_3 + (4 + j3)\tilde{I}_4 = -0.5\tilde{V}.$$

We can find the voltage $\tilde{V}$ as

$$\tilde{V} = -8\tilde{I}_2 = -8(2\underline{/30°}) = -16\underline{/30°} \text{ V}.$$

We can plug the value of $\tilde{V}$ back into the mesh 4 equation to get

$$-4\tilde{I}_3 + (4 + j3)\tilde{I}_4 = 8\underline{/30°}.$$

Rearranging them in matrix form, we get

$$
\begin{bmatrix}
6 - j8 & -6 & 4 & -4 \\
0 & 1 & 0 & 0 \\
1 & 0 & -1 & 0 \\
0 & 0 & -4 & 4 + j3
\end{bmatrix}
\begin{bmatrix}
\tilde{I}_1 \\
\tilde{I}_2 \\
\tilde{I}_3 \\
\tilde{I}_4
\end{bmatrix}
=
\begin{bmatrix}
3 + j4 \\
2\underline{/30°} \\
1\underline{/60°} \\
8\underline{/30°}.
\end{bmatrix}
$$

Solving the system of equations yields

$$\begin{bmatrix} \tilde{I}_1 \\[2ex] \tilde{I}_2 \\[2ex] \tilde{I}_3 \\[2ex] \tilde{I}_4 \end{bmatrix} = \begin{bmatrix} 2.29\underline{/70.576°} \\[2ex] 2\underline{/30°} \\[2ex] 1.32\underline{/78.57°} \\[2ex] 2.43\underline{/12.14°} \end{bmatrix} \text{ A.}$$

The voltage drop across the inductor can be calculated using

$$\tilde{V}_L = Z_L \tilde{I}_4 = j3(2.43\underline{/12.14°}) = 7.29\underline{/102.14°} \text{ V.}$$

The voltage across the capacitor is calculated in the same manner as

$$\tilde{V}_C = -Z_C \tilde{I}_1 = -(-j8)(2.29\underline{/70.576°}) = 18.32\underline{/160.576°} \text{ V.}$$

We can now move back to the time domain to find the voltages:

$$v_L(t) = 7.29\cos(100t + 102.14°)\,\text{V},$$
$$v_C(t) = 18.32\cos(100t + 160.576°)\,\text{V}.$$

8.2 Superposition

Although our circuits in this chapter are driven by sinusoidal sources, our elements are still linear. As a result, we can still apply superposition and source transformation to our AC circuits. Superposition is particularly important in SSS circuits because it allows us to analyze circuits driven by sources at different frequencies. This is best demonstrated with an example circuit. Figure 8.7 shows a circuit that is driven by three different sources. Two sources are sinusoidal sources that oscillate at two different frequencies, while the third one is a DC source (zero frequency). The sinusoidal sources in the circuit have the form

$$v_{s_1}(t) = V_{m_1}\cos(\omega_1 t + \theta_1)$$

and

$$i_{s_2}(t) = I_{m_2}\cos(\omega_2 t + \theta_2).$$

We would like to find the current $i(t)$ in the circuit. In order to transfer our circuit to the phasor domain, we have to satisfy the condition that all sources *must* have the same frequency. Clearly, our circuit in Figure 8.7 does not qualify. This is where we can use superposition: we can solve our circuit by considering each source separately, analyzing each case separately, and finally adding the partial results. To be more specific, let us briefly review how to apply superposition:

1. Turn off all sources except for one. As we know, voltage sources become short circuits when they are turned off. On the other hand, current sources become open circuits when they are turned off.

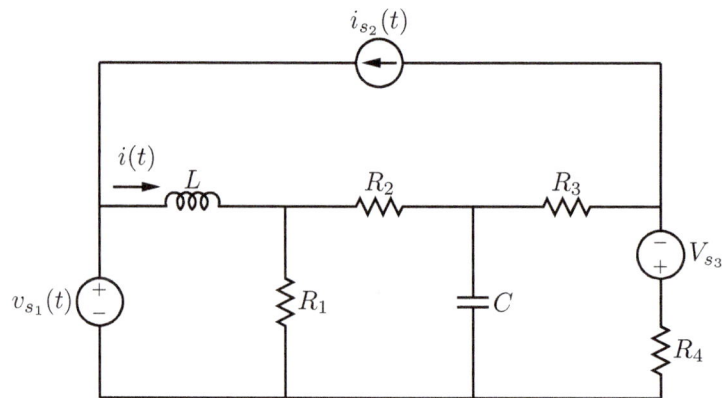

Figure 8.7: An example circuit excited by multiple sources.

2. Analyze the circuit with only one active source at a time and find the desired unknowns—for example, $i_x(t)$.

3. Repeat for as many sources as needed.

4. Find the final result by adding the partial results in the time domain. For example, $i(t) = i_x(t) + i_y(t) + \ldots$.

Let us apply this process to our circuit in Figure 8.7. We have three sources, so we will have three different partial circuits. We will start by keeping $v_{s_1}(t)$ active. All other sources are turned off, as seen in Figure 8.8. The same figure shows the circuit in the phasor domain as well.

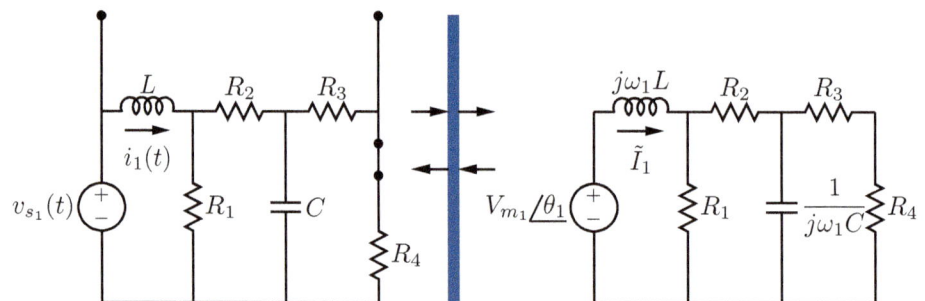

Figure 8.8: The circuit from Figure 8.7 with only v_{s_1} turned on shown in both the time and phasor domains.

We can see from the phasor domain circuit in Figure 8.8 that we can easily calculate the current $\tilde{I}_1$ as

$$\tilde{I}_1 = \frac{\tilde{V}_{s_1}}{Z_{\mathrm{eq}_1}} = \frac{V_{m_1}\angle\theta_1}{Z_{\mathrm{eq}_1}},$$

where $Z_{\mathrm{eq}S_1}$ can be found by a series-parallel combination of the impedances as follows:

$$Z_{\mathrm{eq}_1} = j\omega_1 L + [R_1 \parallel (R_2 + (\frac{1}{j\omega_1 C} \parallel (R_3 + R_4)))] = |Z_{\mathrm{eq}_1}|\angle\theta_{\mathrm{eq}_1}.$$

We would particularly like to notice here that the impedances $j\omega_1 L$ and $\frac{1}{j\omega_1 C}$ are functions of frequency. Thus, every time you change the frequency, these impedances change as well. We can now transfer the current to the time domain $i_1(t)$ as

$$i_1(t) = \frac{V_{m_1}}{|Z_{\text{eq}_1}|} \cos(\omega_1 t + (\theta_1 - \theta_{\text{eq}_1})) = I_1 \cos(\omega_1 t + \theta_{i_1}).$$

Let us now consider the second source $i_{s_2}(t)$ by itself. We have redrawn the circuit in the time and phasor domains as shown in Figure 8.9.

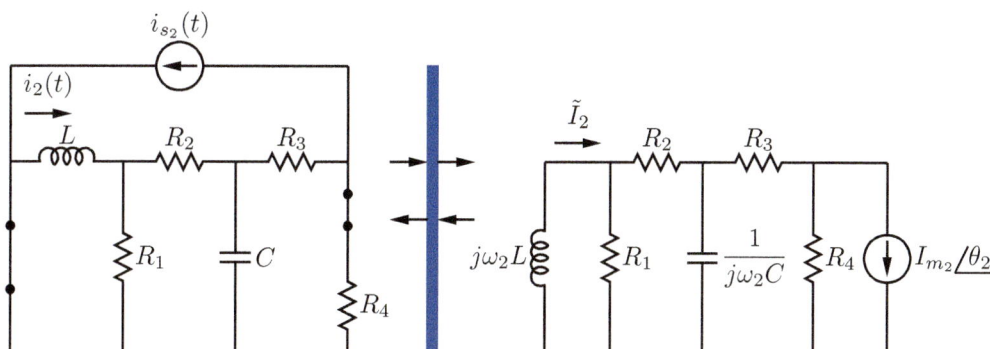

Figure 8.9: The circuit from Figure 8.7 with only i_{s_2} turned on in both the time and phasor domains.

We can proceed with source transformation in the phasor domain starting from the far right of the circuit. We will end up with a simpler circuit, as seen in Figure 8.10. Using current division, we can find $\tilde{I}_2$ as

$$\tilde{I}_2 = -\frac{Z_{\text{eq}_2}}{j\omega_2 L + Z_{\text{eq}_2}} I_{\text{eq}_2} \underline{/\theta_{i_{\text{eq}2}}} = I_2 \underline{/\theta_{i_2}}.$$

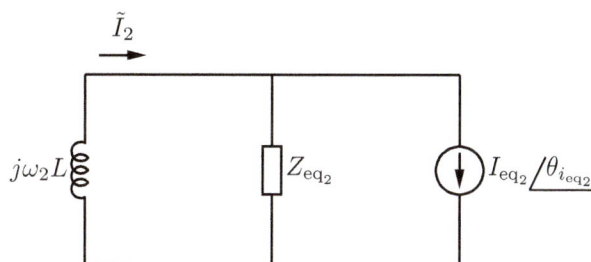

Figure 8.10: The phasor domain circuit of Figure 8.9.

We will revisit this circuit in the source transformation section to see the procedure in detail. Transferring the current back to the time domain yields

$$i_2(t) = I_2 \cos(\omega_2 t + \theta_{i_2}).$$

The last source is a DC source. A DC source is equivalent to a situation where our frequency is zero. Hence, our capacitor and inductor will be treated as open and short circuits, respectively. Let us look at this closely. For the inductor, we have

$$L \to Z_L = j\omega L = j(0)L = 0 \to \text{short circuit.} \qquad (8.16)$$

Similarly,

$$C \to Z_C = \frac{1}{j\omega C} = \frac{1}{j(0)C} = \infty \to \text{open circuit.} \qquad (8.17)$$

Of course, with zero frequency we are dealing with a DC circuit (no phasor domain treatment is needed). The circuit is shown in Figure 8.11. The resistor R_1 is shorted, and we can use

$$I_3 = \frac{V_{s_3}}{R_2 + R_3 + R_4}.$$

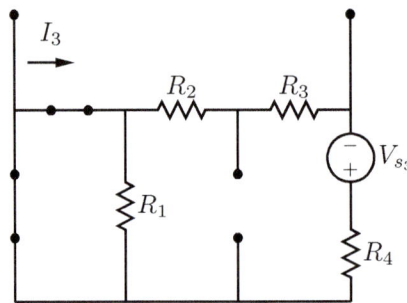

Figure 8.11: The circuit from Figure 8.7 with only V_{s_3} turned on and L, C drawn in DC conditions.

Using superposition, the current $i(t)$ in Figure 8.7 can be calculated as

$$i(t) = i_1(t) + i_2(t) + I_3$$
$$= I_1 \cos(\omega_1 t + \theta_{i_1}) + I_2 \cos(\omega_2 t + \theta_{i_2}) + I_3.$$

In closing this section, we would like to re-emphasize our main point: adding phasors referenced to different frequencies is *not* allowed. If a circuit is excited by sources at different frequencies, each frequency *must* be considered separately. Partial results can only be added in the time domain.

Example 8.2.1

For the circuit shown in Figure 8.12, $v_{s_1}(t) = 100\cos(10t + 40°) + 25\,\text{V}$ and $v_{s_2}(t) = 50\cos(20t + 30°)\,\text{V}$. Find the voltage drop across the 24 Ω resistor $v(t)$.

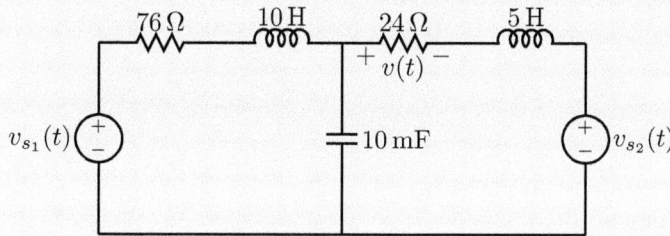

Figure 8.12: The circuit for Example 8.2.1 shown in the time domain.

Since both sources oscillate at different frequencies and since $v_{s_1}(t)$ has a DC component, we will have to use superposition in order to solve the circuit in the frequency domain. We will need to split our circuit into three different circuits. Let us start with the DC portion of the circuit. Figure 8.13 shows the DC circuit.

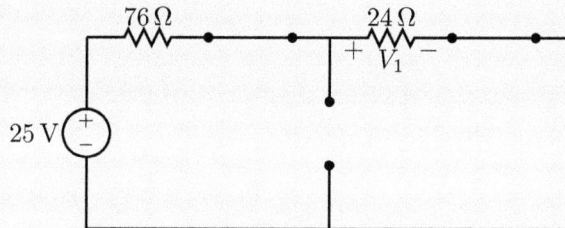

Figure 8.13: The circuit for Example 8.2.1 with only the DC component of $v_{s_1}(t)$.

Remember that in DC, inductors are short circuits and capacitors are open circuits. We can use voltage division to find V_1 as

$$V_1 = \frac{24}{76 + 24} 25 = 6 \, \text{V}.$$

Now we can analyze the circuit with the sinusoidal portion of $v_{s_1}(t)$.

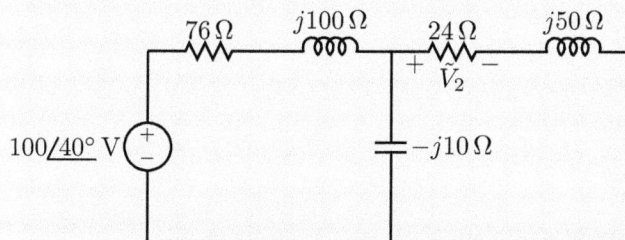

Figure 8.14: The circuit for Example 8.2.1 with only the sinusoidal portion of $v_{s_1}(t)$.

We can use source transformation and then use current division to find the current through the 24 Ω resistor. Since we technically have not covered source transformation yet, we will use voltage division in two stages. We will first use voltage division between the $76 + j100 \, \Omega$ impedance and the parallel combination of $-j10 \, \Omega$ and $24 + j50 \, \Omega$. This is

$$Z_1 = (-j10) \parallel (24 + j50) = 11.89\underline{/-84.68^\circ}\,\Omega,$$

$$\tilde{V}_{Z_1} = \frac{Z_1}{76 + j100 + Z_1}100\underline{/40^\circ} = 10.15\underline{/-93.51^\circ}\,\text{V}.$$

We can use voltage division again to find the desired voltage. Hence,

$$\tilde{V}_2 = \frac{24}{24 + j50}\tilde{V}_{Z_1} = \frac{24}{24 + j50}10.15\underline{/-93.51^\circ} = 4.39\underline{/-157.87^\circ}\,\text{V}.$$

We need to transfer our phasor back to the time domain. Thus we have

$$v_2(t) = 4.39\cos(10t - 157.87^\circ)\,\text{V}.$$

Once you study source transformation, come back to this circuit and solve for $\tilde{V}_2$ using source transformation. It is good to practice multiple ways of approaching a solution.

Let us now move on to the last circuit with only $v_{s_2}(t)$ active.

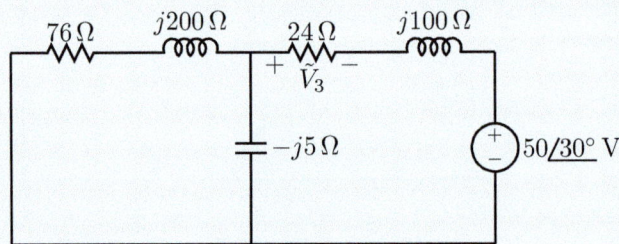

Figure 8.15: The circuit for Example 8.2.1 with $v_{s_2}(t)$.

We will use voltage division again, but first we will find the parallel combination of $-j5\,\Omega$ and $76 + j200\,\Omega$. This is

$$Z_1 = (-j5) \parallel (76 + j200) = 5.11\underline{/-89.51^\circ}\,\Omega = 0.0434 - j5.11\,\Omega.$$

Now we can use voltage division to find the desired voltage. We have

$$\tilde{V}_3 = \frac{-24}{24 + j100 + 0.0434 - j5.11}50\underline{/30^\circ} = 12.26\underline{/134.22^\circ}\,\text{V}.$$

We need to transfer our phasor back to the time domain as

$$v_3(t) = 12.26\cos(20t + 134.22^\circ)\,\text{V}.$$

The total voltage can now be found as

$$v(t) = V_1 + v_2(t) + v_3(t),$$

$$v(t) = 4.39\cos(10t - 157.86^\circ) + 12.26\cos(20t + 134.22^\circ) + 6\,\text{V}.$$

8.3 Source Transformation

Source transformation can easily be applied to our circuits in the phasor domain. Since all elements become impedances, we can use the same techniques we used in DC circuits. Figure 8.16 demonstrates the application of source transformation in the phasor domain. The figure shows the transformation of a voltage source connected in series with an impedance to a current source connected in parallel to the same impedance. Remember that the impedance Z can be a resistor, capacitor, inductor, or a combination of them.

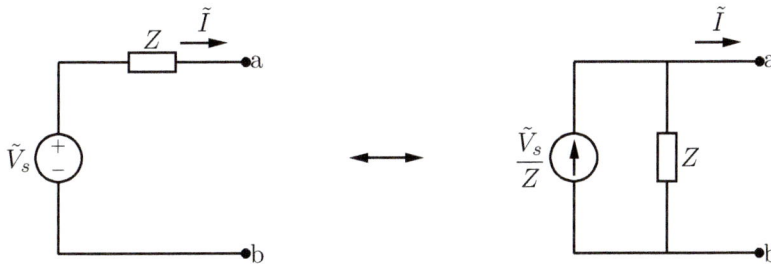

Figure 8.16: Source transformation applied in the phasor domain: voltage source to current source.

Of course here we will be dealing with complex numbers, which will require us to be a little bit more careful. Obviously, we can also use source transformation when we have a current source in parallel with an impedance, as seen in Figure 8.17. As discussed when solving DC circuits, it is important to pay attention to the polarity of sources to ensure you apply this theorem correctly. As an example, let us revisit the circuit in Figure 8.9 and give it some numbers.

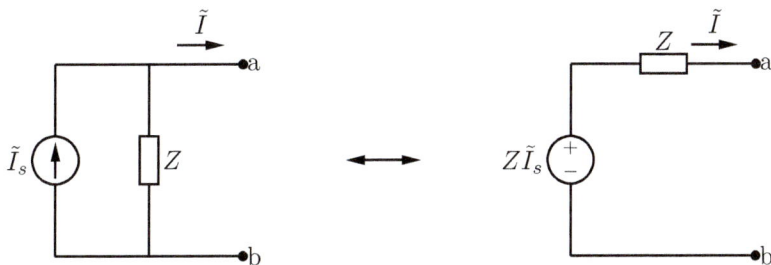

Figure 8.17: Source transformation applied in the phasor domain: current source to voltage source.

Example 8.3.1

For the circuit in Figure 8.18, find $\tilde{I}$.

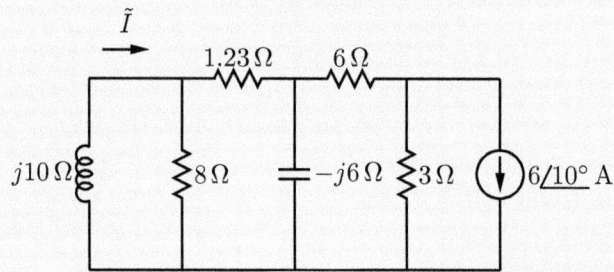

Figure 8.18: The circuit for Example 8.3.1 in the phasor domain.

We can use source transformation from the far right of the circuit and work our way to the unknown current.

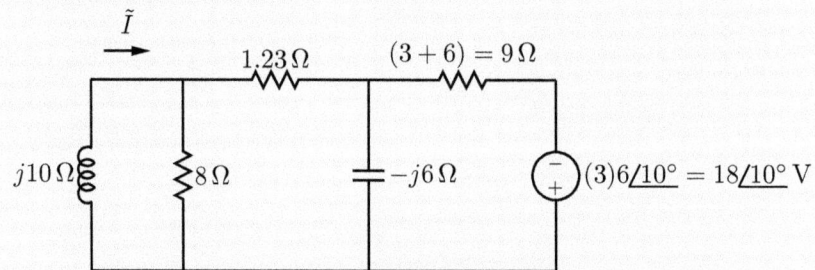

Figure 8.19: Applying source transformation to the circuit in Figure 8.18.

We can perform another source transformation, as seen in Figure 8.20.

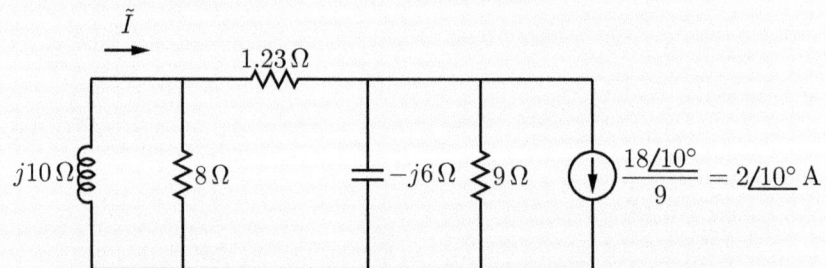

Figure 8.20: Applying source transformation to the circuit in Figure 8.18.

We will need to find the parallel combination of $-j6\,\Omega$ and $9\,\Omega$ as

$$Z_{eq_1} = \frac{9(-j6)}{9-j6} = 5\underline{/-56.3°}\,\Omega = 2.77 - j4.2\,\Omega.$$

Now we can go back to using source transformation, as seen in Figure 8.21.

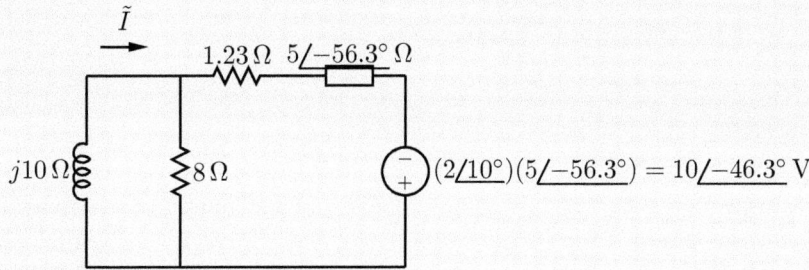

Figure 8.21: Applying source transformation to the circuit in Figure 8.18.

We can add Z_{eq_1} to the $1.33\,\Omega$ resistor and then do another transformation

$$Z_{eq_2} = 1.23 + 2.77 - j4.2 = 4 - j4.2\,\Omega = 5.8\underline{/-46.4^\circ}\,\Omega.$$

We can perform one more source transformation, as seen in Figure 8.22.

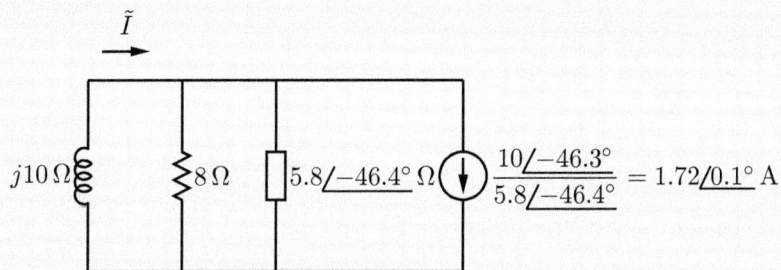

Figure 8.22: Applying source transformation to the circuit in Figure 8.18.

We have two choices at this point: the first is to apply current division between multiple impedances in parallel; the second is to keep using source transformation until the end. We will use the current division formula here, and you can do the source transformation option as an exercise. The general current division equation between N impedances connected in parallel is

$$\tilde{I}_1 = \frac{\frac{1}{Z_1}}{\frac{1}{Z_1} + \frac{1}{Z_2} + \ldots + \frac{1}{Z_N}}\tilde{I}_s.$$

Hence, the current $\tilde{I}$ can be calculated as

$$\tilde{I} = \frac{\frac{1}{j10}}{\frac{1}{j10} + \frac{1}{8} + \frac{1}{5.8\underline{/-46.4^\circ}}}(1.72\underline{/0.1^\circ}) = 0.7\underline{/-95.7^\circ}\,\text{A}.$$

8.4   Thévenin and Norton Equivalent Circuits in the Phasor Domain

Just like superposition and source transformation, the Thévenin and Norton theorems are also applicable in the phasor domain. We will use the same techniques we learned in section 4.6; the only difference is that we will be dealing with impedances instead of resistances. We can use source transformation on the Thévenin equivalent circuit to find the Norton equivalent circuit and vice versa, as seen in Figure 8.23.

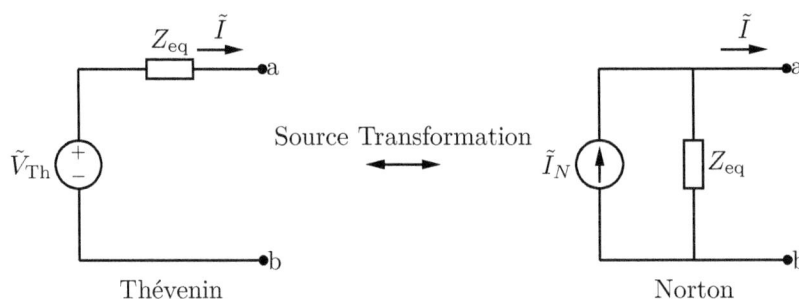

Figure 8.23: Thévenin and Norton equivalent circuits via the source transformation technique.

To find the Thévenin or Norton equivalent for an AC circuit, we will have to first transfer the circuit to the phasor domain. Once we have the circuit in the phasor domain, we can follow the same steps in section 4.6. To find the Thévenin or Norton equivalent circuits, we need to find two of the following three quantities:

- The equivalent impedance Z_{eq} at the desired port.

- The open-circuit voltage ($\tilde{V}_{Th}$) across the desired port.

- The short-circuit current ($\tilde{I}_N$) through the desired port.

It does not matter what we are asked to find; we should always evaluate the circuit and find out the two easiest values to find because

- If you have Z_{eq} and $\tilde{V}_{Th}$, then

$$\tilde{I}_N = \frac{\tilde{V}_{Th}}{Z_{eq}}.$$

- If you have Z_{eq} and $\tilde{I}_N$, then

$$\tilde{V}_{Th} = \tilde{I}_N Z_{eq}.$$

- If you have $\tilde{I}_N$ and $\tilde{V}_{Th}$, then

$$Z_{eq} = \frac{\tilde{V}_{Th}}{\tilde{I}_N}.$$

A summary of how to find any of the above is shown below.

- Equivalent impedance (Z_{eq}): Turn off all independent sources (dependent sources remain unchanged) and calculate the resulting impedance at the desired port. Notice that you may have to apply the $i-v$ test if impedances cannot be combined through series and parallel connections, or if the circuit includes dependent sources. In several books and papers, Z_{eq} is also called Thévenin equivalent impedance Z_{Th} or Norton equivalent impedance (Z_N). In other words, $Z_{eq} = Z_{Th} = Z_N$.

- Open-circuit (Thévenin) voltage ($\tilde{V}_{Th}$): Leave the desired port open-circuited (i.e., no load connected) and find the voltage across it.

- Short-circuit (Norton) current ($\tilde{I}_N$): Short-circuit the desired port (i.e., connect a short circuit across the port) and find the current through it.

Example 8.4.1

For the circuit in Figure 8.24, find the Thévenin equivalent circuit at port a-b.

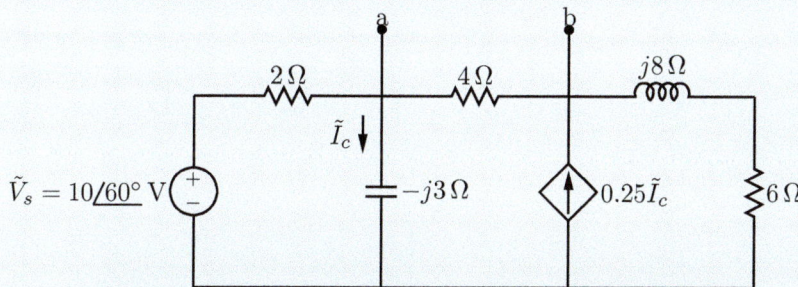

Figure 8.24: The circuit for Example 8.4.1 in the phasor domain.

Since we have a dependent source, we know that we will have to use the $i-v$ test method if we want to find the equivalent impedance. We turn off all independent sources and apply a test source at the port. We will use a current test source with $1\,\text{A}$ value, as seen in Figure 8.25.

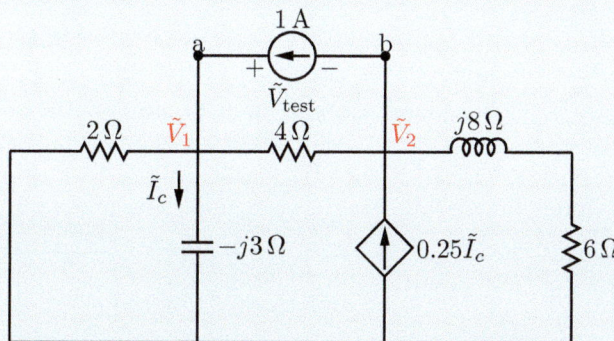

Figure 8.25: Applying the $i-v$ test method to the circuit in Figure 8.24.

We can use nodal analysis to find the nodal voltages. The final matrix is going to be

$$
\begin{bmatrix}
\dfrac{1}{2} - \dfrac{1}{j3} + \dfrac{1}{4} & -\dfrac{1}{4} \\[2ex]
-\dfrac{1}{4} + \dfrac{0.25}{j3} & \dfrac{1}{4} + \dfrac{1}{6 + j8}
\end{bmatrix}
\begin{bmatrix}
\tilde{V}_1 \\[2ex]
\tilde{V}_2
\end{bmatrix}
=
\begin{bmatrix}
1 \\[2ex]
-1
\end{bmatrix}.
$$

Solving the system of equations, we get

$$
\begin{bmatrix}
\tilde{V}_1 \\[2ex]
\tilde{V}_2
\end{bmatrix}
=
\begin{bmatrix}
0.26 - j0.44 \\[2ex]
-2.65 - j0.97
\end{bmatrix}
\text{V.}
$$

Now we can find Z_{eq} by

$$
Z_{\text{eq}} = \frac{\tilde{V}_{\text{test}}}{1} = \frac{\tilde{V}_1 - \tilde{V}_2}{1} = 2.9 + j0.53\,\Omega = 2.95\underline{/10.36^\circ}\,\Omega.
$$

Now we have to find the open-circuit voltage or the short-circuit current. If we find the short-circuit current, we will have only one node to work with. If we find the open-circuit voltage, we will have two nodal voltages. We will find the open-circuit voltage here and you can try to find the short-circuit current as an exercise.

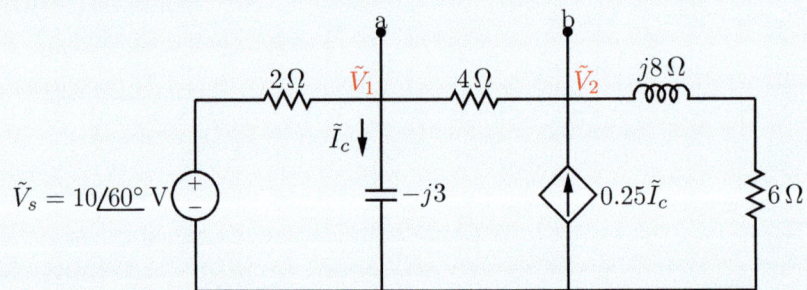

Figure 8.26: Finding the open-circuit voltage for the circuit in Figure 8.24.

We will use nodal analysis again. Notice that the circuit we have here is almost the same as the circuit in Figure 8.25. The only thing that changed is the independent source. This means that the admittance matrix will remain the same; the source vector will be the only thing that is going to change. The final matrix is going to be

$$
\begin{bmatrix}
\dfrac{1}{2} - \dfrac{1}{j3} + \dfrac{1}{4} & -\dfrac{1}{4} \\[2ex]
-\dfrac{1}{4} + \dfrac{0.25}{j3} & \dfrac{1}{4} + \dfrac{1}{6 + j8}
\end{bmatrix}
\begin{bmatrix}
\tilde{V}_1 \\[2ex]
\tilde{V}_2
\end{bmatrix}
=
\begin{bmatrix}
\dfrac{10\underline{/60^\circ}}{2} \\[2ex]
0
\end{bmatrix}.
$$

Solving the system of equations yields

$$\begin{bmatrix} \tilde{V}_1 \\ \\ \tilde{V}_2 \end{bmatrix} = \begin{bmatrix} 6.28 + j5.09 \\ \\ 2.07 + j6.33 \end{bmatrix} \text{V}.$$

Therefore, the open-circuit voltage or the Thévenin voltage can be calculated using

$$\tilde{V}_{\text{Th}} = \tilde{V}_1 - \tilde{V}_2 = 4.22 - j1.24 \, \text{V} = 4.4\underline{/-16.4°} \, \text{V}.$$

If you try to find the Norton current $\tilde{I}_N$, you can check your result with

$$\tilde{I}_N = \frac{\tilde{V}_{\text{Th}}}{Z_{\text{eq}}} = \frac{4.4\underline{/-16.4°}}{2.95\underline{/10.36°}} = 1.5\underline{/-26.7°} \, \text{A}.$$

8.5   Power Calculations in the Phasor Domain

Now that we have a reliable method to calculate voltages and currents in SSS circuits, let us focus on what matters the most in real life: power.

8.5.1   Instantaneous Power

We remember from our basics chapter that instantaneous power is calculated as

$$p(t) = v(t)i(t). \tag{8.18}$$

In this chapter, our voltages and currents are of sinusoidal form and oscillate at the same frequency. Let us focus for now on our simple AC circuit, shown again in Figure 8.27 for convenience.

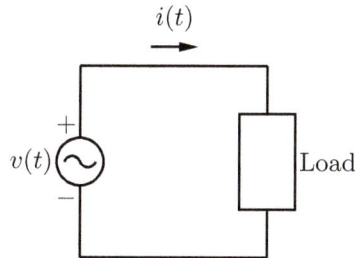

Figure 8.27: Simple AC circuit.

The voltage $v(t)$ is expressed as

$$v(t) = V_m \cos(\omega t + \theta_v) \, (\text{V}). \tag{8.19}$$

The current will also have a similar form:

$$i(t) = I_m \cos(\omega t + \theta_i)\,(\text{A}). \qquad (8.20)$$

The instantaneous power then will be

$$p(t) = V_m I_m \cos(\omega t + \theta_v) \cos(\omega t + \theta_i). \qquad (8.21)$$

We can use the following identity for a cosine product:

$$\cos(\alpha)\cos(\beta) = \frac{\cos(\alpha + \beta) + \cos(\alpha - \beta)}{2}. \qquad (8.22)$$

Using this identity, we get

$$
\begin{aligned}
p(t) &= \frac{V_m I_m}{2}\left(\cos(\omega t + \theta_v + \omega t + \theta_i) + \cos(\omega t + \theta_v - \omega t - \theta_i)\right) \\
&= \frac{V_m I_m}{2}\left(\cos(2\omega t + \theta_v + \theta_i) + \cos(\theta_v - \theta_i)\right) \\
&= \frac{V_m I_m}{2}\cos(2\omega t + \theta_v + \theta_i) + \frac{V_m I_m}{2}\cos(\theta_v - \theta_i). \qquad (8.23)
\end{aligned}
$$

It is clear that the instantaneous power we got in (8.23) has two parts

$$p(t) = \underbrace{\frac{V_m I_m}{2}\cos(2\omega t + \theta_v + \theta_i)}_{a} + \underbrace{\frac{V_m I_m}{2}\cos(\theta_v - \theta_i)}_{b}. \qquad (8.24)$$

Part (a) is a cosine wave but with twice the source's frequency, which is actually one of the problems with AC generation and distribution. This problem was solved by three-phase generation and distribution, but we will not discuss this topic in this text. Part (b) is a constant or what we would call a DC shift. We also call Part (b) of (8.24) the average power. Why? Because if you calculate the average power in one period, you will find that the time varying term in (8.24) yields a zero value. You can see that the sinusoidal portion of the equation will result in positive power for half the cycle and negative power with the same magnitude for the other half cycle. Only part (b) of (8.24) will contribute to non-zero average power. Consequently, this is the critical term for power. The next section shows this calculation in detail.

Example 8.5.1

For the circuit in Figure 8.28, find the total instantaneous power delivered by the voltage source.

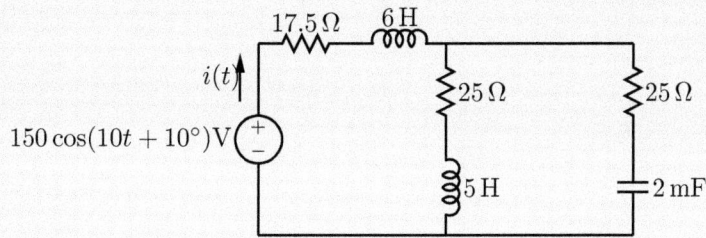

Figure 8.28: The circuit for Example 8.5.1 in the time domain.

We have to first transform the circuit into the phasor domain using $\omega = 10\,\text{rad/s}$. The transformed circuit is shown in Figure 8.29.

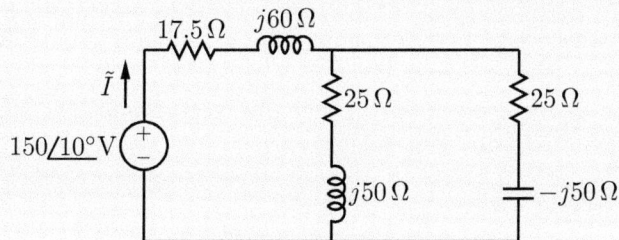

Figure 8.29: The circuit for Example 8.5.1 in the phasor domain.

We can now find the current leaving the voltage source. We will start by finding the parallel combination between the $25 + j50\,\Omega$ and the $25 - j50\,\Omega$ impedances. The resulting circuit is shown in Figure 8.30 and the needed impedance is

$$Z_{\text{eq}} = 25 + j50\,\Omega \parallel 25 - j50\,\Omega = 62.5\,\Omega.$$

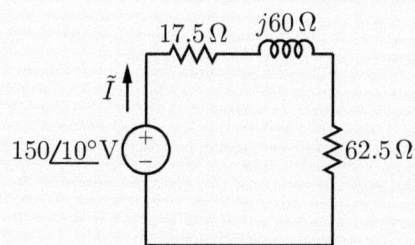

Figure 8.30: The circuit for Example 8.5.1 in the phasor domain.

The current can be found by Ohm's law as

$$\tilde{I} = \frac{150\underline{/10^\circ}}{17.5 + 62.5 + j60} = \frac{150\underline{/10^\circ}}{100\underline{/36.87^\circ}} = 1.5\underline{/-26.87^\circ}\text{A}.$$

Transferring the total current into the time domain yields

$$i(t) = 1.5\cos(10t - 26.87^\circ)\text{A}.$$

The instantanous power is calculated using (8.24),

$$
\begin{aligned}
p(t) &= \frac{V_m I_m}{2} \cos(2\omega t + \theta_v + \theta_i) + \frac{V_m I_m}{2} \cos(\theta_v - \theta_i) \\
&= \frac{150 \times 1.5}{2} \cos(20t + 10 - 26.87^\circ) + \frac{150 \times 1.5}{2} \cos(10 + 26.87^\circ) \\
&= 112.5 \cos(20t - 16.87^\circ) + 90\,\text{W}.
\end{aligned}
$$

8.5.2   Average (Real) Power Calculations

If you take a look at your energy (electric) bill or take a peek at your power meter outside your residence, you will notice that you are being billed for your kilowatt-hour (kWh) power consumption. This means that you are billed for the average (real) power you consume during an hour, even though you are delivered instantaneous power of the form shown in (8.24). Let us see how to calculate this average power. Recall that the average value of a periodic signal $x(t)$ with a period T can be calculated as

$$
X = \frac{1}{T} \int_{t_0}^{t_0+T} X(t) \, \mathrm{d}t. \tag{8.25}
$$

Consequently, the average power is given by

$$
\begin{aligned}
P &= \frac{1}{T} \int_{t_0}^{t_0+T} p(t) \, \mathrm{d}t \\
&= \frac{1}{T} \int_{t_0}^{t_0+T} \left[\frac{V_m I_m}{2} \cos(2\omega t + \theta_v + \theta_i) + \frac{V_m I_m}{2} \cos(\theta_v - \theta_i) \right] \, \mathrm{d}t \\
&= \underbrace{\frac{1}{T} \int_{t_0}^{t_0+T} \frac{V_m I_m}{2} \cos(2\omega t + \theta_v + \theta_i) \, \mathrm{d}t}_{=0} + \frac{1}{T} \int_{t_0}^{t_0+T} \frac{V_m I_m}{2} \cos(\theta_v - \theta_i) \, \mathrm{d}t \\
&= \frac{1}{T} \int_{t_0}^{t_0+T} \frac{V_m I_m}{2} \cos(\theta_v - \theta_i) \, \mathrm{d}t = \frac{V_m I_m}{2} \cos(\theta_v - \theta_i). \tag{8.26}
\end{aligned}
$$

To summarize, the **average power** consumed by a general complex load (impedance) can be found by

$$
\boxed{ P = \frac{V_m I_m}{2} \cos(\theta_v - \theta_i). } \tag{8.27}
$$

While we have calculated average power starting from time-domain expressions, it would be more convenient to do so using phasors. We can indeed do so by noticing the following facts. If we multiply the phasor representation of the voltage in (8.19) by the complex conjugate of the phasor representation of the current in (8.20), we will get

$$
\frac{1}{2} \tilde{V} \tilde{I}^* = \frac{1}{2} V_m \underline{/\theta_v}\, I_m \underline{/-\theta_i} = \frac{V_m I_m}{2} \underline{/\theta_v - \theta_i}. \tag{8.28}
$$

In rectangular form, this becomes

$$\frac{1}{2}\tilde{V}\tilde{I}^* = \frac{V_m I_m}{2}\cos(\theta_v - \theta_i) + j\frac{V_m I_m}{2}\sin(\theta_v - \theta_i). \tag{8.29}$$

From (8.29), we notice that

$$\boxed{P = \Re\left[\frac{1}{2}\tilde{V}\tilde{I}^*\right] = \frac{V_m I_m}{2}\cos(\theta_v - \theta_i).} \tag{8.30}$$

Hence, the real part of the product $\frac{1}{2}\tilde{V}\tilde{I}^*$ represents the **real power** consumed by a device in SSS. The imaginary part of the same product is called the **reactive power**. While it plays an important role in distribution networks and power transfer, it will not be covered in this text.

What Absorbs Average Power?

Let us examine (8.27) for different types of loads. We can easily see that the average power depends on the relative phase difference between voltage and current. We will start by assuming that the load in Figure 8.31 is purely resistive. For a resistive load, the voltage and current are in-phase,

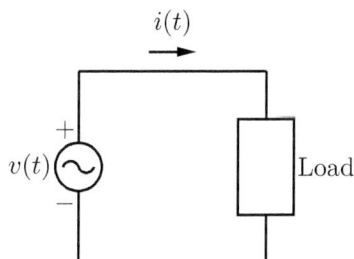

Figure 8.31: Simple AC circuit.

$$\theta_v = \theta_i.$$

Therefore, the average power absorbed by a resistor is

$$P_R = \frac{V_m I_m}{2}\cos(\theta_v - \theta_i) = \frac{V_m I_m}{2}\cos(0) = \frac{V_m I_m}{2}. \tag{8.31}$$

We can also write the average power absorbed by the resistor in terms of the resistance. Remember that

$$\tilde{V} = \tilde{I}R = RI_m\underline{/\theta_i} \tag{8.32}$$

and

$$\tilde{I} = \frac{\tilde{V}}{R} = \frac{V_m}{R}\underline{/\theta_v}. \tag{8.33}$$

Using (8.32) and (8.33) in (8.31) yields

$$P_R = \frac{RI_m^2}{2} = \frac{1}{2}R|\tilde{I}|^2 \tag{8.34}$$

and

$$P_R = \frac{V_m^2}{2R} = \frac{|\tilde{V}|^2}{2R}. \tag{8.35}$$

On the other hand, if the load in Figure 8.31 is purely inductive or capacitive, we will have

$$|\theta_v - \theta_i| = 90°.$$

This will make the average consumed power zero, as seen from

$$P_L = P_C = \frac{V_m I_m}{2}\cos(\theta_v - \theta_i) = \frac{V_m I_m}{2}\cos(\pm 90°) = 0. \tag{8.36}$$

This is of course consistent with our understanding that (ideal) inductors and capacitors do not absorb real power. However, this does not mean that their role is insignificant. *They are in dynamic balance with the rest of the circuit.* For part of the AC cycle they store energy and for the remaining part they deliver it back to the circuit. This dynamic balance is very useful in applications and is exploited to achieve a variety of useful functions. For instance, wireless energy transfer is based on this dynamic balance. We will review this further in the next chapter (magnetically coupled circuits).

In most practical cases, the load in Figure 8.31 is a combination of resistors, inductors, and/or capacitors. It is important to remember that only the resistive part of the load absorbs real power.

Example 8.5.2

For the circuit in Figure 8.32, find the power absorbed by the 6 Ω resistor and the total average power absorbed by the entire circuit.

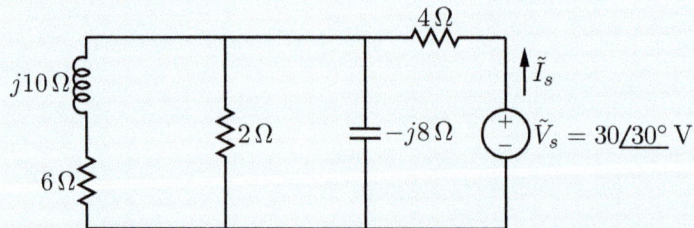

Figure 8.32: The circuit for Example 8.5.2 in the phasor domain.

The power absorbed by the 6 Ω resistor can be calculated by using

$$P_R = \frac{1}{2}R|\tilde{I}|^2 = \frac{|\tilde{V}|^2}{2R}.$$

We can first find the equivalent impedance Z_{eq} as shown in Figure 8.33. Subsequently, we can find the voltage across Z_{eq} to eventually find the current through the 6 Ω resistor. In addition, once we know Z_{eq}, it will be easy to find the power absorbed by the whole circuit. This is the basic plan outlined in detail below.

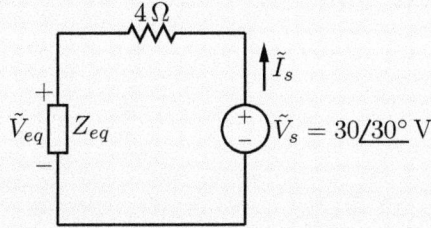

Figure 8.33: The circuit for Example 8.5.2 in the phasor domain.

The equivalent impedance is

$$Z_{eq} = (6 + j10) \parallel (2) \parallel (-j8) = 1.83\underline{/-5.4^\circ}\,\Omega = 1.82 - j0.17\,\Omega.$$

The total current $\tilde{I}_s$ can be calculated using

$$\tilde{I}_s = \frac{\tilde{V}_s}{Z_{eq} + 4} = \frac{30\underline{/30^\circ}}{1.82 - j0.17 + 4} = 5.15\underline{/31.67^\circ}\,\text{A}.$$

The total power absorbed by the circuit can be calculated by

$$P_{\mathrm{T}} = \frac{|\tilde{V}_s||\tilde{I}_s|}{2}\cos(\theta_v - \theta_i) = \frac{30(5.15)}{2}\cos(30^\circ - 31.67^\circ) = 77.22\,\text{W}.$$

We can use voltage division to find the voltage $\tilde{V}_{eq}$ as

$$\tilde{V}_{eq} = \frac{Z_{eq}}{Z_{eq} + 4}\tilde{V}_s = \frac{1.83\underline{/-5.4^\circ}}{1.82 - j0.17 + 4}30\underline{/30^\circ} = 9.43\underline{/26.3^\circ}\,\text{V}.$$

We can use voltage division again to find the voltage across the 6 Ω resistor or find the total current through the resistor by

$$\tilde{I}_{6\,\Omega} = \frac{\tilde{V}_{eq}}{6 + j10} = 0.81\underline{/-32.74^\circ}\,\text{A}.$$

The total power absorbed by the resistor is

$$P_{6\,\Omega} = \frac{1}{2}6\,\Omega|\tilde{I}_{6\,\Omega}|^2 = 3(0.81)^2 = 1.97\,\text{W}.$$

8.6   Root Mean Square (RMS) or Effective Value

Let us examine (8.27) again for the circuit in Figure 8.27. The equation and the figure are repeated here for convenience:

$$P = \frac{V_m I_m}{2}\cos(\theta_v - \theta_i). \tag{8.37}$$

We can rewrite (8.37) again as

$$P = \left(\frac{V_m}{\sqrt{2}}\right)\left(\frac{I_m}{\sqrt{2}}\right)\cos(\theta_v - \theta_i). \tag{8.38}$$

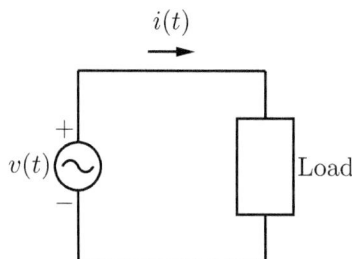

Figure 8.34: Simple AC circuit

If we were in DC conditions (i.e., DC voltage supply with magnitude V_m), the voltage and current would have been constants, which would make the instantaneous power constant. The average power would have been simply given by

$$P = VI = V_m I_m. \tag{8.39}$$

In DC conditions, the load would feel the full value of the applied voltage V_m all the time. On the other hand, in the case of the sinusoidal source (or any other periodic signal) the load experiences the maximum voltage V_m only for a small fraction of the time (twice per cycle). We can intuitively say that the load experiences an "effective" voltage value that is smaller than V_m. To be more precise, we can say that there is an effective *constant* value V_{eff} that would result in the same power consumption if applied across the same load. Looking at (8.38) through this lens, we can say that the effective voltage and current values are

$$V_{\text{eff}} = \frac{V_m}{\sqrt{2}}, \tag{8.40}$$

$$I_{\text{eff}} = \frac{I_m}{\sqrt{2}}. \tag{8.41}$$

We can actually prove (although we will not do it here) that the effective value for a periodic signal $x(t)$ can be calculated by taking its root mean square or RMS value given by

$$\boxed{X_{\text{eff}} = X_{\text{rms}} = \sqrt{\frac{1}{T}\int_{t_0}^{t_0+T} x^2(t)\,\mathrm{d}t} } \tag{8.42}$$

where T is its period and t_0 is any arbitrary time.

For a sinusoidal voltage

$$v(t) = V_m\cos(\omega t + \theta_v)\,(\mathrm{V}), \tag{8.43}$$

the RMS value is calculated as

$$V_{\text{rms}} = \sqrt{\frac{1}{T} \int_{t_0}^{t_0+T} V_m^2 \cos^2(\omega t + \theta_v)\, dt}$$

$$= \sqrt{\frac{V_m^2}{T} \int_{t_0}^{t_0+T} \cos^2(\omega t + \theta_v)\, dt}. \tag{8.44}$$

Remember that

$$\cos^2(x) = \frac{1}{2} + \frac{\cos(2x)}{2}. \tag{8.45}$$

Using (8.45) in (8.44) yields

$$V_{\text{rms}} = \sqrt{\frac{V_m^2}{T} \int_{t_0}^{t_0+T} \left(\frac{1}{2} + \frac{\cos(2x)}{2} \right) dt}$$

$$= \sqrt{\frac{V_m^2}{T} \left(\int_{t_0}^{t_0+T} \frac{1}{2}\, dt + \underbrace{\int_{t_0}^{t_0+T} \frac{\cos(2x)}{2}\, dt}_{=0} \right)}$$

$$= \sqrt{\frac{V_m^2}{T} \int_{t_0}^{t_0+T} \frac{1}{2}\, dt} = \sqrt{\frac{V_m^2}{2T}(t_0 + T - t_0)} = \frac{V_m}{\sqrt{2}}. \tag{8.46}$$

This proves that the RMS value for a sinusoidal signal is equal to its effective value as defined above. The same can be proven for all periodic signals.

Going back to power, we can now write the power as a function of the RMS values

$$\boxed{P = V_{\text{rms}} I_{\text{rms}} \cos(\theta_v - \theta_i).} \tag{8.47}$$

Due to the importance of RMS values in calculating consumed power, in practice we almost always use RMS values for AC voltages and currents. Voltage and current magnitudes are practically never used in real life. Consider, for example, the voltage value at your house. If you live in the United States, you probably know that your home appliances need to be powered by 120 V. This rating actually refers to the RMS value of the voltage. The peak voltage is 120 V × $\sqrt{2}$, which is about 170 V. While in this text we will be careful and always distinguish between RMS and peak values, in real life this is almost never done. All AC values should be interpreted as RMS values unless otherwise specified.

Let us now look at some examples of other periodic functions and see how to find their root mean square (effective) values.

Example 8.6.1

Figure 8.35 shows a nonideal current source that is modeled by an ideal source with a parallel resistance to compensate for losses. The current source is periodic and its plot is shown in Figure 8.36. Find the value of the load resistance that will make the total average power (by both R_L and R_s) absorbed to be 26 W.

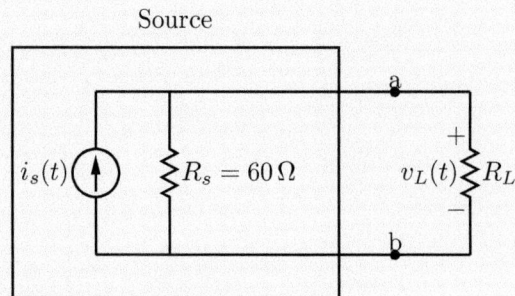

Figure 8.35: The circuit for Example 8.6.1 in the time domain.

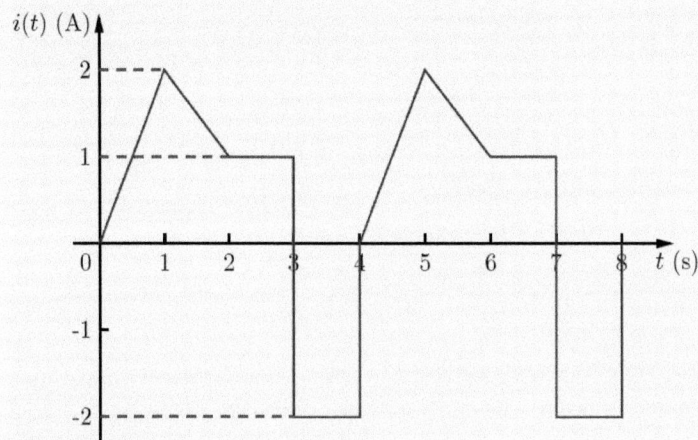

Figure 8.36: The current source plot for the circuit in Figure 8.35.

Since we have a periodic function we will need to find the RMS value of the current using

$$I_{\text{rms}} = \sqrt{\frac{1}{T} \int_{t_0}^{t_0+T} i^2(t)\, dt}.$$

The total absorbed average power is

$$P_T = I_{\text{rms}}^2 R_T$$

where

$$R_T = R_s \parallel R_L.$$

We already know the total power, so we will be able to find the load resistance R_L. We will start by finding the RMS value of the current source. The time period T is 4 s and the current will have to be divided into four separate functions.

$$i_s(t) = \begin{cases} 2t & 0 < t < 1, \\ -t + 3 & 1 < t < 2, \\ 1 & 2 < t < 3, \\ -2 & 3 < t < 4 \end{cases} \text{A.}$$

We will have to break up the integration:

$$I_{\text{rms}} = \sqrt{\frac{1}{T}\left(\int_0^1 (2t)^2\,\mathrm{d}t + \int_1^2 (-t+3)^2\,\mathrm{d}t + \int_2^3 (1)^2\,\mathrm{d}t + \int_3^4 (-2)^2\,\mathrm{d}t\right)}.$$

Evaluating the integrals separately yields:

$$\int_0^1 (2t)^2\,\mathrm{d}t = \left.\frac{4t^3}{3}\right|_0^1 = \frac{4}{3}\,\mathrm{A}^2\cdot\mathrm{s}$$

$$\int_1^2 (-t+3)^2\,\mathrm{d}t = \int_1^2 t^2 - 6t + 9\,\mathrm{d}t = \left.\frac{t^3}{3} - \frac{6t^2}{2} + 9t\right|_1^2 = \frac{7}{3}\,\mathrm{A}^2\cdot\mathrm{s}$$

$$\int_2^3 (1)^2\,\mathrm{d}t = \left.t\right|_2^3 = 1\,\mathrm{A}^2\cdot\mathrm{s}$$

$$\int_3^4 (-2)^2\,\mathrm{d}t = \left.4t\right|_3^4 = 4\,\mathrm{A}^2\cdot\mathrm{s}.$$

Now we can go back to evaluating the RMS current:

$$I_{\text{rms}} = \sqrt{\frac{1}{4}\left(\frac{4}{3} + \frac{7}{3} + 1 + 4\right)} = 1.472\,\mathrm{A}.$$

The total resistance is

$$R_T = \frac{P_T}{I_{\text{rms}}^2} = \frac{26}{1.472^2} = 12\,\Omega.$$

The load resistance can be found by using

$$R_L = \frac{R_s R_T}{R_s - R_T} = \frac{60(12)}{60 - 12} = 15\,\Omega.$$

Note: It would be fun to refresh your memory and derive the R_L equation.

8.7 Summary for Power Calculations

In this section we present a summary table that displays all the ways we can calculate real (average) power (Table 8.1).

Let us take a look at a simple example.

Amplitude	RMS	When to Use
$P = \Re\left[\frac{1}{2}\tilde{V}_L \tilde{I}_L^*\right]$	$P = \Re\left[\tilde{V}_{L\mathrm{rms}} \tilde{I}_{L\mathrm{rms}}^*\right]$	Valid all the time
$P = \frac{V_L I_L}{2}\cos(\theta_v - \theta_i)$	$P = V_{L\mathrm{rms}} I_{L\mathrm{rms}}\cos(\theta_v - \theta_i)$	Valid all the time
$P = \frac{1}{2}R_L|\tilde{I}_L|^2$	$P = R_L|\tilde{I}_{L\mathrm{rms}}|^2$	Valid all the time
$P = \frac{1}{2}|\tilde{V}_L|^2 \frac{R_L}{R_L^2 + X_L^2}$	$P = |\tilde{V}_{L\mathrm{rms}}|^2 \frac{R_L}{R_L^2 + X_L^2}$	Valid all the time
$P = \frac{1}{2}\frac{|\tilde{V}_L|^2}{R_L}$	$P = \frac{|\tilde{V}_{L\mathrm{rms}}|^2}{R_L}$	Only for purely resistive circuits $X_L = 0$

Table 8.1: Power equations.

Example 8.7.1

For the circuit in Figure 8.37, find the average power delivered by the source, the average power absorbed by the line, and the average power absorbed by the load.

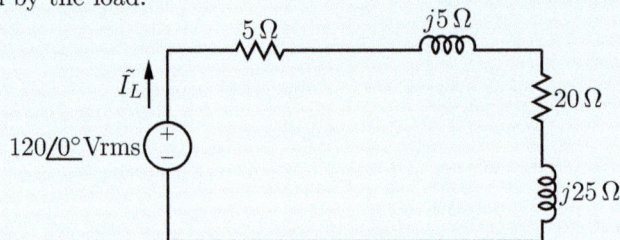

Figure 8.37: The circuit for Example 8.7.1 in the phasor domain in RMS.

We can use Ohm's law to find the load current $\tilde{I}_L$ first:

$$\tilde{I}_L = \frac{120}{Z_{\text{total}}} = \frac{120}{25 + j30} = 3.073\underline{/-50.19°}\,\text{A}.$$

Now that we have the current through all the elements, we can use our power equation that relates the resistance and current magnitude to the power. The average power absorbed by the load is calculated as

$$P_{\text{load}} = R_L|\tilde{I}_{L\mathrm{rms}}|^2 = 20(3.073)^2 = 188.74\,\text{W}.$$

The average power dissipated by the line is calculated as

$$P_{\text{line}} = R_{\text{line}}|\tilde{I}_{L\mathrm{rms}}|^2 = 5(3.073)^2 = 47.22\,\text{W}.$$

The total average power delivered by the source can be calculated using conservation of power:

$$P_{\text{source}} = P_{\text{line}} + P_{\text{load}} = 235.96\,\text{W}.$$

8.8   Maximum Power Transfer

In many applications, we need to transfer power from one place to another since
the source of power is physically far from the load. Consider for instance a cell
phone antenna receiving a wireless signal. The received power needs to be transferred to the detector so the signal can be decoded. As an additional example,
consider the case of a GPS receiver. The signal needs to be transferred from
the antenna to the signal decoder. In these cases, maximum power transfer is of
paramount importance since the received signal could be very weak (sometimes
quite a bit below a nW). Successfully receiving such signals requires us to design
our system carefully so that we transfer the maximum amount of power from
the source to the intended load. This is the problem we will address in this
section.

Before we derive the solution, keep in mind that there are also practical
cases in which maximum power transfer is not desired or feasible. For example,
consider your house, which is connected to the grid. The objective here is clearly
different: rather than trying to transfer the maximum power from the power
station to your house, we design our system to minimize distribution losses
and maximize power transfer efficiency. While we will not discuss this problem
here, it is important to note that the two goals (maximum power transfer and
maximum power efficiency) cannot be achieved at the same time. One can
optimize only one of the two. Consequently, it is important to understand what
the objective really is to design the best system possible.

Let us now focus on the problem at hand. Similarly to the DC circuits,
a practical AC source can always be modeled with its Thévenin or Norton
equivalent circuit. Figure 8.38 shows a source that is modeled by its phasor
domain Thévenin equivalent circuit and attached to its load impedance Z_{Load}.
The goal is to find the load impedance value that absorbs the maximum amount
of power.

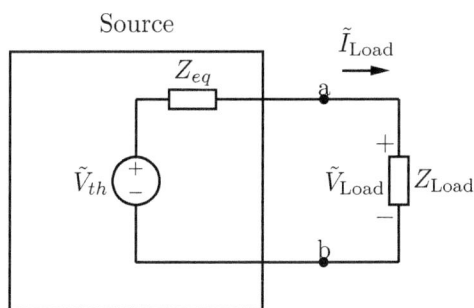

Figure 8.38: The phasor domain Thévenin equivalent circuit of a source with an attached
load.

The load impedance Z_{Load} can be written in rectangular form as

$$Z_{\text{Load}} = R_{\text{Load}} + jX_{\text{Load}}. \tag{8.48}$$

Similarly, the source's equivalent impedance can also be written in rectangular form as

$$Z_{eq} = R_{eq} + jX_{eq}. \tag{8.49}$$

Real power is only absorbed by the resistive part of the load impedance Z_L and can be calculated using

$$P_{\text{Load}} = \frac{1}{2} R_{\text{Load}} |\tilde{I}_{\text{Load}}|^2. \tag{8.50}$$

Since we are in the phasor domain, we can find the load current using Ohm's law as

$$\tilde{I}_{\text{Load}} = \frac{\tilde{V}_{th}}{Z_{eq} + Z_{\text{Load}}} = \frac{\tilde{V}_{th}}{(R_{eq} + R_{\text{Load}}) + j(X_{eq} + X_{\text{Load}})}. \tag{8.51}$$

For (8.50) we need to find the magnitude of the load current $|\tilde{I}_{\text{Load}}|$ as

$$|\tilde{I}_{\text{Load}}| = \frac{|\tilde{V}_{th}|}{\sqrt{(R_{eq} + R_{\text{Load}})^2 + (X_{eq} + X_{\text{Load}})^2}}. \tag{8.52}$$

To make things easier we can find $|\tilde{I}_{\text{Load}}|^2$ as

$$|\tilde{I}_{\text{Load}}|^2 = \frac{|\tilde{V}_{th}|^2}{(R_{eq} + R_{\text{Load}})^2 + (X_{eq} + X_{\text{Load}})^2}. \tag{8.53}$$

Plugging (8.53) into (8.50) yields

$$P_{\text{Load}} = \frac{1}{2} \frac{R_{\text{Load}} |\tilde{V}_{th}|^2}{(R_{eq} + R_{\text{Load}})^2 + (X_{eq} + X_{\text{Load}})^2}. \tag{8.54}$$

Remember that we are trying to find the load impedance value that absorbs the maximum real power. In order to find the maximum or minimum of a function, we need to take the derivative of the function and equate it to zero. The problem here is that the power equation we have in (8.54) depends on both the real and imaginary parts of our load impedance. This means we will have to use partial derivatives. Specifically, we need to get

$$\frac{\partial P_{\text{Load}}}{\partial R_{\text{Load}}} = 0 \tag{8.55}$$

and

$$\frac{\partial P_{\text{Load}}}{\partial X_{\text{Load}}} = 0. \tag{8.56}$$

Finding the partial derivative of (8.56) seems a bit easier than finding the partial derivative of (8.55). So let us start with this equation. Since we will be taking the partial derivative with respect to X_{Load}, everything else becomes a constant including R_{Load}, which means that we can rewrite the equation to make things a little bit simpler. Rewriting (8.54) yields

$$\frac{\partial P_{\text{Load}}}{\partial X_{\text{Load}}} = \frac{\partial}{\partial X_{\text{Load}}} \left(\frac{A}{B + (X_{eq} + X_{\text{Load}})^2} \right) \tag{8.57}$$

where

$$A = \frac{1}{2} R_{\text{Load}} |\tilde{V}_{th}|^2, \tag{8.58}$$

$$B = (R_{eq} + R_{\text{Load}})^2. \tag{8.59}$$

Taking the partial derivative now yields

$$\begin{aligned}
\frac{\partial P_{\text{Load}}}{\partial X_{\text{Load}}} &= \frac{\partial}{\partial X_{\text{Load}}} \left(\frac{A}{B + (X_{eq} + X_{\text{Load}})^2} \right) \\
&= -A(B + (X_{eq} + X_{\text{Load}})^2)^{-2} \left(2(X_{eq} + X_{\text{Load}}) \right)(1) \\
&= \frac{-A(2X_{eq} + 2X_{\text{Load}})}{(B + (X_{eq} + X_{\text{Load}})^2)^2}.
\end{aligned} \tag{8.60}$$

We can now equate (8.60) to zero as

$$\frac{\partial P_{\text{Load}}}{\partial X_{\text{Load}}} = \frac{-A(2X_{eq} + 2X_{\text{Load}})}{(B + (X_{eq} + X_{\text{Load}})^2)^2} = 0. \tag{8.61}$$

Solving for X_{Load} yields

$$\boxed{X_{\text{Load}} = -X_{eq}.} \tag{8.62}$$

We can plug our solution from (8.62) into (8.54) before we proceed to finding $\frac{\partial P_{\text{Load}}}{\partial R_{\text{Load}}}$. This results in

$$P_{\text{Load}} = \frac{1}{2} \frac{R_{\text{Load}} |\tilde{V}_{th}|^2}{(R_{eq} + R_{\text{Load}})^2 + (X_{eq} - X_{eq})^2} = \frac{1}{2} \frac{R_{\text{Load}} |\tilde{V}_{th}|^2}{(R_{eq} + R_{\text{Load}})^2}. \tag{8.63}$$

We can rewrite (8.63) as

$$P_{\text{Load}} = \frac{1}{2} |\tilde{V}_{th}|^2 R_{\text{Load}} (R_{eq} + R_{\text{Load}})^{-2}. \tag{8.64}$$

We can use the product or chain rule here to find $\frac{\partial P_{\text{Load}}}{\partial R_{\text{Load}}}$. As a refresher, the chain rule is

$$\frac{\mathrm{d}}{\mathrm{d}x} (f(x)g(x)) = \frac{\mathrm{d}f(x)}{\mathrm{d}x} (g(x)) + \frac{\mathrm{d}g(x)}{\mathrm{d}x} (f(x)). \tag{8.65}$$

Applying the chain rule yields

$$\begin{aligned}
\frac{\partial P_{\text{Load}}}{\partial R_{\text{Load}}} &= \frac{\partial}{\partial R_{\text{Load}}} \left(\frac{1}{2} |\tilde{V}_{th}|^2 R_{\text{Load}} (R_{eq} + R_{\text{Load}})^{-2} \right) \\
&= \frac{|\tilde{V}_{th}|^2}{2} \left((1)(R_{eq} + R_{\text{Load}})^{-2} - 2R_{\text{Load}}(R_{eq} + R_{\text{Load}})^{-3}(1) \right) \\
&= \frac{|\tilde{V}_{th}|^2}{2} \left(\frac{1}{(R_{eq} + R_{\text{Load}})^2} - \frac{2R_{\text{Load}}}{(R_{eq} + R_{\text{Load}})^3} \right).
\end{aligned} \tag{8.66}$$

Equating (8.66) to zero yields

$$\frac{\partial P_{\text{Load}}}{\partial R_{\text{Load}}} = 0,$$

$$1 - \frac{2R_{\text{Load}}}{(R_{eq} + R_{\text{Load}})} = 0,$$

$$2R_{\text{Load}} = R_{\text{Load}} + R_{eq},$$

$$\boxed{R_{\text{Load}} = R_{eq}.} \tag{8.67}$$

From (8.62) and (8.67), we have now found that the impedance that absorbs the maximum possible power is given by

$$Z_{\text{Load}} = R_{eq} - jX_{eq} \tag{8.68}$$

or, in other words,

$$\boxed{Z_{\text{Load}} = Z_{eq}^*.} \tag{8.69}$$

We can obtain the maximum absorbed power by plugging $Z_{\text{Load}} = Z_{eq}^*$ into (8.54). Hence,

$$P_{\text{Load}} = \frac{1}{2}\frac{R_{\text{Load}}|\tilde{V}_{th}|^2}{(R_{eq} + R_{\text{Load}})^2 + (X_{eq} + X_{\text{Load}})^2} = \frac{1}{2}\frac{R_{eq}|\tilde{V}_{th}|^2}{(R_{eq} + R_{eq})^2 + (X_{eq} - X_{eq})^2}. \tag{8.70}$$

This makes the maximum absorbed power

$$\boxed{P_{\text{Load}}(max) = \frac{|\tilde{V}_{th}|^2}{8R_{eq}}.} \tag{8.71}$$

Example 8.8.1

For the circuit in Figure 8.39, find the value of the load impedance Z_L that absorbs the maximum power and find this power as well.

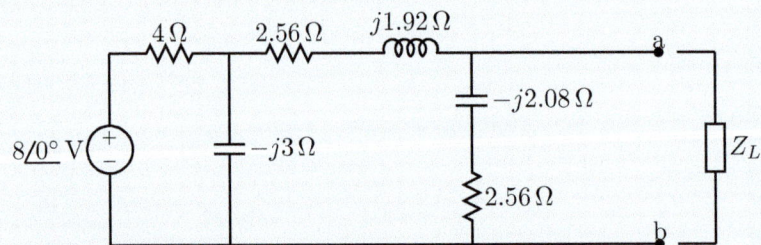

Figure 8.39: The circuit for Example 8.8.1 in the phasor domain.

Maximum power transfer to the load requires

$$Z_L = Z_{ab}^*.$$

As a result we need to find the Thévenin equivalent circuit with respect to port a-b. The maximum power will be given by

$$P_{\max} = \frac{|\tilde{V}_{th}|^2}{8R_{ab}}.$$

One way to find the Thévenin equivalent circuit is to use source transformation. We encourage you to also find it by using the formal method and comparing your results. Let us start from the far left and work our way to port a-b.

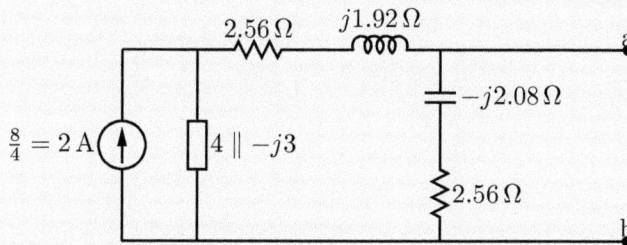

Figure 8.40: Applying source transformation to the circuit shown in Figure 8.39.

$$4 \parallel -j3 = 2.4\underline{/-53.13°}\,\Omega = 1.44 - j1.92\,\Omega.$$

Applying source transformation again and redrawing the circuit as seen in Figure 8.41, we get

$$\tilde{V} = (4 \parallel -j3)(2) = 4.8\underline{/-53.13°}\,\text{V} = 2.88 - j3.84\,\text{V}.$$

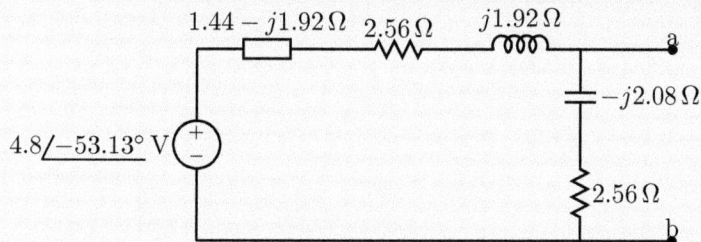

Figure 8.41: Applying source transformation to the circuit shown in Figure 8.39.

We can sum the series impedances together to get

$$Z = 1.44 - j1.92 + 2.56 + j1.92 = 4\,\Omega.$$

Apply another source transformation now as

$$\tilde{I} = \frac{4.8\underline{/-53.13°}}{4} = 1.2\underline{/-53.13°}\,\text{A} = 0.72 - j0.96\,\text{A}.$$

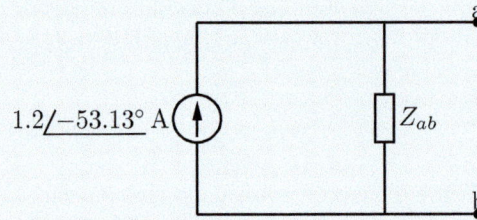

Figure 8.42: The Norton equivalent circuit for the circuit shown in Figure 8.39.

The equivalent impedance is

$$Z_{ab} = 4 \parallel (2.56 - j2.08) = 1.78 - j0.7\,\Omega = 1.92\underline{/-21.5°}\,\Omega.$$

The optimal load impedance for maximum power transfer needs to be

$$Z_L = Z_{ab}^* = 1.78 + j0.7\,\Omega = 1.92\underline{/21.5°}\,\Omega.$$

To find the maximum power, we need to find $\tilde{V}_{th}$ first:

$$\tilde{V}_{th} = Z_{ab}\tilde{I}_N = (1.92\underline{/-21.5°})(1.2\underline{/-53.13°}) = 2.3\underline{/-74.63°}\,\text{V}.$$

The maximum power delivered is given then as

$$P_{\max} = \frac{|\tilde{V}_{th}|^2}{8R_{ab}} = \frac{2.3^2}{8(1.78)} = 370\,\text{mW}.$$

Example 8.8.2

For the circuit in Figure 8.43, find the load impedance Z_L that absorbs the maximum possible power.

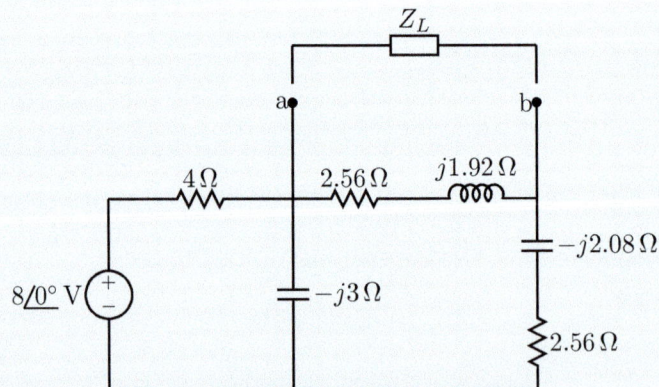

Figure 8.43: The circuit for Example 8.8.2 in the phasor domain.

In order for us to have maximum power transfer, we need to achieve

$$Z_L = Z_{ab}^*.$$

This time, since the power is not needed, we can find the equivalent impedance quickly by turning off the source and combining impedances, as seen in Figure 8.44.

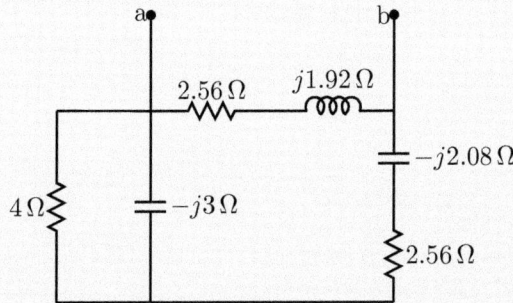

Figure 8.44: Turning off the source and finding the equivalent impedance at port a-b for the circuit of Example 8.8.2.

Hence,

$$Z_{ab} = (2.56 + j1.92) \parallel (2.56 - j2.08 + (4 \parallel -j3))$$
$$= (2.56 + j1.92) \parallel (5.67\underline{/-45°}) = 2.63\underline{/9.47°}\ \Omega = 2.6 + j0.43\ \Omega.$$

8.9   Maximum Power versus Maximum Efficiency

From our previous section, we found out that in order to obtain maximum power transfer to the load in Figure 8.45, the load needs to be equal to the complex conjugate of the source's Thévenin equivalent:

$$Z_L = Z_{eq}^* = R_{eq} - jX_{eq}. \tag{8.72}$$

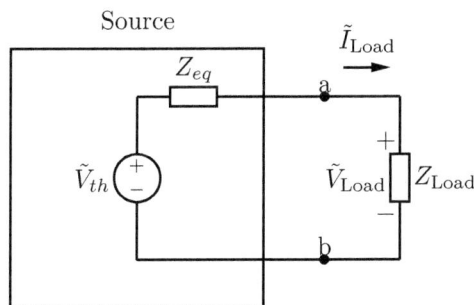

Figure 8.45: The phasor domain Thévenin equivalent circuit of a source with an attached load.

Now that maximum power transfer has been achieved, let us check on power efficiency. Remember that power efficiency is calculated using

$$\eta = \frac{P_{\text{Load}}}{P_{\text{Source}}} \tag{8.73}$$

In our case, the output power is the power absorbed by the load and the input power is the power delivered by the source:

$$P_{\text{Source}} = \frac{1}{2}|\tilde{I}_{\text{L}}|^2(R_{eq} + R_{eq}) = |\tilde{I}_{\text{L}}|^2 R_{eq}, \tag{8.74}$$

$$P_{\text{Load}} = \frac{1}{2}|\tilde{I}_{\text{L}}|^2 R_{eq}. \tag{8.75}$$

The power efficiency is

$$\eta = \frac{P_{\text{Load}}}{P_{\text{Source}}} = \frac{\frac{1}{2}|\tilde{I}_{\text{L}}|^2 R_{eq}}{|\tilde{I}_{\text{L}}|^2 R_{eq}} = 50\%. \tag{8.76}$$

We can see that although maximum power transfer is achieved, it is not the most efficient. You might be wondering, "Isn't this bad?" Unfortunately, you cannot have both maximum efficiency and maximum power transfer. It will depend on the application. Real-life applications will clarify this concept further.

A case where maximum power is needed is our cell phone antennas. Cell phone antennas operate at the nanowatt level, which means that the RF receiver needs to capture as much power as possible received by the antenna.

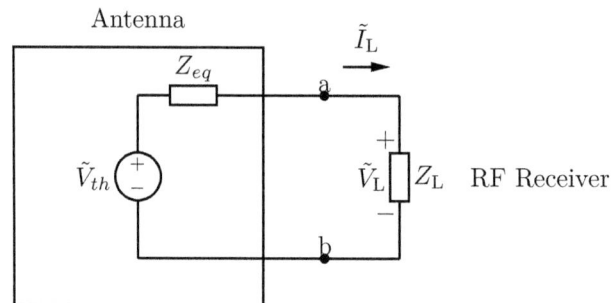

Figure 8.46: Cell phone antenna and RF receiver circuit.

Maximum power efficiency is picked over maximum power transfer in the case of our distribution networks. Let us discover why. Figure 8.47 shows a simplified power distribution system.

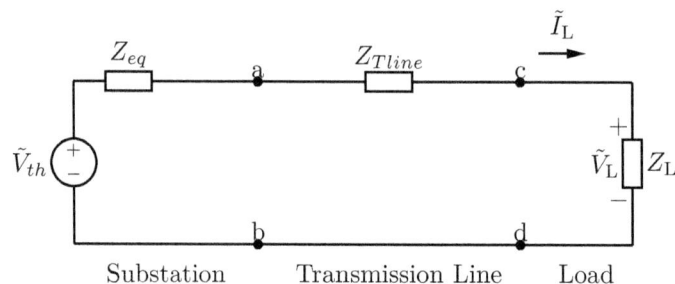

Figure 8.47: Power distribution network.

Unlike a regular wire, transmission lines are miles long and have significant impedance that can lead to power loss. The power delivered by the substation is

$$P_{\text{source}} = P_{\text{Tline}} + P_{\text{Load}} \tag{8.77}$$

We would like to minimize the power lost in the line, so we need to look at P_{Tline}.

$$P_{\text{Tline}} = \frac{1}{2}|\tilde{I}_L|^2 R_{\text{Tline}}. \tag{8.78}$$

The only thing we can control in this equation is the current drawn by the load. The load current is known. How? Loads in the distribution system are usually known and represented by real power absorption and a power factor angle $(\theta_v - \theta_i)$. The voltage across the load is also known because the power distributor promises a certain voltage magnitude—that is, 120 V. This leads to

$$|\tilde{I}_L| = \frac{P}{|\tilde{V}_L|\cos(\theta_v - \theta_i)}. \tag{8.79}$$

The voltage $|\tilde{V}_L|$ is constant and the power required by the load is also constant. The only thing we can control here is the power factor $\cos(\theta_v - \theta_i)$. Yes, we might not be able to change the load itself, but power providers can install parallel loads to act as a power factor corrector. This is a topic to be discussed in other courses. The only thing that we need to know here is that by making the power factor closer to unity, we minimize the magnitude of the load current, which in turn minimizes the loss in the line. This ultimately leads to maximum power efficiency.

Some numbers can help solidify this concept even more, so let us take a look at a simple example.

Example 8.9.1

For the circuit in Figure 8.48, $|\tilde{V}_L| = 120\,\text{V}$, and $P_{\text{load}} = 2400\,\text{W}$. Find the magnitude of the load current $|\tilde{I}_L|$ if the load Z_L has the following power factors:

1. $\cos(\theta_v - \theta_i) = 1$.

2. $\cos(\theta_v - \theta_i) = 0.5$.

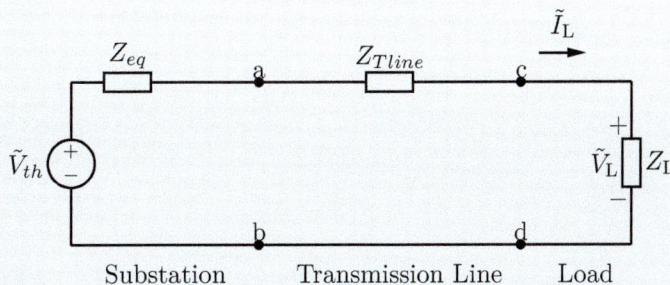

Figure 8.48: Power distribution system.

Recall that

$$|\tilde{I}_L| = \frac{P_{\text{Load}}}{|\tilde{V}_L| \cos(\theta_v - \theta_i)}.$$

For $\cos(\theta_v - \theta_i) = 1$, we can find the current as

$$|\tilde{I}_L| = \frac{P_{\text{Load}}}{|\tilde{V}_L| \cos(\theta_v - \theta_i)} = \frac{2400}{120(1)} = 20 \text{ A}.$$

Calculating the current again for $\cos(\theta_v - \theta_i) = 0.5$,

$$|\tilde{I}_L| = \frac{P_{\text{Load}}}{|\tilde{V}_L| \cos(\theta_v - \theta_i)} = \frac{2400}{120(0.5)} = 40 \text{ A}.$$

Notice how the drop in power factor resulted in higher current. Higher current will mean more line losses, and that is why we try our best to correct the load's power factor.

8.10 Frequency Response (RLC Circuits)

So far, we have analyzed our circuits in the phasor domain using the source's single frequency. In this section, we would like to see what happens to the circuit as a function of frequency. We will pick a circuit and examine it under all ranges of frequency, from DC ($f = 0$) to very high frequencies ($f = \infty$).

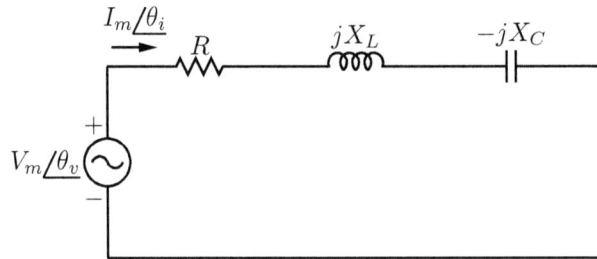

Figure 8.49: Series RLC circuit with a variable frequency AC source in the phasor domain.

Let us start by looking at the series RLC circuit shown in Figure 8.49. The total impedance seen by the source is

$$Z_{\text{eq}} = R + j(X_L - X_C). \tag{8.80}$$

Recall that

$$X_L = \omega L = 2\pi f L, \tag{8.81}$$

$$X_C = \frac{1}{\omega C} = \frac{1}{2\pi f C}. \tag{8.82}$$

A plot of each reactance in (8.81) and (8.82) is shown in Figure 8.50.

From the plot, we see that the inductor acts as a short circuit at DC and very low frequencies and as an open circuit at very high frequencies. On the

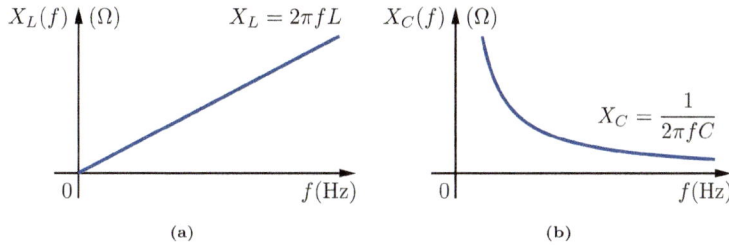

Figure 8.50: Reactance verus frequency for (a) an inductor and (b) a capacitor.

other hand, the capacitor acts as an open circuit at DC and very low frequencies and a short circuit at very high frequencies. Logically, we can deduce that no current will be flowing through the circuit at DC, very low frequencies, or very high frequencies due to the open circuits caused by the inductor and capacitor. We will investigate this further in a bit.

Now, in order to maximize the current, we need to minimize the impedance. More specifically, we must find the frequency at which the impedance is a minimum. There are multiple ways to tackle this problem. We can look at (8.80) and quickly see that the total impedance is the smallest when

$$X_L - X_C = 0, \tag{8.83}$$

or

$$X_L = X_C. \tag{8.84}$$

The frequency where this occurs can be found using the definition of the reactances:

$$X_L = X_C, \tag{8.85}$$

$$\omega_r L = \frac{1}{\omega_r C}, \tag{8.86}$$

$$\omega_r = \frac{1}{\sqrt{LC}}, \tag{8.87}$$

$$f_r = \frac{1}{2\pi\sqrt{LC}}. \tag{8.88}$$

The frequency in (8.88) is a special frequency called the resonance frequency.

Resonance frequency: A frequency where the total impedance in the circuit is a minimum or the total current through the circuit is a maximum.

We can also look at a plot of both reactances and find the point of intersection where $X_L = X_C$, as shown in Figure 8.51.

Figure 8.52 shows a plot of the total impedance's magnitude with respect to frequency.

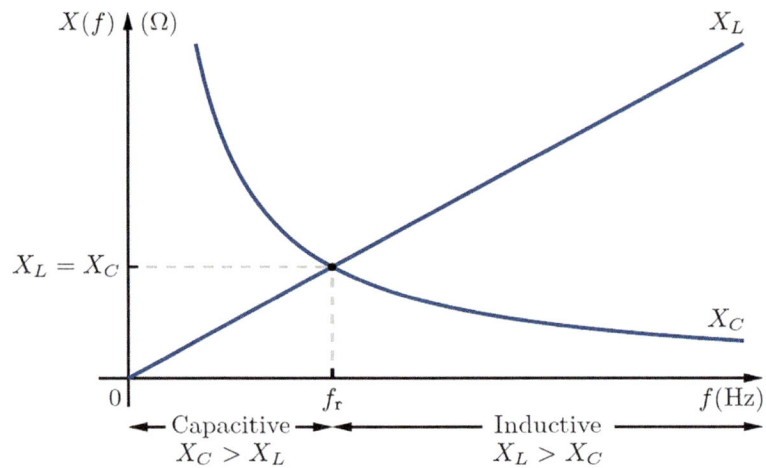

Figure 8.51: RLC series X_L and X_C versus f.

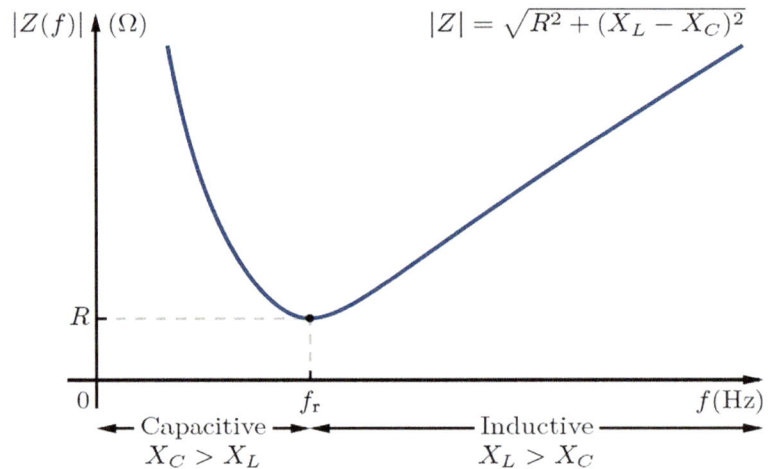

Figure 8.52: RLC series Z versus f.

$$|Z(\omega)| = \sqrt{R^2 + (\omega L - \frac{1}{\omega C})^2}. \tag{8.89}$$

As expected, the impedance is the lowest at the resonance frequency. At resonance, the circuit is purely resistive, and the magnitude of the current flowing through the circuit only depends on the value of that resistor. We can also see from the plot that our series RLC circuit is heavily capacitive at very low frequencies and heavily inductive at higher frequencies.

Let us move on to the current, since it is the main reason for our analysis. Figure 8.53 shows a plot of our current's magnitude versus frequency.

The current's magnitude is calculated using Ohm's law:

$$|\tilde{I}| = \frac{|\tilde{V}|}{|Z|} = \frac{V_m}{\sqrt{R^2 + (\omega L - \frac{1}{\omega C})^2}}. \tag{8.90}$$

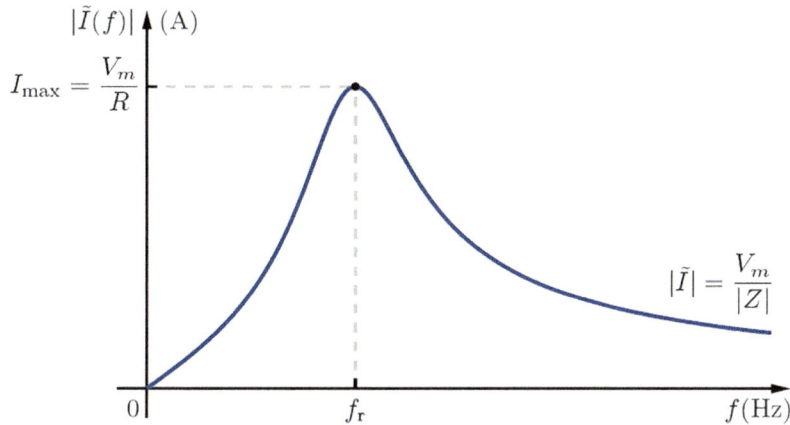

Figure 8.53: RLC series I versus f.

The maximum current does occur at the resonance frequency and is virtually zero at both extreme points—that is, DC and ∞.

An interesting observation can be made looking at the bell curve current in Figure 8.54. There is a range of frequency where the current is substantially high. We call this frequency range the bandwidth of the circuit.

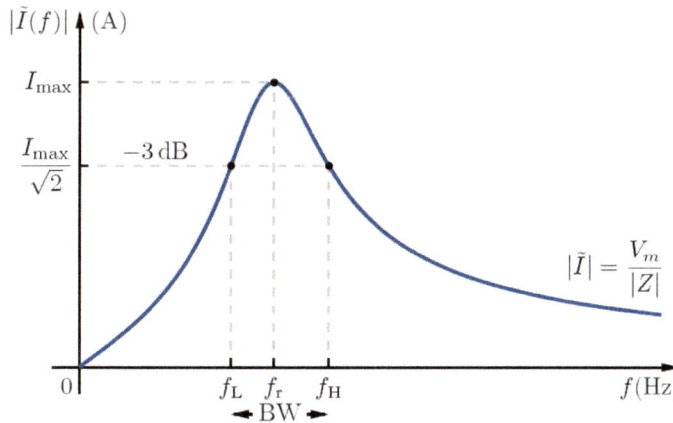

Figure 8.54: RLC series I versus f bandwidth.

Bandwidth: A range of frequencies in which resonance still has a strong dominance.

The bandwidth of the circuit is usually taken at -3 dB, which is the frequency at which the current is at about 30% less of its maximum value.

So far we saw that we can control the frequency at which our resonance can occur by changing the values of the inductor and/or the capacitor. What about changing the value of the resistor in the circuit? Figure 8.55 shows the circuit's current magnitude for different resistor values.

The strength of the resonance is what's affected by the change in resistance. We can see from the plot that small resistance leads to strong resonance or a

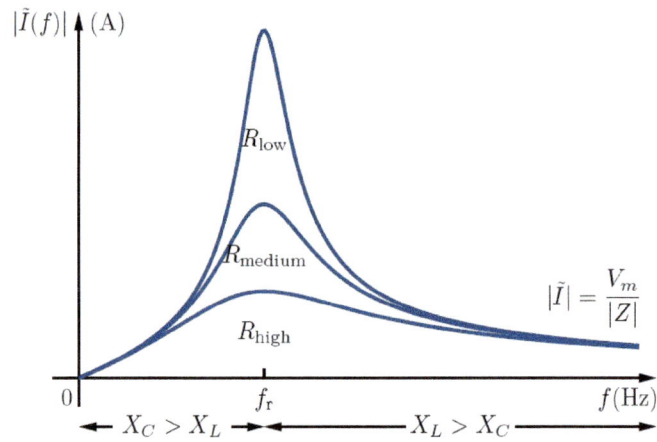

Figure 8.55: RLC series I versus f with R.

narrow bandwidth. An application for that is your cell phone filter. We wouldn't want our cell phone to pick up every frequency around it. Instead, we would want it to be highly selective and focus only on the small range of desired frequencies.

A large resistor will lead to weak resonance or a very wide bandwidth. An application for that is your car suspension. A very wide bandwidth allows for the fast dispersion of any change in terrain, such as going over speed bumps or potholes.

8.11 Exercises

Section 8.1.1

8.1. For the circuit in Figure 8.56, use nodal analysis to find $\tilde{I}$.

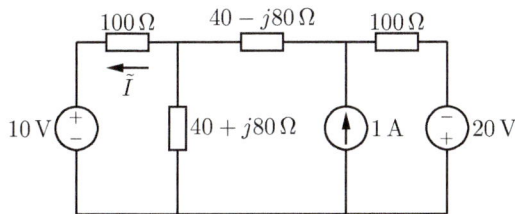

Figure 8.56

8.2. For the circuit in Figure 8.57, use nodal analysis to find $\tilde{V}_z$.

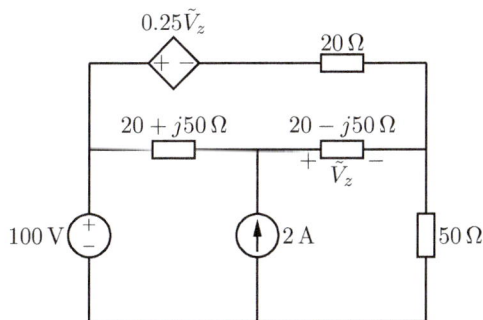

Figure 8.57

8.3. For the circuit in Figure 8.58, use nodal analysis to find $\tilde{I}$.

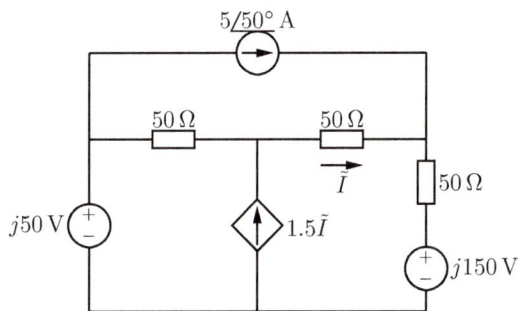

Figure 8.58

8.4. For the circuit in Figure 8.59, use nodal analysis to find $\tilde{I}$.

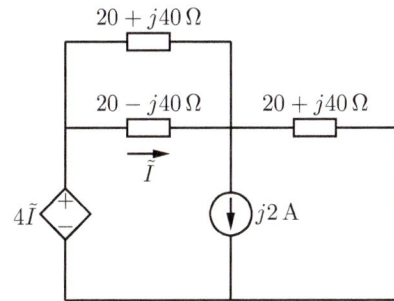

Figure 8.59

Section 8.1.2

8.5. For the circuit in Figure 8.60, find the mesh currents.

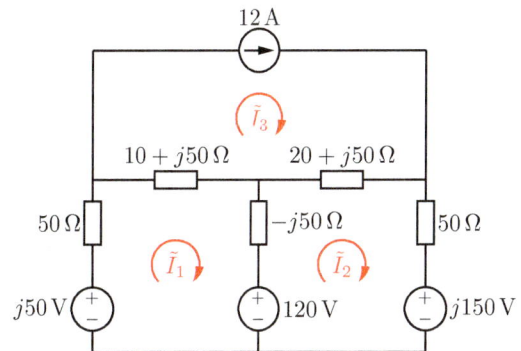

Figure 8.60

8.6. For the circuit in Figure 8.61, use mesh analysis to find $\tilde{V}_z$.

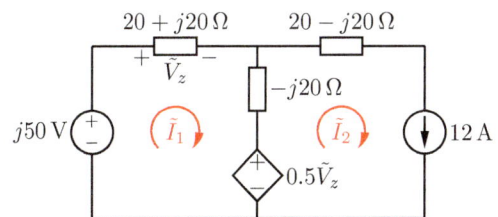

Figure 8.61

8.7. For the circuit in Figure 8.62, use mesh analysis to find $\tilde{V}_z$.

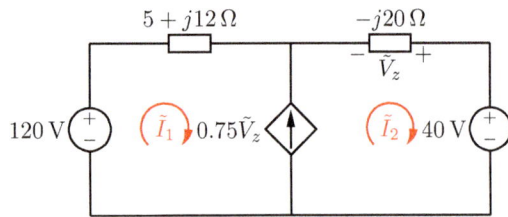

Figure 8.62

Section 8.2

8.8. For the circuit in Figure 8.63, the voltage $v_s(t) = 1.53\cos(50t + 78.69°) + 0.1\,\text{kV}$. Find $i(t)$. (Hint: The voltage has a DC component.)

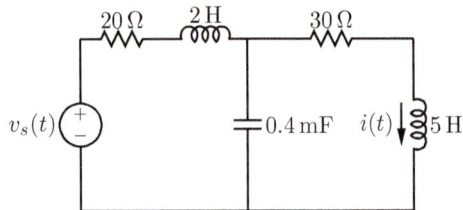

Figure 8.63

8.9. For the circuit in Figure 8.64, the current $i_s(t) = 2\cos(10t + 10°) + 4\,\text{A}$. Find $i(t)$. (Hint: The current has a DC component.)

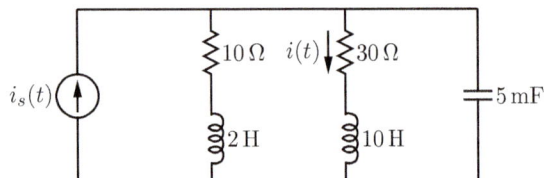

Figure 8.64

8.10. For the circuit in Figure 8.65, the voltage $v_s(t) = 559.02\cos(10t + 63.43°) + 30\,\text{V}$ and the current $i_s(t) = 2\cos(20t + 45°)\,\text{A}$. Find $v(t)$.

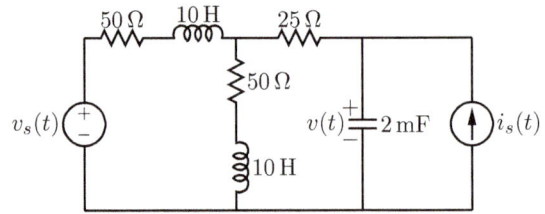

Figure 8.65

Sections 8.3 and 8.4

8.11. For the circuit in Figure 8.66, use source transformation to find the Thévenin equivalent circuit at port a-b.

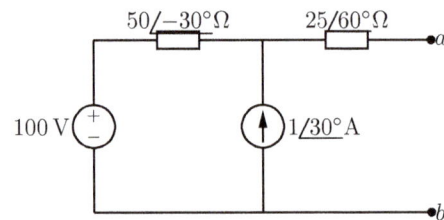

Figure 8.66

8.12. For the circuit in Figure 8.67, use source transformation to find the Norton equivalent circuit at port a-b.

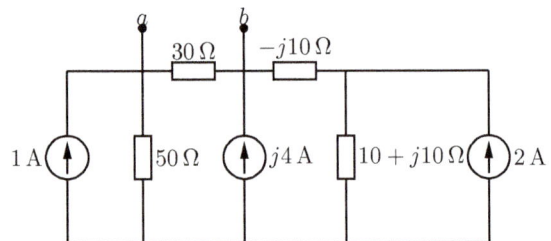

Figure 8.67

8.13. For the circuit in Figure 8.68, use source transformation to find the Norton equivalent circuit at port c-d.

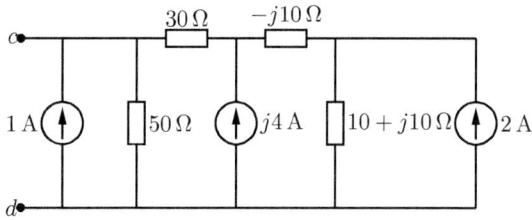

Figure 8.68

8.14. For the circuit in Figure 8.69, use source transformation to find the Norton equivalent circuit at port a-b.

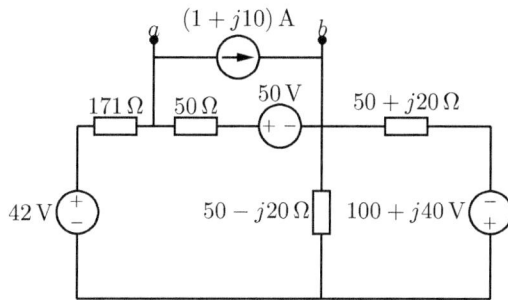

Figure 8.69

8.15. For the circuit in Figure 8.70, find the Thévenin equivalent circuit at port a-b.

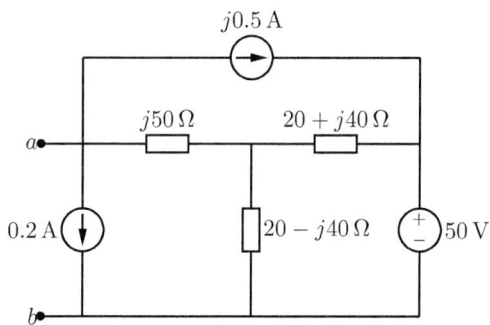

Figure 8.70

8.16. For the circuit in Figure 8.71, find the Thévenin equivalent circuit at port a-b.

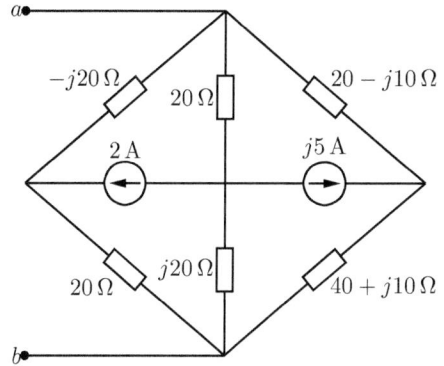

Figure 8.71

8.17. For the circuit in Figure 8.72, find the Thévenin equivalent circuit at port a-b.

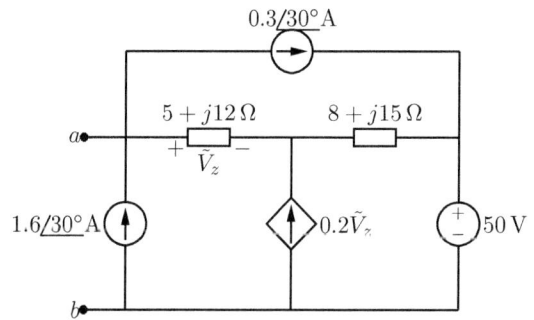

Figure 8.72

8.18. For the circuit in Figure 8.73, find the Thévenin equivalent circuit at port a-b.

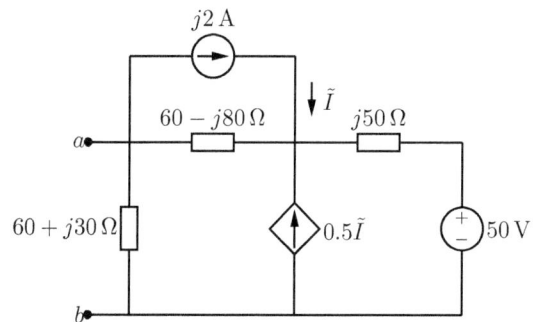

Figure 8.73

Section 8.5

8.19. For the circuit in Figure 8.74, $i(t) = 5\cos(20t - 15°)$ A. Find the instantaneous power delivered by the source and the total absorbed average power.

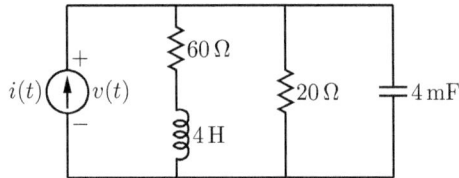

Figure 8.74

8.20. For the circuit in Figure 8.75, find the total instantaneous power delivered to the circuit and the total absorbed average power.

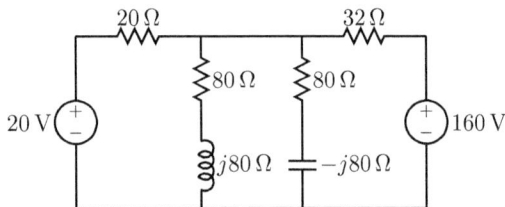

Figure 8.75

8.21. For the circuit in Figure 8.76, the total average power delivered by the source is 20 W. Find the equivalent resistance seen by the that source.

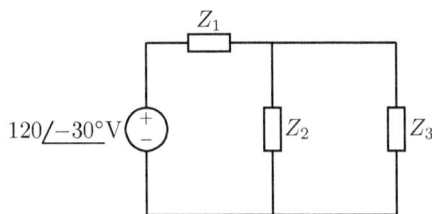

Figure 8.76

8.22. For the circuit in Figure 8.77, the total average power delivered by the source is 81 W. Find the equivalent resistance seen by the source.

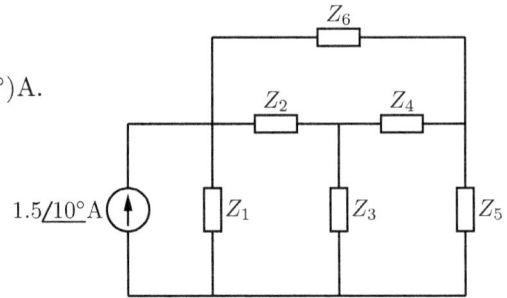

Figure 8.77

8.23. For the circuit shown in Figure 8.78, find the average power absorbed by the 15 Ω resistor.

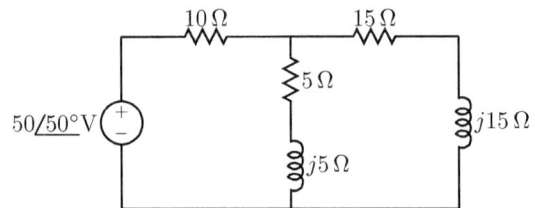

Figure 8.78

8.24. For the circuit in Figure 8.79, find the total average power.

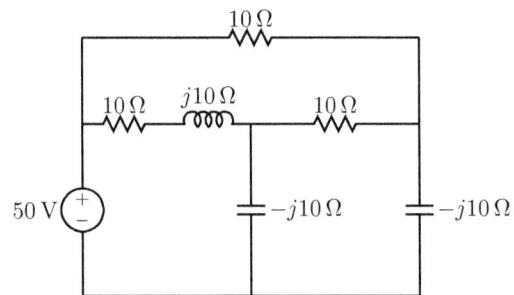

Figure 8.79

8.25. For the circuit in Figure 8.80, find the total average power.

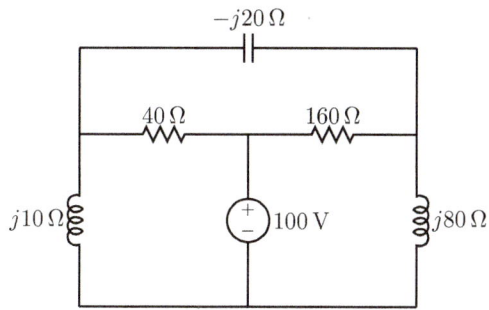

Figure 8.80

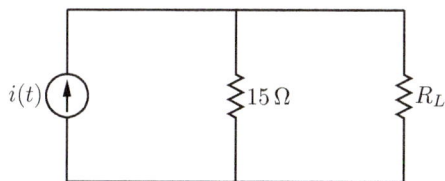

Figure 8.83

Section 8.6

8.26. The plot of the source current in Figure 8.81 is shown in Figure 8.82. Find the value of the load resistance that will make the total average power delivered by the source 312.5 W.

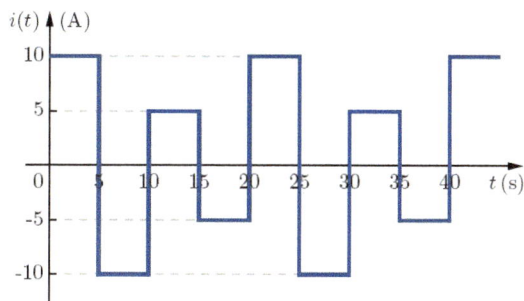

Figure 8.84

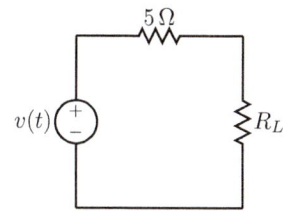

Figure 8.81

8.28. The plot of the source current through a 30 Ω is shown in Figure 8.85. Find the average power dissipated by the resistor.

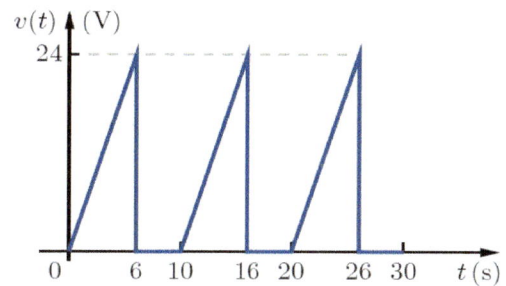

Figure 8.85

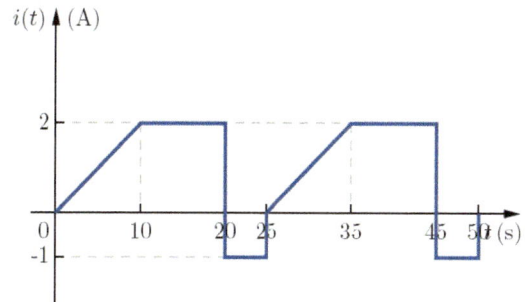

Figure 8.82

Section 8.8

8.29. For the circuit in Figure 8.86, find the value of the load impedance Z_L to achieve maximum power transformation, then find the maximum power delivered to the load.

8.27. The plot of the source voltage in Figure 8.83 is shown in Figure 8.84. Find the value of the load resistance that will make the total average power delivered by the source 5.76 W.

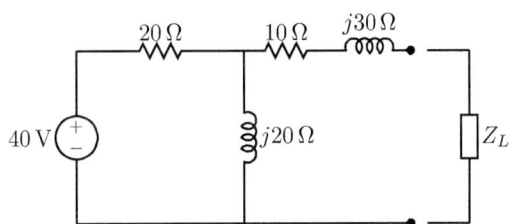

Figure 8.86

8.30. For the circuit in Figure 8.87, find the value of the load impedance Z_L to achieve maximum power transformation, then find the maximum power delivered to the load.

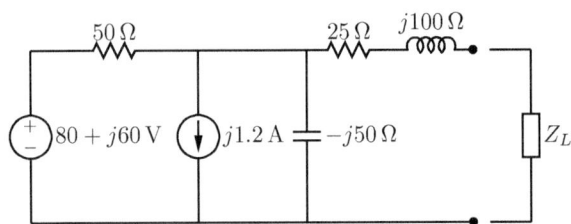

Figure 8.87

8.31. For the circuit in Figure 8.88, find the value of the load impedance Z_L to achieve maximum power transformation, then find the maximum power delivered to the load.

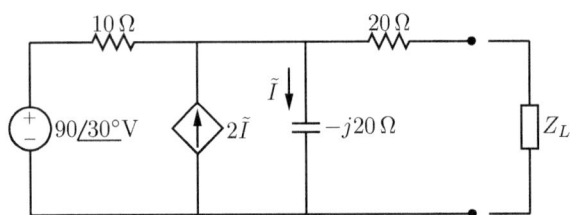

Figure 8.88

Chapter 9

Analysis and Applications of Magnetically Coupled Circuits

Introduction

So far we have discussed self-inductance, where we found the flux flowing through a structure that generated that flux. In real life, it is also important to consider cases where a structure is affected by a flux generated by other nearby structures. This allows structures to be magnetically linked or coupled through what we will define as mutual inductance. This is a very important concept, since it allows devices to "talk" to other seemingly unconnected devices through free space. Sometimes we purposely maximize the coupling between two structures: (nearly) all flux generated from one structure is going through the neighboring structure. This is the case for transformers. Other times, minimum coupling of < 0.1 is all that is needed to link two structures. This is the case in RFID applications. We will begin this chapter by exploring mutual inductance, followed by introducing its circuit representation. We will also talk about the coupling coefficient k and explore applications of both its upper and lower limits.

<table>
<tr><td>(a) Single phase transformer.</td><td>(b) Three phase transformer.</td></tr>
</table>

Figure 9.1: Transformers used in power distribution.

9.1 Mutual Inductance

When we first introduced inductors, we studied the concept of self-inductance in section 5.3.1. We started with a current loop, as seen in Figure 9.2, and used the flux to define the self-inductance as

$$L = \frac{\Phi}{I}. \tag{9.1}$$

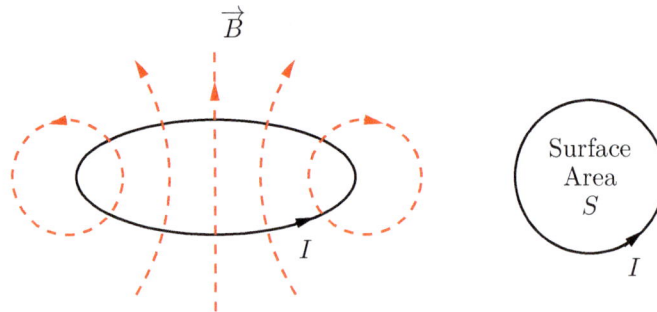

Figure 9.2: Magnetic field due to current in a closed loop.

Let us now try to figure out mutual inductance. Just like before, we will start with the same loop of Figure 9.2 and add another loop at a close distance, as shown in Figure 9.3. This second loop is not excited by any current.

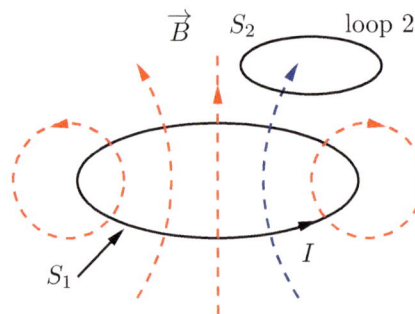

Figure 9.3: Mutual inductance

The flux that flows through the second loop due to the flux generated by the first loop can be calculated as

$$\Phi_{12} = \int_{S_2} \vec{B_1} \cdot \vec{dS_2}. \tag{9.2}$$

The mutual inductance between the two loops is defined by

$$L_{12} = \frac{N_2 \Phi_{12}}{I_1} \tag{9.3}$$

where N_2 is the number of turns in loop 2. We conclude from this that a large mutual inductance implies strong magnetic linkage and vice versa. So if the mutual inductance $L_{12} = 0$, loop 2 is not "seen" by the first loop.

Here are some additional notes to keep in mind.

- Since we use the symbol L_{12} for mutual inductance, we sometimes use the symbol L_{11} for self-inductance.

- Self-inductance is always higher than mutual inductance.

$$L_{11} > L_{12}$$

since

$$\Phi_{11} > \Phi_{12}.$$

- Due to reciprocity, $L_{12} = L_{21}$. Reciprocity is a fundamental property and it means that if you exchange the source and observation points, the field values will not change. Therefore, if you need to find the mutual inductance between two structures, you typically excite with current the structure whose magnetic field is the easiest to calculate.

9.2 Mutual Inductance Circuit Representation and the Dot Convention

Now that we understand the existence of mutual inductance, let us take a look at the simple circuit shown in Figure 9.4. The current source $i(t)$ through the first inductor generates flux that reaches the neighboring source-free inductor.

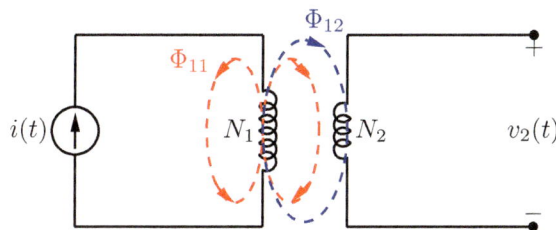

Figure 9.4: Circuit representation of mutual inductance.

The circuit in Figure 9.4 can be further simplified by removing the flux lines and adding a different notation that represents the existence of mutual inductance between the two inductors (loops) in the circuit. A double-ended arrow is used to point to the two inductors that are coupled; the value of the mutual inductance is shown above that arrow, as seen in Figure 9.5.

The next question to ask is how to define the polarity of the induced voltage in the circuit. The polarity depends on the direction of the windings. For example, it is possible to wind the coils such that an increase in the current of the first inductor leads to an increase of the induced voltage in the second inductor. Alternatively, we can wind them so that increasing the current of the first inductor will lead to a decrease of the voltage in the second inductor. Consequently, polarity depends on geometry and somehow we need to include this in our circuit model. In the literature, it is common to do this by introducing the dot convention. The dot convention is shown in Figure 9.6. We use the

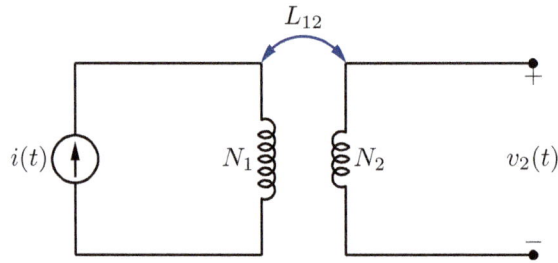

Figure 9.5: Circuit representation of mutual inductance.

dot position along with the current direction to determine the polarity of the induced voltage.

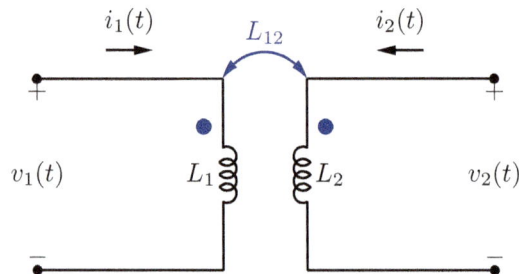

Figure 9.6: The dot convention.

For the circuit in Figure 9.6, if we did not have mutual inductance and wanted to find the voltage $v_1(t)$, we would have

$$v_1(t) = L_1 \frac{di_1(t)}{dt}. \tag{9.4}$$

Due to the mutual inductance, though, the voltage $v_1(t)$ is the sum of both the voltage induced by the current $i_1(t)$ (self-inductance) and the voltage induced by the current $i_2(t)$ going through L_2 (mutual inductance). This is where we introduce the **passive sign convention**.

The current entering the dotted terminal of one coil (inductor) induces a positive voltage oriented toward the dotted terminal of the other coil.

Since the current $i_2(t)$ is entering the dotted terminal of L_2, a positive voltage is induced and $v_1(t)$ will be

$$v_1(t) = L_1 \frac{di_1(t)}{dt} + L_{12} \frac{di_2(t)}{dt}. \tag{9.5}$$

This is called the passive sign convention. The same process can be used to find the voltage $v_2(t)$:

$$v_2(t) = L_2 \frac{di_2(t)}{dt} + L_{12} \frac{di_1(t)}{dt}. \tag{9.6}$$

Before we go any further, let us contemplate this thought. We say that a voltage is induced in a part of the circuit due to a current through another element in the circuit. This should sound very familiar; remember the definition of a current-controlled voltage source (CCVS)? We can replace the dot convention along with the mutual inductance arrow with a current-controlled voltage source. Let us take a look at how we can do this.

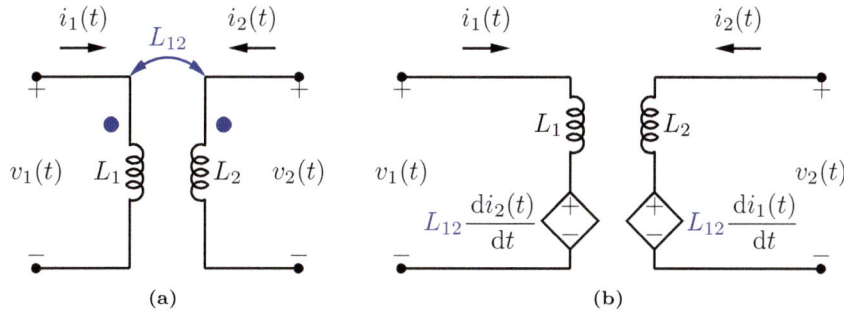

Figure 9.7: Simulating the dot convention using a CCVS.

We start by writing the voltage equations for Figure 9.7b:

$$v_1(t) = L_1 \frac{di_1(t)}{dt} + L_{12} \frac{di_2(t)}{dt} \tag{9.7}$$

and

$$v_2(t) = L_2 \frac{di_2(t)}{dt} + L_{12} \frac{di_1(t)}{dt}. \tag{9.8}$$

We can see that we get the exact same equations. Let us try other dot current combinations. The circuit in Figure 9.8 has the dot at the bottom of the inductor L_2, which means the current i_2 is leaving the dot, which will result in a negative induced voltage.

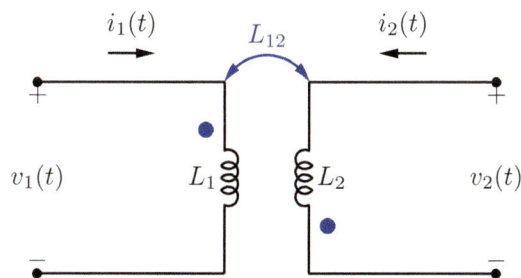

Figure 9.8: The dot convention.

In order to translate the dot convention to a CCVS, we should follow the following rules:

1. Draw a dependent voltage source in series with the inductors with dots across them.

2. The polarity of the dependent source will follow the dot. If the dot was at the top of the inductor, the positive sign of the voltage source will be at the top.

3. The value of the CCVS will be the mutual inductance multiplied by the derivative of the current responsible for the induced voltage.

4. The sign of the induced voltage will depend on whether the current enters or leaves the dot. If the current enters the dot, the voltage will be positive. Otherwise it will be negative.

Let us try to apply these steps to Figure 9.8. We will start with the CCVS in series with L_1. The CCVS source will have its positive terminal at the top, as seen in Figure 9.9b. The value of the CCVS is calculated using the current in the second loop. The current $i_2(t)$ is leaving the dot, so the voltage will be negative:

$$v_{M_1} = -L_{12}\frac{di_2(t)}{dt}.$$

The second CCVS is added in series with L_2. Since the dot is at the bottom of L_2, the negative terminal of the CCVS will be on top, as shown in Figure 9.9b. The voltage is induced by the current $i_1(t)$ entering the dotted terminal of L_1. Therefore, the induced voltage is positive and is calculated by

$$v_{M_2} = L_{12}\frac{di_1(t)}{dt}.$$

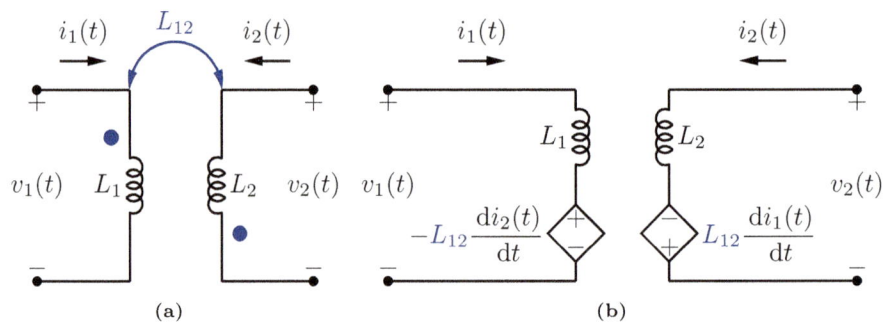

Figure 9.9: Simulating the dot convention using a CCVS.

We have two other possible combinations. We show them here in Figures 9.10 and 9.11.

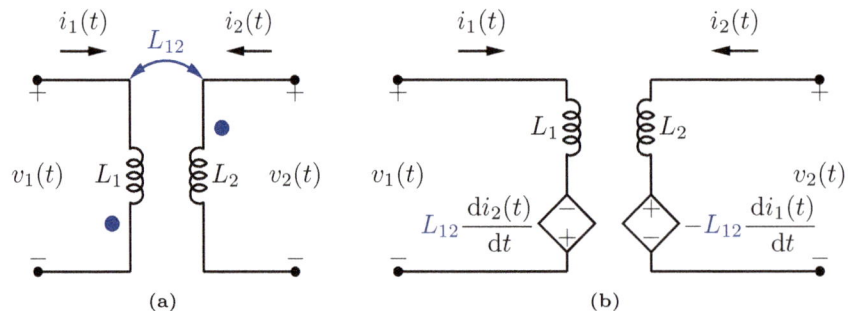

Figure 9.10: Simulating the dot convention using a CCVS.

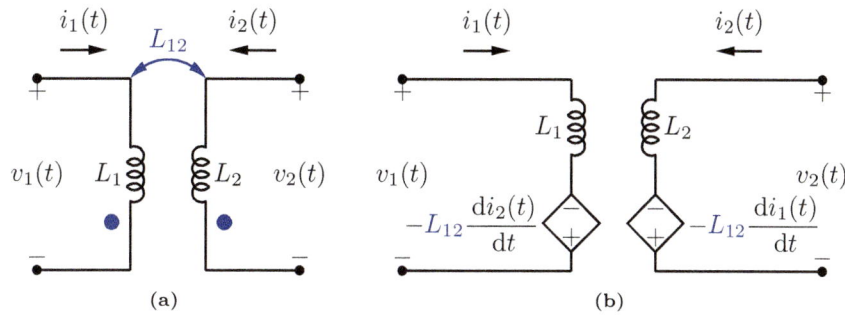

Figure 9.11: Simulating the dot convention using a CCVS.

If our circuit is driven by a sinusoidal source, we can transfer our magnetically coupled circuit to the phasor domain. Figure 9.12 demonstrates the magnetically coupled circuit in the phasor domain with the dot convention and the translation of the dot convention. Another thing to notice here is that mesh analysis is often the best solution method, since the currents through the coupled inductors are very important. The best way to solidify this new concept is to demonstrate it in an example.

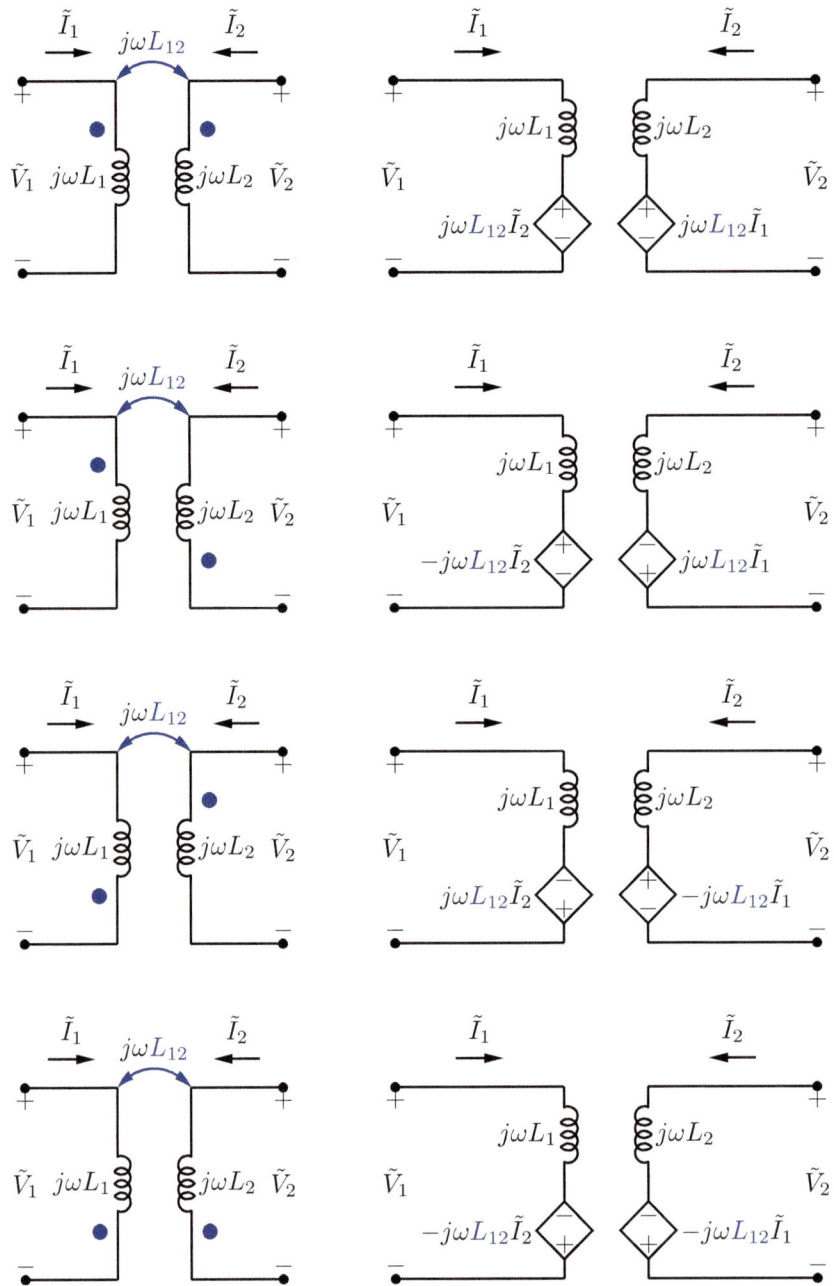

Figure 9.12: Magnetically coupled circuit in the phasor domain.

Example 9.2.1

For the circuit in Figure 9.13, find the equations for the mesh currents $\tilde{I}_1$ and $\tilde{I}_2$.

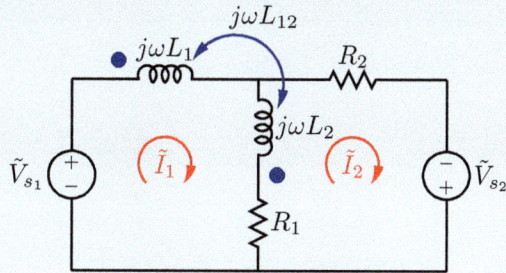

Figure 9.13: Example 9.2.1.

We can start writing the equations right away, or we can translate the circuit to the CCVS model and then write the equations. The redrawn circuit is shown in Figure 9.14.

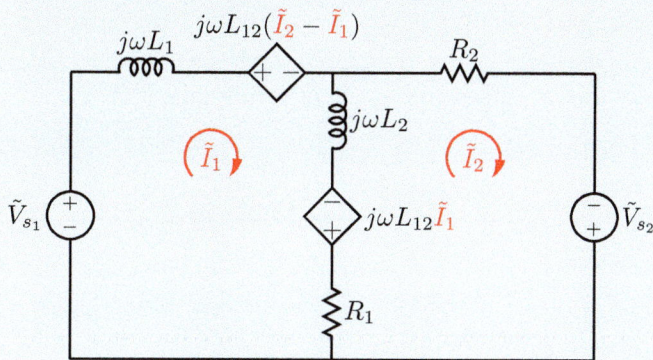

Figure 9.14: Example 9.2.1.

Starting with mesh 1, we have

$$\tilde{V}_{s_1} = j\omega L_1 \tilde{I}_1 + j\omega L_{12}(\tilde{I}_2 - \tilde{I}_1) + (R_1 + j\omega L_2)(\tilde{I}_1 - \tilde{I}_2) - j\omega L_{12}\tilde{I}_1$$
$$\tilde{V}_{s_1} = (R_1 + j\omega L_1 + j\omega L_2 - 2j\omega L_{12})\tilde{I}_1 - (R_1 + j\omega L_2 - j\omega L_{12})\tilde{I}_2$$

Moving to mesh 2 yields

$$\tilde{V}_{s_2} = R_2\tilde{I}_2 + (R_1 + j\omega L_2)\left(\tilde{I}_2 - \tilde{I}_1\right) + j\omega L_{12}\tilde{I}_1$$
$$\tilde{V}_{s_2} = -\left(R_1 + j\omega L_2 - j\omega L_{12}\right)\tilde{I}_1 + (R_1 + R_2 + j\omega L_2)\tilde{I}_2$$

In matrix form, the equations become

$$\begin{bmatrix} R_1 + j\omega L_1 + j\omega L_2 - 2j\omega L_{12} & -R_1 - j\omega L_2 + j\omega L_{12} \\ -R_1 - j\omega L_2 + j\omega L_{12} & R_1 + R_2 + j\omega L_2 \end{bmatrix}\begin{bmatrix}\tilde{I}_1 \\ \tilde{I}_2\end{bmatrix} = \begin{bmatrix}\tilde{V}_{s_1} \\ \tilde{V}_{s_2}\end{bmatrix}$$

Example 9.2.2

For the circuit in Figure 9.15, find the mesh currents $\tilde{I}_1$, $\tilde{I}_2$, and $\tilde{I}_3$.

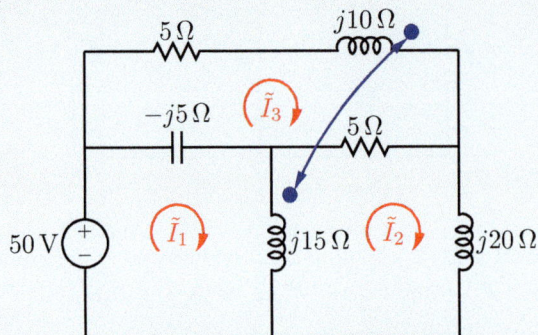

Figure 9.15: Example 9.2.2.

We can convert the dot convention and mutual inductance into the dependent source representation, as seen in Figure 9.16.

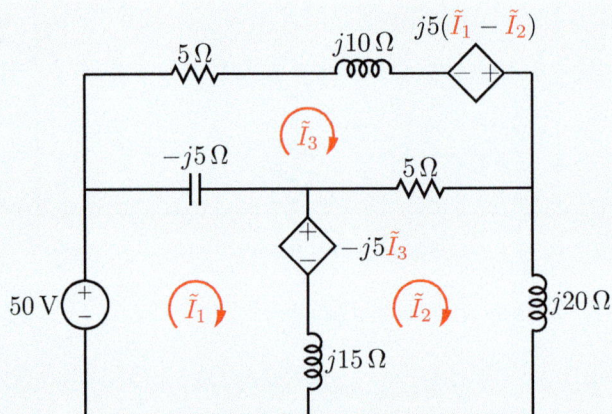

Figure 9.16: Example 9.2.2.

From here we can apply mesh analysis to find the mesh currents. The final matrix representation of the mesh equations is shown below.

$$\begin{bmatrix} j10 & -j15 & 0 \\ -j15 & 5+j35 & -5+j5 \\ 0 & -5+j5 & 10+j5 \end{bmatrix} \begin{bmatrix} \tilde{I}_1 \\ \tilde{I}_2 \\ \tilde{I}_3 \end{bmatrix} = \begin{bmatrix} 50 \\ 0 \\ 0 \end{bmatrix}$$

Solving for the mesh currents results in

$$\begin{bmatrix} \tilde{I}_1 \\ \tilde{I}_2 \\ \tilde{I}_3 \end{bmatrix} = \begin{bmatrix} 2.45 - j10.78 \\ 1.63 - j3.85 \\ -1.98 - j1.75 \end{bmatrix} = \begin{bmatrix} 11.05\underline{/-77.18°} \\ 4.18\underline{/-67.01°} \\ 2.65\underline{/-138.57°} \end{bmatrix} \text{ A}$$

9.3   Energy Stored in the Coupling Field

As we have discussed several times before, calculating power and energy (consumed or delivered) is the most important reason for finding currents and voltages in a circuit. Let us now calculate the energy stored in a circuit with two magnetically coupled inductors (Figure 9.17). As a reminder, the $i - v$ characteristics of the circuit are

$$v_1(t) = L_1 \frac{di_1(t)}{dt} + L_{12} \frac{di_2(t)}{dt} \tag{9.9}$$

$$v_2(t) = L_2 \frac{di_2(t)}{dt} + L_{12} \frac{di_1(t)}{dt}. \tag{9.10}$$

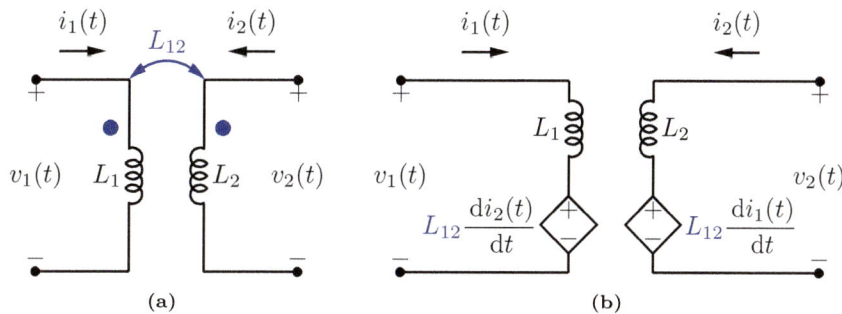

Figure 9.17: Simulating the dot convention using a CCVS.

In general, the equation for computing instantaneous power in any circuit is

$$p(t) = v(t)i(t). \tag{9.11}$$

To simplify the calculation, we will make the following assumptions:

1. Assume that at time zero ($t = 0$), both currents are zero ($i_1(0) = i_2(0) = 0$).

2. Assume that there is no energy stored in the circuit at $t = 0$.

It is also simpler to increase each current separately. First let us increase $i_1(t)$ from 0 to some current $i_1(t_1)$ achieved at $t = t_1$. The current $i_2(t)$ is held at zero for the entire time interval $0 \le t \le t_1$. The voltages $v_1(t)$ and $v_2(t)$ can be found as

$$v_1(t) = L_1 \frac{di_1(t)}{dt} \tag{9.12}$$

$$v_2(t) = L_{12} \frac{di_1(t)}{dt}. \tag{9.13}$$

The instantaneous power at any time between 0 and t_1 can be calculated as

$$p(0 < t < t_1) = p_1(t) = v_1(t)i_1(t) + v_2(t)i_2(t) = i_1(t)L_1 \frac{di_1(t)}{dt}. \tag{9.14}$$

The energy stored between the time $t = 0$ and t_1 can be calculated by

$$w_1 = \int_0^{t_1} p_1(t)\,dt = \int_0^{t_1} i_1(t)L_1 \frac{di_1(t)}{dt}\,dt = \int_0^{i_1(t_1)} L_1 i_1(t)\,di_1$$

$$= \frac{1}{2}L_1 i_1^2(t_1). \tag{9.15}$$

Next, let us keep the first current constant—that is, $i_1(t) = i_1(t_1)$—and increase the second current $i_2(t)$ to some value $i_2(t_2)$ achieved at $t = t_2 > t_1$. The voltages again are calculated using our general expressions:

$$v_1(t) = L_1 \frac{di_1(t)}{dt} + L_{12} \frac{di_2(t)}{dt} \tag{9.16}$$

$$v_2(t) = L_2 \frac{di_2(t)}{dt} + L_{12} \frac{di_1(t)}{dt}. \tag{9.17}$$

Since $i_1(t)$ is held constant for $t_1 \le t \le t_2$, its derivative will be zero—that is, $\frac{di_1(t)}{dt} = 0$. Hence,

$$v_1(t) = L_{12} \frac{di_2(t)}{dt} \tag{9.18}$$

$$v_2(t) = L_2 \frac{di_2(t)}{dt}. \tag{9.19}$$

The instantaneous power at any time between t_1 and t_2 can be calculated as

$$p(t_1 < t < t_2) = p_2(t) = v_1(t)i_1(t) + v_2(t)i_2(t)$$

$$= i_1(t_1)L_{12} \frac{di_2(t)}{dt} + i_2(t)L_2 \frac{di_2(t)}{dt}. \tag{9.20}$$

The energy stored between $t = t_1$ and $t = t_2$ can be calculated as

$$w_2 = \int_{t_1}^{t_2} p_2(t)\,dt = \int_{t_1}^{t_2} i_1(t_1)L_{12} \frac{di_2(t)}{dt}\,dt + \int_{t_1}^{t_2} i_2(t)L_2 \frac{di_2(t)}{dt}\,dt$$

$$= \int_0^{i_2(t_2)} i_1(t_1)L_{12}\,di_2 + \int_0^{i_2(t_2)} i_2(t)L_2\,di_2(t)$$

$$= L_{12}i_1(t_1)i_2(t_2) + \frac{1}{2}L_2 i_2^2(t_2). \tag{9.21}$$

Therefore, the total energy stored at $0 < t < t_2$ is

$$w_T = w_1 + w_2 = \frac{1}{2}L_1i_1^2(t_1) + \frac{1}{2}L_2i_2^2(t_2) + L_{12}i_1(t_1)i_2(t_2). \qquad (9.22)$$

While the above equation was derived by sequentially increasing the currents at two time intervals $0 \leq t \leq t_1$ and $t_1 \leq t \leq t_2$, it is important to recognize that t_1, t_2, $i_1(t_1)$, and $i_2(t_2)$ were completely arbitrary. Hence, (9.22) is generally applicable.

If one of the currents leaves the dot as seen in Figure 9.18, then the voltage equations will differ slightly:

$$v_1(t) = L_1\frac{di_1(t)}{dt} - L_{12}\frac{di_2(t)}{dt} \qquad (9.23)$$

$$v_2(t) = L_2\frac{di_2(t)}{dt} - L_{12}\frac{di_1(t)}{dt}. \qquad (9.24)$$

This in return will introduce a negative sign at the mutual inductance part of the equation, so the total stored energy will be

$$w_T = \frac{1}{2}L_1i_1^2(t_1) + \frac{1}{2}L_2i_2^2(t_2) - L_{12}i_1(t_1)i_2(t_2). \qquad (9.25)$$

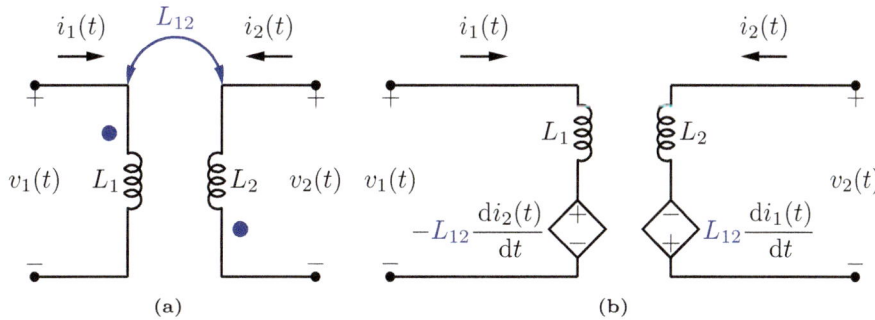

Figure 9.18: Simulating the dot convention using a CCVS.

In general, the total energy stored in a magnetically coupled circuit can be calculated using

$$\boxed{w_T = \frac{1}{2}L_1i_1^2 + \frac{1}{2}L_2i_2^2 \pm L_{12}i_1i_2.} \qquad (9.26)$$

9.3.1 The Coupling Coefficient (k)

As we have discussed before, mutual inductance is always smaller than self-inductance. Specifically, we know that if two inductors are physically far from each other, their fields will not interact strongly and their mutual inductance is practically zero. On the other hand, if they physically overlap each other, they will be nearly perfectly coupled. Consequently, the magnetic coupling can range from 0 to 100%. We would like to mathematically describe this by defining a coupling coefficient k that will range from 0 to 1—that is,

$$0 \leq k \leq 1. \tag{9.27}$$

To do so, we define the coupling coefficient k as

$$\boxed{k = \frac{L_{12}}{\sqrt{L_1 L_2}}.} \tag{9.28}$$

Let us now show that (9.27) is true from an energy perspective. To start, let us look at the energy equation we just established in (9.26). Since this network includes only *passive* elements, the energy stored can *never* be negative. Based on this, we can find the maximum value of L_{12} in (9.26) that will never allow the energy stored w_T to be negative. To do so, we will work on the energy equation with the negative sign

$$w_T = \frac{1}{2} L_1 i_1^2 + \frac{1}{2} L_2 i_2^2 - L_{12} i_1 i_2 \tag{9.29}$$

and find the value of L_{12} to have

$$w_T \geq 0. \tag{9.30}$$

Adding and subtracting $\sqrt{L_1 L_2} i_1 i_2$ from (9.29) to complete the square yields

$$w_T = \frac{1}{2} (\sqrt{L_1} i_1 - \sqrt{L_2} i_2)^2 + \sqrt{L_1 L_2} i_1 i_2 - L_{12} i_1 i_2 \geq 0. \tag{9.31}$$

Rearranging this yields

$$w_T = \underbrace{\frac{1}{2} (\sqrt{L_1} i_1 - \sqrt{L_2} i_2)^2}_{\geq 0} + (\sqrt{L_1 L_2} - L_{12}) i_1 i_2 \geq 0. \tag{9.32}$$

For a nonnegative energy, we need

$$L_{12} \leq \sqrt{L_1 L_2}. \tag{9.33}$$

Example 9.3.1

For the circuit in Figure 9.19, find the mesh currents $\tilde{I}_1$ and $\tilde{I}_2$ for $k = 0.15$ and $k = 0.9$.

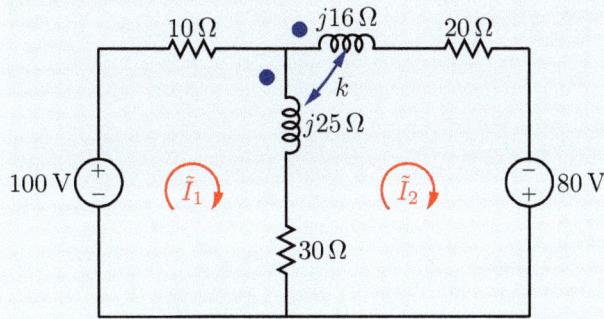

Figure 9.19: Example 9.3.1.

First, we will need to find the impedance of the mutual in terms of the coupling coefficient k.

$$j\omega L_{12} = jk\omega\sqrt{L_1 L_2} = jk\sqrt{25 \times 16} = j20k.$$

Now, we can convert the dot convention and mutual inductance into the dependent source representation, as seen in Figure 9.20.

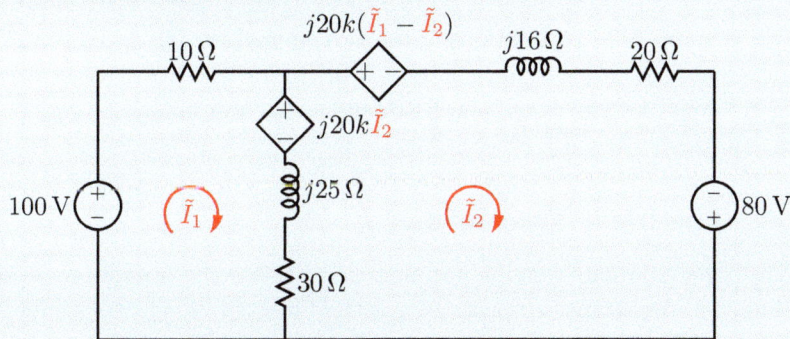

Figure 9.20: Example 9.3.1.

From here we can apply mesh analysis to find the mesh currents. The simplest form of the mesh equations is shown below in matrix form.

$$\begin{bmatrix} 40 + j25 & -30 + j20k - j25 \\ -30 + j20k - j25 & 50 + j41 - j40k \end{bmatrix} \begin{bmatrix} \tilde{I}_1 \\ \tilde{I}_2 \end{bmatrix} = \begin{bmatrix} 100 \\ 80 \end{bmatrix}$$

We can now solve the matrix by plugging in the different values of k. For $k = 0.15$, we get

$$\begin{bmatrix} 40 + j25 & -30 - j22 \\ -30 - j22 & 50 + j35 \end{bmatrix} \begin{bmatrix} \tilde{I}_1 \\ \tilde{I}_2 \end{bmatrix} = \begin{bmatrix} 100 \\ 80 \end{bmatrix}$$

Solving for the mesh currents results in

$$\begin{bmatrix} \tilde{I}_1 \\ \tilde{I}_2 \end{bmatrix} = \begin{bmatrix} 5.39 - j2.69 \\ 4.39 - j2.32 \end{bmatrix} = \begin{bmatrix} 6.02\underline{/-26.52°} \\ 4.97\underline{/27.82°} \end{bmatrix} \text{ A}$$

For $k = 0.9$, we get

$$\begin{bmatrix} 40 + j25 & -30 - j7 \\ -30 - j7 & 50 + j5 \end{bmatrix} \begin{bmatrix} \tilde{I}_1 \\ \tilde{I}_2 \end{bmatrix} = \begin{bmatrix} 100 \\ 80 \end{bmatrix}$$

Solving for the mesh currents results in

$$\begin{bmatrix} \tilde{I}_1 \\ \tilde{I}_2 \end{bmatrix} = \begin{bmatrix} 4.12 - j3.1 \\ 4.33 - j1.72 \end{bmatrix} = \begin{bmatrix} 5.16\underline{/-36.96°} \\ 4.67\underline{/21.66°} \end{bmatrix} \text{ A}$$

9.4   Applications

We will examine here applications of the two ends of the spectrum of the coupling coefficient k. We will take a look at a the ideal transformer as an example of perfect coupling $k = 1$ and the RFID circuit as an example of weaker coupling.

9.4.1   The Ideal Transformer $k = 1$

A transformer is a four-terminal element that utilizes the benefits of magnetic coupling. An ideal transformer is characterized by containing two or more coils tightly wound around a highly permeable iron core. The high permeability allows an easy path for the flux lines. That is how we achieve perfect coupling $k = 1$.

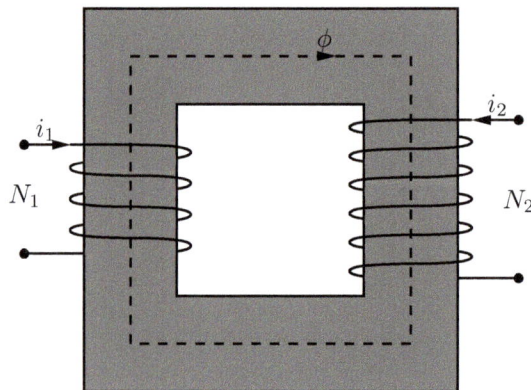

Figure 9.21: The iron-core transformer.

An ideal transformer is also characterized by the large number of turns per coil.

Remember that the inductance is directly proportional to the number of turns, which means the larger the number of turns, the higher the inductance. Ideally, the large number of turns will lead to infinite inductance. The circuit symbol for an ideal transformer is shown in Figure 9.22.

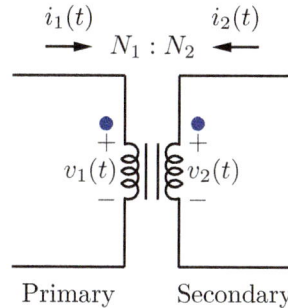

Figure 9.22: The iron-core transformer circuit representation.

The transformer, in general, has a primary side and a secondary side. The primary side is where the source ($i_1(t)$) is connected, while the secondary side is where the load ($i_2(t)$) is connected. Since both coils experience the same flux, as seen in Figure 9.21, the voltages $v_1(t)$ and $v_2(t)$ can be calculated using the following formula:

$$v(t) = \frac{d\lambda}{dt} = N\frac{d\Phi}{dt}. \tag{9.34}$$

Applying (9.34) to our transformer voltages while paying attention to the dots,

$$v_1(t) = N_1\frac{d\Phi}{dt} \tag{9.35}$$

and

$$v_2(t) = N_2\frac{d\Phi}{dt}. \tag{9.36}$$

We relate the primary to the secondary voltages by finding their ratio:

$$\frac{v_2(t)}{v_1(t)} = \frac{N_2\frac{d\Phi}{dt}}{N_1\frac{d\Phi}{dt}} \tag{9.37}$$

$$\boxed{\frac{v_2(t)}{v_1(t)} = \frac{N_2}{N_1}.} \tag{9.38}$$

This is how we define the turns ratio n:

$$\boxed{n = \frac{N_2}{N_1}.} \tag{9.39}$$

If $N_2 > N_1$, we have a step-up transformer, which implies that the secondary voltage is higher than the input voltage. A step-up transformer can be used in transmitting power over long distances: for the same amount of power to be transferred, a higher voltage will result in a smaller current which will in turn result in minimizing power loss. If $N_1 > N_2$, we have a step-down transformer, which means that the output voltage is smaller than the input. A step-down transformer is used to step down the voltage just before it reaches your house. If $N_1 = N_2$, we have a buffer transformer. This kind of transformer can be used between systems to isolate unwanted harmonics from entering or leaving a network.

The current, on the other hand, is opposite to the voltage. The secondary to primary current ratio is

$$\frac{i_2(t)}{i_1(t)} = -\frac{N_1}{N_2} = -\frac{1}{n}. \tag{9.40}$$

Example 9.4.1

The ideal transformer doesn't store or generate energy. Is this statement true? Provide an explanation.

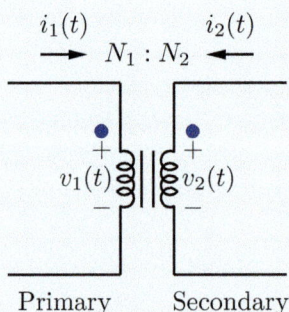

Figure 9.23: Example 9.4.1.

In order to find the energy stored or generated by the transformer, we need to first find the instantaneous power because

$$w(t) = \int_{t_i}^{t_f} p(x)\,\mathrm{d}x.$$

The instantaneous power in general is

$$p(t) = v(t)i(t).$$

Since the transformer is a two-terminal device, the instantaneous power will be the sum of the powers at its two ports,

$$p(t) = v_1(t)i_1(t) + v_2(t)i_2(t).$$

We know that the primary and secondary voltages and currents are related to each other by the turns ratio. Using equations (9.38) and (9.40),

$$p(t) = v_1(t)i_1(t) + nv_1(t)\frac{i_1(t)}{-n} = 0.$$

This indicates that the energy stored/generated by the transformer is

$$w(t) = \int_{t_i}^{t_f} p(x)\,\mathrm{d}x = \int_{t_i}^{t_f} 0\,\mathrm{d}x = 0.$$

The statement we started with is true: an ideal transformer neither generates nor stores energy.

In practice, the transformer is used in an input-output configuration, where we have an input voltage and current in the primary side and the output current leaving the transformer from the secondary side. The practical configuration of the transformer is shown in Figure 9.24.

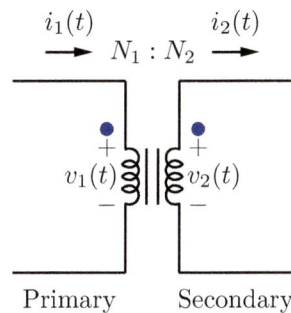

Figure 9.24: The ideal transformer with input-output configuration.

The secondary to primary current ratio for this configuration is

$$\boxed{\frac{i_{\text{out}}(t)}{i_{\text{in}}(t)} = \frac{1}{n}.}$$
(9.41)

Other ideal transformer dot configurations are shown in Figure 9.25.

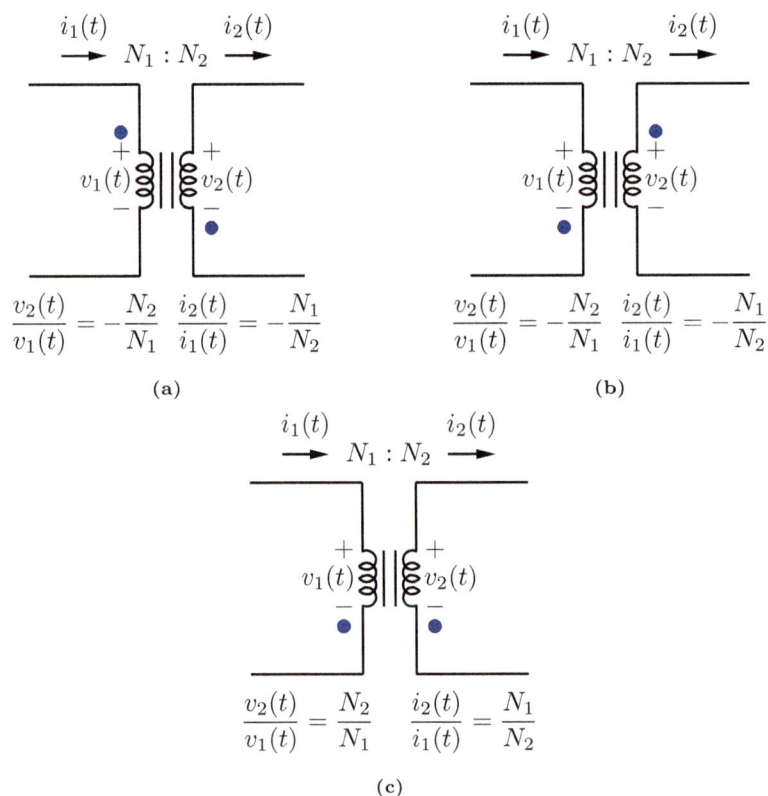

Figure 9.25: Different dot configurations for a practical ideal transformer.

If the transformer was driven by a sinusoidal source, which is the case in most applications, then we can take it to the phasor domain. Figure 9.26 shows a general ideal transformer circuit in the phasor domain.

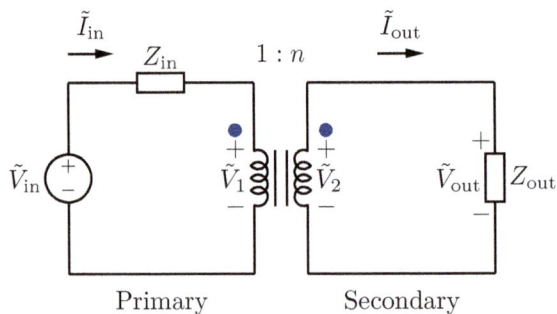

Figure 9.26: General input-output configuration transformer circuit in the phasor domain.

Let us try to write the KVL equation to both the primary and secondary sides of the circuit in 9.26.

$$\tilde{V}_{\text{in}} = Z_{\text{in}}\tilde{I}_{\text{in}} + \tilde{V}_1 \tag{9.42}$$

$$\tilde{V}_2 = Z_{\text{out}}\tilde{I}_{\text{out}}. \tag{9.43}$$

Using (9.38) and (9.41),

$$\tilde{V}_1 = \frac{\tilde{V}_2}{n} \tag{9.44}$$

$$\tilde{I}_{\text{out}} = \frac{\tilde{I}_{\text{in}}}{n}. \tag{9.45}$$

We can now substitute the output current in (9.43) with (9.45):

$$\tilde{V}_2 = Z_{\text{out}} \frac{\tilde{I}_{\text{in}}}{n}. \tag{9.46}$$

Rewriting (9.42),

$$\tilde{V}_{\text{in}} = Z_{\text{in}}\tilde{I}_{\text{in}} + \frac{\tilde{V}_2}{n} = Z_{\text{in}}\tilde{I}_{\text{in}} + \frac{Z_{\text{out}}\frac{\tilde{I}_{\text{in}}}{n}}{n}$$

$$= \left(Z_{\text{in}} + \frac{Z_{\text{out}}}{n^2} \right) \tilde{I}_{\text{in}}. \tag{9.47}$$

We can see from (9.47) that the primary side of the transformer sees a reflection of the output impedance. The turns ratio allows us to transfer impedances between the sides of the transformer. The circuit in Figure 9.26 can be redrawn with everything referenced to the primary side.

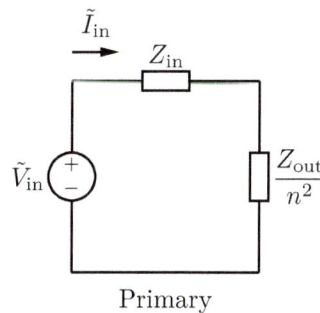

Figure 9.27: The circuit in Figure 9.26.

Example 9.4.2

For the ideal transformer in Figure 9.28, find the equivalent input impedance, the input current $\tilde{I}_{\text{in}}$, and the output voltage $\tilde{V}_{\text{out}}$.

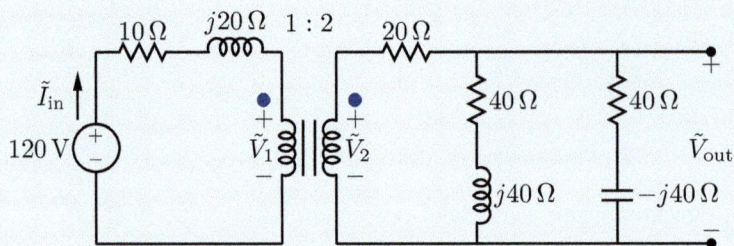

Figure 9.28: Example 9.4.2.

We can start by simplifying the circuit a little bit by combining the parallel impedances at the output together, as seen in Figure 9.29.

$$Z_{\text{out}} = \frac{(40 + j40)(40 - j40)}{40 + j40 + 40 - j40} = 40\,\Omega.$$

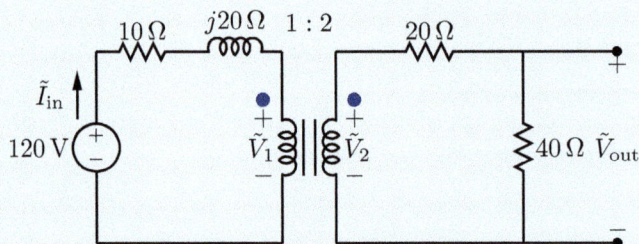

Figure 9.29: Example 9.4.2.

We can now transfer the total secondary impedance to the primary side of the transformer, as seen in Figure 9.30.

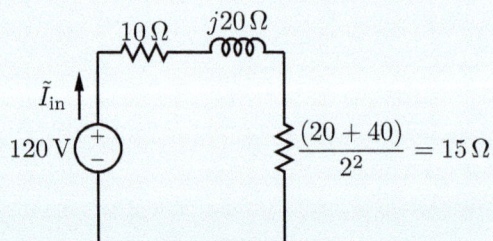

Figure 9.30: Example 9.4.2.

The equivalent input impedance is

$$Z_{\text{eq}}(in) = 25 + j20\,\Omega.$$

The input current is calculated using our frequency domain Ohm's law,

$$\tilde{I}_{\text{in}} = \frac{120}{25 + j20} = 3.75\underline{/-38.66°}\,\text{A}.$$

To find $\tilde{V}_{\text{out}}$, we can transfer the input current we just calculated to the secondary side of the transformer and then use Ohm's law again to find the the output voltage as follows:

$$\tilde{I}_2 = \frac{\tilde{I}_{\text{in}}}{n} = \frac{3.75\underline{/-38.66°}}{2} = 1.875\underline{/-38.66°}\,\text{A}.$$

Then

$$\tilde{V}_{\text{out}} = 40\tilde{I}_2 = 75\underline{/-38.66°}\,\text{V}.$$

Example 9.4.3

For the ideal transformer circuit in Figure 9.28, $|\tilde{V}_2| = 50\,\text{V}$, and $|\tilde{I}_s| = 0.9\,\text{A}$. Find the equivalent input impedance and the input currents $|\tilde{V}_s|$, $|\tilde{I}_{R_1}|$, and $|\tilde{I}_{R_2}|$.

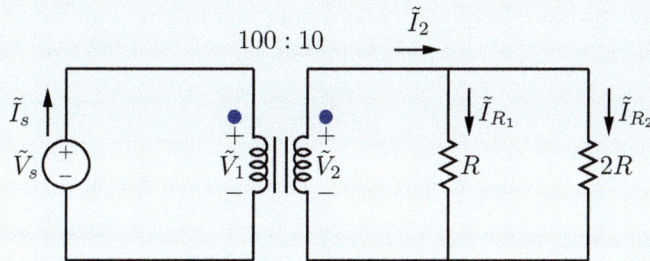

Figure 9.31: Example 9.4.3.

We can find the source voltage right away by transferring the secondary voltage to the primary side of the transformer.

$$|\tilde{V}_s| = |\tilde{V}_1| = n|\tilde{V}_2| = \frac{100}{10}(50) = 500\,\text{V}.$$

Transferring the source current to the secondary side of the transformer,

$$|\tilde{I}_2| = n|\tilde{I}_1| = \frac{100}{10}0.9 = 9\,\text{A}.$$

Using current division, we can find the currents through the resistors R_1 and R_2:

$$|\tilde{I}_{R_1}| = \frac{2R}{R + 2R}|\tilde{I}_2| = \frac{2}{3}(9) = 6\,\text{A},$$

and

$$|\tilde{I}_{R_2}| = \frac{R}{R + 2R}|\tilde{I}_2| = \frac{1}{3}(9) = 3\,\text{A}.$$

9.4.2 Radio-frequency Identification (RFID) $k < 0.1$

Have you ever wondered how this small little sticker on the items you buy can trip an alarm if not deactivated properly before leaving the store? Have you ever wondered why you have to pass by the two pillars at the door for it to activate? How close or far do you have to be to trip the alarm?

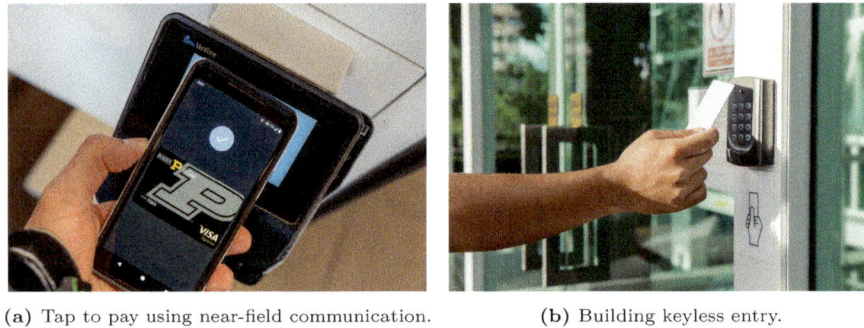

(a) Tap to pay using near-field communication.　　(b) Building keyless entry.

Figure 9.32: Applications of RFIDs.

The RFIDs depend on coupling to provide power. Just like in the ideal transformer with a primary and a secondary side, the RFIDs are made of two main parts, the reader and the transponder. The transponder is your card, the sticker on the product you are about to buy, or anything that needs to be read or sensed. The beauty of this system is that there is no need to attach a power source to the transponder, which means:

1. Products can stay on shelves for a long time without worrying about recharging or changing batteries. Stick it and forget it.

2. Sourceless transponders make lighter items. Can you imagine every card in your wallet having a battery?

3. False alarms are minimized because the power is supplied by the reader, which needs to be within a certain distance from the transponder.

4. Unlike bar codes, the transponder (tag) can be hidden or concealed.

In the case of the ideal transformer, we were able to achieve maximum coupling by physically attaching the winding to the same core. In the case of RFIDs, we will have to settle for weaker coupling due to the distance between the reader and transponder. Figure 9.33 shows a simple circuit of a reader/transponder circuit.

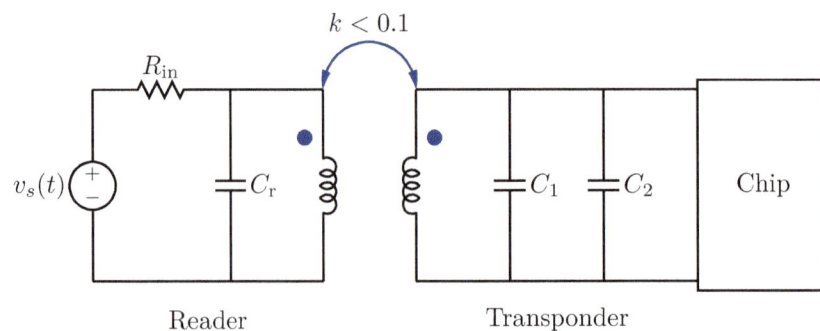

Figure 9.33: A simple circuit of a reader/transponder circuit.

The reader powers the transponder circuit due to coupling and allows it to send the information stored. The information can vary from a simple single bit to a stream of bits. Just remember that the more information you want, the larger the circuit, which in turn will require more power or closer distance.

9.5 Exercises

Section 9.1

9.1. For the circuit in Figure 9.34, find the output voltage $\tilde{V}_o$.

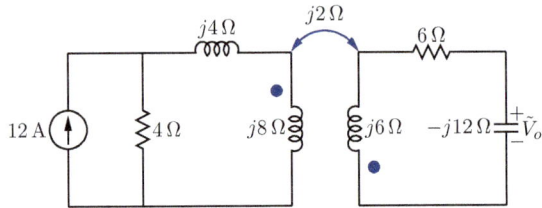

Figure 9.34

9.2. For the circuit in Figure 9.35, find the output current $\tilde{I}_o$.

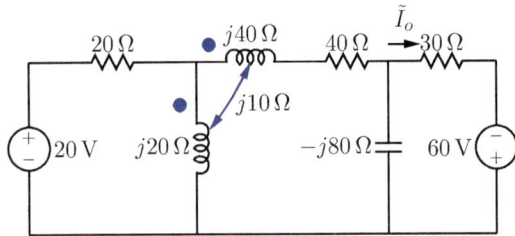

Figure 9.35

9.3. For the circuit in Figure 9.36, find the capacitor's current $\tilde{I}_C$.

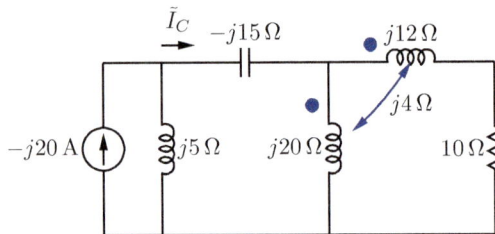

Figure 9.36

9.4. For the circuit in Figure 9.37, find the capacitor's voltage $\tilde{V}_C$.

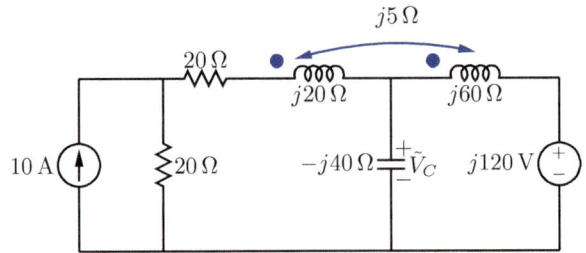

Figure 9.37

9.5. For the circuit in Figure 9.38, find the resistor's voltage $\tilde{V}_R$.

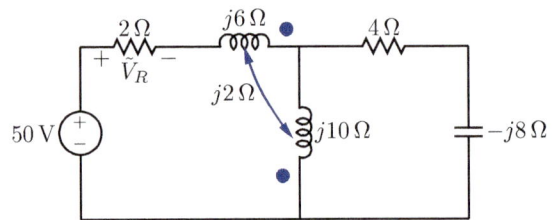

Figure 9.38

Section 9.3.1

9.6. For the circuit in Figure 9.39, find the coupling coefficient k, the output voltage $\tilde{V}_o$, and the power absorbed by the $10\,\Omega$ resistor.

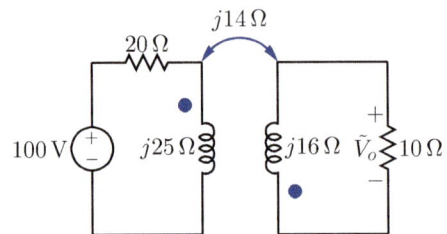

Figure 9.39

9.7. For the circuit in Figure 9.40, find the output current $\tilde{I}_o$ for $k = 0.9$, $k = 0.5$, and $k = 0.1$.

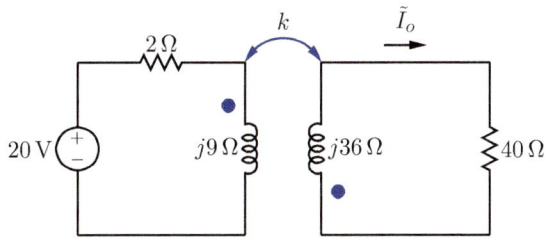

Figure 9.40

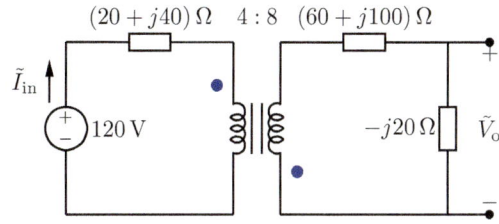

Figure 9.43

Section 9.4.1

9.8. The circuit in Figure 9.41 has the following impedances: $Z_1 = (25.5 + j4.5)\,\Omega$, $Z_2 = (43 + j62)\,\Omega$, $Z_3 = 20\,\Omega$, $Z_4 = (12 - j30)\,\Omega$, and $Z_5 = (48 + j30)\,\Omega$. Find the output voltage $\tilde{V}_o$.

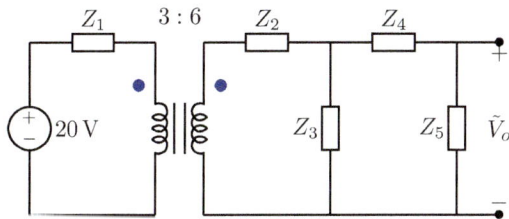

Figure 9.41

9.9. For the circuit in Figure 9.42, find the output current $\tilde{I}_o$ and the equivalent impedance seen by the source at the primary side of the ideal transformer.

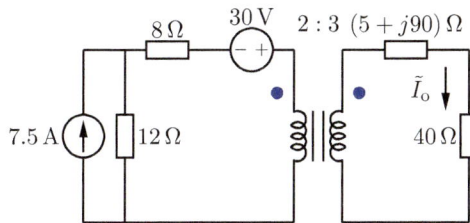

Figure 9.42

9.10. For the circuit in Figure 9.43, find the output voltage $\tilde{V}_o$ and the equivalent impedance seen by the source at the primary side of the ideal transformer.

9.11. The circuit in Figure 9.44 has a voltage source

$$\tilde{V}_s = \frac{120}{\sqrt{2}}\underline{/45°}\,\text{V}.$$

The impedance values are as follows:

$$Z_1 = (50 + j50)\,\Omega,$$
$$Z_2 = (50 - j50)\,\Omega,$$
$$Z_3 = (25 - j250)\,\Omega,$$
$$Z_4 = 400\,\Omega,$$
$$Z_5 = (100 + j1000)\,\Omega.$$

Find the output current $\tilde{I}_o$ and the equivalent impedance seen by the source at the primary side of the ideal transformer.

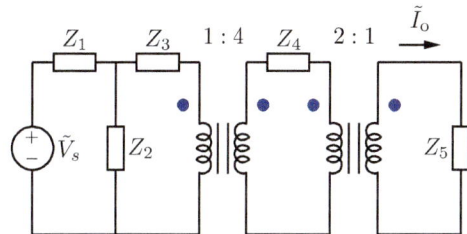

Figure 9.44

Chapter 10

Semiconductors

Introduction

Semiconductor materials are foundational to the construction of nearly all modern electronic devices. They can be used to make basic circuit elements like resistors or capacitors and are essential to the advanced components in a wide variety of applications such as computer logic, switching, amplifiers, sensors, lighting, and solar panels. The name "semiconductor" comes from the fact that their conductivities generally fall in a range between those of insulators, like glass, and those of conductors, like copper.

We begin our study of semiconductors with a review of the quantum mechanical model of atoms and atomic bonding in sections 10.1 and 10.2, respectively. These two sections provide supplemental material that ties the semiconductor properties presented in the rest of the chapter to principles typically introduced in chemistry and physics courses. If you are already familiar with these concepts (or are in a hurry), you can safely skip straight to section 10.3.

In section 10.3, we introduce the crystal structure of silicon (the most widely used semiconductor) and how this structure gives rise to discrete bands of allowable energy states for the electrons within it. Section 10.4 examines the effect of exciting electrons across the band-gap and introduces the electron's complementary pseudo particle, the hole. This section concludes with a discussion of how the available energy states influence the conductivity of a material. Finally, in section 10.5, we show how the concentration of carriers within a semiconductor can be changed by intentionally adding impurities known as dopants, how those carriers are distributed with respect to energy, and how they move in response to concentration gradients and electric fields.

10.1 The Quantum Mechanical Model of Atoms

10.1.1 Atomic Orbitals

Atoms are composed of protons, neutrons, and electrons. The protons and neutrons of an atom are held tightly together in its center, forming a small dense nucleus, while the electrons occupy a much larger volume of space around the nucleus known as the electron cloud. For a sense of the difference in scale, the radius of an atom's nucleus is about 13,000 to 30,000 times smaller than the radius of its electron cloud. For example, if an atom were enlarged such that the electron cloud became the size of a world cup soccer stadium ($\approx 130\,\text{m}$ radius), the atom's nucleus would still be much smaller than a golf ball ($\approx 2.13\,\text{cm}$ radius).

The exact location of the electrons around an atom are unknown, but the Schrödinger equation predicts their probability distribution. We won't cover the Schrödinger equation in any detail, but it tells us not only where we are likely to find electrons of a given energy but also that electrons within an atom are restricted to a set of discrete energy levels. Electrons can move between levels by gaining or losing energy, but they must do so by gaining or losing the precise amount corresponding to the difference in energy between two levels.

These energy levels are categorized into principal energy levels of increasing energy by a principle quantum number $n = 1, 2, 3, \cdots$ etc., and then are further divided into sublevels based on their shapes. The sublevels are designated in increasing energy as s, p, d, and f, and have one, three, five, and seven **atomic orbitals**, respectively. The first principle energy level, $n = 1$, has only the s sublevel. The second, $n = 2$, has s and p sublevels. The third, $n = 3$, has s, p, and d sublevels, and this pattern continues into the higher principle energy levels. The resulting sublevels are named $1s$, $2s$, $2p$, $3s$, $3p$, $3d$, and so on. A chart of the relative energy of the various levels and the number of orbitals in each level is shown in Figure 10.1.

Theoretically, the pattern of energy levels, sublevels, and their corresponding orbitals can be continued to higher levels as needed; however, principle energy levels beyond $n = 7$ and sublevels beyond f are rarely used. In fact, for our present purposes, where we will be focusing on silicon-based semiconductors, we will only be concerned with orbitals up to $3p$.

10.1.2   Electron Configurations

An electron configuration specifies to which atomic orbital each electron in an atom belongs. This can be conveniently illustrated with an Aufbau diagram like that of Figure 10.2. In the diagram, energy increases toward the top and each atomic orbital is represented as a box positioned at its relative energy level. By convention, we define a free electron to have zero energy. If that electron is captured by an atom, it loses energy while descending through the possible energy levels, getting closer and closer to the atomic nucleus. Due to this convention, the energy levels represented in the Aufbau diagram are negative and increase toward zero as we move up the diagram.

The electron configuration of an atom can be determined by applying three basic principles that govern how electrons fill atomic orbitals. These are the Aufbau principle, the Pauli exclusion principle, and Hund's rule. The Aufbau principle states that electrons fill the lowest energy orbitals first. These are the lowest boxes in the Aufbau diagram. The Pauli exclusion principle limits each orbital to having a maximum of two electrons, and those electrons must have opposite spin, either up or down. We indicate the spin of an electron with either an up arrow or down arrow drawn within the orbital's box and only allow one of each per box. Finally, Hund's rule states that equal energy orbitals are filled

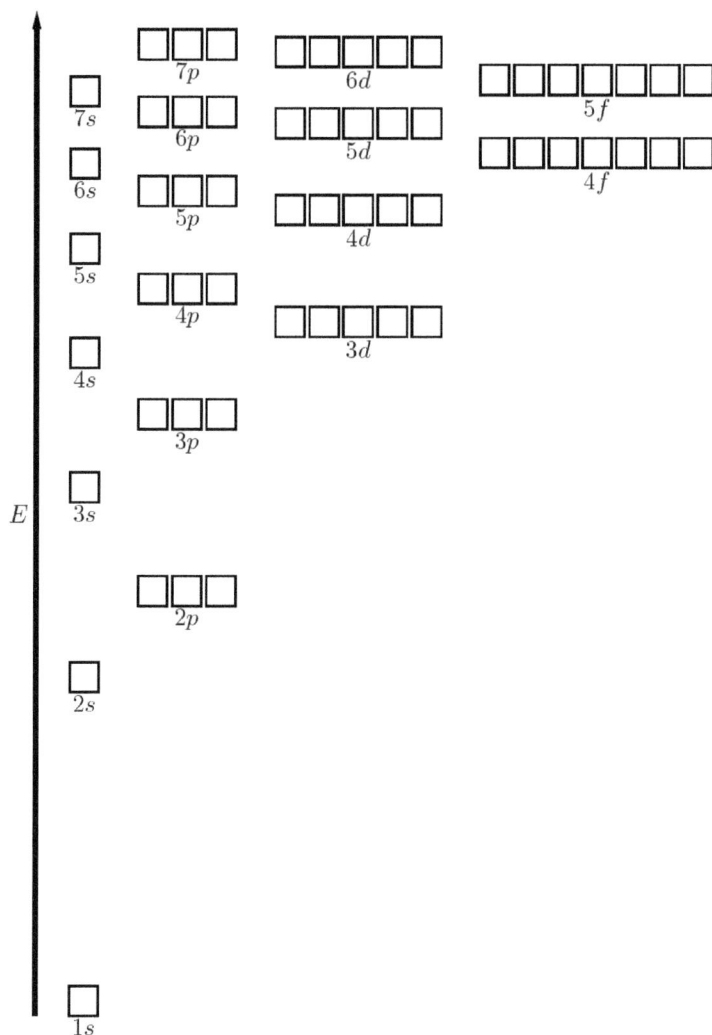

Figure 10.1: Relative energy levels of atomic orbitals with each orbital represented as a square.

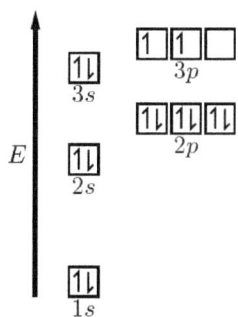

Figure 10.2: An Aufbau diagram illustrating the electron configuration of a neutral silicon atom.

with electrons of parallel spin before any orbital at that energy level gets a second electron of opposite spin. For example, each of the three $2p$ orbitals are filled with one electron with up spin before any of the $2p$ orbitals receive an electron with down spin. These three principles are summarized as follows:

Aufbau principle: The lowest energy orbitals are filled first.

Pauli exclusion principle: There is a maximum of two electrons per orbital, each of opposite spin.

Hund's rule: Equal energy orbitals are filled with one electron each in parallel spins before any of them gets a second electron of opposite spin.

To see how these principles are applied, consider the electron configuration of a neutral silicon atom. The silicon atom has fourteen protons in its nucleus. Each proton has a positive charge, so in order for the atom to be charge neutral, it must be balanced by the negative charges of fourteen electrons. The electron configuration can now be found by adding each of the fourteen electrons to an Aufbau diagram, one at a time, while following the preceding three principles. By the Aufbau principle, the first electron will go into the lowest energy orbital. That is the $1s$ orbital. The second electron will also go into the $1s$ orbital, but per the Pauli exclusion principle, it must have opposite spin of the first electron and it causes the $1s$ orbital to be completely filled. Now the third and fourth electrons will go into the $2s$ orbital. Again they will have opposite spin and will completely fill that orbital. Then, by Hund's rule, each of the three $2p$ orbitals will accept one of the fifth, sixth, and seventh electrons which will all have the same spin. Next the eighth, ninth, and tenth electrons will fill the three $2p$ orbitals with electrons of opposite spin so that the $2p$ orbitals are all completely full. Again, by the Aufbau and Pauli exclusion principles, the eleventh and twelfth electrons will fill the $3s$ orbital. Finally, by Hund's rule, the last two electrons will go into two of the three $3p$ orbitals and will have matching spin. The resulting completed Aufbau diagram for a neutral silicon atom is shown in Figure 10.2.

10.2   Energy of Atomic Bonds

10.2.1   Valence Electrons

When atoms interact with each other, such as when forming atomic bonds, they do so through interactions between their outermost electrons. Without going into the details of the shapes of atomic orbitals, higher energy orbitals, like $3s$, tend to be larger than lower energy orbitals, like $2s$ or $2p$. This means that the electrons in higher energy orbitals will, on average, tend to be further from the atomic nucleus than electrons in lower energy orbitals. As a consequence, when isolated atoms are brought close enough to each other to interact, their interactions involve their outermost, highest energy electrons. We refer to these electrons as **valence electrons**.

In silicon atoms, the four electrons in the $3s$ and $3p$ orbitals are valence electrons and are available to interact with other atoms. The ten electrons in the $1s$, $2s$, and $2p$ orbitals are held much more tightly by the atomic nucleus and are not available for interatomic interaction. The valence electrons of an atom are often depicted as dots surrounding the element's symbol. This type of depiction is known as an electron dot structure. As an example, the electron dot structure of a neutral silicon atom is shown below, with four dots surrounding the symbol Si.

$$\cdot \overset{\textstyle \cdot}{\underset{\textstyle \cdot}{\text{Si}}} \cdot$$

The electron dot structure of neutral hydrogen, which has only one valence electron, is drawn with one dot next to the symbol H.

$$\text{H} \cdot$$

10.2.2 Atomic Bonds

An atom can interact with another by forming bonds which are either ionic or covalent. In an ionic bond, one atom essentially steals an electron from the other and the resulting ions are held together by the electrostatic attraction of oppositely charged particles. In a covalent bond, two atoms share a pair of electrons. In either case, stable arrangements tend to be those in which the atoms involved achieve the electron configuration of a nearby noble gas.

A hydrogen atom can obtain a stable electron configuration by losing its only electron in the case of an ionic bond, or sharing a pair of electrons in a covalent bond. By sharing a pair of electrons, the hydrogen atom is able to fill its $1s$ orbital, and so obtains the electron configuration of the noble gas helium. Silicon can achieve a stable electron configuration by forming four covalent bonds. With each bond, it shares one of its valence electrons with another atom and receives a shared valence electron from the other atom. These eight shared electrons fill the silicon atom's $3s$ and $3p$ orbitals, allowing the atom to obtain the electron configuration of the noble gas argon.

If a silicon atom forms covalent bonds with four hydrogen atoms, it results in the five-atom molecule SiH_4, known as silane. In this molecule, each hydrogen atom has obtained the electron configuration of helium, and the silicon atom has obtained the electron configuration of argon. The electron dot structure for the resulting SiH_4 molecule is shown in the reaction below. Each pair of valence electrons in the molecule are shared between the silicon atom and one of the hydrogen atoms.

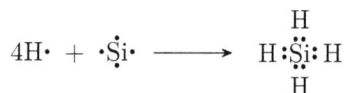

$$4\text{H}\cdot \; + \; \cdot \overset{\textstyle \cdot}{\underset{\textstyle \cdot}{\text{Si}}} \cdot \quad \longrightarrow \quad \text{H} \overset{\textstyle \text{H}}{\underset{\textstyle \text{H}}{:\overset{\textstyle \cdot\cdot}{\text{Si}}:}} \text{H}$$

It is important to note that the electron dot structure for silane shown above is only a two-dimensional representation of the molecule. Specifically, it does not represent the molecule's three-dimensional shape, which is a tetrahedral with the silicon atom at its center, as shown in Figure 10.3. The four negatively charged covalent bonds are repelled from each other but are attracted to the positively charged silicon nucleus. These forces act to spread the bonds around the silicon nucleus such that they are as far away from each other as they can get, which results in its tetrahedral structure.

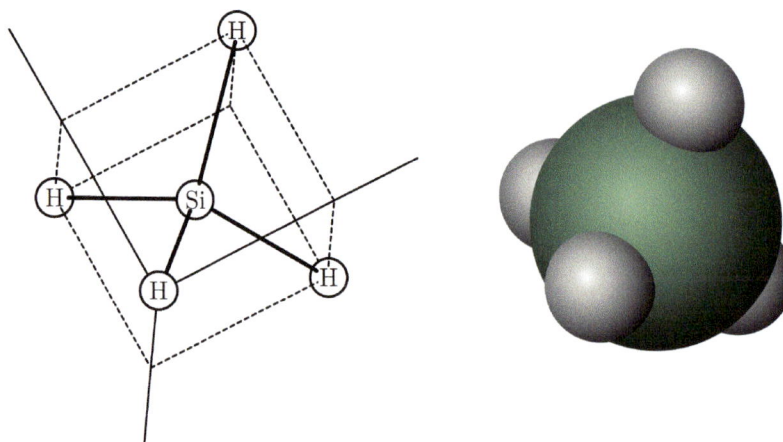

Figure 10.3: Tetrahedral shape of a silane molecule illustrated via a ball-and-stick model on the left and a space-filling model on the right.

10.2.3   Molecular Orbitals

When a pair of electrons form a covalent bond between two atoms, the two atomic orbitals the electrons formerly occupied are blended together to form two new **molecular orbitals**. One of these new molecular orbitals, called the **bonding molecular orbital**, is at a lower energy than the other. When electrons occupy this lower energy orbital, they contribute to a stable bond. If electrons occupy the higher energy molecular orbital, called the **antibonding molecular orbital**, then the bond is weakened.

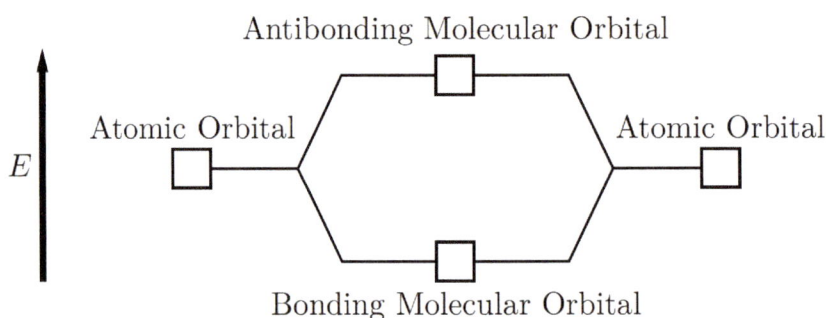

Figure 10.4: Molecular orbital energy levels relative to atomic orbital energy levels.

Consider a covalent bond between two hydrogen atoms to form the molecule H_2. Each atom contributes one electron to the bond, and by sharing electrons, each hydrogen atom is able to obtain the electron configuration of the nearby noble gas helium.

$$2H\cdot \longrightarrow H\!:\!H$$

By bonding, the $1s$ atomic orbitals of the hydrogen atoms are converted into a bonding and antibonding pair of molecular orbitals, as shown in Figure 10.5.

The electrons tend to occupy the lower energy bonding molecular orbital, leaving the antibonding molecular orbital empty. As a result, H_2 is a stable molecule.

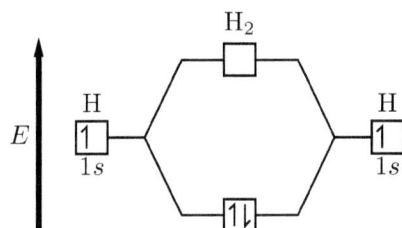

Figure 10.5: Molecular orbitals resulting from a H–H bond.

In contrast to the stable H_2 molecule, consider trying to form a molecule from two helium atoms. Neutral helium atoms possess two electrons. When unexcited, these electrons fill the atom's $1s$ orbital. As two of these atoms are brought closer together, their atomic orbitals separate in energy to form bonding and antibonding molecular orbitals just as we saw with H_2. However, in this case both the bonding and antibonding orbital must be filled, as shown in Figure 10.6. By requiring the antibonding molecular orbital to be filled, the resulting He_2 molecule is unstable and can only exist under special circumstances.

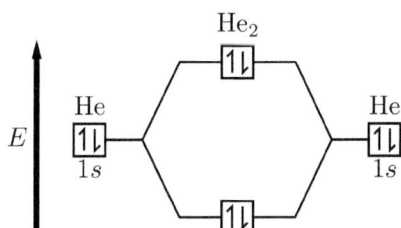

Figure 10.6: Molecular orbitals resulting from a He–He bond.

The energy levels of the molecular orbitals in a bond depend on the distance between the bonded atoms, which is referred to as their bond length. The relationship between bond length and molecular orbital energy is shown in Figure 10.7. Starting from isolated hydrogen atoms on the right of the figure, the energy levels of the bonding and antibonding molecular orbitals are equivalent to the atom's atomic orbitals. As the atoms are brought closer together, their valence electrons begin to interact and the energy levels of the molecular orbitals separate. The energy level of the antibonding molecular orbital increases as the atoms are brought closer together while the energy level of the bonding molecular orbital decreases until reaching a minimum at the equilibrium bond length, and then increases if the atoms are pushed even closer together.

The total number of orbitals available in the H_2 molecule is equal to the sum of the atomic orbitals in two isolated hydrogen atoms. In general, the number of molecular orbitals in a molecule will be equal to the number of atomic orbitals that are combined to form bonds. In the case of silane, each of the four hydrogen

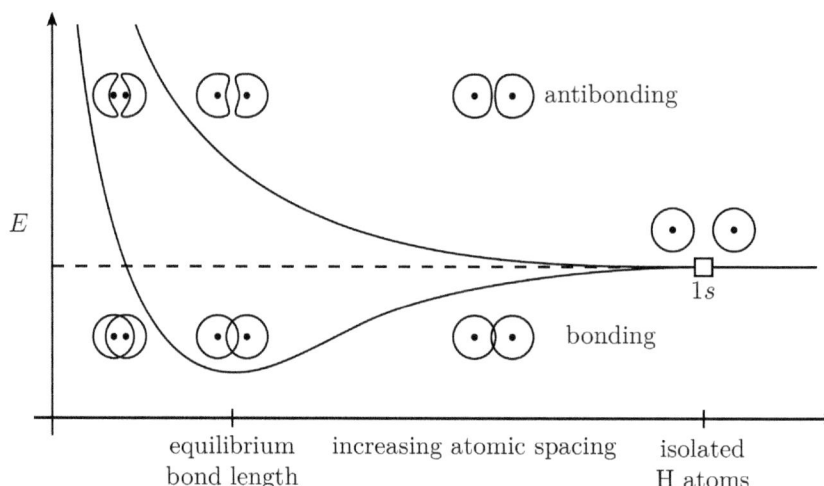

Figure 10.7: Molecular orbital energy and orbital shape versus bond length for a pair of hydrogen atoms.

atoms will contribute their one $1s$ orbital, while the silicon atom contributes its one $3s$ orbital and three $3p$ orbitals. These eight atomic orbitals result in eight molecular orbitals when the atoms are combined to form a molecule. Four of these will be bonding, while the other four will be antibonding. Since the eight total valence electrons all fit into the four bonding molecular orbitals, we can deduce that silane is a stable molecule.

10.3   Silicon Crystal

10.3.1   Crystal Lattice Structure

In a silicon crystal, a typical silicon atom shares its four valence electrons to form covalent bonds with four other silicon atoms. As with silane (SiH_4), the four bonds around each silicon atom are spaced as far away from each other as they can get, such that each atom sits at the center of a tetrahedral. This arrangement forms a repeating cubic structure known as the diamond lattice. The smallest repeating unit of the diamond lattice, called a unit cell, is shown in Figure 10.8a.

We sometimes use the bonding model, as shown in Figure 10.8b, to help visualize the movements of electrons within a crystal lattice. The bonding model is a two-dimensional representation of the atoms and bonds of the diamond lattice structure. Each atom in the model is depicted as a circle, and each bond is depicted by connecting two atoms with a pair of lines. Unlike the structural formulas of chemical compounds often used in chemistry where bonding pairs of electrons are depicted with a single line, in the bonding model, we draw a line for each electron in the bond. As is true in the diamond lattice, each atom in the bonding model has bonds to its four nearest neighbors. However, the two-dimensional grid layout of the bonding model is merely for convenience in drawing and does not represent the crystal's three-dimensional structure.

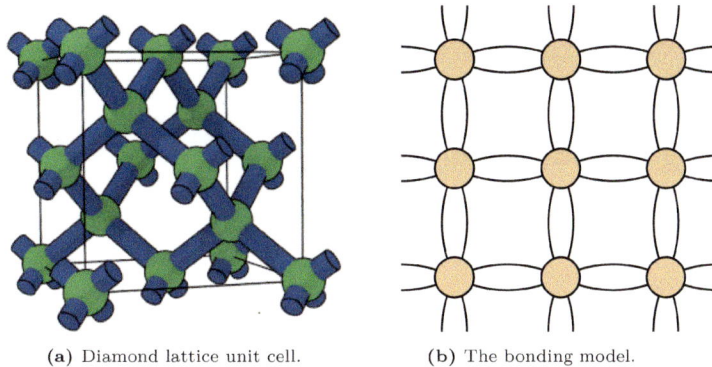

(a) Diamond lattice unit cell.   (b) The bonding model.

Figure 10.8: The diamond lattice versus the bonding model.

10.3.2 Energy Bands

If we count carefully, we will find that the diamond lattice unit cell contains eight atoms. One eighth of an atom comes from each of the unit cell's eight corners. One half of an atom comes from each of its six sides. Four whole atoms are within its interior. Adding these all together yields eight atoms per unit cell. For crystalline silicon, the length of the unit cell's sides is $5.43\,\text{Å}$ (5.43×10^{-8} cm). With eight atoms per unit cell, that means there are approximately 5×10^{22} silicon atoms per cubic centimeter of silicon crystal. All of the silicon atoms in the crystal are bonded to other atoms in the crystal, and so the entire crystal is a single large molecule.

The number of molecular orbitals in a molecule is equal to the number of atomic orbitals that are combined to form bonds. Each silicon atom contributes four atomic orbitals (one $3s$ and three $3p$). As a result, one cubic centimeter of silicon crystal has around $4 \cdot 5 \times 10^{22} = 2 \times 10^{23}$ molecular orbitals. Half (1×10^{23}) of those will be bonding molecular orbitals, and the other half will be antibonding. However, unlike in the case of H_2 where the bond is alone in the molecule, the many bonds in the silicon crystal lattice interact and influence each other. As a result of these interactions, the energy levels of the molecular orbitals are spread out into bands of possible energies. The antibonding orbitals form a band known as the **conduction band**, and the band of bonding molecular orbitals is known as the **valence band**.

As with the H_2 molecule, the molecular orbital energies in silicon crystal are dependent on bond length. If we could vary the lattice spacing, and thus the length of the bonds in the crystal, the interactions between bonds would increase as the bond length decreased because this brings the bonds closer together. As these interactions increase, the band of energies over which the molecular orbitals are spread gets wider such that the s and p orbitals blend into a single band. At the same time, decreasing the bond length causes the energy of the antibonding molecular orbitals to increase while the average energies of the bonding molecular orbitals decrease to a minimum at the equilibrium bond length, and then increase if the bonds are shortened further. This is the same relationship we saw in the energies of the antibonding and bonding molecular orbitals of the H_2 molecule, except that the many orbitals of each type in silicon crystal are spread out into bands. The relationship between lattice spacing and orbital energies, including the development of energy bands, is illustrated in Figure 10.9. The allowed energy levels within each band are technically discrete, but because there are so many orbitals and thus energy levels within each band, we approximate the densely packed discrete energy levels within each band as a continuous distribution of allowable energies.

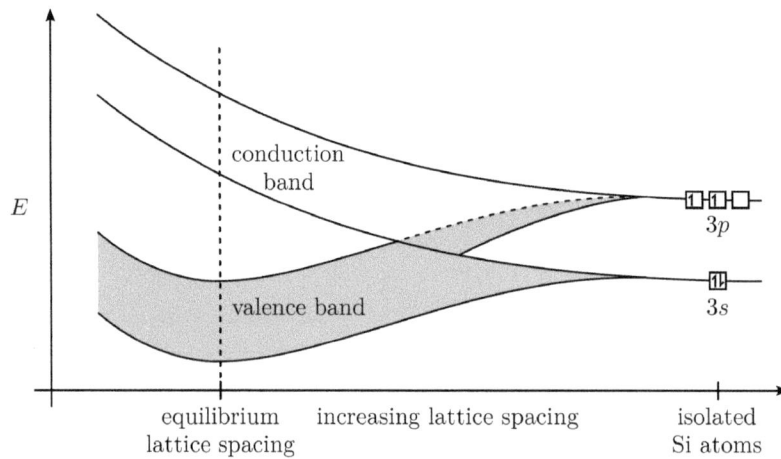

Figure 10.9: Silicon crystal orbital energy versus lattice spacing.

10.3.3   Energy Band Diagram

The allowed electron states of Figure 10.9 apply at every location within the crystal. In other words, since the crystal has the same structure throughout, the allowed electron states in the crystal do not change with position. Choosing some arbitrary direction x, a plot of the allowed energy levels along the x direction, known as an energy band diagram, is shown in Figure 10.10

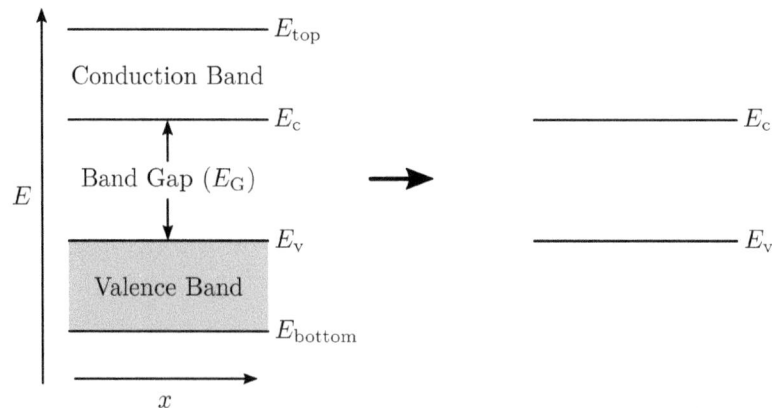

Figure 10.10: Detailed energy band diagram (left) and its simplified counterpart (right).

Electrons can occupy any of the available states within the conduction band or valence band; however, these bands are separated by a gap in the allowed states. This **band gap** is a range of energies that electrons within the crystal are unable to obtain. It is a sort of forbidden zone that electrons cannot enter. Of all the energy states that the bonding electrons can occupy, the lowest in energy is the bottom of the valence band, which we denote as E_{bottom}. The highest energy state is the top of the conduction band, denoted as E_{top}. We then denote the top of the valence band as E_{v} and the bottom of the conduction band as E_{c}. The band gap, E_{G}, is the energy difference between the bottom of the conduction band and the top of the valence band ($E_{\text{G}} = E_{\text{c}} - E_{\text{v}}$).

We are often only concerned with the behavior of electrons near the edges of the band gap, and so in practice, the detailed energy band diagram, as shown on the left side of Figure 10.10, is often simplified to that shown on the right side of the figure. All of the features shown on the left-hand side still exist in the right-hand figure, but they are implied rather than drawn explicitly.

10.4 Charge Carriers

Each silicon atom in silicon crystal contributes four valence electrons to molecular orbitals of the crystal. This results in 2×10^{23} electrons per cubic centimeter. There are also 2×10^{23} molecular orbitals per cubic centimeter available in the crystal and each orbital can hold two electrons. Hence, in pure silicon crystal, there are only enough electrons to fill half of the available orbitals. Half of the orbitals are in the valence band and the other half are in the conduction band. As the temperature decreases, fewer and fewer electrons have enough energy to reach the conduction band. In fact, as the temperature approaches absolute zero (0 K), the valence band will be completely filled, leaving the conduction band completely empty. With no electrons in the conduction band, it cannot carry any current because there are simply no charge carriers available. Perhaps less apparent is that the valence band, by being completely filled with electrons, is also unable to carry any current. Without any vacancies in the valence band, there is nowhere an electron can go unless another electron simultaneously moves in the opposite direction. The result is that, although the electrons in the valence band are free to move from one bond to another, their net movements are exactly balanced by electrons moving in the opposing direction and, thus, the net current in the valence band must also be exactly zero.

10.4.1 The Electron

The situation within the crystal becomes more interesting if some electrons are excited from the valence band up to the conduction band. As illustrated with the aid of a band diagram in Figure 10.11a, this requires energy equal to or greater than the band gap energy E_{G}. This same event is illustrated via the bonding model in Figure 10.11b as the freeing of an electron from a bond and allowing it to roam the crystal's lattice. In both diagrams, the electron is depicted as a black circle centered on a white minus sign, and the electron's initial location is marked with a gray circle. Note that the two gray circles in the band diagram represent the same spatial location.

An electron excited to the conduction band is not a "free" electron because it does not have sufficient energy to escape the crystal entirely. Instead it is constrained to the antibonding molecular orbitals of the conduction band. However, the vast sea of empty orbitals available within the conduction band allow it to move freely within the confines of the crystal. This freedom of movement allows the electrons in the conduction band to carry current.

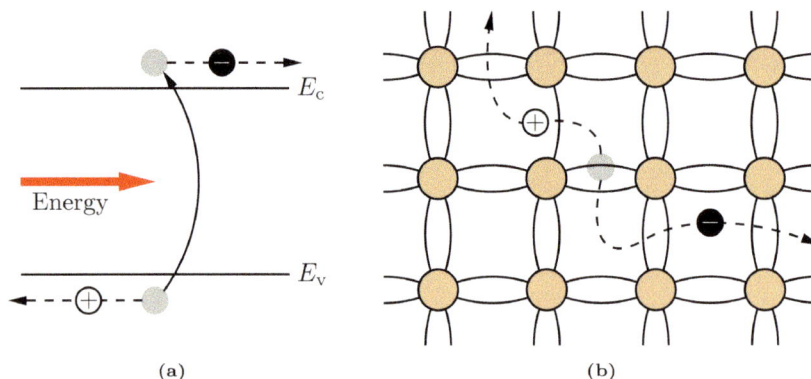

Figure 10.11: An electron excited from the valence band up to the conduction band shown in (a) the band diagram, and (b) the bonding model.

10.4.2   The Hole

As a negative charge carrier, an electron in the conduction band is capable of carrying a current, but when an electron is elevated to the conduction band, a positively charged vacancy is opened in a molecular orbital of the valence band. Another valence band electron may move to fill this vacancy, but doing so opens a new vacancy behind it. In this way, the vacancy in the valence band may move about like a bubble in water. As a result, the net movements of the electrons in the valence band are no longer exactly zero, and so, like the conduction band, the valence band is also able to carry a current.

Instead of considering the movements of all of the valence band electrons in the role of carrying a current, it is conceptually simpler to consider the vacancy as a positively charged quasiparticle known as a **hole**. We depict holes in band diagrams and in the bonding model as white circles centered on a black plus sign. A hole in the valence band is the direct counterpart to an electron in the conduction band. Just as the negatively charged electron can carry a current in the conduction band, the positively charged hole can carry a current in the valence band.

When an electron gains energy, it moves up the energy band diagram; however, when a hole gains energy, it moves down the energy band diagram. The water and bubble analogy of electrons and holes may help you remember this relationship. If the valence electrons were a pool of water, it would require positive energy input to lift a drop of water (an electron) up and out of the pool. Similarly, it requires positive energy input to push an air bubble (a hole) down beneath the water's surface.

10.4.3   Material Classification

The conductivity of silicon crystal depends directly on the concentration of charge carriers in the material (electrons in the conduction band and holes in the valence band). When more carriers are available, current can flow more easily and the material has a higher conductivity. Conversely, when fewer carriers are available, it is more difficult for current to flow and the material has a lower conductivity. The concentration of carriers depends on the energy required for an electron to jump from the valence band into the conduction band. If the band gap were smaller, more electrons could make the jump, and so conductivity would be higher. Of course, if the band gap were larger, fewer electrons could jump to the conduction band, and so the material's conductivity would be lower. This relationship between band gap and conductivity can be seen in how materials are classified as either insulators, semiconductors, or conductors, as shown in Figure 10.12.

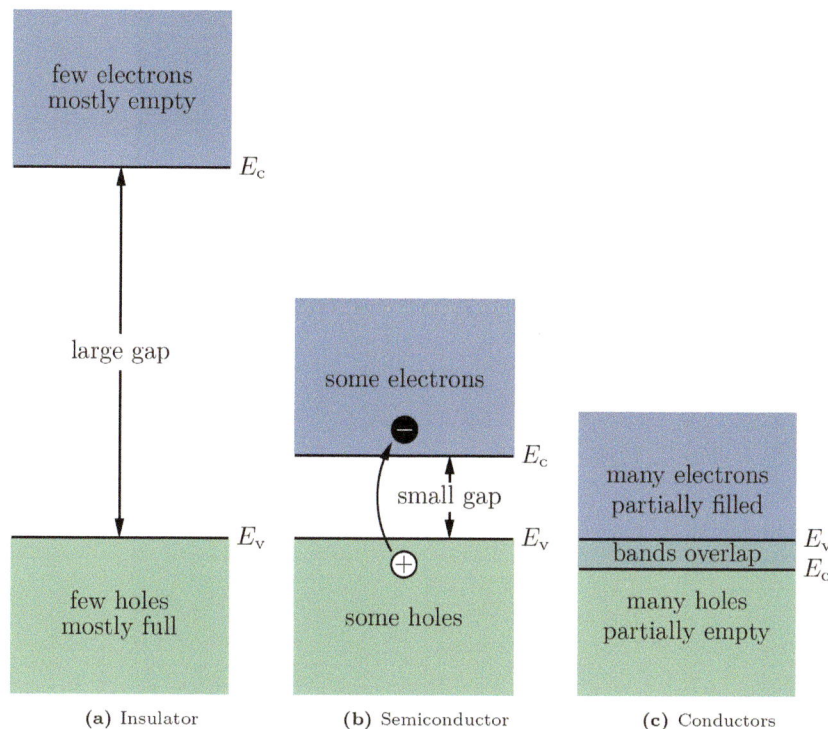

Figure 10.12: Relative band gap and material classifications.

While the energy band diagram of Figure 10.10 was developed with silicon crystal in mind, all materials possess a band structure of available electron states due to their bonding and antibonding molecular orbitals. The key difference between, insulators, semiconductors, and conductors is in the size of their band gap. Insulators, like silicon dioxide ($E_G \approx 8\,\mathrm{eV}$), have a large band gap and so very few charge carriers are available for conduction. Semiconductors, like silicon crystal ($E_G \approx 1.12\,\mathrm{eV}$), have a smaller band gap and can have widely ranging conductivities based on other conditions, like temperature or incident lighting. Conductors, like metals, have overlapping energy bands such that they have no band gap at all. Since there are energy levels in the conduction band that are at a lower energy than the highest energy levels in the valence band, the conduction band in metals is always partially filled with electrons and the valence band is always partially vacant (filled with holes). This results in an abundance of charge carriers and in high conductivity.

10.5   Doping

The periodic table of elements is organized such that elements with similar chemical properties are in the same column. The chemical properties of elements are determined by the way each element's atoms interact with other atoms, which depends upon the number of valence electrons in each atom. Since the number of valence electrons determines an element's chemical properties, and the elements are organized by their chemical properties, they are also organized by their number of valence electrons. In the abbreviated periodic table in Figure 10.13, the elements of each column have the same number of valence electrons. For example, the elements of column IV (C, Si, Ge, Sn, Pb) all have four valence electrons. As we have seen before, for silicon these are the four electrons in the $3s$ and $3p$ orbitals. For carbon, these are the four electrons in the $2s$ and $2p$ orbitals. For germanium, they are the four electrons in the $4s$ and $4p$ orbitals and so on down the column.

I							VIII
1 H Hydrogen	II	III	IV	V	VI	VII	2 He Helium
3 Li Lithium	4 Be Beryllium	5 B Boron	6 C Carbon	7 N Nitrogen	8 O Oxygen	9 F Fluorine	10 Ne Neon
11 Na Sodium	12 Mg Magnesium	13 Al Aluminum	14 Si Silicon	15 P Phosphorus	16 S Sulfur	17 Cl Chlorine	18 Ar Argon
19 K Potassium	30 Zn Zinc	31 Ga Gallium	32 Ge Germanium	33 As Arsenic	34 Se Selenium	35 Br Bromine	36 Kr Krypton
37 Rb Rubidium	48 Cd Cadmium	49 In Indium	50 Sn Tin	51 Sb Antimony	52 Te Tellurium	53 I Iodine	54 Xe Xenon
55 Cs Cesium	80 Hg Mercury	81 Tl Thallium	82 Pb Lead	83 Bi Bismuth	84 Po Polonium	85 At Astatine	86 Rn Radon

Figure 10.13: Selected columns of the periodic table.

Up to this point, we have only considered properties of pure silicon crystal, but something interesting happens if we replace a silicon atom in the crystal with an atom that has a different number of valence electrons. There are a variety of methods for intentionally introducing such impurities into semiconductor crystals. Collectively, these methods are referred to as **doping**, and the impurities are called **dopants**. For a specific example, let us consider replacing a silicon atom with phosphorus. Phosphorus is in column V of the periodic table and has five valence electrons. As shown in the bonding model Figure 10.14b, the phosphorus atom forms four covalent bonds with its neighboring silicon atoms, and then has one electron left over.

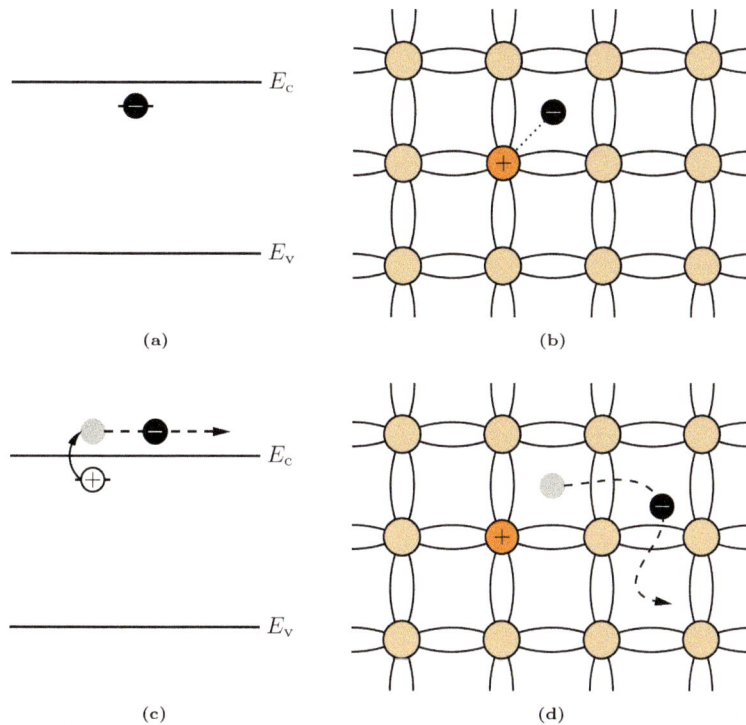

Figure 10.14: A silicon crystal doped with a donor atom (phosphorus), shown in (a) the band diagram and (b) the bonding model.

Very little energy is required to excite this extra electron from its atomic orbital around the phosphorus atom up to one of the vacant orbitals of the conduction band. In fact, in the case of phosphorus, this only requires 0.045 eV, which is very small compared to the 1.12 eV band gap of silicon. In the band diagram, as shown in Figure 10.14a, the addition of this impurity is illustrated by adding a line to represent an energy level in the band gap near the bottom of the conduction band. This new level only exists in the vicinity of the phosphorus atom. To account for this in the band diagram, the line is kept short to span a small Δx.

Since this new energy level rests so close to the conduction band, there is sufficient energy at room temperature to essentially guarantee that the electron within it will be excited into the conduction band. In this process, the electron becomes mobile and is free to move around the crystal conducting current, while a vacancy is opened in an orbital around the phosphorus atom. There are no broken bonds associated with this vacancy, and unlike the holes in the valence band, which are free to move, it is fixed to the phosphorus atom, which is firmly bonded in place inside the crystal. Hence, replacing a silicon atom in the silicon crystal with a phosphorus atom results in an additional mobile electron and a fixed positive charge at the location of the phosphorus atom.

This is illustrated via the band diagram and bonding model in Figures 10.14c and 10.14d, respectively.

Now let us consider replacing a silicon atom with boron instead of phosphorus. Boron is in column III of the periodic table and has three valence electrons. When it bonds with its four neighboring silicon atoms in the crystal, it is one electron short of completing the four bonds. This missing electron is represented as a hole bound to the boron atom in the bonding model of Figure 10.15b.

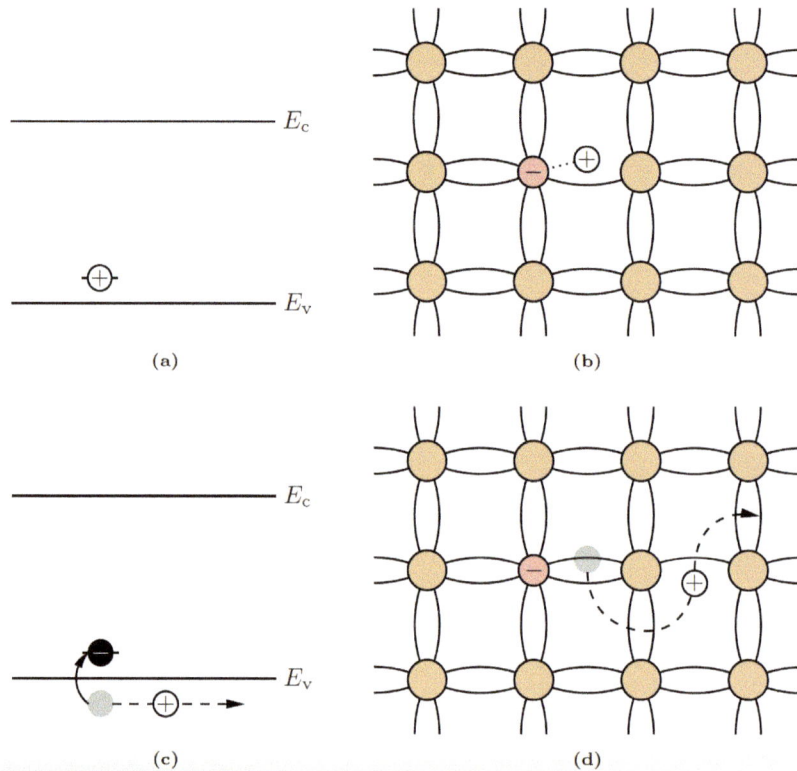

Figure 10.15: A silicon crystal doped with an acceptor atom (boron), shown in (a) the band diagram and (b) the bonding model.

Just like the extra electron of the phosphorus atom, the binding energy of this hole to the boron atom is only 0.045 eV. In other words, the hole only needs 0.045 eV to jump down to the valence band, which is exactly the same as saying an electron in the valence band (i.e., in a bond between silicon atoms) only needs to gain 0.045 eV to complete the fourth covalent bond between the boron atom and one of its silicon neighbors. Hence, the addition of the boron impurity adds a new energy level to the band diagram in the band gap near the top of the valence band, as shown in Figure 10.15a. As with the addition of the phosphorus atom, this new energy level only exists in the vicinity of the boron atom, and so we draw only a short line at the dopant site.

At room temperature, there is near certainty that an electron in the valence band will have sufficient energy to fill this new energy level and leave behind a hole in the valence band. In this case, the hole is now mobile and free to move about the crystal conducting current while the electron has become stuck in a bond at the dopant site, creating a fixed negative charge. This scenario is illustrated via the band diagram and bonding model in Figures 10.15c and 10.15d.

When the addition of a dopant results in additional mobile electrons, it is called a **donor** because it *donates* extra electrons. When its addition results in additional mobile holes, it is called an **acceptor** because it *accepts* extra electrons. In silicon crystal, phosphorus is the most commonly used donor and boron is the most commonly used acceptor. Other common donors for silicon crystal include arsenic and antimony, while common acceptors include aluminum, gallium, and indium. Note that the common donors all come from column V of the periodic table while the common acceptors all come from column III. In each case, the column V donors have one extra electron that they can easily donate to the conduction band, and the column III acceptors are short one electron but can easily accept one from the valence band, resulting in a valence band hole. In some semiconductors, such as gallium arsenide, it is not as easy to work out whether a particular dopant will be considered a donor or an acceptor. In general, regardless of a dopant's position in the periodic table, those that result in extra electrons are donors and those that result in extra holes are acceptors.

10.5.1 Carrier Concentration

In undoped semiconductors, every electron excited into the conduction band leaves behind a hole in the valence band. Thus, the concentration of electrons n and the concentration of holes p are equal. These undoped semiconductors are called **intrinsic** semiconductors and the intrinsic carrier density is denoted n_i. Hence, in an intrinsic semiconductor,

$$n = p = n_i. \tag{10.1}$$

The intrinsic carrier concentration increases with temperature. The general trend is plotted in Figure 10.16. At absolute zero, n_i will also be zero, but for silicon at 300 K, $n_i = 1 \times 10^{10} \, \text{cm}^{-3}$.

Note that each electron-hole pair represents a broken bond in the crystal. So at 300 K there are 1×10^{10} broken bonds per cubic centimeter of silicon crystal. (Is that a lot?) For perspective, recall that there are 2×10^{23} total bonds per cubic centimeter, which means that at 300 K, only one in every twenty trillion bonds is broken. This is far too few to threaten the structural integrity of the crystal. At 1687 K, there is enough energy to break all of the bonds and the crystal liquefies.

When a semiconductor is doped with donor atoms, the concentration of electrons in the conduction band is increased beyond the intrinsic level without adding to the concentration of holes in the valence band. As a result, n is larger than p, and we refer to donor-doped semiconductors as being n-type semiconductors. Similarly, in acceptor-doped semiconductors, p is larger than

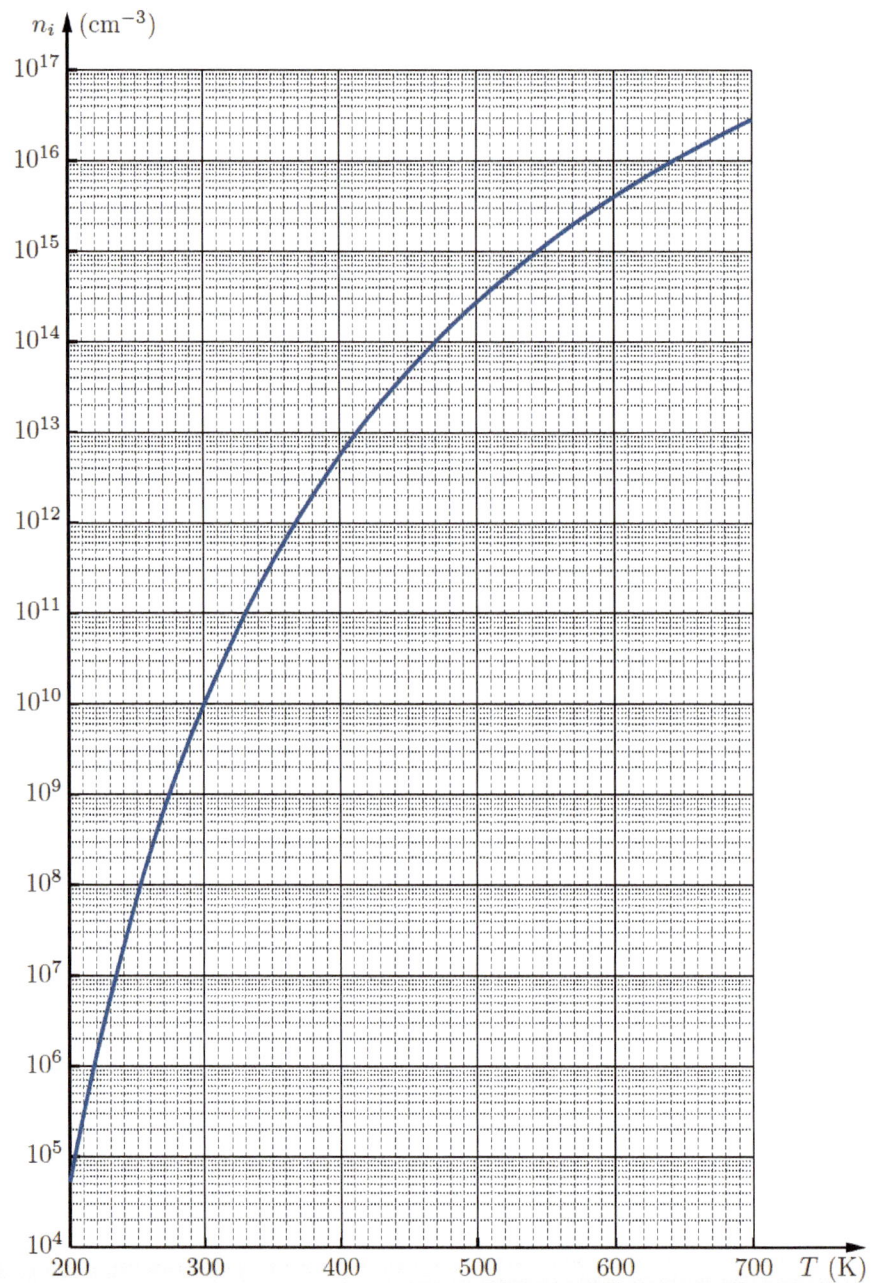

Figure 10.16: Intrinsic carrier concentration in silicon crystal as a function of temperature.

n and these are referred to as p-type semiconductors. In either donor-doped or acceptor-doped semiconductors, the dopant determines the **majority carrier**; either electrons (n) for donors or holes (p) for acceptors. The carrier with the lower concentration is called the **minority carrier**.

The concentration of dopants can be accurately controlled, which allows a device designer to establish desired carrier concentrations and conductivity within the semiconductor. Doping concentrations are denoted N_D for donors and N_A for acceptors. Typical doping concentrations for silicon crystal range from 1×10^{13} dopant atoms per cubic centimeter to 1×10^{18} dopant atoms per cubic centimeter. Since there are 5×10^{22} silicon atoms per cubic centimeter in a silicon crystal, at these doping levels, there are many more silicon atoms than dopant atoms. At the same time, since the intrinsic carrier concentration at room temperature is only about 1×10^{10} carriers per cubic centimeter, there are many more dopant atoms than intrinsic charge carriers (i.e., $N_D \gg n_i$ or $N_A \gg n_i$).

The proximity of the energy levels added by the dopant atoms to the semiconductor's band gap edges means that at room temperature, there is sufficient energy that nearly every dopant atom will ionize and contribute a charge carrier (electrons for donors and holes for acceptors). Thus, it is common to approximate the concentration of additional charge carriers as equal to the concentration of dopant atoms. This simplifying approximation is known as the total ionization assumption. If the dopant concentration is also much larger than the intrinsic carrier concentration, then we can approximate the majority-carrier concentration as being equal to the dopant concentration.

$$n \approx N_D, \text{ for donors} \tag{10.2a}$$

$$p \approx N_A, \text{ for acceptors} \tag{10.2b}$$

As the charge carriers move about a semiconductor, sometimes an electron and a hole will encounter each other and cancel out in an energy-releasing process called **recombination**, as illustrated in Figure 10.17a. Conversely, sometimes just the right amount of energy is absorbed by an electron to produce an electron-hole pair in a process known as **generation**, as illustrated in Figure 10.17b. At equilibrium, the rates of recombination and generation are balanced and a constant concentration of charge carriers is maintained.

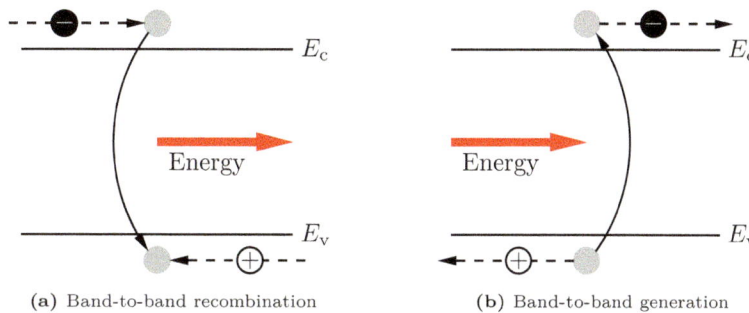

(a) Band-to-band recombination   **(b)** Band-to-band generation

Figure 10.17: Energy band diagram representation of band-to-band recombination and generation processes.

We have stated before that the addition of donor atoms increases the concentration of electrons in the conduction band, but we would also like to know what effect this will have on the concentration of holes. As it turns out, the increased electron concentration results in a reduced concentration of holes. To understand this, consider what would happen if the concentration of holes remained at the intrinsic level after doping with donor atoms increased the electron concentration far above the intrinsic level. The holes roaming the crystal would now be much more likely to encounter an electron, and thus much more likely to be eliminated through recombination. On the other hand, there is no increase in the rate of generation, so the concentration of the holes must decrease until the rate of recombination and generation are again balanced. Note that the concentration of electrons, being approximately equal to the donor concentration, is several orders of magnitude larger than the concentration of holes and can be considered as constant throughout this event. Similarly, in a p-type semiconductor, the concentration of electrons is reduced below the intrinsic concentration. In fact, a detailed analysis of charge carrier concentrations results in a simple relationship between the concentration of electrons and holes known as the np product, which is given by

$$\boxed{np = n_i^2.}$$ (10.3)

This relationship allows one of the carrier concentrations to be found whenever the other is already known. This is very useful when combined with the total ionization assumption, as now both carrier concentrations can be estimated from the dopant concentration and the semiconductor's intrinsic carrier concentration. Approximations for the carrier concentrations in doped semiconductors are summarized below.

$$n \approx N_\mathrm{D}, \quad p \approx \frac{n_i^2}{N_\mathrm{D}}, \quad \text{for } N_\mathrm{D} \gg n_i, \quad N_\mathrm{D} \gg N_\mathrm{A}$$ (10.4)

$$p \approx N_\mathrm{A}, \quad n \approx \frac{n_i^2}{N_\mathrm{A}}, \quad \text{for } N_\mathrm{A} \gg n_i, \quad N_\mathrm{A} \gg N_\mathrm{D}$$ (10.5)

The approximations in (10.4) and (10.5) are valid over a wide range of temperatures. However they cannot be applied if temperatures are too cold or too hot. If the temperature is low enough such that there is not enough thermal energy to ionize the dopant atoms, then the carrier concentration will drop off significantly below the donor concentration. As the temperature approaches absolute zero, the carrier concentration approaches zero as well. At these temperatures, the device is said to be in the "freeze-out temperature region." On the other hand, if the temperature is raised to the point that a significant number of electrons are excited directly from the valence band, then the carrier concentration can be much larger than the donor concentration. At sufficiently high temperatures, the carriers—due to thermal excitation of electrons from the valence band—will completely swamp the concentration of carriers due to the donor atoms. Since at these temperatures the overall carrier concentration is approximately the same as the intrinsic carrier concentration, the device is said to be in the "intrinsic temperature region." Temperatures between these extremes are referred to as the "extrinsic temperature region." A plot of the carrier concentration relative to the donor concentration for a phosphorus-doped $N_\mathrm{D} = 1 \times 10^{16}\,\mathrm{cm}^{-3}$ silicon sample throughout each of these temperature regions is shown in Figure 10.18.

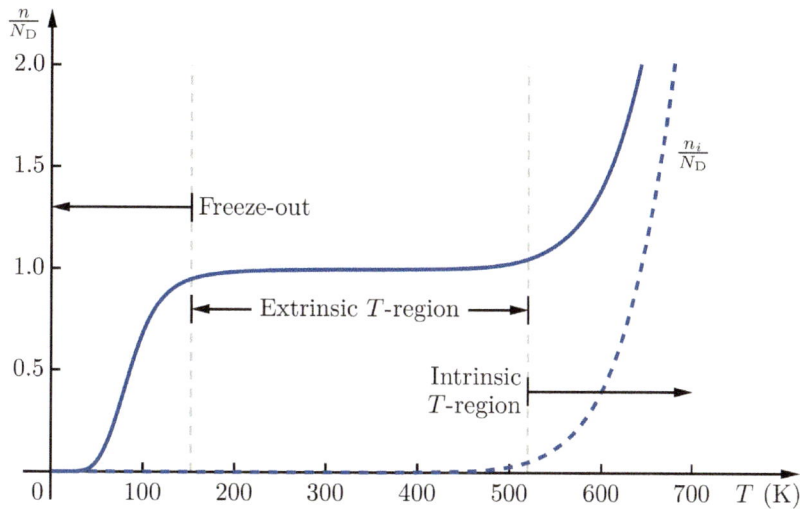

Figure 10.18: Carrier concentration relative to dopant concentration for a phosphorus-doped $N_D = 1 \times 10^{16}$ cm^{-3} silicon sample as a function of temperature compared to the intrinsic silicon carrier concentration (dashed line). The general form of the temperature dependence shown here is typical for majority-carrier concentrations of moderately doped semiconductors.

10.5.2 Carrier Energy Distribution

With (10.4) and (10.5), we can approximate the concentration of electrons in the conduction band and holes in the valence, band but as we will see in chapter 11, we also need to develop an idea about how the charge carriers are distributed with respect to energy within these bands. For electrons, energy increases as we move up in the energy band diagram. There are many closely spaced energy levels available within the conduction band, and so the electrons therein can readily gain or lose any amount of energy that is small enough for them to remain within the band. They naturally settle toward lower energy states by releasing photons.

Analogous reasoning applies to holes in the valence band so long as we note that energy for holes increases as we move downward in the band diagram, causing holes to naturally float up the energy band diagram. Recalling that holes are just vacancies in the bonding molecular orbitals, we can understand holes losing energy and floating up toward the valence band edge as a series of valence band electrons releasing photons and sinking into the vacant orbitals. Each time this happens, the vacancy moves up closer to the valence band edge.

The much larger change in energy required for recombination of holes and electrons across the band gap means that such transitions are much less likely to occur than energy changes within a band. Thus, the electrons and holes accumulate near the band edges, as shown in the subfigures of Figure 10.19. The actual carrier distribution with respect to energy is plotted on the left half of each subfigure, and a schematic representation of this distribution is shown with carrier triangles on the right. Note that the carriers appear in equal numbers in the intrinsic semiconductor of Figure 10.19a, while the n-type and p-type semiconductors of Figures 10.19b and 10.19c have excess electrons and holes respectively. In each figure, care has been taken to satisfy the np product rule.

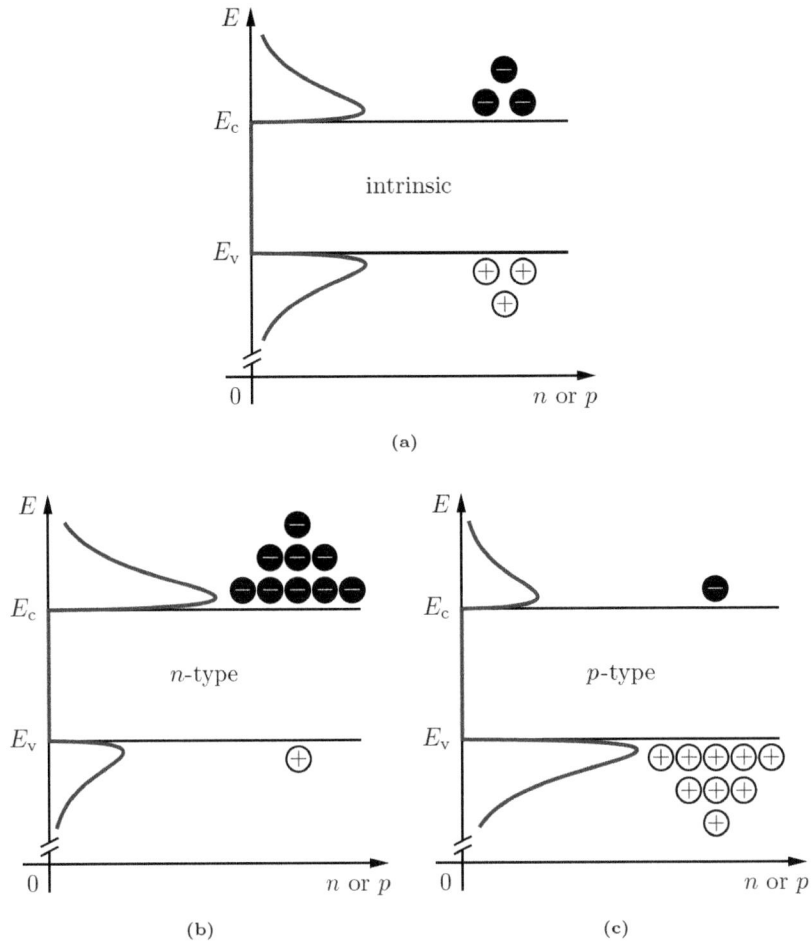

Figure 10.19: Carrier distributions with respect to energy for (a) intrinsic, (b) n-type, and (c) p-type semiconductor.

10.5.3   Diffusion and Drift

To this point, we have assumed any dopant atoms in the semiconductor were evenly distributed throughout the lattice. Further, and without proper explanation, we have assumed that the resulting carrier concentrations were evenly distributed with respect to position. The evenness of the carrier concentration with respect to position is a result of the process of **diffusion**, which we will now examine more closely.

Due to thermal agitation, the charge carriers are in constant motion. On a nanoscopic scale, the thermally induced movements of the individual carriers are random. However, on a macroscopic scale, the cumulative effect of the random movements of the individual carriers results in a net movement from areas of high concentration to areas of low concentration. To gain some insight into this phenomenon, consider a simple thought experiment in which gas-like particles are enclosed in a chamber, as shown in Figure 10.20. Each of the particles moves in a random walk within the chamber. At some initial time, we select four particles from the top half of the chamber and eight particles from the

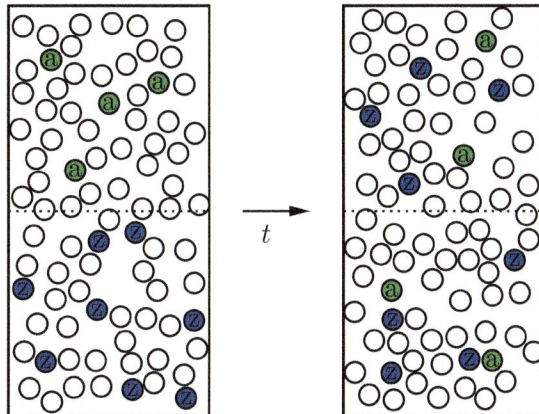

Figure 10.20: Visualisation of diffusion wherein the random movements of particles result in an equalizing of their distribution over time.

bottom half of the chamber. We label the particles in the top half with an "a" and color them green. The particles in the bottom half are labeled with a "z" and colored blue. Then, we monitor these twelve particles as time progresses. Their random movement can carry them anywhere within the chamber and after sufficient time has passed, the four from the top half have just as much chance of being found in the bottom half as they do of remaining in the top. Likewise, the eight from the bottom half have an equal chance of being found in the top as they do of remaining in the bottom. If the particles are distributed according to these odds, in each half we will find two of the four we selected from the top and four of the eight we selected from the bottom. Hence, we now find six of the originally selected particles in each half of the chamber. The random movements of the individual particles have replaced the initial concentration imbalance with an even distribution.

This tendency to even out changes in concentration is what is meant by the word "diffusion." In semiconductor devices, any unevenness in the concentration of holes or electrons results in the diffusion of these particles away from areas of high concentration and toward areas of lower concentration. Quantitatively, if we limit our consideration to one dimension, the diffusion of particles can be expressed as

$$\mathscr{F} = -D\frac{\mathrm{d}\eta}{\mathrm{d}x},^{[1]} \tag{10.6}$$

which states that the particle flux density, $\mathscr{F}$, is proportional to the derivative of the particle concentration $\frac{\mathrm{d}\eta}{\mathrm{d}x}$. The particle flux density is the number of particles crossing some unit of area in some unit of time. In other words, it is the rate of particles passing through a given cross section of a device. Note that the proportionality constant is $-D$ where D, which is known as the diffusion coefficient, is always positive. This ensures that when the concentration of particles increases with x, the resulting particle flux will be negative (directed

---

[1]This relation, known as Fick's law, is typically written as

$$\mathscr{F} = -D\nabla\eta, \tag{10.7}$$

where ∇ is the gradient operator, which represents a generalization of the concept of a derivative to three dimensions.

toward decreasing x), which is in agreement with our qualitative assessment that the particles should be directed away from areas of higher concentration and toward areas of lower concentration.

Suppose we have a piece of silicon, uniform in all directions except that the concentration of electrons increases as a function of x. Referencing (10.6), we can express the rate of electrons per unit area flowing in the x direction as

$$\mathscr{F}_{\mathrm{N}} = -D_{\mathrm{N}}\frac{\mathrm{d}n}{\mathrm{d}x}, \tag{10.8}$$

where $\mathscr{F}_{\mathrm{N}}$ is known as the electron flux density, D_{N} is the diffusion coefficient for electrons, and $\frac{\mathrm{d}n}{\mathrm{d}x}$ is the derivative of the electron concentration. Since electrons are charged particles, their movement due to diffusion produces a current. This diffusion current is an important aspect of semiconductor device behavior. If we multiply the electron flux density by the charge of an electron, $-q$, we get the current density due to diffusion $J_{\mathrm{N}|\mathrm{diff}}$.

$$J_{\mathrm{N}|\mathrm{diff}} = -q\mathscr{F}_{\mathrm{N}}. \tag{10.9}$$

We could now calculate the electron diffusion current, $I_{\mathrm{N}|\mathrm{diff}}$, by multiplying this result by the cross-sectional area of the device, but by using current density instead of current, our analysis can be generalized to devices of any size. To complete the expression for electron diffusion current density, we replace the electron flux density in (10.9) with the right-hand side of (10.8) to get

$$J_{\mathrm{N}|\mathrm{diff}} = qD_{\mathrm{n}}\frac{\mathrm{d}n}{\mathrm{d}x}. \tag{10.10}$$

We can write an analogous equation for the current density due to diffusion of holes as

$$J_{\mathrm{P}|\mathrm{diff}} = -qD_{\mathrm{p}}\frac{\mathrm{d}p}{\mathrm{d}x}. \tag{10.11}$$

Recall that the charge on electrons and holes is $-q$ and q, respectively, and thus an electron flux directed toward negative x results in electron current flowing toward positive x, whereas hole flux and hole current flow in the same direction.

Diffusion currents explain how a uniform carrier concentration is maintained in a uniformly doped semiconductor because any variation that appears in the carrier concentration will result in diffusion currents that act to cancel it back out. However, what if the doping concentration were not uniform? Let us consider an n-type semiconductor where the dopant concentration is increased from one end to the other, as shown in Figure 10.21a. The same donor variation is depicted with a charge model in Figure 10.21b. In the charge model, the donor atoms are drawn as a circular negative charge and a square positive charge. The circular negative charge represents the extra electron the donor carries into the lattice. This electron will become mobile when the donor atom ionizes. The square positive charge represents the extra positive charge of the donor atom's nucleus. This positive charge remains fixed in place when the atom ionizes. Later on, we will use an analogous representation for acceptor atoms using a circular positive charge to represent a potentially mobile hole and a square negative charge to represent a fixed acceptor atom. Note that whether the dopant atoms are donors or acceptors, before ionization, the dopant atoms are charge neutral.

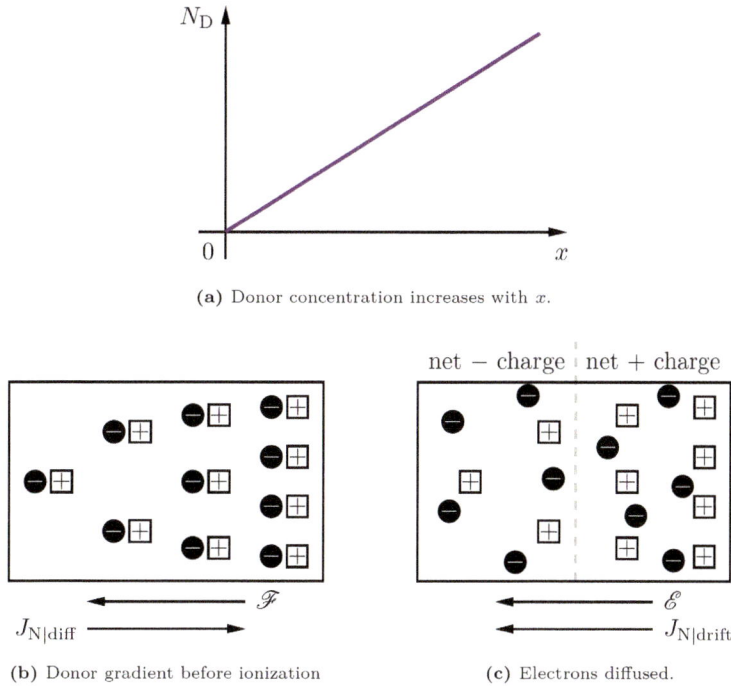

(a) Donor concentration increases with x.

(b) Donor gradient before ionization

(c) Electrons diffused.

Figure 10.21: A variation in donor concentration results in a diffusion of electrons toward lower concentration and a diffusion current toward higher concentration. The redistribution of charge due to diffusion results in a drift current to balance the diffusion current.

Immediately after the donor atoms in Figure 10.21b ionize, the electron concentration will increase as we move toward the right in the positive x direction. This increasing concentration results in a diffusion of electrons to the left and, thus, a current to the right. We know that if acting alone, diffusion will result in an even electron concentration across the semiconductor, as shown in Figure 10.21c. However, because electrons are negatively-charged particles and the positively-charged donor sites are fixed and cannot diffuse, the charge distribution shown in Figure 10.21c results in an electric field pointed from the area of net positive charge on the right to the area of net negative charge on the left.

This electric field exerts a force on the electrons, pulling them back to the right. From the definition of an electric field, the force on each electron is given by

$$F_x = -q\mathscr{E}_x. \tag{10.12}$$

By pulling some electrons back toward the right, the electric field is responsible for a current directed toward the left. We call this current a **drift** current, and for electrons, the drift current density can be expressed quantitatively as

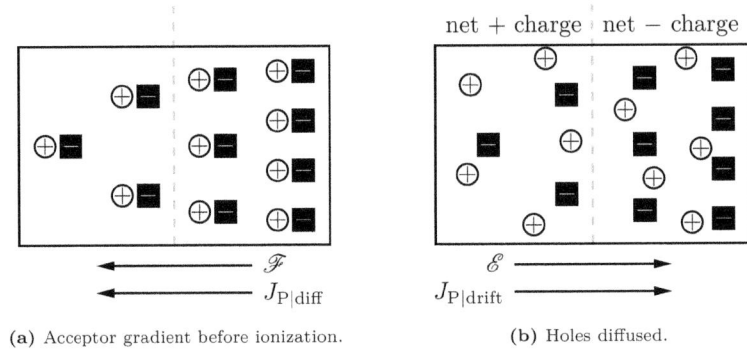

(a) Acceptor gradient before ionization.

(b) Holes diffused.

Figure 10.22: An acceptor gradient results in a diffusion current of holes toward lower concentration. The redistribution of charge due to diffusion results in a drift current to balance the diffusion current.

$$J_{\mathrm{N|drift}} = q\mu_n n \mathscr{E}_x \tag{10.13}$$

where q is the magnitude of the electron's charge, n is the electron concentration, and $\mathscr{E}_x$ is the electric field in the x direction. The quantity μ_n is known as the electron mobility and it quantifies how easily an electron can move through the by taking into account the scattering that occurs when the electrons bump into thermally agitated lattice atoms and ionized dopant atoms.

If instead of an n-type semiconductor we consider a p-type semiconductor with an acceptor concentration that increases with x, as shown in Figure 10.22a, we will find that once the acceptors ionize, the extra holes will diffuse to the left, producing a hole diffusion current toward the left. The charge imbalance due to diffusion results in an electric field pointed toward the right, as shown in Figure 10.22b. This electric field exerts a force pushing the holes back toward the right and thus establishes a hole drift current to the right. The drift current density for holes can be expressed as

$$J_{\mathrm{P|drift}} = q\mu_p p \mathscr{E}_x. \tag{10.14}$$

Notice that the drift current direction for both holes and electrons is in the same direction as the electric field, while the holes move in the direction of the field and the electrons move against it.

Summing the drift and diffusion components of the current density for electrons and holes, we find that

$$J_{\mathrm{N}} = J_{\mathrm{N|drift}} + J_{\mathrm{N|diff}} = q\mu_n n \mathscr{E}_x + qD_n \frac{dn}{dx} \tag{10.15a}$$

$$J_{\mathrm{P}} = J_{\mathrm{P|drift}} + J_{\mathrm{P|diff}} = q\mu_p p \mathscr{E}_x - qD_p \frac{dp}{dx}. \tag{10.15b}$$

The total current density accounting for both electrons and holes is then given by

$$J = J_{\mathrm{N}} + J_{\mathrm{P}}. \tag{10.16}$$

The diffusion coefficients and carrier mobilities found in the equations for diffusion and drift currents are related by a simple equation known as the **Einstein relation**. Under common conditions, the Einstein relation is given by

$$\frac{D_n}{\mu_n} = \frac{kT}{q} \tag{10.17}$$

for electrons, and by

$$\frac{D_p}{\mu_p} = \frac{kT}{q} \tag{10.18}$$

for holes. The T in the above relations is temperature, and the k is the Boltzmann constant ($8.617 \times 10^{-5}\,\text{eV K}^{-1}$). At room temperature ($300\,\text{K}$),

$$kT = 0.0259\,\text{eV} = (1.60 \times 10^{-19})(0.0259)\text{J} \tag{10.19}$$

and

$$\frac{kT}{q} = \frac{(1.60 \times 10^{-19})(0.0259)\text{J}}{1.60 \times 10^{-19}\,\text{C}} = 0.0259\,\text{V}. \tag{10.20}$$

The quantity kT/q shows up so often in device analysis that its value at room temperature is worth remembering.

10.6   Exercises

Section 10.1

10.1. How many atomic orbitals are in each of the following energy levels?

   a) 1s

   b) 2s

   c) 2p

   d) 3s

   e) 3p

10.2. Indicate if an electron transitioning between the following orbitals gains or loses energy.

   a) from 2p to 3s   b) from 3s to 3p

   c) from 4d to 3d   d) from 3d to 4s

10.3. The nucleus of an oxygen atom has eight protons. Draw an Aufbau diagram illustrating the electron configuration for a neutrally charged oxygen atom.

Section 10.2

10.4. Atoms of aluminum, oxygen, and phosphorus have 13, 8, and 15 protons respectively. If atoms of each are neutrally charged, how many valence electrons do they have?

10.5. Two atomic orbitals are combined to form an antibonding and bonding molecular orbital. Rank the original atomic orbitals, the antibonding molecular orbital, and the bonding molecular orbital from lowest to highest energy level.

Section 10.3

10.6. Under the right conditions, carbon atoms can crystallize into a diamond lattice structure, producing the diamonds for which the structure is named. Given that the length of each side of a unit cell in a diamond is 3.57 Å, how many carbon atoms are contained within a cubic centimeter of diamond?

10.7. How many bonding molecular orbitals are present in one cubic centimeter of diamond? (Hint: see Exercise 10.6.)

10.8. Draw a detailed energy band diagram for a semiconductor. Clearly label E_c, E_v, E_G, E_{top}, and E_{bottom}. Also label the conduction band and the valence band in your diagram.

Section 10.4

10.9. If the temperature of a piece of pure silicon crystal is increased, will its conductivity increase, decrease, or stay the same? Explain your answer.

10.10. At room temperature, the band gap energies of pure diamond, silicon, and germanium crystals are approximately 5.4 eV, 1.12 eV, and 0.66 eV respectively. Rank these materials from most to least conductive.

Section 10.5

10.11. Boron (B) impurities in naturally formed diamond result in a blue coloration, as seen in the famous Hope diamond. This blue coloring can also be seen in lab-grown diamonds, which use boron as a dopant to increase conductivity. Based on its position in the periodic table, predict whether the boron dopant results in extra holes or extra electrons and state whether it acts as a donor or an acceptor in diamond crystal.

10.12. In the compound semiconductor gallium arsenide (GaAs), beryllium (Be) impurities can be added such that they replace some of the gallium atoms in the semiconductor's crystal structure. Judging by their position in the periodic table, predict whether substituting beryllium atoms in place of gallium atoms results in extra holes or extra electrons and state whether beryllium acts as a donor or an acceptor.

10.13. In the compound semiconductor gallium arsenide (GaAs), silicon (Si) impurities can be added such that they replace some of the gallium atoms in the semiconductor's crystal structure. Judging by their position in the periodic table, predict whether substituting silicon atoms in place of gallium atoms results in extra holes or extra electrons and state whether silicon acts as a donor or an acceptor.

10.14. An intrinsic semiconductor is found to have 1×10^{14} holes in its valence band. How many electrons are in its conduction band?

10.15. An intrinsic silicon crystal is heated to 380 K. Estimate the number of electrons in its conduction band (Hint: see Figure 10.16).

10.16. A silicon wafer at room temperature ($\approx$ 300 K) is doped with aluminum atoms at a concentration of $1 \times 10^{17}\,\mathrm{cm}^{-3}$.

   a) Which carrier (holes or electrons) is the majority carrier?

   b) What is the mobile electron concentration n?

   c) What is the mobile hole concentration p?

10.17. A silicon wafer at room temperature ($\approx$ 300 K) is doped with phosphorus atoms at a concentration of $1 \times 10^{15}\,\mathrm{cm}^{-3}$.

   a) Which carrier (holes or electrons) is the majority carrier?

   b) What is the mobile electron concentration n?

   c) What is the mobile hole concentration p?

10.18. In a nitrogen-doped diamond crystal ($E_\mathrm{g} \approx$ 5.4 eV), the energy level of the nitrogen atoms' extra electron is about 4 eV below the bottom of the conduction band. Predict whether n will be greater than, less than, or approximately equal to N_D and explain why the total ionization assumption cannot be used to estimate carrier concentrations.

10.19. A boron-doped silicon sample is heated to 450 K. If the dopant concentration is $1 \times 10^{16}\,\mathrm{cm}^{-3}$, determine n, p, and n_i.

10.20. A phosphorus-doped silicon sample is heated to 500 K. If the dopant concentration is $1 \times 10^{18}\,\mathrm{cm}^{-3}$, determine n, p, and n_i.

10.21. Determine if the mobile electron concentration, n, of an n-type semiconductor is greater than, less than, or approximately equal to the dopant concentration, N_D, if the semiconductor's temperature is

   a) in the freeze-out region

   b) in the intrinsic region

   c) in the extrinsic region

10.22. Determine if the mobile electron concentration, n, of an n-type semiconductor is greater than, less than, or approximately equal to the intrinsic concentration, n_i, if the semiconductor's temperature is

   a) in the freeze-out region

   b) in the intrinsic region

   c) in the extrinsic region

10.23. Some particles are distributed with a density given by
$$\eta = \alpha x^2$$
where $\alpha = 2\,\mathrm{cm}^{-5}$. If the particles are subject to diffusion with a diffusion coefficient of $5\,\mathrm{cm}^2\,\mathrm{s}^{-1}$, calculate the particle flux density $\mathscr{F}$ in terms of x using (10.6) and clearly specify the direction of particle flow.

10.24. Find the diffusion current, $I_{\mathrm{N|diff}}$, through a piece of silicon with a cross-sectional area of $0.25\,\mathrm{cm}^2$ if the electron flux density through the piece is $1.25 \times 10^{19}\,\mathrm{C}/(\mathrm{cm}^2\,\mathrm{s})$. (Hint: The charge of an electron is $1.60 \times 10^{-19}\,\mathrm{C}$.)

10.25. If electron concentration in a semiconductor decreases with x, what will be the direction of the electron flux density and what will be the direction of electron diffusion current density?

10.26. If hole concentration in a semiconductor decreases with x, what will be the direction of the hole flux density and what will be the direction of hole diffusion current density?

10.27. If an electric field directed in the positive x direction is present inside a semiconductor,

   a) in which direction will the electron drift current density, $J_{\mathrm{N|drift}}$, flow?

   b) in which direction will the hole drift current density, $J_{\mathrm{P|drift}}$, flow?

Chapter 11

The *pn* Junction

Introduction

Now we turn our attention to the *pn* junction, which is an essential component of nearly every semiconductor device. A *pn* junction occurs wherever a semiconductor transitions from *p*-type to *n*-type. In this chapter, we first consider the behavior of charge carriers near such a transition and work out the electrostatic variables of the *pn* junction in equilibrium. This prepares us for the subsequent development of the *pn* junction's $i - v$ characteristics, which concludes the chapter.

11.1   Charge Distribution

To simplify our investigation into the behavior of charge carriers within a *pn* junction, we introduce an idealized junction as shown in Figure 11.1. The left side of this junction has been uniformly doped with acceptors to form a region of *p*-type material, and the right side has been uniformly doped with donors to form an *n*-type region. The boundary in the middle is assumed to be flawless. In other words, there are no defects in the semiconductor at the boundary. We define the x direction to be increasing toward the right with $x = 0$ at the boundary between the two types. Further, we assume the device is uniform in the other directions and large enough to neglect any edge effects. The difference between the donor concentration and the acceptor concentration, known as the doping profile, is shown in Figure 11.1b. Typical *pn* junctions have a more gradual transition from *p*-type to *n*-type than the stepped profile shown here. However, this profile, known as the step junction, is often used as a first approximation in the analysis of actual *pn* junctions.

As soon as such a junction is formed, the difference in carrier concentrations between the two sides results in diffusion. Holes diffuse from the *p*-side to the *n*-side while electrons defuse from the *n*-side to the *p*-side. Far from the junction, the majority carrier concentrations are readily replenished via carrier generation, which keeps the charges of the ionized dopants balanced. However, in the region near the junction, the rapid depletion of majority carriers due to diffusion across the boundary reduces the majority carrier concentrations.

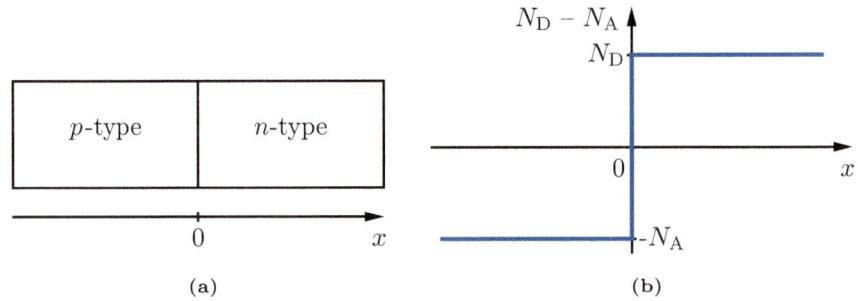

Figure 11.1: (a) Schematic of an idealized *pn* step junction and (b) its corresponding doping profile.

As the majority carriers from each side diffuse across the junction, they leave behind fixed ionized dopant atoms, resulting in a charge imbalance. This region of reduced majority carrier concentration near the junction is known as the **depletion region** or sometimes as the space charge region. The charge imbalance here creates an electric field, which results in drift currents that pull electrons back toward the *n*-side and holes back toward the *p*-side of the junction. As more carriers diffuse across the junction, these drift currents increase with the increasing electric field until an equilibrium is reached where the electron diffusion current is balanced by the electron drift current and the hole diffusion current is balanced by the hole drift current.

$$J_{\mathrm{N}} = J_{\mathrm{N}|\mathrm{drift}} + J_{\mathrm{N}|\mathrm{diff}} = 0 \tag{11.1a}$$

$$J_{\mathrm{P}} = J_{\mathrm{P}|\mathrm{drift}} + J_{\mathrm{P}|\mathrm{diff}} = 0 \tag{11.1b}$$

The equilibrium situation inside a *pn* junction is illustrated with a charge model in Figure 11.2. In this figure, ionized dopant atoms are drawn as white squares with a plus sign for donors and black squares with a minus sign for acceptors. The majority carriers are drawn as black circles with a minus sign for electrons and white circles with a plus sign for holes. Minority carriers are not drawn because their concentrations are several orders of magnitude less than the concentrations of the dopant ions and majority carriers throughout the device. This permits their contribution to the overall charge to be neglected. Within the depletion region, the concentration of majority carriers drops off rapidly. Thus, in this region, the total charge concentration is dominated by ionized dopant atoms and the majority carriers are not drawn.

The total charge concentration throughout the device is expressed as the sum of the contributions from each of the charged entities inside; namely, the holes, electrons, ionized donors, and ionized acceptors. Hence,

$$\frac{\mathrm{charge}}{\mathrm{cm}^3} = qp - qn + qN_{\mathrm{D}}^{+} - qN_{\mathrm{A}}^{-} \tag{11.2}$$

or

$$\rho = q(p - n + N_{\mathrm{D}}^{+} - N_{\mathrm{A}}^{-}) \tag{11.3}$$

where ρ is the total charge density and N_{D}^{+} and N_{A}^{-} are the concentrations of ionized donors and acceptors, respectively. Note that in this discussion, the

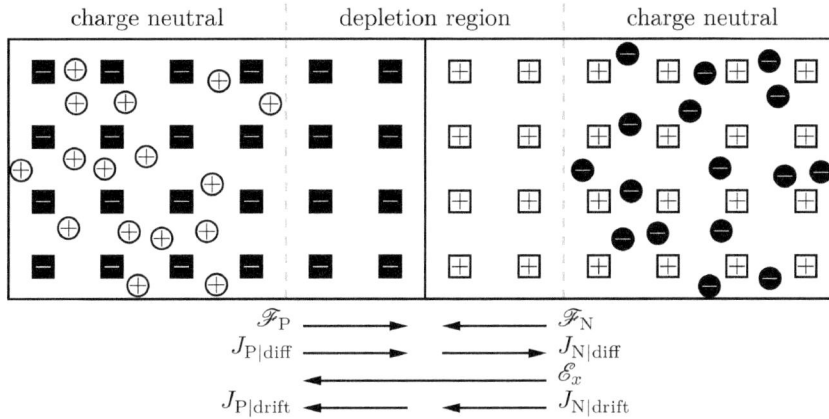

Figure 11.2: Schematic of the charge distribution and associated diffusion and drift currents inside a *pn* step junction at equilibrium.

terms "charge density" and "charge concentration" are both used to denote charge per unit volume. We have already stated that in silicon at room temperature, there is sufficient energy for nearly all of the dopants to ionize. In fact, this is true for shallow energy level dopants in any semiconductor, and so by applying the total ionization assumption, we have

$$N_D{}^+ = N_D \tag{11.4}$$

and

$$N_A{}^- = N_A. \tag{11.5}$$

By substituting (11.4) and (11.5) back into (11.3), we arrive at a simple expression for the total charge density in the device:

$$\rho = q(p - n + N_D - N_A). \tag{11.6}$$

11.2 Electrostatics

Using our knowledge of the charge imbalance in the *pn* junction in equilibrium and the junction's doping profile (repeated in Figure 11.3a), we can develop an intuitive understanding of the form of the total charge density as a function of position shown in Figure 11.3b. Outside of the depletion region, the total charge density is zero because the charges of the ionized dopant atoms are balanced by the charges of the majority carriers. As we enter the depletion region from the *p*-side, the concentration of holes decreases exponentially. As the hole concentration declines, an ever-larger portion of the negatively charged acceptor ions are uncovered. As a result, the total charge concentration is soon dominated by the acceptor concentration. We see a similar trend as we approach the junction from the *n*-side. When we enter the depletion region, the electron concentration drops off rapidly, thereby uncovering positively charged donor ions. The total charge concentration on the *n*-side of the depletion region is soon dominated by the donor concentration. Due to the stepped doping profile, there is an abrupt change from negative charge to positive charge at the junction.

(a)

(b)

(c)

(d)

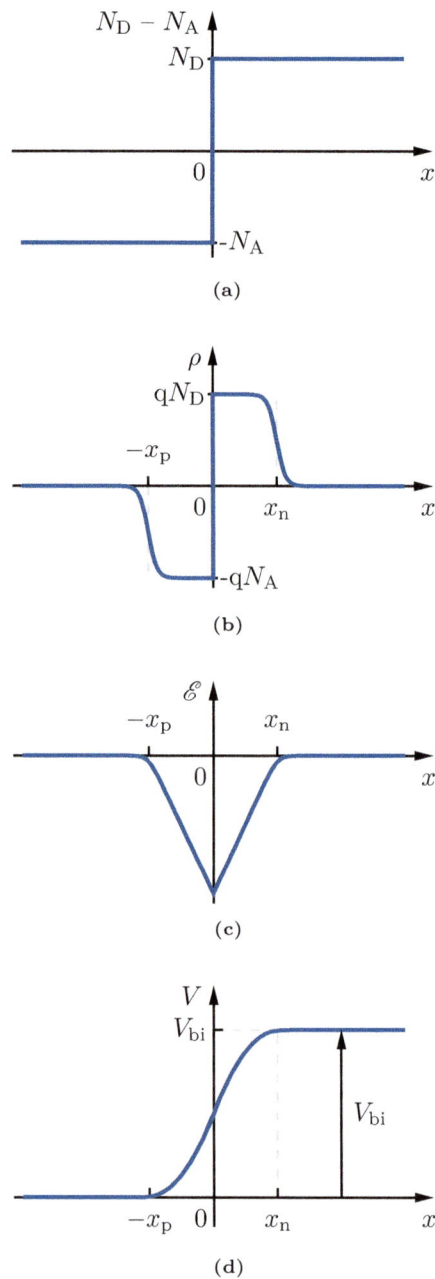

Figure 11.3: A *pn* step junction doping profile and the resulting electrostatic variables in equilibrium. (a) Doping profile, (b) charge density, (c) electric field, and (d) electric potential as a function of position.

Given the total electric charge density ρ, we can establish the electric field from Gauss's law.[1] In our idealized junction in which edge effects are neglected,

---

[1] The three-dimensional version of Gauss's law in differential form is

$$\nabla \cdot \mathscr{E} = \frac{\rho}{\epsilon_r \epsilon_0} \tag{11.7}$$

where $\nabla \cdot \mathscr{E}$ represents the divergence of the electric field $(\frac{\partial \mathscr{E}}{\partial x} + \frac{\partial \mathscr{E}}{\partial y} + \frac{\partial \mathscr{E}}{\partial z})$

the electric field only has a component in the x direction. Thus, by only considering the x dimension, Gauss's law can be expressed as

$$\frac{\mathrm{d}\mathscr{E}_x}{\mathrm{d}x} = \frac{\rho}{\epsilon_r \epsilon_0} \tag{11.8}$$

where ϵ_r is the relative permittivity of the semiconductor[2] and ϵ_0 is the permittivity of free space.[3]

From (11.8), we can see that the slope of the electric field is proportional to the charge density. Then, since the charge density is approximately zero outside of the depletion region, the electric field there must be constant. In fact, the electric field a small distance from the outside edges of the depletion region is exactly zero. This is because without external stimulus to maintain an imbalance, the mobile charge carriers will quickly move via drift currents to cancel out any nonzero field that arises in the uniformly doped portion of the device. Thus, by defining $-x_\mathrm{p}$ and x_n to be the left and right edges of the depletion region, the electric field at any point f within this region can be found by integrating (11.8) from $-x_\mathrm{p}$ to f:

$$\mathscr{E}_x(f) = \int_{-x_\mathrm{p}}^{f} \frac{\rho}{\epsilon_r \epsilon_0} \, \mathrm{d}x. \tag{11.9}$$

The resulting electric field is shown in Figure 11.3c.

Having established the electric field within the junction, the change in voltage with increasing x can be calculated as[4]

$$V(f) - V(i) = -\int_{i}^{f} \mathscr{E}_x \, \mathrm{d}x. \tag{11.10}$$

Since the electric field is zero outside of the depletion region, the voltage there must be constant. We chose arbitrarily to set the voltage on the p-side of the depletion region to zero. Then, to determine the voltage throughout the rest of the device, we carry out the integration in (11.10) starting from $-x_\mathrm{p}$ at the left edge of the depletion region. Since we are starting from a position with an initial electric potential of zero volts, (11.10) simplifies to

$$V(f) = -\int_{-x_\mathrm{p}}^{f} \mathscr{E}_x \, \mathrm{d}x. \tag{11.11}$$

Solving for $V(f)$ as we move f along in the x direction results in the electric potential shown in Figure 11.3d.

---

[2] In some texts, the relative permittivity of the semiconductor is denoted as K_s and is called the semiconductor dielectric constant.

[3] In general, the permittivity of a material ϵ_s is equal to the product of its relative permittivity ϵ_r and the permittivity of free space ϵ_0. Hence, $\epsilon_s = \epsilon_r \epsilon_0$.

[4] In general, the voltage along any path s through an electric field can be calculated as

$$V(f) - V(i) = -\int_{i}^{f} \mathscr{E} \cdot \mathrm{d}\vec{s}.$$

Since our chosen path is parallel to the electric field, we can simplify this equation to (11.10).

The potential change across the depletion region under equilibrium conditions is known as the **built-in potential** V_{bi}. This built-in potential can be calculated from (11.11) as

$$V_{bi} = V(x_n) - V(-x_p) = -\int_{-x_p}^{x_n} \mathscr{E}_x \, dx. \tag{11.12}$$

With a little work, we can derive an expression for V_{bi} from (11.1a) that depends only on the semiconductor's intrinsic level, the dopant concentrations, temperature, and physical constants. We begin by recalling that in equilibrium, the electron drift and diffusion currents sum to zero:

$$J_{N|drift} + J_{N|diff} = 0. \tag{11.13}$$

Expanding the current components using (10.15a), we have

$$q\mu_n n \mathscr{E}_x + q D_n \frac{dn}{dx} = 0. \tag{11.14}$$

Solving for $\mathscr{E}_x$ results in

$$\mathscr{E}_x = \frac{-D_n}{\mu_n} \frac{1}{n} \frac{dn}{dx}. \tag{11.15}$$

Employing the Einstein relation from (10.17) to replace the first quotient yields an expression for $\mathscr{E}_x$ that depends only on physical constants, temperature, and the electron concentration:

$$\mathscr{E}_x = \frac{-kT}{q} \frac{1}{n} \frac{dn}{dx}. \tag{11.16}$$

Plugging this expression for the electric field into (11.12) yields

$$V_{bi} = \frac{kT}{q} \int_{-x_p}^{x_n} \frac{1}{n} \frac{dn}{dx} \, dx. \tag{11.17}$$

Then, with a change of variables of integration, we have

$$V_{bi} = \frac{kT}{q} \int_{n(-x_p)}^{n(x_n)} \frac{1}{n} \, dn, \tag{11.18}$$

which can be solved to yield

$$V_{bi} = \frac{kT}{q} \ln\left(\frac{n(x_n)}{n(-x_p)}\right). \tag{11.19}$$

For the step junction under consideration, n at x_n is approximated using the total ionization assumption to be

$$n(x_n) = N_D. \tag{11.20}$$

Similarly,

$$p(-x_p) = N_A. \tag{11.21}$$

Then from the np product rule, we find n at $-x_\mathrm{p}$ to be

$$n(-x_\mathrm{p}) = \frac{n_i^2}{p(-x_\mathrm{p})} = \frac{n_i^2}{N_\mathrm{A}}. \tag{11.22}$$

Substituting these results back into (11.19), we arrive, as promised, at an expression for V_bi that depends only upon the semiconductor's intrinsic level, the device's doping concentrations, temperature, and physical constants.

$$\boxed{V_\mathrm{bi} = \frac{kT}{q} \ln\left(\frac{N_\mathrm{A} N_\mathrm{D}}{n_i^2}\right)} \tag{11.23}$$

Example 11.2.1

For a given temperature and semiconductor, (11.23) allows us to compute the built-in voltage of a pn step junction from the doping concentrations of each side. For example, consider a silicon step junction at room temperature with a donor concentration of $1 \times 10^{16}\,\mathrm{cm^{-3}}$ and an acceptor concentration of $1 \times 10^{16}\,\mathrm{cm^{-3}}$. Plugging these values into (11.23), we have

$$V_\mathrm{bi} = (0.0259\,\mathrm{V}) \ln\left(\frac{10^{16} \times 10^{16}}{(10^{10})^2}\right) = 0.7\,\mathrm{V}. \tag{11.24}$$

We have found the built-in voltage of this pn junction to be about $0.7\,\mathrm{V}$, which is a typical result.

11.3 Depletion Width

In the previous section, we were able to solve for the electric field $\mathscr{E}_x(x)$ and the electric potential $V(x)$ throughout the pn junction by assuming that the charge density $\rho(x)$ was already known. However, in general, the charge density within the depletion region is not known ahead of time because it depends on the carrier concentrations p and n, which in turn depend on the electric potential.

$$\frac{\mathrm{d}\mathscr{E}_x}{\mathrm{d}x} = \frac{\rho}{\epsilon_r\epsilon_0} = \frac{q}{\epsilon_r\epsilon_0}\left(p - n + N_\mathrm{D} - N_\mathrm{A}\right) \tag{11.25}$$

In some cases, we may be able to write expressions for p and n as a function of x, which allow for a closed-form solution to (11.25), but more often we apply a simple approximation to the charge carriers known as the **depletion approximation**. The depletion approximation assumes that the charge carriers perfectly balance the ionized dopants outside of the depletion region and are negligible within.

Thus, under the depletion region approximation, the total charge density ρ of a pn step junction takes on a squared-off form relative to its exact value, as shown in Figure 11.4. Specifically, the charge density is zero outside of the depletion region and equal to the unit charge multiplied by the doping profile within. Stated formally:

$$\rho = \begin{cases} 0 & \text{if } x \leq -x_\mathrm{p} \text{ or } x \geq x_\mathrm{n} \\ -qN_\mathrm{A} & \text{if } -x_\mathrm{p} \leq x \leq 0 \\ qN_\mathrm{D} & \text{if } 0 \leq x \leq x_\mathrm{n} \end{cases} \quad . \qquad (11.26)$$

Since the doping profile $(N_\mathrm{D} - N_\mathrm{A})$ is assumed to be known and q is a constant, the only unknowns in these expressions for charge density are the depletion region widths x_p and x_n. Therefore, in order to completely specify the depletion approximation, we need to establish these boundaries on each side of the junction.

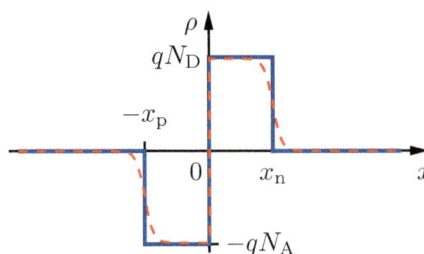

Figure 11.4: Comparison between the exact solution for charge density (dashed) and the depletion approximation (solid).

We will begin by first working out the widths x_p and x_n relative to each other. In the previous section, we considered a *pn* junction in which the depletion region was evenly split between the *p*-side and the *n*-side of the device $(x_\mathrm{p} = x_\mathrm{n})$. However, the width of the depletion region on each side of the junction is dependent on the relative doping level between the two sides such that the *p*-side and *n*-side widths will only be equal if the device's donor and acceptor doping concentrations are also equal.

To relate their relative widths, we note that the complete device is charge neutral, by which we mean that even though there is a charge imbalance in the depletion region, the sum of all charges within the device must be equal to zero. This must be true because the device is assumed to be charge neutral when constructed, and the movements of carriers as the device approaches equilibrium neither create nor destroy charge. Hence, the excess negative charge on the *p*-side of the junction must precisely balance the excess positive charge on the *n*-side of the junction:

$$Q_\mathrm{p}^- = Q_\mathrm{n}^+. \qquad (11.27)$$

Due to the depletion approximation, these excess charges can be expressed as the charge density multiplied by the volume of the depletion region on each side of the junction:

$$qN_\mathrm{A}V_\mathrm{p} = qN_\mathrm{D}V_\mathrm{n}. \qquad (11.28)$$

The volume of the depletion region on a given side of the junction is equal to the depletion width of that side times the cross-sectional area of the device, which is assumed to be constant:

$$V_\mathrm{p} = x_\mathrm{p}A, \quad V_\mathrm{n} = x_\mathrm{n}A. \qquad (11.29)$$

By dropping the area from each side of the equation, the charge balance between each side can be expressed as

$$qN_A x_p = qN_D x_n. \tag{11.30}$$

Finally, we can relate the relative widths of the depletion region on each side of the junction by solving for x_n in terms of x_p, which results in

$$x_n = \frac{N_A}{N_D} x_p. \tag{11.31}$$

Recall that N_A and N_D can vary over several orders of magnitude, and so the depletion widths x_p and x_n can also vary by several orders of magnitude. In particular, note that most of the depletion region will appear on the side of the junction with the lower doping level.

Example 11.3.1

Determine the ratio of the n-side to p-side depletion region widths (x_n/x_p) for a pn step junction with $N_A = 1 \times 10^{17}\,\text{cm}^{-3}$ and $N_D = 1 \times 10^{14}\,\text{cm}^{-3}$:

$$\frac{x_n}{x_p} = \frac{N_A}{N_D} = \frac{1 \times 10^{17}\,\text{cm}^{-3}}{1 \times 10^{14}\,\text{cm}^{-3}} = 1{,}000. \tag{11.32}$$

This result tells us that for the specified doping levels, the depletion region extends 1,000 times farther into the n-side of the junction than into the p-side.

At this point, we have established one equation relating x_p and x_n (11.30), but because we have two unknowns, we require a second independent equation relating them in order to locate both endpoints of the depletion region. A second equation relating x_p and x_n can be found by considering the electric potential within the junction. In order to determine the electric potential, we will first need to establish the electric field, which can be calculated from the charge distribution.

We begin by substituting the charge density we found using the depletion approximation into the one-dimensional version of Gauss's law in differential form from (11.25):

$$\frac{d\mathscr{E}_x}{dx} = \frac{\rho}{\epsilon_r \epsilon_0} = \begin{cases} 0 & \text{if } x \leq -x_p \text{ or } x \geq x_n \\ -\dfrac{qN_A}{\epsilon_r \epsilon_0} & \text{if } -x_p \leq x \leq 0 \\ \dfrac{qN_D}{\epsilon_r \epsilon_0} & \text{if } 0 \leq x \leq x_n \end{cases} . \tag{11.33}$$

We have previously established that the electric field is zero and constant outside of the depletion region since the charge density here is zero. Thus, the electric field must be zero at the depletion region edges $-x_p$ and x_n. Then for the region between $-x_p$ and 0, we have

$$\frac{d\mathscr{E}_x}{dx} = -\frac{qN_A}{\epsilon_r\epsilon_0}. \tag{11.34}$$

We solve for $\mathscr{E}_x(x)$ in the interval by separating variables and integrating each side. The lower bounds are fixed at $\mathscr{E}_x = 0$, $x = -x_p$, while the upper bounds vary through the extent of the interval:

$$\int_0^{\mathscr{E}_x(x)} d\dot{\mathscr{E}}_x = \int_{-x_p}^x \frac{-qN_A}{\epsilon_r\epsilon_0} d\dot{x}. \tag{11.35}$$

Completing the integration yields

$$\mathscr{E}_x(x) = \frac{-qN_A}{\epsilon_r\epsilon_0}(x + x_p), \quad \text{for } -x_p \leq x \leq 0. \tag{11.36}$$

Next, for the region between 0 and x_n, we have

$$\frac{d\mathscr{E}_x}{dx} = \frac{qN_D}{\epsilon_r\epsilon_0}. \tag{11.37}$$

Again separating variables and integrating each side, but this time fixing the upper bounds at $\mathscr{E}_x = 0$, $x = x_n$ and letting the lower bounds vary throughout the interval, we have

$$\int_{\mathscr{E}_x(x)}^0 d\dot{\mathscr{E}}_x = \int_x^{x_n} \frac{qN_D}{\epsilon_r\epsilon_0} d\dot{x}. \tag{11.38}$$

Completing the integration yields

$$\mathscr{E}_x(x) = \frac{-qN_D}{\epsilon_r\epsilon_0}(x_n - x), \quad \text{for } 0 \leq x \leq x_n. \tag{11.39}$$

From the resulting expressions in (11.36) and (11.39), we see that the electric field in the depletion region is always negative and that it decreases linearly from $x = -x_p$ until $x = 0$ and then increases linearly until $x = x_n$. If we assume that the electric field at the inflection point ($x = 0$) is continuous, this means that (11.36) and (11.39) must be equal at $x = 0$. Hence,

$$\frac{-qN_A}{\epsilon_r\epsilon_0}x_p = \frac{-qN_D}{\epsilon_r\epsilon_0}x_n, \tag{11.40}$$

which can be simplified to

$$qN_Ax_p = qN_Dx_n. \tag{11.41}$$

We recognize that (11.41), is the same as (11.30) which was derived from the requirement that the junction as a whole be charge neutral. Thus, assuming that the electric field is continuous at $x = 0$ is the same as assuming the device to be charge neutral and is in agreement with our initial assumptions. The resulting electric field, shown as a blue line in Figure 11.5, is in close agreement with the exact solution, which is shown as a dashed red line in the same figure. The exact solution is also shown separately in Figure 11.3c.

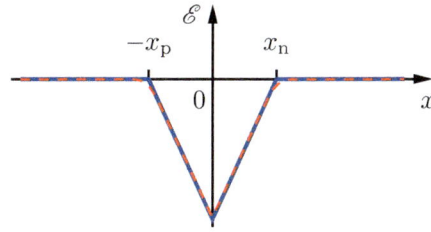

Figure 11.5: Comparison between the exact solution for electric field (dashed) and the depletion approximation (solid).

To remain consistent with our analysis from section 11.2, we define the electric potential in the $x \leq -x_{\mathrm{p}}$ region to be zero volts. Then, the electric potential in the $x \geq x_{\mathrm{n}}$ region will be equal to the built-in potential V_{bi}, which can be found from (11.23). Next, the potential within the depletion region can be found by applying (11.10) and using the established voltages $V(-x_{\mathrm{p}}) = 0$ and $V(x_{\mathrm{n}}) = V_{\mathrm{bi}}$ as boundary conditions. Due to the piecewise linear nature of our solution for $\mathscr{E}_x$, we break the integral of (11.10) into intervals. For the interval on the p-side of the junction, we let the upper bound vary through the extent of the interval, resulting in

$$V(x) - V(-x_{\mathrm{p}}) = V(x) - 0 = \int_{-x_{\mathrm{p}}}^{x} \frac{qN_{\mathrm{A}}}{\epsilon_r \epsilon_0} (x_{\mathrm{p}} + \dot{x}) \, \mathrm{d}x, \qquad (11.42)$$

which upon completing the integration provides

$$V(x) = \frac{qN_{\mathrm{A}}}{2\epsilon_r \epsilon_0} (x_{\mathrm{p}} + x)^2, \quad \text{for } -x_{\mathrm{p}} \leq x \leq 0. \qquad (11.43)$$

For the interval on the n-side of the junction, we vary the lower bound throughout the interval such that

$$V(x_{\mathrm{n}}) - V(x) = V_{\mathrm{bi}} - V(x) = \int_{x}^{x_{\mathrm{n}}} \frac{qN_{\mathrm{D}}}{\epsilon_r \epsilon_0} (x_{\mathrm{n}} - \dot{x}) \, \mathrm{d}x, \qquad (11.44)$$

which upon completing the integration provides

$$V(x) = V_{\mathrm{bi}} - \frac{qN_{\mathrm{D}}}{2\epsilon_r \epsilon_0} (x_{\mathrm{n}} - x)^2, \quad \text{for } 0 \leq x \leq x_{\mathrm{n}}. \qquad (11.45)$$

From (11.43), we see that the electric potential starts from zero and increases quadratically throughout the p-side depletion region. It is continuous at the boundary due to the assumed charge distribution.[5] Then, from (11.45), we see that the electric potential continues to increase but in a quadratically decreasing manner until reaching V_{bi} at $x = x_{\mathrm{n}}$. The resulting electric potential closely matches the exact solution, as shown in Figure 11.6.

---

[5]The electric potential could be discontinuous if the charge distribution included a charge layer with zero thickness—that is, a sheet charge.

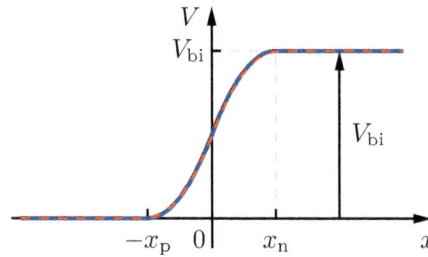

Figure 11.6: Comparison between the exact solution for electric potential (dashed) and the depletion approximation (solid).

Since the voltage is continuous, we can set (11.43) equal to (11.45) at $x = 0$ to arrive at a second equation relating x_p and x_n:

$$\frac{qN_\mathrm{A}}{2\epsilon_r\epsilon_0}x_\mathrm{p}{}^2 = V_\mathrm{bi} - \frac{qN_\mathrm{D}}{2\epsilon_r\epsilon_0}x_\mathrm{n}{}^2. \tag{11.46}$$

Using (11.31) and (11.46), we can now express the p-side and n-side depletion widths (x_p and x_n) in terms of dopant levels. We begin by substituting the solution for x_n in terms of x_p from (11.31) into (11.46), resulting in

$$\frac{qN_\mathrm{A}}{2\epsilon_r\epsilon_0}x_\mathrm{p}{}^2 = V_\mathrm{bi} - \frac{qN_\mathrm{D}}{2\epsilon_r\epsilon_0}\frac{N_\mathrm{A}{}^2}{N_\mathrm{D}{}^2}x_\mathrm{p}{}^2. \tag{11.47}$$

With a bit more algebra, we can solve for x_p to find

$$x_\mathrm{p} = \left[\frac{2\epsilon_r\epsilon_0}{q}\left(\frac{N_\mathrm{D}}{N_\mathrm{A}(N_\mathrm{A} + N_\mathrm{D})}\right)V_\mathrm{bi}\right]^{1/2}. \tag{11.48}$$

Then, by using (11.31) again, we can also solve for x_n to find

$$x_\mathrm{n} = \frac{N_\mathrm{A}}{N_\mathrm{D}}x_\mathrm{p} = \left[\frac{2\epsilon_r\epsilon_0}{q}\left(\frac{N_\mathrm{A}}{N_\mathrm{D}(N_\mathrm{A} + N_\mathrm{D})}\right)V_\mathrm{bi}\right]^{1/2}. \tag{11.49}$$

By expressing the depletion width on each side of the junction in terms of the dopant levels and physical parameters in (11.48) and (11.49), we have completely specified the depletion approximation.

The total depletion region width, W can be defined as the sum of the p-side and n-side depletion region widths:

$$W = x_\mathrm{p} + x_\mathrm{n}. \tag{11.50}$$

By choosing a common denominator in the expressions for x_p and x_n, and combining like terms, (11.50) can be written as

$$W = \left[\frac{2\epsilon_r\epsilon_0}{q}\right]^{1/2}\left(\frac{N_\mathrm{D} + N_\mathrm{A}}{[N_\mathrm{A}N_\mathrm{D}(N_\mathrm{A} + N_\mathrm{D})]^{1/2}}\right)V_\mathrm{bi}{}^{1/2}. \tag{11.51}$$

Then, by canceling a factor of $\sqrt{N_A + N_D}$ and grouping all the terms under same square root, we arrive at

$$W = \left[\frac{2\epsilon_r \epsilon_0}{q} \left(\frac{N_A + N_D}{N_A N_D} \right) V_{bi} \right]^{1/2}, \qquad (11.52)$$

which gives us the total width of the depletion region in an isolated pn step junction at equilibrium.

11.4 Band Bending

We can gain a deeper insight into the pn junction with the aid of the band diagram. In section 10.5.1, we established band diagrams for p-type and n-type materials. They were distinguished by their distribution of carriers with respect to energy as illustrated in Figures 10.19b and 10.19c. Now we set out to construct the band diagram of a pn junction. We will begin by simply connecting the band diagrams of the p-type and n-type materials together, conduction band edge to conduction band edge and valence band edge to valence band edge, as shown in Figure 11.7.

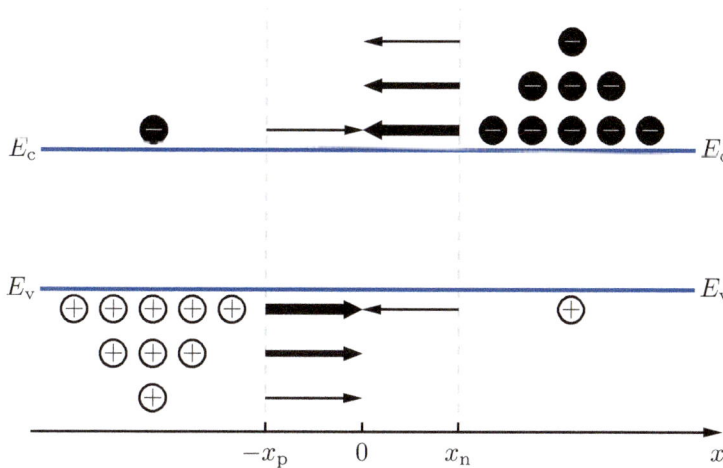

Figure 11.7: Nonequilibrium energy band diagram of a pn junction with zero electric field. The difference in carrier concentrations results in diffusion currents that, due to the lack of an electric field, are not balanced by drift currents.

This arrangement could represent the carrier distribution with respect to energy the moment the junction is formed, but as described previously via the charge distribution model (Figure 11.2), the uneven distribution of carriers immediately results in diffusion currents. A net flow of electrons moves from the n-side into the p-side and a net flow of holes moves from the p-side into the n-side. Fixed charges at the dopant sites, which are uncovered by diffusion, produce an electric field that begins to pull electrons back toward the n-side and holes back toward the p-side, resulting in drift currents. As more dopant sites are uncovered by diffusion, the electric field grows and the resulting drift currents increasingly counteract the diffusion currents until an equilibrium is reached.

As we saw in Figure 11.3d, at equilibrium there is a built-in voltage difference V_{bi} between the n-side and p-side of the junction. Now we would like to determine how this built-in voltage is reflected in the band diagram. To begin, we relate voltage to energy by noting that the electric potential V is equivalent to the electric potential energy (P.E.) per unit of charge at a given point in an electric field. Hence,

$$V = \frac{\text{P.E.}}{q} \tag{11.53}$$

where q is the elementary charge (1.6×10^{-19} C). For an electron with charge $-q$, we can write its electrical potential energy in terms of the electric potential as

$$\text{P.E.} = -qV. \tag{11.54}$$

With this relationship in mind, we can see that for electrons, an increase in electric potential (voltage) corresponds to a decrease in electric potential energy. For example, if the voltage in a semiconductor increases from V_a as some point x_a to V_b as another point x_b, as shown in Figure 11.8a, the electric potential energy of an electron moving from x_a to x_b will change by $-q(V_b - V_a)$. This reduction in energy for an electron is illustrated in Figure 11.8c.

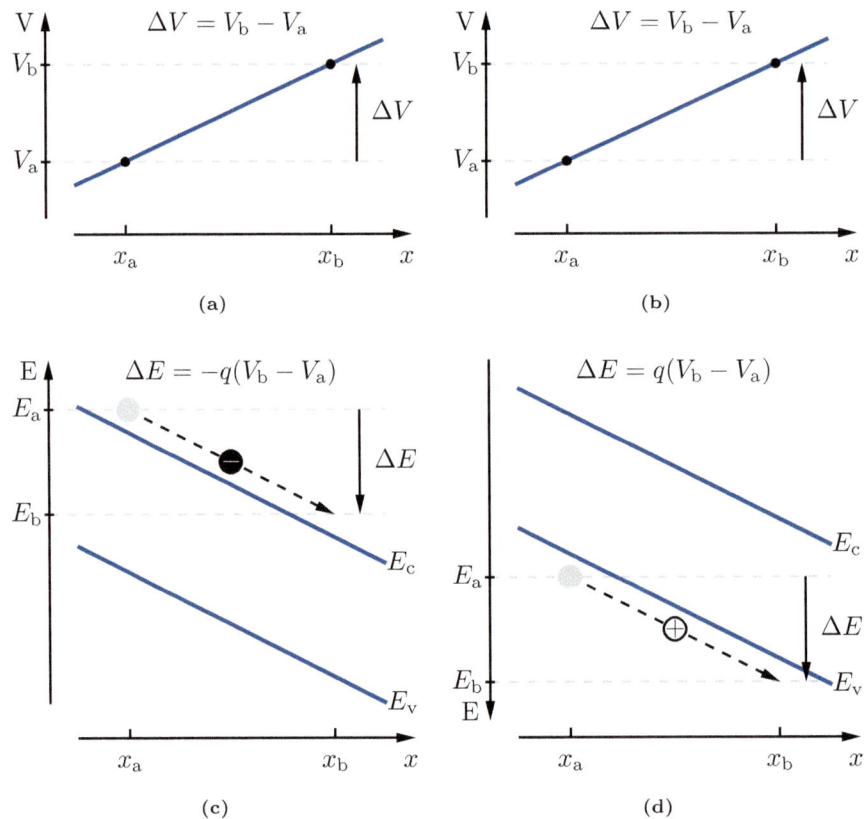

Figure 11.8: When moving from the lower voltage V_a to the higher voltage V_b as shown in (a), the electron *loses* energy ($\Delta E = -q(V_b - V_a)$) as shown in (c). When moving from the lower voltage V_a to the higher voltage V_b, as shown in (b), the hole *gains* energy ($\Delta E = q(V_b - V_a)$) as shown in (d).

If a hole moves from x_a to x_b through the same semiconductor as shown in Figure 11.8b, its electric potential energy changes by $q(V_b - V_a)$. This represents an increase in energy for the hole, as illustrated in Figure 11.8d. Recall that energy increases in the downward direction for holes in an energy band diagram.

In the *pn* junction, by moving to achieve equilibrium, the carriers have increased the voltage on the *n*-side of the junction by V_{bi} relative to the *p*-side. From (11.54), we see that this corresponds to a change in potential energy of $-qV_{bi}$. In other words, by coming to equilibrium, the electric potential energy of the electrons on the *n*-side of the junction have been reduced by qV_{bi}. The fact that the electrons are at a lower potential energy when in equilibrium should not be surprising because dynamic systems in general lose energy as they approach stable equilibria.

The energy of the carriers in the energy band diagram represents their total energy, including both potential and kinetic energy. Therefore, since their potential energy has decreased by qV_{bi} and their kinetic energy (thermal energy) has not changed, the energy levels on the *n*-side of the junction must be qV_{bi} lower than on the *p*-side as shown in Figure 11.9. Graphically, the energy levels are shaped like an upside-down plot of the voltage across the device.

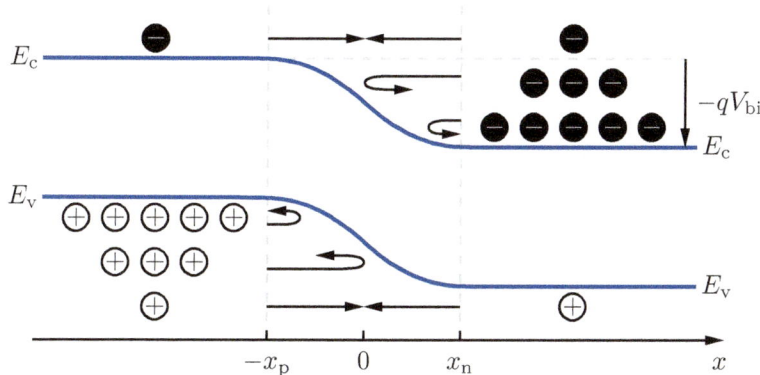

Figure 11.9: Energy band diagram of a *pn* junction in equilibrium.

If the idea that increasing the voltage results in lowering the energy of the electrons seems counter intuitive, consider the simple circuit of Figure 11.10. Just as water lowers its gravitational potential energy by flowing downhill, electrons lower their electrical potential energy by flowing from the negative terminal of the source toward the higher voltage positive terminal. Stated another way, electrons lower their potential energy by moving toward higher electrical potential. These results are due to the fact that electrons have negative charge.

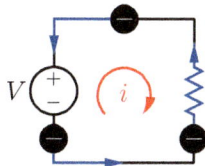

Figure 11.10: Direction of electron flow relative to the current direction in a simple resistive circuit.

Taking a close look at the equilibrium conditions shown in Figure 11.9, we see that an energy barrier or hill has formed in the depletion region, preventing most of the electrons in the *n*-side from wandering across the junction into the *p*-side. Simultaneously, any electrons that wander into the depletion region from the *p*-side are readily swept down this energy hill and into the *n*-side. Analogously, very few holes on the *p*-side have enough energy to overcome the energy barrier in the depletion region and travel to the *n*-side while any holes that enter the depletion region from the *n*-side readily float up and across the junction into the *p*-side. This difference in energy between the *n* and *p* sides of the junction is commonly referred to as **band bending**. This band bending gives us another way to visualize the motion of charge carriers within a device. In particular, we can see how lowering the energy of the *n*-side of the junction by qV_{bi} relative to the *p*-side brings the drift and diffusion current for both holes and electrons into balance and thus brings the junction into equilibrium.

11.5 $i - v$ Characteristics

In order to determine how the *pn* junction behaves in a circuit, we will need to relate the current through the device to the voltage across it. In other words, we need to establish the *pn* junction's $i - v$ characteristics. As we will soon see, the *pn* junction acts as a **diode**, which is a device that limits current flowing in one direction while allowing it to flow easily in the other. This one-way valve-like behavior of *pn* junction diodes has many practical applications, some of which will be presented in the next chapter.

We will begin our investigation into the *pn* junction's $i - v$ characteristics by considering what happens when the ends of the junction are connected to each other with a conducting wire, as shown in Figure 11.11. Will any current flow through this wire? If we assume the circuit is in thermodynamic equilibrium and is isolated from electromagnetic radiation, we can answer this question with an energy argument. First, we note that the wire will have some small but non-zero resistance R. Therefore, any current I flowing through the wire will dissipate energy in the form of heat at a rate of I^2R. Thanks to the law of conservation of energy, we know that this energy must come from somewhere, but due to our previous assumptions (equilibrium, no EM radiation), there is no source of energy in the circuit. Thus the energy dissipated in the wire and the current through it must be zero.

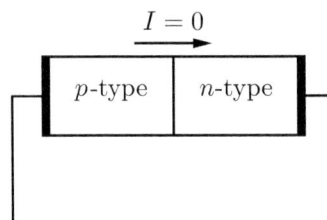

Figure 11.11: A wire connecting both ends of a *pn* junction in equilibrium will carry no current.

You may be wondering about the junction's built-in potential and why the voltage difference between the p- and n-sides of the junction in equilibrium doesn't lead to current through the wire. The explanation is that there will be built-in potential differences at the interfaces between the wire and each side of the pn junction, just as there is between the p-side and the n-side of the junction. Known as contact potentials, the potential differences across the wire-semiconductor junctions precisely cancel the built-in potential of the pn junction. As a result, there is no potential difference between each end of the wire, and so no current flows through it, in agreement with the previous energy argument.

So far, we have established only that the pn junction's $i-v$ characteristic passes through the origin $(v, i = 0, 0)$. Now we set out to develop a qualitative understanding of the junction's forward- and reverse-biased characteristics. Consider the simple circuit of Figure 11.12, in which an independent voltage

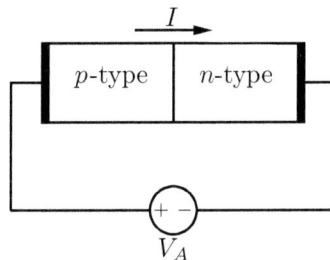

Figure 11.12: Voltage V_A is applied to a pn junction via an independent voltage source.

source V_A is connected to the pn junction. The positive terminal of the source is connected to the p-side of the junction and the negative terminal to the n-side. Applying the passive sign convention, current flowing through the device from the p-side to the n-side will be considered positive. If the wire contacts are well made, they will have very low resistance such that we can neglect any drop in the applied voltage across these contacts. Further, if the resulting current I is not too large, the relatively abundant carriers in the neutral portions of the junction provide a low resistance path up to the depletion region edges. Thus we can also neglect any drop in the applied voltage across the neutral portions of the p-side and n-side of the junction. Conversely, the shortage of charge carriers in the depletion region results in a much larger resistance, such that essentially all of the applied voltage is dropped here.

Since we will assume that all of the applied voltage is dropped across the depletion region, when a positive voltage V_A is applied, the potential on the p-side of the junction will increase by V_A, and when a negative V_A is applied, the potential on the p-side of the junction will decrease by V_A. Regardless of the voltage applied, we will take the equilibrium potential of the p-side of the junction when $V_A = 0$ as our reference voltage. Then the n-side of the junction remains at V_{bi} for all applied voltages, as shown in Figure 11.13.

The difference in potential between the n-side and the p-side of the junction is given by $V_{bi} - V_A$, from which we see that a positive applied voltage reduces the potential difference between the two sides, whereas a negative applied voltage increases it. Also notice in Figure 11.13 that the width of the depletion

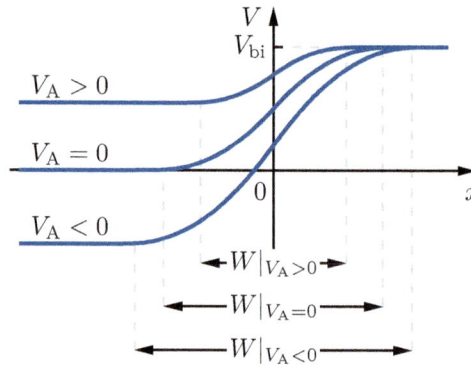

Figure 11.13: The influence of applied voltage on the depletion width and the electrostatic potential within a *pn* junction.

region is influenced by the applied voltage. A positive applied voltage decreases the depletion width, whereas a negative applied voltage increases it. Equations (11.48), (11.49), and (11.52) for depletion widths in an isolated *pn* junction can be generalized to account for the applied voltage by replacing V_{bi} in each equation with $V_{bi} - V_A$ such that

$$x_p = \left[\frac{2\epsilon_r \epsilon_0}{q} \left(\frac{N_D}{N_A(N_A + N_D)} \right) (V_{bi} - V_A) \right]^{1/2}, \qquad (11.55)$$

$$x_n = \left[\frac{2\epsilon_r \epsilon_0}{q} \left(\frac{N_A}{N_D(N_A + N_D)} \right) (V_{bi} - V_A) \right]^{1/2}, \qquad (11.56)$$

and

$$W = \left[\frac{2\epsilon_r \epsilon_0}{q} \left(\frac{N_A + N_D}{N_A N_D} \right) (V_{bi} - V_A) \right]^{1/2}. \qquad (11.57)$$

Note that we must restrict V_A to be less than V_{bi} to prevent imaginary results in the preceding equations. The reason for this is that when V_A exceeds V_{bi}, we will have a large current, and our assumption that the voltage drops across the neutral portions of the device are negligible will no longer be valid.

To work out the effect of the applied voltage on the movements of charge carriers within the junction, we turn our attention to the junction's energy band diagram. From (11.54), we determined that the energy levels in the energy band diagram are shaped like upside-down copies of the electric potential across the junction. Hence, when the voltage of the *p*-side of the junction goes up, the energy bands on the *p*-side move down, and when the voltage goes down, the energy bands move up. When a positive voltage is applied to the *p*-side of the junction, the device is said to be **forward biased**. Under this condition, the energy levels of the *p*-side of the junction will be lowered relative to their equilibrium values, as shown in Figure 11.14. The energy hill that was keeping the drift and diffusion currents in balance has now been lowered as well. By reducing the energy barrier between the two sides, now more of the electrons on the *n*-side of the junction (and holes on the *p*-side) have enough energy to diffuse across the depletion region. On the other hand, there is no change in the rate of minority carriers wandering into the depletion region and being swept across to the other side by the electric field. As a consequence, the drift and diffusion currents for both holes and electrons are no longer balanced. The excess holes diffusing to the right and the excess electrons diffusing to the left both contribute to a net positive current flowing to the right through the device.

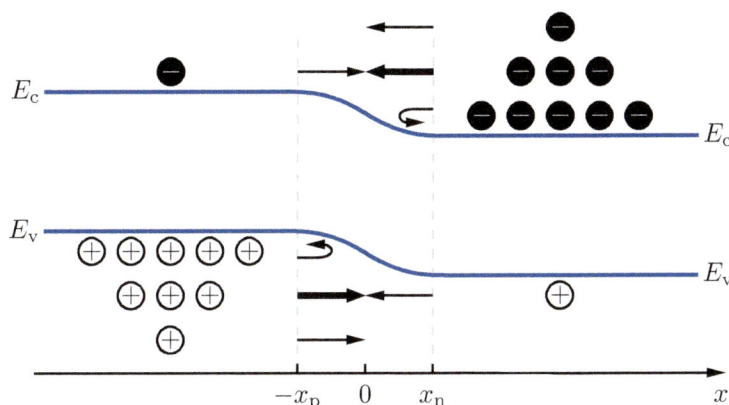

Figure 11.14: Energy band diagram of a forward-biased *pn* junction.

That the application of a positive voltage produces positive current flowing through the device is unsurprising; all passive elements respond in this way. One of the features that differentiate the *pn* junction from the passive elements we have already studied is that the current through it increases exponentially with increasing applied voltage. The charge carriers on either side of the depletion region are concentrated near the band edges and their concentration with respect to energy declines exponentially as we move away from the band edge (see Figure 10.19). Thus, as the applied voltage increases from zero and the energy barrier in the depletion region is reduced, the number of majority carriers with sufficient energy to cross the barrier grows exponentially. The result is an exponential increase in current through the device as the forward-biased voltage increases.

If we **reverse bias** the junction by applying a negative voltage, the height of the energy barrier in the depletion region is increased. An energy band diagram for the reverse-biased *pn* junction is shown in Figure 11.15. As the magnitude of the reverse-biased voltage increases, the number of majority carriers with sufficient energy to cross the depletion region decreases exponentially toward zero. Meanwhile, the rate at which minority carriers wander into the depletion region and get swept across to the other side by the electric field remains unchanged because it is not influenced by the height of the energy barrier, or by the strength of the electric field in the depletion region. Rather, it depends only on the minority carrier concentration in the neutral regions of the junction. Thus, as the magnitude of the reverse-biased voltage increases, the flow of majority carriers diminishes and the current saturates toward a small negative value due solely to the flow of minority carriers. The minority electrons flowing from the *p*-side to the *n*-side and minority holes flowing from the *n*-side to the *p*-side each contribute to a net current flowing from the *n*-side to the *p*-side of the junction. Given the assigned current polarity, this current is negative and its magnitude, which is called the **saturation current**, is commonly denoted I_0.

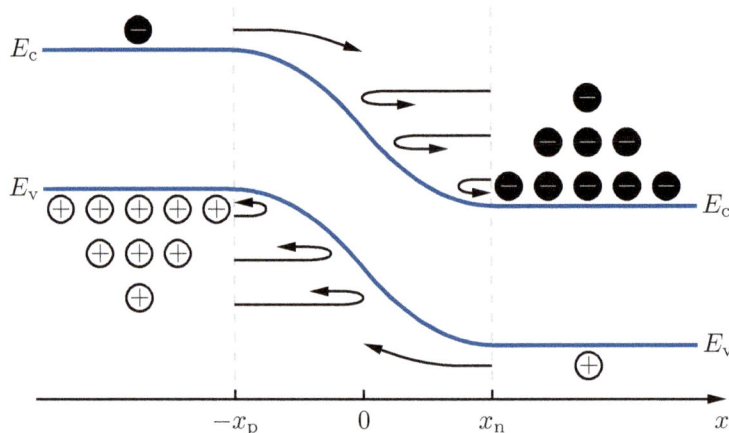

Figure 11.15: Energy band diagram of a reverse-biased *pn* junction.

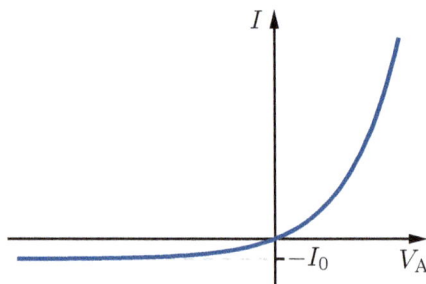

Figure 11.16: The $i - v$ characteristic curve of an ideal *pn* junction diode.

Mathematically, the complete $i - v$ characteristic curve for an ideal *pn* junction diode is given by

$$I = I_0(e^{qV_A/kT} - 1), \tag{11.58}$$

which is plotted in Figure 11.16. Note how the current I given by (11.58) is in agreement with the preceding qualitative analysis. First, when the applied voltage V_A is zero, the current I is also zero as required. Next, the current I increases exponentially with the applied voltage V_A, as predicted. Finally, for V_A less than zero, the current quickly saturates at $-I_0$ as V_A decreases.

11.6 Exercises

Section 11.1

11.1. A semiconductor device is constructed with the doping profile $(N_D - N_A)$ shown in Figure 11.17. For each region of the device (R_1, R_2, R_3), determine if the region is n-type or p-type.

11.2. Assuming a device with the doping profile of Figure 11.17 is in electrostatic equilibrium, determine the direction of the following quantities at $x = 0$.

 a) electron flux density $\mathscr{F}_N$

 b) hole flux density $\mathscr{F}_P$

 c) electron diffusion current density $J_{N|diff}$

 d) hole diffusion current density $J_{P|diff}$

 e) electron drift current density $J_{N|drift}$

 f) hole drift current density $J_{P|drift}$

 g) electric field $\mathscr{E}_x$

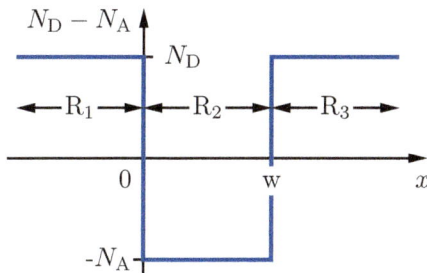

Figure 11.17: The doping profile $(N_D - N_A)$ of an unspecified semiconductor device.

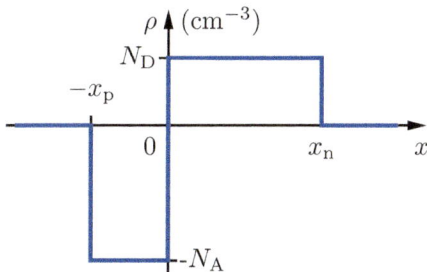

Figure 11.18: The depletion approximation of the charge density of a pn step junction in electrostatic equilibrium.

Section 11.2

11.3. Given the charge density, ρ, of a pn junction in electrostatic equilibrium, as shown in Figure 11.18,

 a) sketch a plot of $\mathscr{E}_x(x)$ within the junction and

 b) sketch a plot of $V_x(x)$ within the junction.

11.4. Assuming total ionization, determine the built-in voltage of a germanium step junction at $300\,\mathrm{K}$. The junction's doping concentrations are $N_A = 6 \times 10^{17}\,\mathrm{cm}^{-3}$ and $N_D = 4 \times 10^{15}\,\mathrm{cm}^{-3}$. Hint: The intrinsic carrier concentration, n_i, of germanium at $300\,\mathrm{K}$ is about $2.2 \times 10^{13}\,\mathrm{cm}^{-3}$.

11.5. A pn step junction is constructed from silicon, with a phosphorus concentration of $1 \times 10^{15}\,\mathrm{cm}^{-3}$ on one side and a boron concentration of $2 \times 10^{16}\,\mathrm{cm}^{-3}$ on the other. If the junction is operating at $400\,\mathrm{K}$, find the intrinsic carrier concentration, n_i, and the built-in voltage, V_{bi}.

11.6. A silicon pn step junction is doped with an acceptor concentration of $3 \times 10^{17}\,\mathrm{cm}^{-3}$. Assuming total ionization, determine the necessary donor concentration if the junction is to have a built-in voltage of $0.8\,\mathrm{V}$ at $300\,\mathrm{K}$.

Section 11.3

11.7. A pn step junction with doping concentrations of $N_A = 2 \times 10^{14}\,\mathrm{cm}^{-3}$ and $N_D = 6 \times 10^{14}\,\mathrm{cm}^{-3}$, has a p-side depletion region width of $x_p = 1.62\,\mu\mathrm{m}$.

 a) Use the total ionization assumption and the fact that the excess charge on each side of the junction must be equal to determine the n-side depletion region width x_n.

 b) Use the depletion approximation to plot the charge density $\rho(x)$ in equilibrium. Clearly label N_D, $-N_A$, x_n, and $-x_p$ in the plot. Finally, compare the areas of the negatively and positively charged regions of the device.

11.8. Find x_p, x_n, and the depletion width, W, of a silicon pn step junction at $300\,\mathrm{K}$, given that the ionized dopant concentrations of the donors and acceptors are $1 \times 10^{15}\,\mathrm{cm}^{-3}$ and $5 \times 10^{14}\,\mathrm{cm}^{-3}$

respectively. Hint: The relative permittivity of silicon is 11.68, and the permittivity of free space is $8.854 \times 10^{-14}\,\mathrm{F\,cm^{-3}}$.

11.9. Using the depletion approximation, make a plot of the equilibrium charge density for the *pn* junction in Exercise 11.8. Clearly label $-x_{\mathrm{p}}$, x_{n}, N_{D}, and $-N_{\mathrm{A}}$.

11.10. Find the depletion width, W, for a silicon *pn* step junction with $N_{\mathrm{A}} = N_{\mathrm{D}} = 1 \times 10^{16}\,\mathrm{cm^{-3}}$ at $T = 250\,\mathrm{K}$ and $T = 350\,\mathrm{K}$, assuming the dopants are totally ionized. How does the depletion width change with temperature?

Section 11.4

11.11. Given the plots of electrical potential versus position within a semiconductor device shown in Figures 11.19a and 11.19b, sketch each device's corresponding energy band diagram.

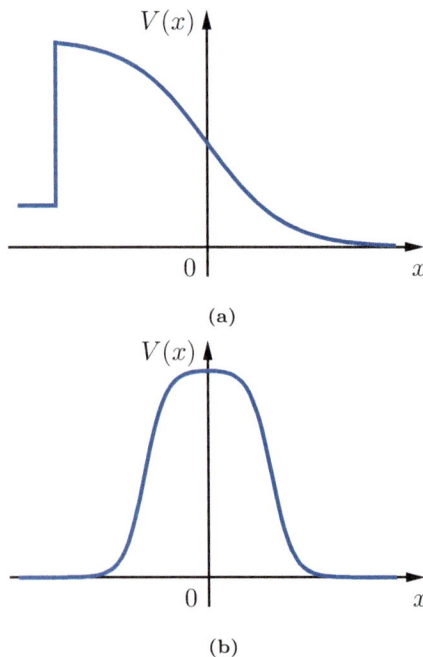

(a)

(b)

Figure 11.19: Electrostatic potentials in unspecified semiconductor devices.

11.12. Given the semiconductor device energy band diagrams shown in Figures 11.20a and 11.20b, sketch plots of their corresponding electric potential versus position.

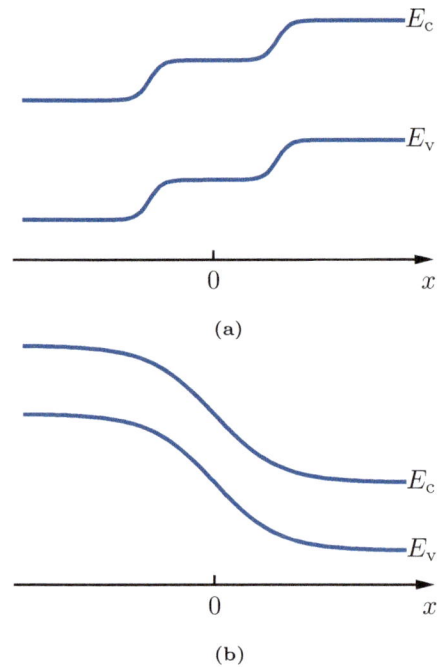

(a)

(b)

Figure 11.20: Energy band diagrams for unspecified semiconductor devices.

Section 11.5

11.13. Calculate the width of the depletion region, W, after a voltage of $0.3\,\mathrm{V}$ is applied from the *p*-side to the *n*-side of the *pn* junction in Exercise 11.8. Repeat this calculation for an applied voltage of $-0.3\,\mathrm{V}$.

11.14. A bias voltage, V_{A}, is applied to a *pn* junction diode that has a reverse-biased saturation current, I_0, of $30\,\mathrm{nA}$. Determine the current through the device if

a) $V_{\mathrm{A}} = 0.5\,\mathrm{V}$ and $T = 300\,\mathrm{K}$.

b) $V_{\mathrm{A}} = 0.5\,\mathrm{V}$ and $T = 400\,\mathrm{K}$.

c) $V_{\mathrm{A}} = -0.5\,\mathrm{V}$ and $T = 300\,\mathrm{K}$.

d) $V_{\mathrm{A}} = -0.5\,\mathrm{V}$ and $T = 400\,\mathrm{K}$.

11.15. Determine the reverse-biased saturation current, I_0, necessary for a *pn* junction diode to allow a forward-biased current of $1\,\mathrm{A}$ when $0.5\,\mathrm{V}$ is applied at room temperature.

11.16. Two *pn* junction diodes, identical to the one in Exercise 11.14, are connected back-to-back in a circuit, as shown in Figure 11.21. If V_s is 5 V, approximate I.

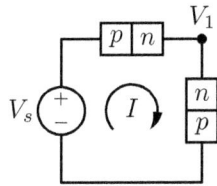

Figure 11.21: Back-to-back *pn* junction diode circuit.

Chapter 12

Diodes

12.1 Diodes in Circuits

Diodes allow current to flow easily in one direction, called the forward direction, while presenting a large resistance against current flowing in the opposite or reverse direction. You can think of them as one-way valves for current that essentially only allow current to flow in the forward direction. The terminal through which current in the forward direction enters a diode is called the **anode**, while the terminal through which current in the forward direction leaves a diode is called the **cathode**. Both the anode and the cathode are types of **electrodes**.

Early diodes (from the late 1800s until the mid 1900s) relied on thermionic emission from a hot cathode and were constructed in vacuum tubes. In these devices, there is a gap between the cathode and the anode within the evacuated tube. The cathode material is heated until some electrons gain enough energy to escape the cathode, at which point they become attracted to the nearby positively charged anode. Thus, the current, which flows in the opposite direction of the electrons, flows in the forward direction through the device from the anode to the cathode. When the polarity is reversed, the cathode becomes positively charged and the negatively charged electrons accumulate on the anode. Since the anode is colder, electrons there don't have enough energy to escape from this electrode, and so current cannot flow in the reverse direction.

In modern pn junction diodes like those studied in chapter 11, the forward direction for current is from the p-side to the n-side, and so the p-side of the junction is the anode while the n-side is the cathode.

12.1.1 Diode Circuit Symbol

The circuit symbol for a diode consists of a triangle with one vertex pointing toward the direction in which large positive current is allowed to flow, and a line through that vertex orthogonal to the diode's connecting wire to remind us that current coming from the opposite direction is blocked. This is illustrated in Figure 12.1. You can imagine the diodes as a funnel that helps large currents flow through the circuit in the positive direction while making it difficult for current to flow in the negative direction.

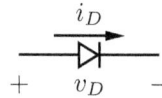

Figure 12.1: The circuit symbol of a diode.

12.1.2   Simple Logic Application

The two-diode circuit of Figure 12.2a is an implementation of a digital logic OR gate. This circuit, like all of the circuits presented so far in this book, is an analog circuit. However, when analyzing digital logic, the focus is usually on the logic itself, rather than the implementation details of a particular gate. Thus, in digital circuits, each type of logic gate (AND, OR, NAND, NOR, etc.) is given its own symbol. The digital circuit symbol for an OR gate is shown in Figure 12.2b. An OR gate has one output variable, labeled Q, and two or more input variables. In this case, the input variables are labeled A and B. Each of the variables can take on one of two logical values: either 1 or 0.

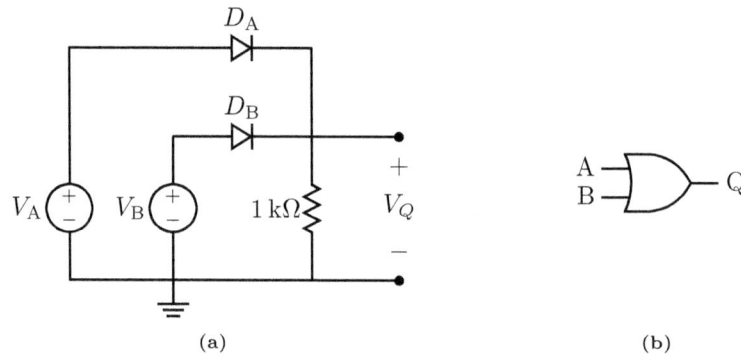

Figure 12.2: A two-input digital logic OR gate implemented with diodes on the left (a) and the digital circuit symbol for an OR gate on the right (b). The voltage sources V_A and V_B on the left supply the gate's input signals, labeled A and B, on the right. The output voltage V_Q provides the logic gate's output signal Q.

When constructing digital logic gates from analog circuit components, as we have done using diodes in Figure 12.2a, voltages are used to represent the logical variables. We could arbitrarily assign voltage ranges to the logical values of 1 and 0, but they are commonly assigned according to the positive logic convention[1] in which a high voltage range (e.g., 2.8 V to 5 V) is used to represent a logical 1, and a low voltage range (e.g., 0 V to 0.5 V) is used to represent a logical 0. The precise voltage ranges for 1 and 0 are dependent on the technology being used, but in general we can refer to the ranges as *high* denoted with an "H" and *low* denoted with an "L."

The logic of an OR gate is such that its output will be a logical 1 whenever one or both of its inputs are 1. This means that the output voltage of the circuit should be high when at least one of its input voltages is high, and the output voltage should be low when both of the input voltages are low.

---

[1] As opposed to negative logic convention, which prescribes a logical 1 to low voltages and a logical 0 to high voltages.

These requirements are conveniently expressed using **truth tables**, as shown in Table 12.1a and Table 12.1b.

A	B	Q
0	0	0
0	1	1
1	0	1
1	1	1

(a)

V_A	V_B	V_Q
L	L	L
L	H	H
H	L	H
H	H	H

(b)

Table 12.1: Truth tables for a digital OR gate. Logical variables are indicated on the left (a) and relative voltage levels in positive logic are indicated on the right (b).

Taking the voltages V_A and V_B to be logic signals, we can work out how this circuit functions as a logical OR gate using the pn junction $i-v$ characteristics established in the section 11.5. Recall that the current through an ideal diode is given by

$$I = I_0(e^{qV_A/kT} - 1). \tag{12.1}$$

By design, the reverse-biased saturation current, I_0, will be several orders of magnitude smaller than relevant circuit currents to ensure that the reverse-biased current can be neglected. Typical values of I_0 for silicon diodes are around 1×10^{-12} A to 1×10^{-15} A. As a consequence, the forward-biased current for low voltages is also very low and is insignificant as long as the voltage across the diode remains less than a few kT/q.

If both input voltages are low, providing logical 0s, the voltage across and current through the diodes will be approximately zero, and the voltage dropped across the 1 kΩ resistor, which is the output voltage V_Q, will be low as well, producing a logical 0 output value. However, if both input voltages are high, providing logical 1s, the diodes allow current to easily flow through them such that the output voltage is also high producing a logical 1 output value. Finally, if one of the input voltages is high while the other is low, the current through the forward-biased diode on the high side causes the output voltage to also be high and produces a logical 1 output value, while the diode on the low side becomes reverse biased and prevents current from flowing back toward the low input.

These results match the input and output values found in the truth tables for a digital OR gate shown in Table 12.1, and so the preceding qualitative analysis suggests that the circuit of Figure 12.2a does, in fact, implement an OR gate. Yet, we also know that the voltage drop across a forward-biased diode is not exactly zero, and so not all of a high-input voltage will be available at the output. Thus, the "high"-output voltage will be lower than the high-input voltage. This may not be a problem if we only have a single logic stage, but in more complicated circuits where many logic gates may be chained together, the successive voltage drops could cause the "high"-output voltage to fall below the voltage range assigned to a logical 1. At this point, we would no longer know how to interpret the output of our digital circuit. We begin to address this problem with a more careful quantitative analysis of a simple diode circuit in the next section.

12.1.3   Simple Diode Circuit

In order to make the quantitative analysis of a diode circuit as straightforward as possible, we will consider only a subcircuit of the diode OR gate, as shown in Figure 12.3. By applying KVL to this circuit loop, we establish

$$V_\mathrm{A} = V_\mathrm{D} + I_\mathrm{D}R_\mathrm{D}. \tag{12.2}$$

Then, recalling the $i-v$ characteristic of the pn junction diode from (11.58), we can express I_D in terms of the diode voltage V_D such that

$$V_\mathrm{A} = V_\mathrm{D} + I_0(e^{qV_\mathrm{D}/kT} - 1)R_\mathrm{D}. \tag{12.3}$$

Note that the only unknown in this equation is V_D—and yet, try as we might, we cannot solve it with algebraic techniques. The difficulty here arises from the inclusion of the transcendental function e^x. Consider what would happen if we set the input voltage and all of the other constant parameters in (12.3) to one. We would then be able to rewrite it as

$$2 = V_\mathrm{D} + e^{V_\mathrm{D}}, \tag{12.4}$$

which makes the difficulty in solving for V_D algebraically more clear. There is no algebraic method to solve transcendental equations like that of (12.4).

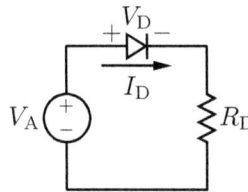

Figure 12.3: This simple diode circuit is difficult to solve using the exponential diode equation.

So what are we to do? In general, we will have to settle for an approximate solution (although we can approximate to arbitrary precision). Approximate solutions can be found through graphical analysis, numerical methods, or analytical approximations.[2] In the next two subsections, we explore the graphical analysis and numerical method approaches. In section 12.2, analytical approximations are introduced.

Graphical Analysis

The graphical approach to solving equations of one variable like (12.3) is to plot each side of the equation as a function of the unknown variable on the same axes and observe where these plots intersect. The value(s) of the unknown variable at the intersection(s) is(are) the solution(s) of the equation. While we could do this with (12.3) in its current form, a few manipulations will make the results

---

[2]Banwell and Jayakumar published an exact analytical solution that relies on the Lambert W-function in 2000.

easier to interpret. We first subtract V_D from each side, and then divide by R_D to yield

$$\frac{V_A - V_D}{R_D} = I_0(e^{qV_D/kT} - 1). \tag{12.5}$$

The left side of (12.5) represents the current through the load resistor, R_D, while the right side is recognized as the exponential current through an ideal pn junction diode. Of course, the current through the resistor and the diode must be equal since they are in series. Each side of (12.5) is plotted in Figure 12.4. The linear plot of the current through the resistor is called the load line and the intersection of the load line with the plot of the diode current is called the quiescent point, or Q point. The voltage V_{DQ} is the solution to (12.3) and (12.5), and the circuit's operating point is found to be the point (V_{DQ}, I_{DQ}).

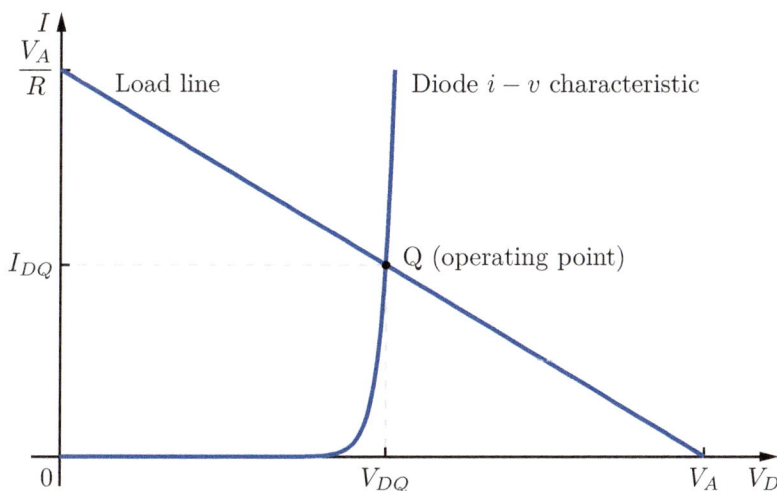

Figure 12.4: Solution of the diode circuit in Figure 12.3 via graphical analysis.

Numerical Analysis

The graphical approach can help us to get a general idea of how a circuit will operate, including helping us to visualize how changes in circuit parameters, such as changing the source voltage V_A, will affect its operating point. However, to get a precise numerical solution, we turn to numerical methods. The basic idea behind numerical methods is to start with some initial guess at the solution and then to iteratively improve upon that guess until some desired level of precision has been achieved. When each successive iteration of a numerical method brings the approximate solution closer to a particular value, the method is said to converge at that value. Convergence is not always assured, and sometimes the convergence of a numerical method depends on the quality of the initial guess.

   To see a numerical method in action, we will add some numbers to the circuit of Figure 12.3, as shown in Figure 12.5. Then we make some initial guess as to the circuit's operating point. Since we can see that the diode is forward biased and we also know that diodes allow current to flow easily in the forward bias direction, we will initially assume the voltage drop across the diode is zero (i.e., $V_D = 0$). This implies that all of the source voltage is applied across the load resistor, from which we determine the current to be

$$I = \frac{V_A - V_D}{R_D} = 1.4\,\text{mA}. \tag{12.6}$$

Thus, our initial guess at the circuit's operating point is $(V_D, I_D) = (0\,\text{V}, 1.4\,\text{mA})$.

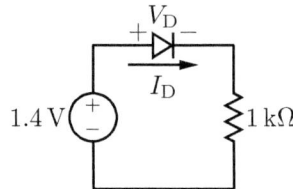

Figure 12.5: Numerical example of a simple diode circuit in which a 1.4 V source is in series with a forward-biased diode and a 1 kΩ resistor.

The next step is to improve upon our initial guess. We can do this by first improving upon our guess for the diode's operating voltage using the ideal *pn* junction diode equation and our previous guess for the diode's current. By solving the right hand side of (12.5) for V_D, we have

$$V_D = \frac{kT}{q} \ln \left(\frac{I_D}{I_0} + 1 \right), \tag{12.7}$$

which, upon plugging in our circuit's parameters (assuming $\frac{kT}{q} = 25.9\,\text{mV}$ and $I_0 = 1\,\text{fA}$), yields $V_D = 0.724\,358\,\text{V}$. Then we use this new voltage in the left-hand side of (12.5) to refine our guess for the current I_D, which yields $0.675\,692\,\text{mA}$. This gives us a new operating point of $(V_D, I_D) = (0.724\,358\,\text{V}, 0.675\,642\,\text{mA})$.

Having established a procedure for improving our guess at the circuit's operating point, we can continue to refine this estimate by repeatedly updating the voltage and then the current estimates in sequence. Each iteration brings us closer to the exact solution. We stop iterating when some predetermined level of precision has been reached. The results of repeating this iterative process are summarized in Table 12.2, in which the process has been repeated until the values for the voltage and current remain the same within the first six digits from one iteration to the next.

#	V_D (V)	I_D (mA)	$\hat{V}_D$ (V)
1	0	1.4	0.724 358
2	0.724 358	0.675 642	0.705 488
3	0.705 488	0.694 512	0.706 202
4	0.706 202	0.693 798	0.706 175
5	0.706 175	0.693 825	0.706 176
6	0.706 176	0.693 824	0.706 176

Table 12.2: Summary of operating point values found for the circuit of Figure 12.5 for each iteration.

In Figure 12.6, the progress of the numerical method is drawn on top of the plots from the graphical method. Starting from our initial guess at $(0\,\text{V}, 1.4\,\text{mA})$ near the top left of the figure, the numerical method can be seen spiraling in toward the operating point at the intersection of the two plot lines where the method converges.

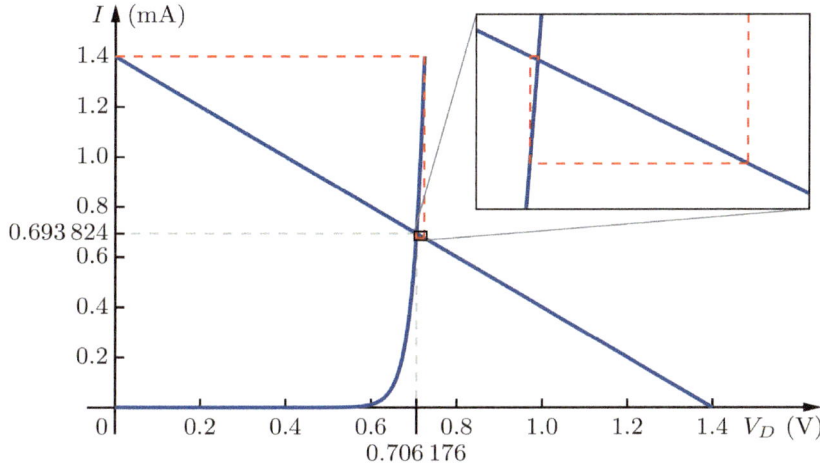

Figure 12.6: The iterative approximate solutions to the circuit in Figure 12.3 are connected with a dashed red line. Note how the iterative solutions close in on the operating point shown in the close-up window.

Even for a simple diode circuit like that of Figure 12.5, graphical and numerical analysis methods can be overly time consuming and might require the aid of a computer. When circuits get more complex, the situation worsens. Consider the two diode OR gate repeated below in Figure 12.7, which was introduced at the beginning of this chapter. Applying Kirchhoff's laws, we can write

$$V_\text{A} - V_\text{DA} = V_\text{B} - V_\text{DB} \tag{12.8}$$

and

$$\frac{V_\text{A} - V_\text{DA}}{R} = I_\text{DA} + I_\text{DB}. \tag{12.9}$$

Then, substituting the ideal pn junction diode equation for the currents I_DA and I_DB, (12.9) becomes

$$\frac{V_\text{A} - V_\text{DA}}{R} = I_0 \left(e^{qV_\text{DA}/kT} + e^{qV_\text{DB}/kT} - 2 \right). \tag{12.10}$$

We now have two equations, (12.8) and (12.10), and two unknowns, V_DA and V_DB. Just as in the single diode circuit, we have a transcendental equation, which requires approximate solutions, but the second diode in the OR gate introduces an additional equation and unknown variable.

If we wish to apply graphical analysis, we must add a third dimension to our plots to account for the additional variable. The operating point will be found by identifying the point where three surfaces intersect. To apply a numerical

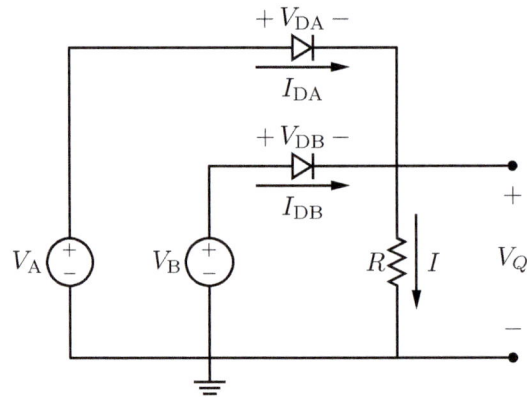

Figure 12.7

Figure 12.8: A two input diode OR gate with diode voltages clearly labeled.

method, we can make some initial guess for the voltage V_{DA}, then use this guess to calculate V_{DB} using (12.8):

$$V_{DB} = V_B - V_A + V_{DA}. \tag{12.11}$$

Next, we would estimate the current through the resistor using the left-hand side of (12.10),

$$I = \frac{V_A - V_{DA}}{R}, \tag{12.12}$$

and then update our initial guess for V_{DA} by using this current and V_{DB} to solve for V_{DA} using the right-hand side of (12.10):

$$V_{DA} = \frac{kT}{q} \ln\left(\frac{I}{I_0} + 2 - e^{qV_{DB}/kT}\right). \tag{12.13}$$

Finally, as we did in the single diode circuit, we would repeat these steps, updating V_{DB}, I, and V_{DA} in sequence, iterating until the solution converges to a sufficiently accurate operating point.

Note how the addition of another diode increased the complexity of both the graphical and numerical methods. Additional diodes in the circuit will make these methods entirely unsuitable for hand calculation. The next section introduces simplified diode models that will allow us to find approximate solutions in these situations with much less effort.

Example 12.1.1

For the circuit in Figure 12.9, find the current I and voltage V_o using the ideal diode model.

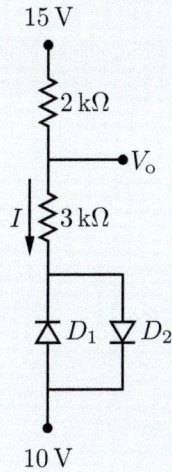

Figure 12.9: Example 12.1.1.

Since we have two diodes, we will have to make an assumption for each one and solve the circuit. Since they are in opposite directions, it might be a waste of time to start with the assumption that they are both on or off. From the voltage source values, we can tell the it is likely to have a positive current in the direction shown in the figure. That will mean that D_1 will be off and D_2 will be on. If that was the case, then

$$I = \frac{15 - 10}{5} = 1\,\mathrm{mA}.$$

This is consistent with our assumption, so we can move on with our solution. Using KVL,

$$V_\mathrm{o} = 3I + 10 = 13\,\mathrm{V}.$$

Example 12.1.2

For the circuit in Figure 12.10, find I and V_o using the ideal diode model.

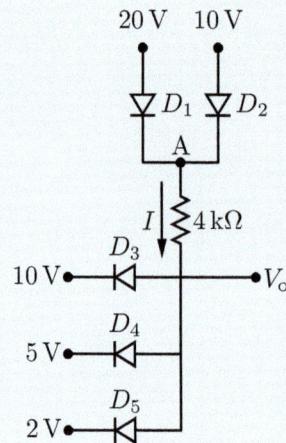

Figure 12.10: Example 12.1.2.

In this circuit, we have five diodes, so we need to look at this efficiently. We can start with D_1 and D_2. We know that both diodes can't be on at the same time, because that would make node A 20 V and 10 V at the same time. If D_1 were on and D_2 were off, then the voltage at node A would be 20 V, which would make D_2 off—so we might be on the right track.

We see the same pattern here in the decision with the rest of the diodes. D_3, D_4, and D_5 can't all be on because that would make the output voltage three different values. The same applies to just two of them: no two could be on at the same time because it would be a conflict. At this point, we know that they can all be off or only one can be on. Now, which one can be on and can keep the other two off? If D_5 were on, that would make the output voltage 2 V. If that were the case, then D_4 and D_3 would be in reverse bias and would be off. This would make the output voltage

$$V_\mathrm{o} = 2\,\mathrm{V}.$$

We can now try to calculate the current with only D_1 and D_5 on:

$$I = \frac{20\,\mathrm{V} - 2\,\mathrm{V}}{4\,\mathrm{k}\Omega} = 4.5\,\mathrm{mA}.$$

The direction of the current is also consistent with our assumption.

Example 12.1.3

For the circuit in Figure 12.11, find I and V_o before and after switching using the ideal diode model.

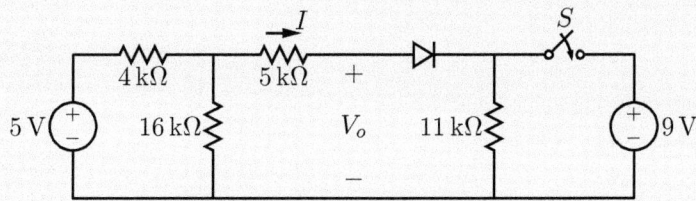

Figure 12.11: Example 12.1.3.

When the switch is open, the 9 V source has no effect on the circuit, and the diode is conducting. We can redraw the circuit as seen in Figure 12.12.

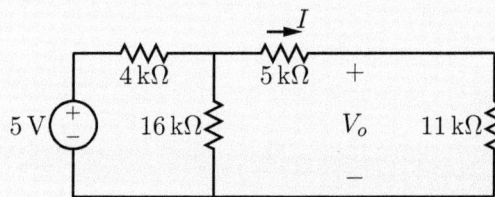

Figure 12.12: Figure 12.11 before switching.

We can use two voltage divisions or a source transformation and a current division; the method is up to you. We now have all the tools we require. We will first find the voltage across the parallel combination of the 16 kΩ resistor and the two resistors in series:

$$V = \frac{16 \parallel (5 + 11)}{4 + (16 \parallel (5 + 11))} 5 = 3.33 \, \text{V}.$$

Then there is one more voltage division to find the output voltage V_o:

$$V_o = \frac{11}{5 + 11} V = \frac{11}{16} 3.33 = 2.29 \, \text{V}.$$

The current I can be calculated using Ohm's law:

$$I = \frac{V}{5 + 11} = \frac{3.33}{16} = 0.208 \, \text{mA}.$$

The current I is positive, which is concomitant with our assumption of the diode on. After the switch closes, the 9 V source is now added to the circuit. This will put the diode in a reverse bias. We can redraw the circuit by replacing the diode with an open circuit. The circuit is shown in Figure 12.13.

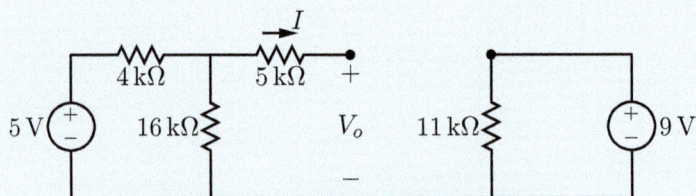

Figure 12.13: Figure 12.11 before switching.

The current I will be zero and the output voltage can be calculated using voltage division as

$$V_o = \frac{16}{16 + 4} 5 = 4\,\mathrm{V}.$$

The diode has $-5\,\mathrm{V}$ drop, which is consistent with the assumption of it being off.

Example 12.1.4

For the circuit in Figure 12.14, find the currents I_D and I_R.

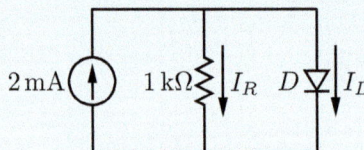

Figure 12.14: Example 12.1.4.

If the diode is off, all the source current will go through the resistor. The voltage drop across the resistor can be calculated using Ohm's law:

$$V_D = 2(1) = 2\,\mathrm{V}.$$

This result is a contradiction to our off diode assumption. The other option is that the diode is on. If the diode is on, all the current will go through the diode and we will have a zero voltage drop across the diode, which is acceptable.

Example 12.1.5

For the circuits in Figure 12.15 and Figure 12.16, find the output voltage V_o for the source voltage shown.

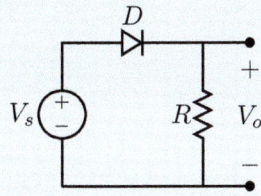

Figure 12.15: Example 12.1.5.

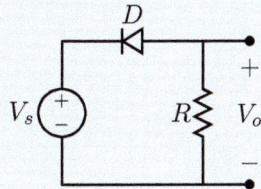

Figure 12.16: Example 12.1.5.

For the circuit in Figure 12.15, the diode is on when V_s is positive and off when the source is negative. The output voltage is plotted in the figure below.

For the circuit in Figure 12.16, the diode is off when V_s is positive and on when the source is negative. The output voltage is plotted in the figure below.

Example 12.1.6

For the circuit in Figure 12.17, using the ideal diode model, find the current I_R and the output voltage V_o if:

1. The diode combination in Figure 12.18a is applied between nodes a and b.

2. The diode combination in Figure 12.18b is applied between nodes a and b.

3. The diode combination in Figure 12.18c is applied between nodes a and b.

4. The diode combination in Figure 12.18d is applied between nodes a and b.

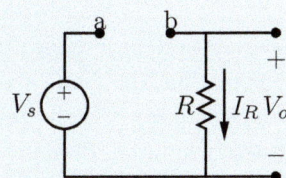

Figure 12.17: Example 12.1.6.

Figure 12.18: Series diode combinations with two diodes directed (a) toward each other, (b) toward the right, (c) away from each other, and (d) toward the left.

For the circuit in Figure 12.18a, the diode near node a is on while the other one is off. This means

$$V_o = 0 \, \text{V},$$

and

$$I_R = 0 \, \text{A}.$$

For the circuit in Figure 12.18b, both diodes are on, which means

$$V_o = V_s,$$

and

$$I_R = \frac{V_s}{R}.$$

For the circuit in Figure 12.18c, the diode near node a is off while the other one is on. This means

$$V_o = 0 \, \text{V},$$

and

$$I_R = 0 \, \text{A}.$$

Finally, for the circuit in Figure 12.18d, both diodes are off, which means

$$V_o = 0 \, \text{V},$$

and

$$I_R = 0 \, \text{A}.$$

12.2   Simplified Diode Models

12.2.1   Ideal Diode Model

The simplest analytical approximation for a diode's operating behavior is shown in Figure 12.19. In this approximation, the diode is modeled as a short circuit in the forward direction, and an open circuit in the reverse direction. The diode's exponential behavior is completely ignored. In many applications, this is how we wish the diode actually behaved, and so we call this approximation the "ideal diode model."

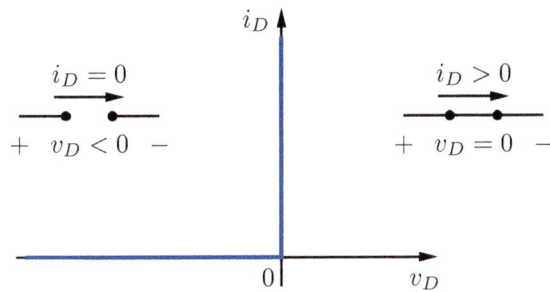

Figure 12.19: A plot of the $i - v$ characteristics of the ideal diode. The circuit equivalent of the reverse-biased ideal diode (open circuit) is shown on the left and the circuit equivalent of the conducting ideal diode (short circuit) is shown on the right.

Consider the circuit in Figure 12.20a. The first step in applying the ideal diode model to solve this circuit is to determine if the diode is biased in the forward or reverse direction. We then apply the appropriate ideal diode model circuit equivalent and solve the circuit. When the bias direction of the diode is not apparent, we will make a guess. After solving the circuit with our assumed diode bias, we must check to see if our guess was correct. If we assumed a forward-biased diode, the current through it should be positive. If we assumed a reverse-biased diode, the voltage across it should be negative. If the resulting current or voltage doesn't agree with our assumed bias, we simply make the opposite assumption and solve the circuit again. With experience, guessing the correct diode bias from the start becomes easier.

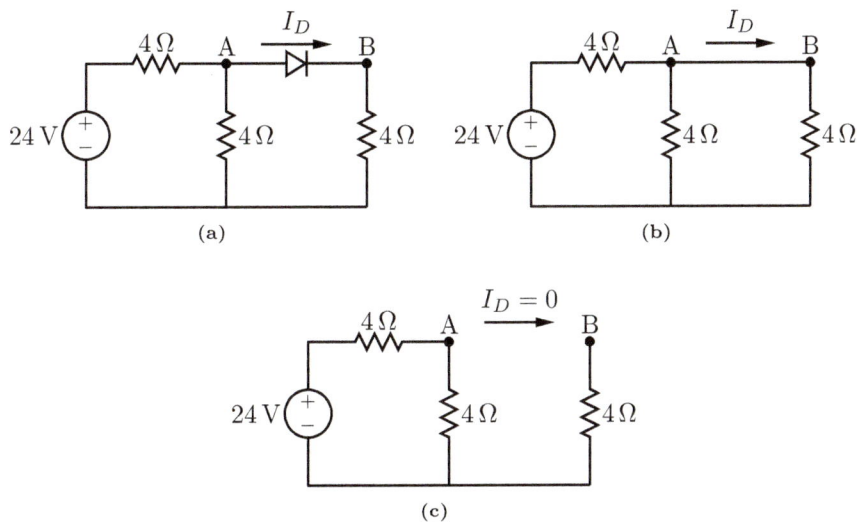

Figure 12.20: On applying the ideal diode model to a diode in circuit (a), it is converted to a short circuit in its equivalent circuit under an assumption of forward bias (b) and to an open circuit under the assumption of reverse bias (c).

Applying this strategy to the circuit in Figure 12.20a, we initially guess that the diode is forward biased. Using the ideal diode model, the forward-biased diode is replaced with a short circuit, allowing us to redraw the circuit as shown in Figure 12.20b. We then apply circuit analysis techniques to find that the voltage at A and B is 8 V and that a positive 2 A current flows from A to B through the diode. These results are consistent with a forward-biased diode, and so our solution is complete. If we had originally assumed that the diode was reverse biased, we would have redrawn the circuit as shown in Figure 12.20c. Upon solving for the node voltages of this circuit, we would find that the voltage of node A is 12 V while the voltage at node B is 0 V. Since the voltage across the diode, V_{AB}, is positive, these results are inconsistent with a reverse-biased diode, which confirms that this diode is in fact forward biased.

We can summarize the steps for solving a circuit containing diodes using the ideal diode model as:

1. Make an initial guess of the bias for each diode.

2. Replace forward-biased diodes with a short circuit and reverse-biased diodes with an open circuit.

3. Solve the circuit for the current through any forward-biased diodes and the voltage across any reverse-biased diodes.

4. Check if the currents and voltages found in step 3 are consistent with the biases assumed in step 1. If they are in agreement, the solution is complete. Otherwise, repeat the process from step 1 with a different assumed bias.

Example 12.2.1

Use the ideal diode model to find the voltage at A in the circuit below.

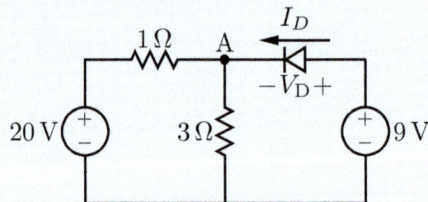

Figure 12.21

Our first step is to guess the bias of the diode. Let's assume it is forward biased. Under the ideal diode model, we replace the forward-biased diode with a short circuit, as show in Figure 12.22. Then, using KCL at A to solve for I_D, we write the sum of the currents, leaving A as

$$\frac{9\,\text{V} - 20\,\text{V}}{1\,\Omega} + \frac{9\,\text{V}}{3\,\Omega} - I_D = 0, \tag{12.14}$$

which results in

$$I_D = -8\,\text{A}. \tag{12.15}$$

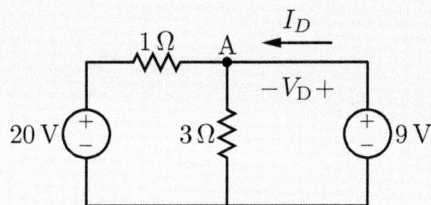

Figure 12.22: The circuit of Figure 12.21 with the diode replaced by a short circuit.

A negative current is inconsistent with our assumption that the diode is forward biased, which informs us that the diode must actually be in reverse bias. Applying the ideal diode model again, we now replace the diode with an open circuit as shown in Figure 12.23 and solve for the voltage across the diode V_D:

$$V_D = 9\,\text{V} - V_A = 9\,\text{V} - \frac{3\,\Omega}{1\,\Omega + 3\,\Omega}20\,\text{V} = -6\,\text{V}. \tag{12.16}$$

A diode voltage of $-6\,$V is consistent with a reverse-biased diode, so our solution is complete.

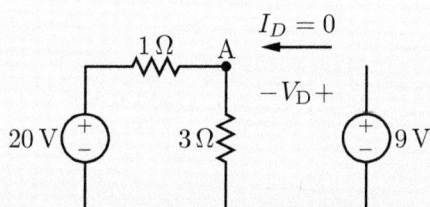

Figure 12.23: The circuit of Figure 12.21 with the diode replaced by an open circuit.

By using the ideal diode model, diode circuits become much easier to solve, but, in exchange, our solution is less accurate than if we had used the exponential diode model. Reconsider the simple diode circuit from Figure 12.5, which we solved using the exponential diode model via graphical and numerical methods in section 12.1.3. The circuit figure is repeated in Figure 12.24a for convenience. If we apply the ideal diode model, the forward-biased diode is replaced by a short circuit, as shown in Figure 12.24b. Then, from Ohm's law, we quickly determine the current I_D to be 1.4 mA. Unfortunately, this is not very close to the value of $0.693\,824$ mA we found using the more accurate exponential model. In fact, it represents an error of greater than 100 %.

This discrepancy illustrates one of the limitations of the ideal diode model, which neglects any forward-biased voltage drop across the diode. If the source voltage in the circuit is ten times larger, the performance of the ideal diode model is much better. To see this, we now solve the circuit again with the source voltage set to 14 V using both the ideal diode model and the exponential model. The diode is still forward biased, and so by applying the ideal diode model, we find the current I_D to be $14\,\text{V}/1\,\text{k}\Omega = 14$ mA. Then, when we solve the circuit using the exponential model via numerical analysis, we find the current I_D to be 13.2 mA (see Table 12.3). This represents an error of a little over 6 %. Thus, with

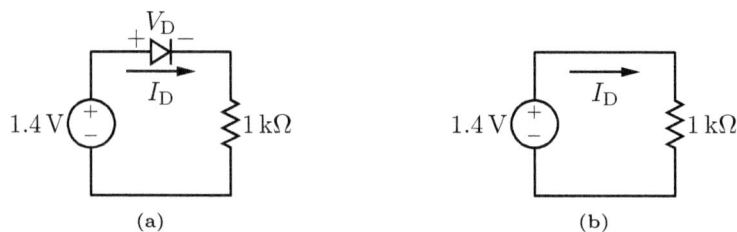

Figure 12.24: A simple forward-biased diode circuit (a) before and (b) after applying the ideal diode model.

the higher source voltage, the error introduced by using the ideal diode model is substantially less than we found in the low voltage circuit. This is because when the source voltage is large, the voltage drop across the forward-biased diode, which the ideal diode model neglects, is relatively small compared to the other voltages in the circuit. Consequently, the error introduced by the ideal diode model is smaller.

#	V_D (V)	I_D (mA)	$\hat{V}_D$ (V)
1	0	14	0.783 995
2	0.783 995	13.216 00	0.782 502
3	0.782 502	13.217 50	0.782 505
4	0.782 505	13.217 49	0.782 505

Table 12.3: Summary of operating point values found for each numerical iteration of the circuit of Figure 12.24a with the source voltage set to 14 V.

When the diode is reverse biased, the ideal diode model requires the current to be zero while the exponential model limits the magnetude of the current to the reverse-biased saturation current I_0. Since I_0 is usually very small (e.g., 1×10^{-12} A to 1×10^{-15} A) compared to currents of interest (e.g., 1 A to 1×10^{-6} A), neglecting the reverse-biased current is usually a good approximation. Although it is not covered in this text, a sufficiently large reverse-biased voltage will cause a breakdown within the diode that results in a short circuit. Some diodes, known as Zener diodes, are designed to break down at a specific voltage in a nondestructive way, which allows the circuit designer to make use of this property.

The point to remember is that sometimes the ideal diode model is a good approximation and sometimes it isn't. In order to be used effectively, we need to know when it is a reasonable approximation. In general, the ideal diode model does well when the diode is reverse biased or when the diode is forward biased and the other voltages in the circuit are large (e.g., > 3 to 5 V) compared to the diode's neglected forward-biased voltage drop, which is often in the order of $0.5 - 0.8$ V. It is also helpful when trying to deduce which diodes in a circuit are forward biased and which are reverse biased before applying a more accurate model to solve the circuit.

12.2.2 Constant Voltage Drop Model

In the preceding sections of this chapter, we solved the circuit of Figure 12.3 with the source voltage set at both 1.4 V (see Table 12.2) and 14 V (see Table 12.3). The operating points of the diode were found to be (0.706 V, 0.694 mA) and (0.782 V, 13.2 mA), respectively. Note how the voltage across the forward-biased diode changes very little despite a large change in the current through it. This is, of course, due to the exponential nature of a diode's $i - v$ characteristic curve expressed in (12.1).

To investigate this effect more closely, we solve for the diode voltage, V_{D}, in terms of I_{D} to produce

$$V_{\mathrm{D}} = \frac{kT}{q} \ln \left(\frac{I_{\mathrm{D}}}{I_0} + 1 \right), \tag{12.17}$$

from which we can readily determine the diode voltage for a given current. Using (12.17), the forward voltage required to drive 1 μA through a diode, assuming $T = 300$ K and $I_0 = 1$ fA, is found to be 0.536 V. Yet the voltage required to drive 1 A through the diode, a current six orders of magnitude larger, is only 0.894 V. Due to the rapid increase in the current through the diode in response to an applied voltage, forward-biased diodes usually operate within a small range of bias voltages between about 0.5 V and 0.9 V. This suggests that we can improve upon the ideal diode model by including a small constant voltage drop in our model when the diode is forward biased. Essentially, we shift the ideal diode model $i - v$ curve, as shown in Figure 12.25. Without knowing more details about the diode or its operating point, we typically use a voltage drop of 0.7 V.

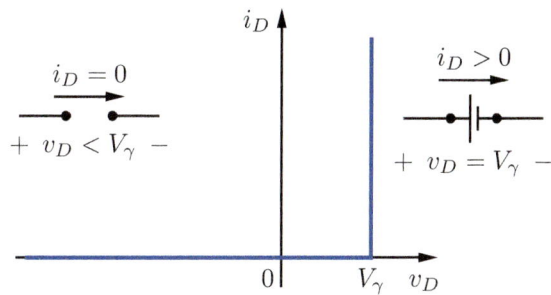

Figure 12.25: A plot of the $i - v$ characteristics of a diode using the constant voltage drop model. The circuit equivalent of the reverse-biased ideal diode (open circuit) is shown on the left and the circuit equivalent of the conducting ideal diode (short circuit) is shown on the right.

12.3   Exercises

12.1. For the circuit in Figure 12.26, find the current I and voltage V_o using the ideal diode model.

10 V

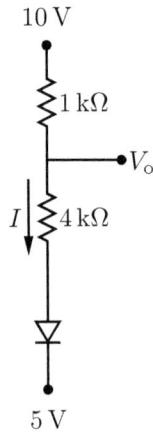

Figure 12.26: Exercise 12.1.

12.2. For the circuit in Figure 12.27, find the current I and voltage V_o using the ideal diode model.

15 V

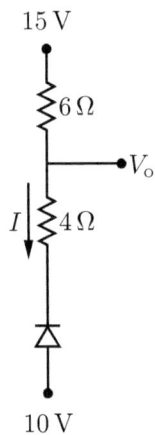

Figure 12.27: Exercise 12.2.

12.3. For the circuit in Figure 12.28, find the current I and voltage V_o using the ideal diode model.

10 V

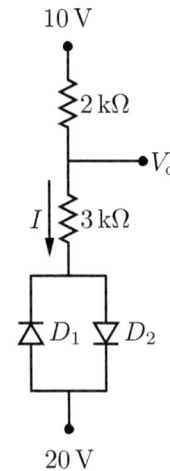

Figure 12.28: Exercise 12.3.

12.4. For the circuit in Figure 12.29, find the current I_R using the ideal diode model.

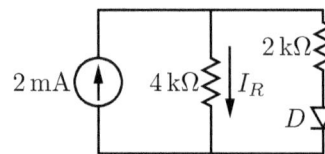

Figure 12.29: Exercise 12.4.

12.5. For the circuits in Figure 12.31, find the output voltage V_o using the ideal diode model.

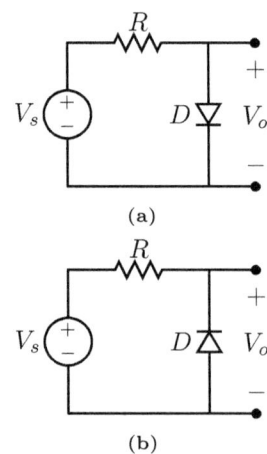

Figure 12.31: Exercise 12.5.

12.6. For the circuit in Figure 12.32, using the ideal diode model, find the current I_R and the output voltage V_o if:

a) The diode combination in Figure 12.33a is applied between nodes a and b.

b) The diode combination in Figure 12.33b is applied between nodes a and b.

c) The diode combination in Figure 12.33c is applied between nodes a and b.

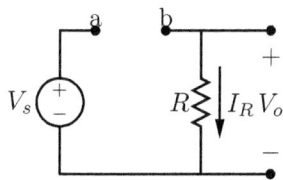

Figure 12.32: Exercise 12.6.

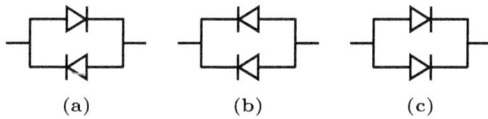

Figure 12.33: Exercise 12.6.

12.7. For the circuit in Figure 12.34, find the node voltages V_1, V_2, V_3, and V_4 using the ideal diode model if:

a) The source voltage $V_s = 5\,\mathrm{V}$.

b) The source voltage $V_s = -5\,\mathrm{V}$.

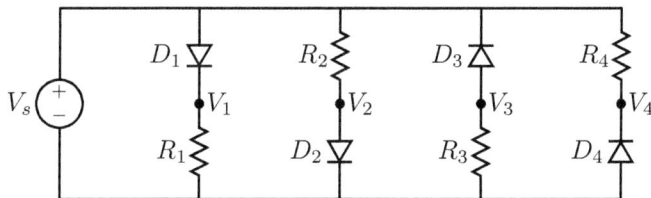

Figure 12.34: Exercise 12.7.

12.8. For the AND gate connection in Figure 12.35, the node a diode is fried (open circuit). Find the new output for all input combinations.

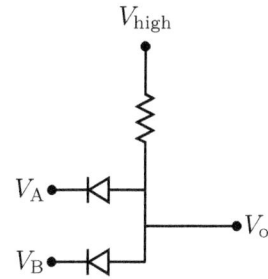

Figure 12.35: Exercise 12.8.

12.9. For the circuit shown in Figure 12.36, fill the truth table below using the ideal diode model.

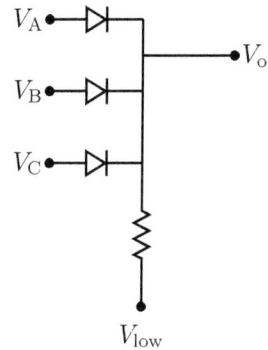

Figure 12.36: Exercise 12.9.

V_A	V_B	V_C	V_o
L	L	L	
L	L	H	
L	H	L	
L	H	H	
H	L	L	
H	L	H	
H	H	L	
H	H	H	

Chapter 13

Transistors

Introduction

From their first commercial production in the early 1950s, transistors, with applications in switching, amplification, and digital logic, have become fundamental components in nearly every electronic device. Originally used in radio signal amplification, transistors became smaller, cheaper, and more reliable alternatives to their vacuum tube predecessors. The first transistor computer, built in 1953, had only about 100 transistors. Since then, transistors have rapidly improved. By 1965, Gordon Moore observed that the number of transistors in integrated circuits, like computer processors, was doubling every year. In 1975, he predicted a continued doubling every two years. This trend, famously known as Moore's law, has held true for more than four decades. As a result, today's large-scale integrated circuits, like those in smart phones, contain billions of transistors.

Unlike most of the circuit components we have studied so far, which have just two terminals, the transistor is a four-terminal device. The fundamental feature of a transistor is that the current through one pair of the transistor's terminals can be controlled by a voltage applied across the other pair of terminals. This behavior is often modeled using the voltage-controlled current source with which we are already familiar. There are many types of transistors available, but the most common type, and the one we will examine in this chapter, is the metal-oxide-semiconductor field-effect transistor (MOSFET). We start with a qualitative description of its operation in section 13.1 and then explore its applications in both analog and digital circuits in sections 13.2 and 13.3, respectively. Many of the operating principles of MOSFETs introduced here apply to other devices as well, which are covered in more advanced courses.

13.1 MOSFET Operating Principles

13.1.1 The MOS-C

We start developing an understanding of the MOSFET by analyzing the metal-oxide-semiconductor capacitor (MOS-C), which is a fundamental component of field-effect devices. As its name implies, a MOS-C is composed of a thin oxide layer (usually SiO_2) between a metal and a semiconductor (usually Si),

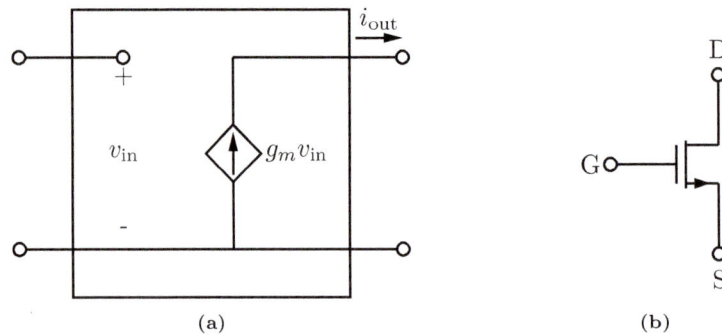

(a)   **(b)**

Figure 13.1: A transistor is a four-terminal device shown as (a) an idealized model using a voltage-controlled current source and as (b) the circuit symbol for a n-type MOSFET. Note that the base terminal is often connected directly to the source terminal and is not drawn in (b).

as shown in Figure 13.2. To gain insight into its operation, we will establish an approximation of the charge carrier distribution and the basic form of the energy band structure within a metal-oxide-semiconductor-capacitor (MOS-C) under different applied voltages.

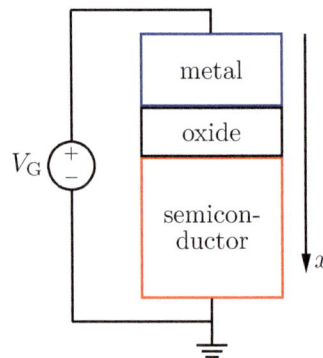

Figure 13.2: The idealized structure of a MOS-C.

Working from top to bottom in Figure 13.2, we start with the metal layer. At room temperature, metals have an abundance of mobile charge carriers, which makes them excellent conductors. Good conductors like metals have either a very small band gap or no band gap at all (see Figure 13.4c). Thus, in the energy band diagram for the metal layer, instead of drawing two energy levels to demarcate a band gap, we draw just a single energy level that is within the wide band of available energy states in the metal. This energy level is chosen such that the available states at this level have an equal chance of being filled as they do of being vacant, or, stated another way, the energy states at this level have a 50% chance of being filled.[1] As a consequence of this choice, the number of electrons above this level is equal to the number of holes below it, as illustrated in Figure 13.3a. The blue line on the right of the figure indicates the boundary between the metal and oxide layers.

The oxide layer is an insulator and accordingly has a wide band gap (see Figure 13.4a). For SiO$_2$ the band gap is about 9 eV at room temperature. With

---

[1] This energy level, known as the Fermi level, is an important concept in solid-state physics and is covered in detail in more advanced texts.

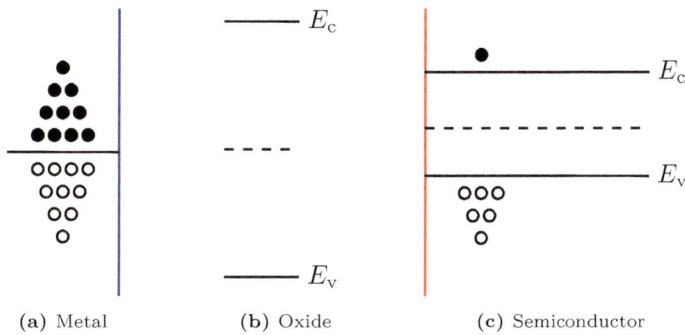

Figure 13.3: Energy bands for the components of a MOS-C in isolation.

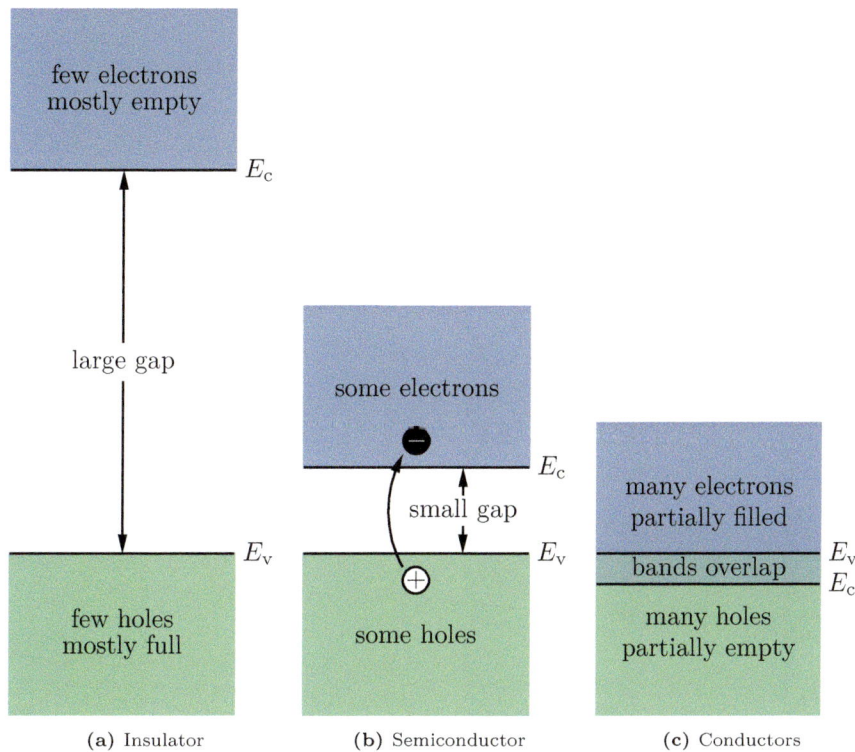

Figure 13.4: Relative band gap of materials (repeated from Figure 10.12 for convenience).

a wide band gap, much more energy is required to create an electron-hole pair in an insulator than in a semiconductor or conductor. Consequently, there are very few conducting electrons or holes. With so few charge carriers available, the insulator has a very low conductivity (high resistance), and practically no current is able to flow through the oxide layer. The band diagram for the oxide layer is shown in Figure 13.3b.

The bottom layer, known as the substrate, is made from a semiconductor. Semiconductors have a medium-sized band gap as shown in Figure 13.4b. For silicon, the band gap is about 1.12 eV at room temperature. The concentration of conducting electrons and holes, and thus the conductivity of the semiconductor, is controlled through doping. The substrate can be either a p-type or n-type semiconductor, but we choose to only analyze the p-type for now. The band diagram for a p-type semiconductor is shown in Figure 13.3c.

In general, when materials are in contact as they are in the MOS-C, an equilibrium will be established that balances the flow of carriers (drift and diffusion currents) just as occurs in the *pn* junction. However, with the MOS-C, there is an insulating oxide layer between the metal and semiconductor. Ideally, this layer is a perfect insulator and as such would prevent the redistribution of carriers. While in the remainder of our analysis we will stick with the perfect insulator assumption, we acknowledge here that a real device has a finite insulator resistance and, if left undisturbed long enough, will reach equilibrium. Thus, just like the energy bands of the *p*-type and *n*-type semiconductors in a *pn* junction, at equilibrium, the energy bands of the metal and semiconductor in a MOS-C will align such that their carrier energies are balanced. The band diagram of an ideal *p*-type MOS-C in equilibrium is shown in Figure 13.5.

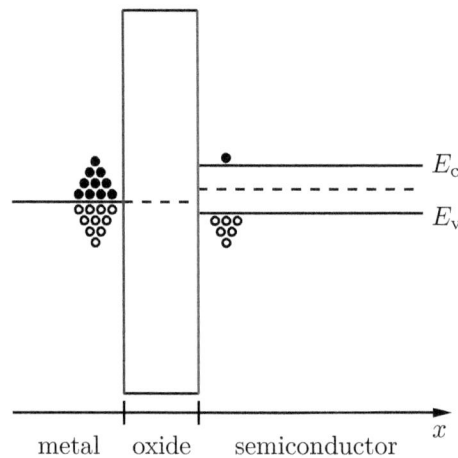

Figure 13.5: The MOS-C band diagram in equilibrium.

Accumulation

Having established the MOS-C in equilibrium with no external inputs, we now seek an understanding of the device when a DC bias voltage is applied. We start by grounding the bottom of the semiconductor layer, sometimes called the **body**, and using it as our reference voltage. Then we apply negative a DC bias voltage V_G to the top of the metal layer, called the **gate**, as shown in Figure 13.6. The metal layer acts in the same way as a metal plate in a parallel

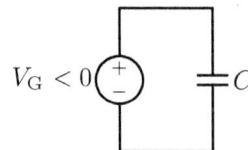

Figure 13.6: A negative voltage applied to a MOS-C.

plate capacitor; as the negative voltage is applied, electrons (negative charge carriers) flow into the metal layer. These electrons are attracted to the grounded body but are unable to cross the barrier formed by the insulating oxide layer, and so a negative charge builds up in the metal.

The electric field produced by this buildup of negative charge attracts holes (positive-charge carriers) in the semiconductor toward the metal. Holes are the majority carrier in the p-type semiconductor, so there are plenty available. The holes are also unable to cross the oxide layer, so they accumulate at the oxide-semiconductor junction until their positive charge balances the negative charge of the electrons. The electrostatic forces attracting the positive holes to the negative electrons pull them into the areas immediately adjacent to the oxide junctions. The resulting charge distribution in the MOS-C is shown, to a first-order approximation, in Figure 13.7a. This biasing mode is known as **accumulation** due to the majority carriers which accumulate near the oxide-semiconductor junction in response to the applied gate voltage.

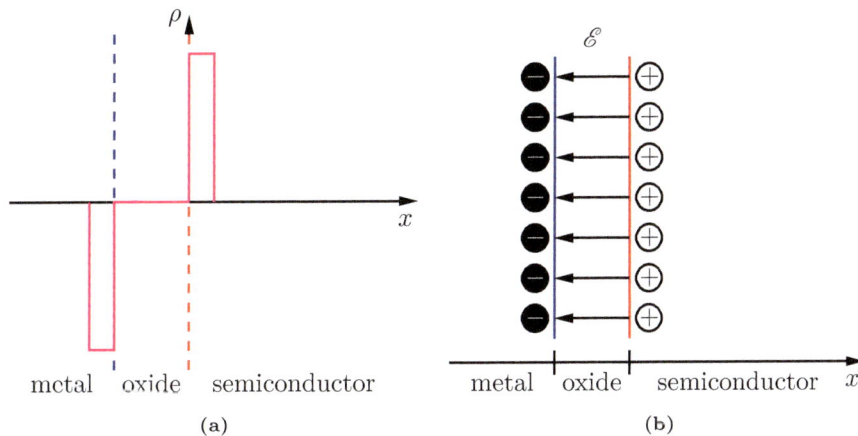

(a)   (b)

Figure 13.7: Block charge diagram and (a) corresponding electric field (b) in a p-type MOS-C in accumulation.

The charge distribution shown in Figure 13.7a results in an electric field $\mathscr{E}_x$ across the oxide between two charged surfaces, as shown in Figure 13.7b. When these surfaces are large compared to the distance between them, the electric field can be found via Gauss's law to be

$$\mathscr{E}_x = \frac{\sigma}{\epsilon}, \tag{13.1}$$

where σ is the surface charge density and ϵ is the permittivity of the oxide. This electric field is constant and directed in the negative x direction away from the positive charges in the semiconductor toward the negative charges in the metal. Since we have defined x as increasing toward the right, we have

$$\mathscr{E}_x < 0. \tag{13.2}$$

The electric field is related to voltage by

$$\mathscr{E}_x = -\frac{\mathrm{d}V}{\mathrm{d}x}, \tag{13.3}$$

and since the electric field is constant and negative throughout the oxide, the electric potential $V(x)$ in the oxide must vary linearly with a positive slope.

We are now ready to work out the shape of the energy bands in the MOS-C. At the gate, the energy level(s) will shift up or down with respect to the body in the same way as the two sides of a pn junction. Recall from section 11.4 that the energy levels in a band diagram are negatively proportional to the electric potential,

$$E \propto -|q|\,V, \tag{13.4}$$

such that increasing the electric potential will lower the energy levels in the band diagram, while decreasing the electric potential will raise them. Thus, applying a negative gate voltage ($V_\mathrm{G} < 0$) to the MOS-C at the gate will raise the energy levels of the metal gate with respect to those of the semiconductor body. In fact,

$$\Delta E_\mathrm{gate} = -|q|\,V_\mathrm{G}. \tag{13.5}$$

As a good conductor, the entire metal layer is an equipotential surface, so a plot of the voltage or the energy level in the metal layer will be completely flat.[2] Since we are only concerned with one energy level on the metal's side of the band diagram, we can complete that side of the diagram all the way to the metal-oxide junction with a flat line, as shown on the left side of Figure 13.8.

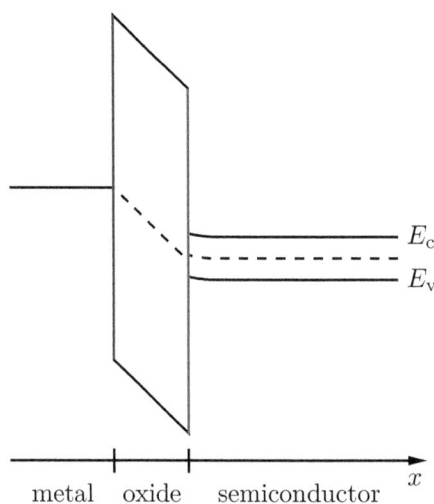

Figure 13.8: The band diagram of a p-type MOS-C in accumulation ($V_\mathrm{G} < 0$).

Within the oxide, we have already worked out that the electric potential will increase linearly, so we can use the relation in (13.4) again to determine that the energy bands within the oxide must decrease linearly, as shown in the middle of Figure 13.8.

Within the semiconductor, the thin layer of positive charges gathered near the oxide-semiconductor interface quickly cancel out the electric field. As a result, the electric potential and energy levels in the semiconductor quickly flatten out as we move with increasing x into the field-free region of the semiconductor. With this information about the shape of the energy bands in the semiconductor, an approximation of the complete band diagram for the MOS-C with a negative voltage applied to the gate can now be drawn, as shown in Figure 13.8.

---

[2]The electrical contact to the body is generally made though another conductive metal layer added beneath it. This provides an equipotential surface at the bottom of the semiconductor to achieve a uniform ground voltage across its entire bottom surface.

Depletion

Now let's consider what happens when a positive gate voltage, $V_G > 0$, is applied as shown in Figure 13.9. The energy level of the metal gate is now lowered

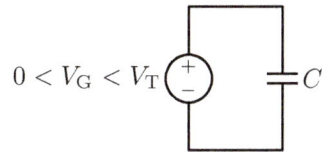

Figure 13.9: A positive voltage applied to a MOS-C.

relative to $V_G = 0$. Positive charge builds up in the metal, creating an electric field that repels the majority carriers of the p-type semiconductor. As the semiconductor's holes are pushed away from the oxide-semiconductor junction, they uncover the fixed negatively charged dopant atoms. In equilibrium, the holes are pushed back far enough that the positive charge in the metal is balanced by the negative charge uncovered in the semiconductor. This biasing mode is called **depletion** because a region of the semiconductor near the oxide-semiconductor junction has been depleted of majority carriers by the electric field. Since the fixed negative charges cannot move, the negatively charged region in depletion mode extends much further into the semiconductor than the positively charged region does in accumulation. However, just as we did for the pn junction, we will assume that the semiconductor layer is sufficiently deep that the depleted region does not extend all the way through it. The resulting block charge diagram and corresponding band diagram are shown in Figures 13.10a and 13.10b. Again, the voltage and energy level in the metal are flat, while the constant

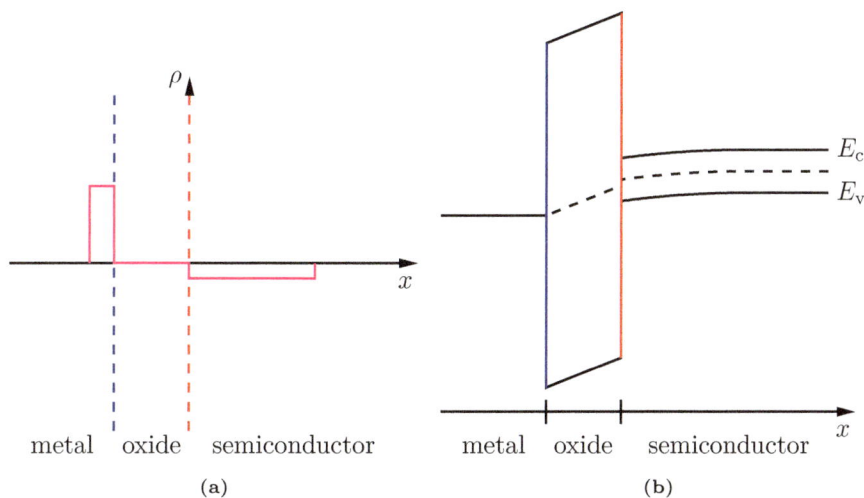

Figure 13.10: Block charge diagram and (a) corresponding band diagram (b) in a p-type MOS-C in depletion ($0 < V_G < V_T$).

electric field across the oxide results in a linear change in voltage and energy level there. In the semiconductor, the energy level gradually flattens as we move further from the oxide-semiconductor junction toward the field-free region of the device.

Inversion

When V_G is positive, it also attracts some of the mobile electrons (minority carriers) in the semiconductor to the oxide-semiconductor junction. If V_G is increased beyond a threshold voltage, V_T, then the concentration of these minority carriers near the junction will exceed the concentration of uncovered dopant atoms. Biasing in this region is known as **inversion** because the concentration of minority carriers near the junction has become greater than the unbiased concentration of majority carriers. Effectively, a thin layer of the p-type semiconductor adjacent to the oxide-semiconductor junction has been converted into an n-type semiconductor by the application of an electric field. As we will see in the next section, the ability to control this inversion layer via the MOS-C gate voltage makes the transistor action of a MOSFET possible. The block charge and energy band diagrams for inversion are shown in Figures 13.11a and 13.11b. The negatively charged region in the semiconductor shows both a wide low-charge density block due to the uncovered negatively charged dopant ions and a narrow, higher-charge density block due to minority carriers that gather in an inversion layer at the junction. Once again, the voltage and energy level in the metal are flat, while the constant electric field across the oxide results in a linear change in voltage and energy level there. In the semiconductor, the energy level begins flattening quickly through the inversion layer, and then finishes flatting out gradually as we move further from the oxide-semiconductor junction toward the field-free region of the device.

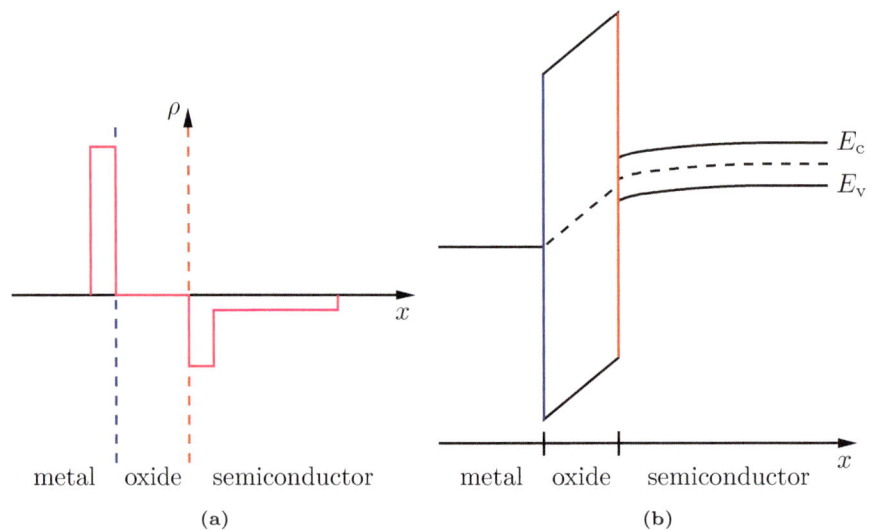

Figure 13.11: Block charge diagram and (a) corresponding band diagram (b) in a p-type MOS-C in inversion ($V_G > V_T$).

13.1.2 MOSFET Operation

The basic structure for an n-channel enhancement mode MOSFET is shown in Figure 13.12. The device consists of a pn junction on either side of a MOS-C. As with the MOS-C, the gate terminal is separated from the p-type substrate by an insulating oxide layer. The two n-type regions on either side are called the **source** and **drain** terminals, while the body terminal connects directly to the substrate. The source and body terminals are typically connected together and to ground. Current flowing between the drain and the source can be controlled by the voltage applied to the gate. When the gate voltage is above the threshold voltage, the transistor is "on," allowing current to flow from the drain to the source. When the gate voltage is below the threshold, the transistor is "off" and the drain current is blocked.

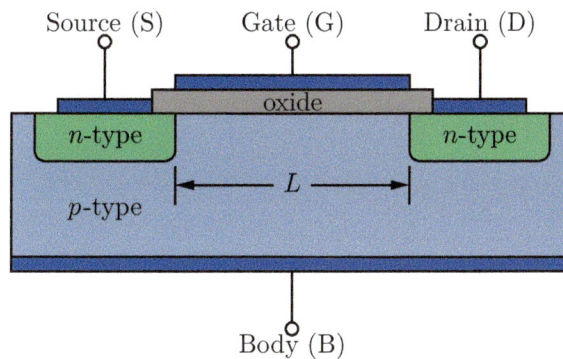

Figure 13.12: Simplified structure of an n-channel enhancement-mode MOSFET.

To see how the gate voltage controls the current through the device, first consider the scenario where the gate voltage is less than the threshold voltage, $V_G < V_T$, and a small positive drain voltage, $V_D > 0$, is applied. When the gate voltage V_G is less than the threshold voltage V_T, then the MOS-C is either in accumulation or depletion (i.e., not in inversion) and there are very few minority carriers available in the substrate. In either case, when a small positive drain voltage V_D is applied, the pn junction between the drain and the substrate will be reverse biased (the substrate/body is grounded). The reverse-biased current is of course very close to zero, so the device is "off." This situation is illustrated in Figure 13.13 with a gate voltage set within the MOS-C's depletion mode $(0 < V_G < V_T)$. Note that the depletion regions around each of the pn junctions are almost entirely within the substrate. This is because, compared to the substrate, the source and drain are much more heavily doped (see Example 11.3). The N^+ notation drawn within the n-type regions of the device is used to indicate the heavy doping.

If the gate voltage is raised above the threshold voltage, $V_G > V_T$, an inversion layer will develop between the source and drain, as shown in Figure 13.14. The extra minority carriers in this layer provide a connection between the source and drain terminals. Now when a small positive drain voltage is applied, current can flow from the drain, through the inversion layer, and into the source, and so the device is "on." In this mode of operation, the inversion layer acts as a resistor. When biased in inversion, the concentration of minority carriers,

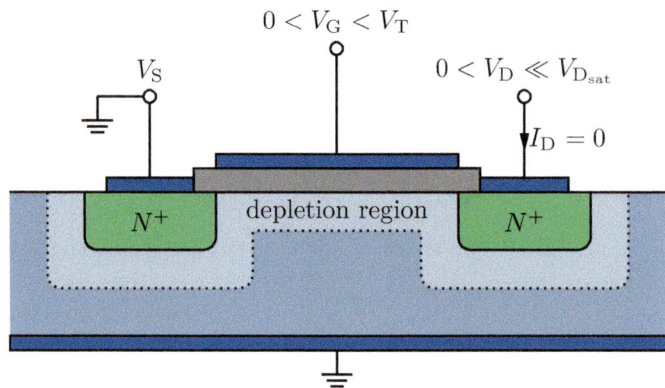

Figure 13.13: An n-channel enhancement-mode MOSFET with a depletion mode gate voltage ($0 < V_\mathrm{G} < V_\mathrm{T}$) and a small positive drain voltage ($0 < V_\mathrm{D} \ll V_{\mathrm{D}_\mathrm{sat}}$).

and thus the conductivity of the inversion layer, can be controlled by the gate voltage. By increasing the gate voltage, the conductivity of the inversion layer is increased and its resistance is lowered. By lowering resistance, the current flowing through the device for a given drain voltage will be increased. Thus, the gate voltage provides control over the current flowing through the device.

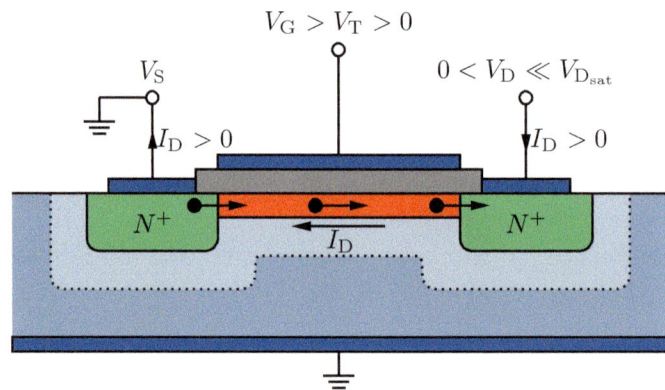

Figure 13.14: An n-channel enhancement-mode MOSFET with an inversion mode gate voltage ($V_\mathrm{G} > V_\mathrm{T}$) and a small positive drain voltage ($0 < V_\mathrm{D} \ll V_{\mathrm{D}_\mathrm{sat}}$).

Given the direction of current flow, from drain to source, one may wonder about the choice of names for these terminals. The key here is that the term *source* is used to refer to the source of primary charge carriers, and the term *drain* refers to the destination of those carriers. In the n-channel device under consideration, electrons are the primary charge carriers, and a positive drain voltage pulls electrons from the source, through the channel, and into the drain. So the *source* is the source of electrons entering the channel, while the *drain* drains the electrons from the channel. Finally, this device is known as an n-channel device because the inversion layer forms a channel with n-type characteristics between the n-type source and n-type drain.

Alternatively, p-channel devices can be constructed using a MOS-C with an n-type substrate. The inversion layer will form a p-type channel connecting a p-type source and drain when a large enough negative voltage is applied to the gate. In these devices, holes are the primary charge carrier, and with the correct bias voltage applied, they will enter the channel from the source and be removed at the drain, just as the electrons in the n-channel device. However, note that in the p-channel MOSFET, the current and the primary charge carriers flow in the same direction.

There are some further constraints on the drain voltage not yet mentioned. As stated earlier, in our n-channel MOSFET, positive drain voltage V_D reverse biases the substrate-drain pn junction. If the drain voltage exceeds the junction's reverse-biased breakdown voltage, a large uncontrolled current will flow from the drain into the substrate, and the device no longer behaves as a transistor. Finally, let's consider what will happen if the drain voltage is less than zero ($V_D < 0$). The apparent symmetry of the device might lead one to predict that it will simply operate in reverse, with current flowing from the grounded source to the negative drain. However, recall that the substrate is also grounded. This means that a negative voltage applied to the drain will forward bias the substrate-drain pn junction. The resulting current flow from the substrate to the drain bypasses the gate and thus also bypasses any control we have over it. Again, under these bias conditions, the device is no longer behaving as a transistor. The takeaway here is that careful consideration of bias voltages is required for proper transistor operation.

13.1.3 $i - v$ Characteristics

Having established that we can control the current I_D through a MOSFET by controlling the gate voltage V_G, we would like to develop a more detailed view of how I_D varies with both the gate voltage V_G and the drain voltage V_D. First, for all V_D between zero and the substrate-drain reverse-biased breakdown voltage, the current I_D will be approximately zero unless the gate voltage is greater than the threshold voltage V_T. The resulting flat line $I_D - V_D$ characteristic is rather uninteresting. However, if we assume that V_G is greater than V_T such that an inversion layer exists beneath the oxide, as shown in Figure 13.15a, then we can vary the conductivity of that layer via the gate voltage and, for *small* values of V_D, the inversion layer behaves as a variable resistor. As shown in Figure 13.15b, increasing the gate voltage increases the slope of a linear $I_D - V_D$ characteristic.

For the remainder of this section, we will stick with a single gate voltage chosen to be larger than the threshold voltage such that an inversion layer is formed beneath the oxide. As the drain voltage V_D increases beyond a few tenths of a volt, the voltage in the substrate near the drain increases as well. This reduces the potential difference between the gate and the substrate and, as a consequence, the concentration of minority carriers in the inversion layer near the drain is also reduced. We represent this reduction in concentration in Figure 13.16a by narrowing the inversion layer at the drain. This *narrowing* of the channel increases its resistance and causes the slope of the $I_D - V_D$ curve to decrease. As V_D is increased further, the $I_D - V_D$ curve continues to slump, as shown in Figure 13.16b.

As the inversion layer near the drain narrows, an increasing portion of the voltage drop between the drain and the source takes place on the drain end of the channel, leaving the source side mostly unaffected. When the drain voltage

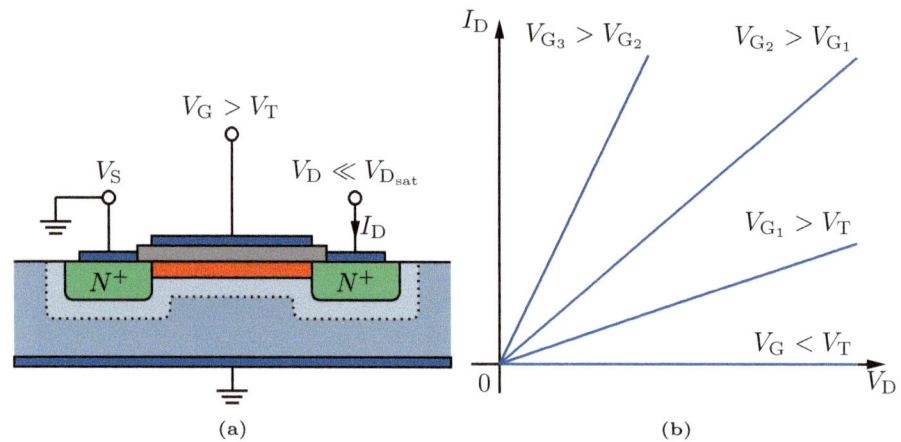

Figure 13.15: An n-channel MOSFET with $V_G > V_T$ and small V_D. (a) Device cross section showing the depletion region and inversion layer. (b) The characteristic $i - v$ curve for small V_D.

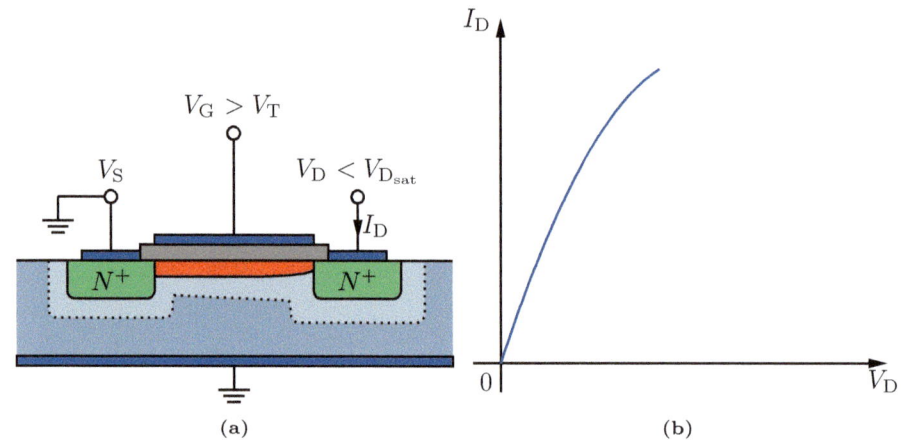

Figure 13.16: An n-channel enhancement-mode MOSFET with $V_G > V_T$ and moderate V_D ($V_D < V_{D_{sat}}$). (a) Device cross section showing the depletion region and inversion layer narrowing. (b) Characteristic $i - v$ curve during inversion layer narrowing.

is increased to the point that the concentration of minority carriers in the inversion layer adjacent to the drain is reduced to the dopant concentration of the substrate, the device is said to have reached *pinch-off*. This is the same concentration of minority carriers associated with the onset of inversion and so pinch-off is the phenomenon at which the inversion layer immediately adjacent to the drain no longer exists. A graphical representation of pinch-off and the resulting influence on the I_D–V_D curve is shown in Figures 13.17a and 13.17b, respectively.

Note that pinch-off of the inversion layer does not stop the current I_D from flowing through the device. Electrons are still flowing from the source into the inversion layer, traveling across the substrate and being pulled into the drain, but as they approach the drain, the electrons accelerate due to the larger voltage drop and corresponding larger electric field present at the drain end of the inversion layer. Since in equilibrium, charge is neither accumulating nor dissipating

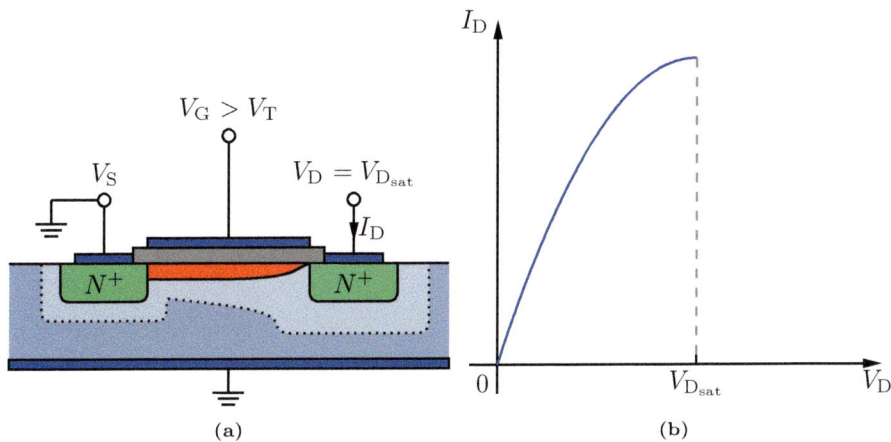

Figure 13.17: An n-channel enhancement-mode MOSFET with $V_\mathrm{G} > V_\mathrm{T}$ and $V_\mathrm{D} = V_{\mathrm{D}_\mathrm{sat}}$. (a) Device cross section showing the depletion region and inversion layer at pinch-off. (b) Characteristic $i - v$ curve at pinch-off.

within the substrate, the current, which is the rate of charge flowing through the device, must be constant with respect to position. Thus, as the electrons accelerate toward the drain, their concentration decreases. Pinch-off is simply the point at which their concentration drops below the dopant concentration.

Example 13.1.1

Suppose people are walking along a track in single file at a speed of $4\,\mathrm{km\,h^{-1}}$. If 2000 people cross the finish line every hour, what is the average spacing between each person?

Figure 13.18: People walking on a track at a speed of 4 km h^{-1}.

If we consider all the people who will cross the finish line in the next hour, the last person to cross will be $4\,\mathrm{km}$ away from the finish line right now. Since 2000 people in total will cross the line during this hour, the average distance between each person will be $4\,\mathrm{km}/2000$ people $= 2\,\mathrm{m}$ per person.

Now suppose that the people on the track are running at $10\,\mathrm{km\,h^{-1}}$ but have increased the space between them such that they are still crossing the finish line at a rate of 2000 per hour. What is the average space between each person at this new speed?

$10\,\mathrm{km\,h^{-1}}$

Finish
Line

Figure 13.19: People running on a track at a speed of $10\,\mathrm{km\,h^{-1}}$.

As before, we consider the last person who will cross the finish line in the next hour. This time they are 10 km away, so the average spacing between each person must be $10\,\mathrm{km}/2000$ people $= 5\,\mathrm{m}$ per person.

We see that the space between each person has increased from 2 m to 5 m. If the people were electrons, the number of people crossing the finish line in a given amount of time would be the current. As the electrons (or people) move faster while the current (or rate across the finish line) stays the same, they must be spaced further apart. In other words, the faster the electrons move for the same current, the lower their concentration.

This is what is happening in the inversion layer at pinch-off. The rate of electrons entering the drain (i.e., the current) stays the same, but the electrons start moving faster as they approach the drain, causing their concentration to decrease. Once the concentration falls below the dopant concentration of the substrate, the inversion layer is pinched off.

From our analysis of the MOS-C, we know that the inversion layer forms when the voltage between the gate and the substrate exceeds V_T. Thus, in the MOSFET, the inversion layer should pinch off at the drain when the difference between the gate voltage and the voltage in the substrate immediately adjacent to the drain drops below the threshold voltage. The drain is heavily doped compared to the substrate, so very little voltage is dropped within the drain. This fact allows us to use the drain voltage as an approximation for the voltage in the substrate immediately adjacent to the drain. Therefore, the inversion layer will reach pinch-off when

$$V_\mathrm{G} - V_\mathrm{D} = V_\mathrm{T}. \tag{13.6}$$

Solving (13.6) for V_D gives us the drain voltage at which pinch-off occurs, called the saturation voltage $V_{\mathrm{D_{sat}}}$.

$$V_{\mathrm{D_{sat}}} = V_\mathrm{G} - V_\mathrm{T} \tag{13.7}$$

Any further increase in the drain voltage beyond $V_{\mathrm{D_{sat}}}$ increases the voltage drop across the channel and pushes the pinch-off point away from the drain, as shown in Figure 13.20a. The channel between the pinch-off point and the drain is now in depletion instead of in inversion. If the depleted portion of the channel is small compared to the total channel length, then most of the voltage in excess of $V_{\mathrm{D_{sat}}}$ is dropped across the depleted portion, and the voltage drop across the rest of the channel is still $V_{\mathrm{D_{sat}}}$. Since the length of the channel in inversion and the voltage across it are approximately unchanged, the current through the channel will be unchanged as well. This results in the flat I_D–V_D current shown in Figure 13.20b. Drain voltages in excess of $V_{\mathrm{D_{sat}}}$ drive the transistor into saturation, where the current through the device, I_D, has only a weak dependence on the drain voltage and is almost entirely determined by the gate voltage.

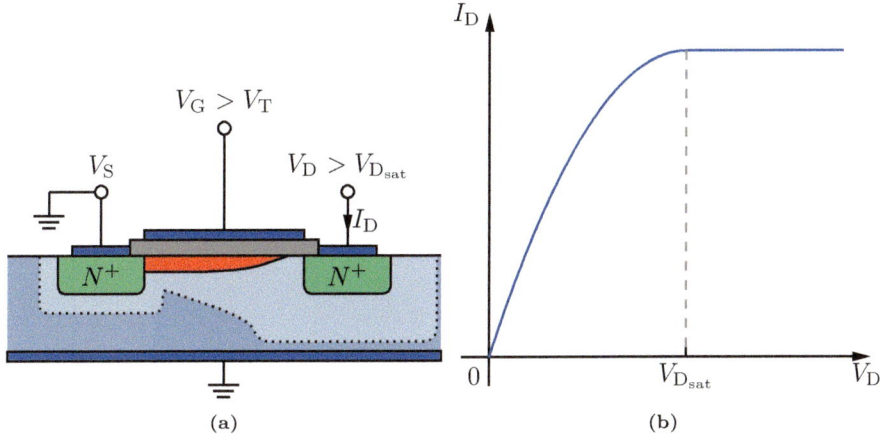

(a)   (b)

Figure 13.20: An n-channel enhancement-mode MOSFET with $V_G > V_T$ and $V_D > V_{D_{sat}}$. (a) Device cross section showing the depletion region and inversion layer post-pinch-off. (b) Characteristic $i - v$ curve after pinch-off.

I_D–V_D versus V_G

To envision the complete $I_D - V_D$ characteristic for this transistor, recall that the current also depends on the gate voltage V_G. Again, the transistor is off when $V_G < V_T$ and on when $V_G > V_T$. As V_G is increased further above V_T, the concentration of minority carriers in the inversion layer is increased. This increases the conductivity of the channel, leading to a larger current I_D for any given drain voltage. A plot showing the relationship of the $I_D - V_D$ curve with respect to V_G is given in Figure 13.21.

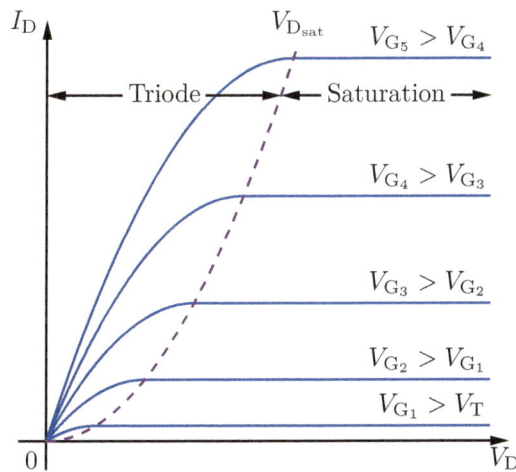

Figure 13.21: Variation of the $I_D - V_D$ characteristic with respect to gate voltage V_G.

A quantitative analysis of the $I_D - V_D$ characteristics of a MOSFET, which can be found in more advanced texts, produces the following equations for the drain current as a function of both the drain voltage and the gate voltage, assuming $V_{\mathrm{GS}} > V_{\mathrm{T}}$.

$$
\begin{aligned}
I_{\mathrm{D}} &= \mu_n C_{\mathrm{ox}} \left(\frac{W}{L} \right) \left[(V_{\mathrm{GS}} - V_{\mathrm{T}}) V_{\mathrm{DS}} - \frac{1}{2} V_{\mathrm{DS}}^2 \right], & \begin{array}{c} V_{\mathrm{DS}} \leq V_{\mathrm{GS}} - V_{\mathrm{T}} \\ V_{\mathrm{D}} \leq V_{\mathrm{G}} - V_{\mathrm{T}} \end{array} && (13.8) \\[2ex]
I_{\mathrm{D}} &= \mu_n C_{\mathrm{ox}} \left(\frac{W}{L} \right) \left[\frac{1}{2} (V_{\mathrm{GS}} - V_{\mathrm{T}})^2 \right], & \begin{array}{c} V_{\mathrm{DS}} \geq V_{\mathrm{GS}} - V_{\mathrm{T}} \\ V_{\mathrm{D}} \geq V_{\mathrm{G}} - V_{\mathrm{T}} \end{array} && (13.9)
\end{aligned}
$$

The first two coefficients in (13.8) and (13.9), μ_n and C_{ox}, are determined by the process and materials used to construct the MOSFET. The coefficient, μ_n, is the electron mobility, which is a measure of how quickly electrons travel through a semiconductor in response to an electric field. The higher their mobility, the faster they move, and thus the higher the resulting current I_{D}. The coefficient, C_{ox}, is the oxide capacitance per unit area across the oxide layer of the gate. As with the parallel plate capacitor, a higher capacitance allows more charge to be stored. Increasing the capacitance increases the current I_{D} by making more charge available to move through the device.

W is the width of the device measured into the page of Figures 13.12 to 13.20a, and L is the length of the channel between the source and drain. Naturally, the current I_{D} will increase as the channel is made wider (increasing W) and will decrease as the channel is made longer (increasing L), just as it would if the channel was a resistor. The quotient, W/L, is the device's **aspect ratio**. Both parameters, W and L, are determined by the MOSFET designer and can be chosen to achieve desired $i - v$ characteristics.

The first four coefficients are often combined into a single coefficient, k, known as the transistor's **transconductance parameter**.

$$
k = \mu_n C_{\mathrm{ox}} \left(\frac{W}{L} \right) \tag{13.10}
$$

Substituting this into (13.8) and (13.9), these equations can be rewritten as

$$
I_{\mathrm{D}} = k \left[(V_{\mathrm{GS}} - V_{\mathrm{T}}) V_{\mathrm{DS}} - \frac{1}{2} V_{\mathrm{DS}}^2 \right], \qquad \begin{array}{c} V_{\mathrm{DS}} \leq V_{\mathrm{GS}} - V_{\mathrm{T}} \\ V_{\mathrm{D}} \leq V_{\mathrm{G}} - V_{\mathrm{T}} \end{array} \tag{13.11}
$$

$$
I_{\mathrm{D}} = \frac{k}{2} (V_{\mathrm{GS}} - V_{\mathrm{T}})^2, \qquad \begin{array}{c} V_{\mathrm{DS}} \geq V_{\mathrm{GS}} - V_{\mathrm{T}} \\ V_{\mathrm{D}} \geq V_{\mathrm{G}} - V_{\mathrm{T}} \end{array} \tag{13.12}
$$

The expressions in square brackets in (13.8) and (13.9) introduce the dependence of the drain current on the device's bias conditions. V_{T} is the threshold voltage required to create an inversion layer in the channel and turn on the device. V_{GS} is the voltage between the gate and the source ($V_{\mathrm{G}} - V_{\mathrm{S}}$), while V_{DS} is the voltage between the drain and the source ($V_{\mathrm{D}} - V_{\mathrm{S}}$). It is still assumed that the source and substrate are connected together, but by referring the gate and drain voltages to the source voltage, the equations (13.8) and (13.9) are valid even when the source voltage is not connected to ground.

Equation (13.8) applies when the device is on ($V_{\mathrm{GS}} > V_{\mathrm{T}}$) and the channel is not pinched off ($V_{\mathrm{DS}} \leq V_{\mathrm{GS}} - V_{\mathrm{T}}$). Notice that for small V_{DS}, V_{DS}^2 will be very small and we can neglect the $-\frac{1}{2}V_{\mathrm{DS}}^2$ term. The remaining relationship between I_{D} and V_{DS} for small V_{DS} becomes linear, as anticipated from the qualitative analysis summarized in Figure 13.15b. As V_{DS} increases, the squared term reduces the drain current from its linear trajectory, producing the slump over shown in Figures 13.16b and 13.17b. We refer to the region of operation up to $V_{\mathrm{DS}_{\mathrm{sat}}}$ as the **triode region**.

Equations (13.8) and (13.9) are equal at the onset of saturation when $V_{\mathrm{DS}} = V_{\mathrm{GS}} - V_{\mathrm{T}} = V_{\mathrm{D}_{\mathrm{sat}}}$. For V_{DS} in excess of $V_{\mathrm{DS}_{\mathrm{sat}}}$, (13.9) applies and the drain current becomes constant with respect to changes in V_{DS}, as shown in Figure 13.20b. We refer to the region of operation beyond $V_{\mathrm{DS}_{\mathrm{sat}}}$ as the **saturation region**.

The transition between triode and saturation occurs when

$$V_{\mathrm{DS}} = V_{\mathrm{GS}} - V_{\mathrm{T}}. \tag{13.13}$$

Solving for I_{D} at this transition voltage using either (13.8) or (13.9) results in

$$I_{\mathrm{D}} = \mu_n C_{\mathrm{ox}} \left(\frac{W}{L}\right)\left[\frac{1}{2}V_{\mathrm{DS}_{\mathrm{sat}}}^2\right] \tag{13.14}$$

which is plotted as the dashed line in Figure 13.21. Every operating point to the left of this line is in the triode region, while every operating point to the right is in the saturation region.

Example 13.1.2

An NMOS transistor is operating at the edge of saturation with $(V_{\mathrm{GS}} - V_{\mathrm{T}}) = 0.5\,\mathrm{V}$ and $I_{\mathrm{D}} = 0.8\,\mathrm{mA}$. Find the following:

1. V_{DS}.

2. The new value of V_{DS} and I_{D} that will keep the transistor at the edge of saturation, if $(V_{\mathrm{GS}} - V_{\mathrm{T}})$ is now halved.

In order for the transistor to be at the edge of saturation, V_{DS} must be equal to $V_{\mathrm{GS}} - V_{\mathrm{T}}$. Hence, for part 1,

$$V_{\mathrm{DS}} = V_{\mathrm{GS}} - V_{\mathrm{T}} = 0.5\,\mathrm{V}.$$

For part 2, we still require V_{DS} to be equal to $V_{\mathrm{GS}} - V_{\mathrm{T}}$ to stay at the edge of saturation, but $(V_{\mathrm{GS}} - V_{\mathrm{T}})$ is now halved. Using the subscript 2 to refer to the new values and 1 to refer to the original values, we have

$$V_{\mathrm{DS}_2} = V_{\mathrm{GS}_2} - V_{\mathrm{T}} = \frac{V_{\mathrm{GS}_1} - V_{\mathrm{T}}}{2} = 0.25\,\mathrm{V}.$$

In saturation, the current I_{D} is given by

$$I_{\mathrm{D}} = \frac{k}{2}(V_{\mathrm{GS}} - V_{\mathrm{T}})^2.$$

However, we do not know the transconductance parameter k, so we will solve for the new current I_{D_2} using its original value I_{D_1}. First we express I_{D_2} in terms of V_{GS_1}:

$$I_{D_2} = \frac{k}{2}(V_{GS_2} - V_T)^2 = \frac{k}{2}\left(\frac{V_{GS_1} - V_T}{2}\right)^2.$$

Then, if we bring the factor of $\frac{1}{2}$ out from squared term, the remaining terms are equal to I_{D_1}. Hence,

$$I_{D_2} = \frac{1}{4}\frac{k}{2}(V_{GS_1} - V_T)^2 = \frac{1}{4}I_{D_1} = \frac{1}{4}0.8\,\text{mA} = 0.2\,\text{mA}.$$

13.1.4 Circuit Symbols

MOSFETs are represented in circuits with either an equivalent circuit model or with a specific circuit symbol for the device, as shown in Figure 13.22. Both representations include three device terminals, which implicitly assumes that the source and body terminals are connected together. The conventional voltage and current polarities have been added for clarity, but in practice, these are frequently omitted.

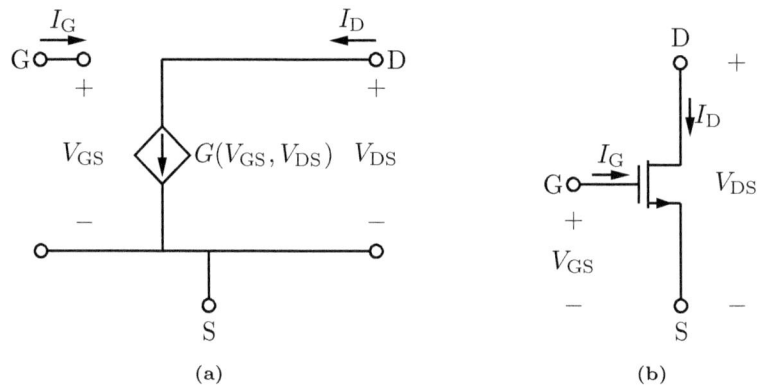

Figure 13.22: Representation of an n-channel MOSFET in circuits using (a) the large-signal equivalent circuit model and (b) its circuit symbol.

The equivalent circuit model represents the behavior of the MOSFET using basic circuit elements. In this case, we can represent the behavior of the device using a voltage-controlled current source. The function $G(V_{GS}, V_{DS})$ determining the current I_D is either (13.8) or (13.9), depending on the bias conditions. This is called a large-signal model because it correctly defines the current over a large set of input conditions. Unfortunately, the large-signal model is difficult to use in circuit analysis because, in general, G is a nonlinear function of two variables.

The circuit symbol shown in Figure 13.22b is specific to n-channel enhancement mode MOSFETs where the source and substrate are connected together. The two parallel lines in the symbol represent the gate and substrate separated by an insulating layer. An arrow placed on the source terminal distinguishes it from the drain terminal. The direction of the arrow indicates the direction of the current I_D.

Example 13.1.3

For the transistor circuit shown in Figure 13.23, when $V_\mathrm{DD} = 5\,\mathrm{V}$ and $V_\mathrm{SS} = -4\,\mathrm{V}$, the drain current $I_\mathrm{D} = 0.6\,\mathrm{mA}$ and the drain voltage $V_\mathrm{D} = 0.8\,\mathrm{V}$. Find the values of R_D and R_S, given that the transistor's parameters are $V_\mathrm{T} = 0.75\,\mathrm{V}$ and $k = 19.2\,\mathrm{mA/V^2}$.

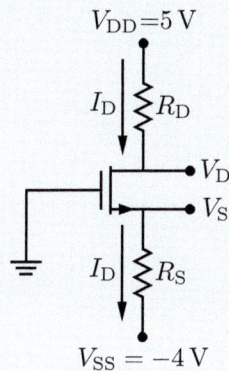

Figure 13.23: Example 13.1.3.

From Ohm's law, we can solve for R_D as

$$R_\mathrm{D} = \frac{V_\mathrm{DD} - V_\mathrm{D}}{I_\mathrm{D}} = \frac{5\,\mathrm{V} - 0.8\,\mathrm{V}}{0.6\,\mathrm{mA}} = 7\,\mathrm{k\Omega}.$$

Similarly, we can express R_S as

$$R_\mathrm{S} = \frac{V_\mathrm{S} - V_\mathrm{SS}}{I_\mathrm{D}}$$

in which V_S is the only unknown. We will use the transistor's current equation to solve for V_S, but we don't know ahead of time if the transistor is operating in saturation or triode mode. To resolve this uncertainty, we will solve for V_S under the assumption that the transistor is in saturation mode, and then check the results to see if they are in agreement with this assumption.

The equation for current through the transistor in saturation is given by

$$I_\mathrm{D} = \frac{k}{2}\left(V_\mathrm{GS} - V_\mathrm{T}\right)^2.$$

Replacing V_GS with $V_\mathrm{G} - V_\mathrm{S}$, we have

$$I_\mathrm{D} = \frac{k}{2}\left(V_\mathrm{G} - V_\mathrm{S} - V_\mathrm{T}\right)^2.$$

With the gate connected to ground, $V_G = 0\,\text{V}$, and so we can solve for V_S to find

$$V_S = V_G - V_T \quad - \sqrt{\frac{2I_D}{k}}$$
$$= 0\,\text{V} - 0.75\,\text{V} \mp 0.25\,\text{V} = -1.0\,\text{V or } -0.5\,\text{V}.$$

We rule out the $V_S = -0.5\,\text{V}$ result by noting that if $V_S = -0.5\,\text{V}$, then $V_{GS} = 0.5\,\text{V}$, which is less than the threshold voltage required to bring the transistor out of cutoff. Since the current I_D is greater than zero, we know the transistor cannot be in cutoff, which leaves only

$$V_S = -1.0\,\text{V}.$$

Now we can check our assumption that the transistor is in saturation by calculating V_{DS} and comparing it to $V_{GS} - V_T$. V_{DS} is given by

$$V_{DS} = V_D - V_S = 0.8\,\text{V} - -1.0\,\text{V} = 1.8\,\text{V},$$

and V_{GS} by

$$V_{GS} = V_G - V_S = 0\,\text{V} - -1.0\,\text{V} = 1.0\,\text{V}.$$

Then, since

$$V_{DS} > V_{GS} - V_T$$
$$1.8\,\text{V} > 1.0\,\text{V} - 0.75\,\text{V},$$

we confirm that the transistor is in saturation.

Finally, we solve for R_S as

$$R_S = \frac{V_S - V_{SS}}{I_D}$$
$$= \frac{-1.0\,\text{V} - (-4\,\text{V})}{0.6\,\text{mA}} = 5\,\text{k}\Omega.$$

Example 13.1.4

For the circuit shown in Figure 13.24 below, find the value of R that results in $V_D = 0.9\,\text{V}$, given that the MOSFET's parameters are $V_T = 0.6\,\text{V}$ and $k = 5\,\text{mA/V}^2$.

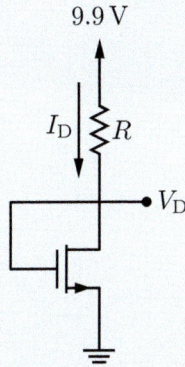

Figure 13.24: Example 13.1.4.

In this circuit, the transistor's gate and drain terminals are connected directly together such that the drain voltage V_{DS} will always be equal to the gate voltage V_{GS}. As a consequence, as long as V_{DS} is greater than V_T, the transistor will be in saturation because V_{DS} will always be greater than $V_{GS} - V_T$. Since V_{GS} equals V_{DS}, the saturation current can be expressed as

$$I_D = \frac{k}{2}(V_{DS} - V_T)^2,$$

which evaluates to

$$I_D = \frac{5\,\text{mA}}{2\,\text{V}^2}(0.9\,\text{V} - 0.6\,\text{V})^2 = 0.225\,\text{mA}.$$

With the voltage across and the current through the resistor established, we can now calculate the appropriate value for R using Ohm's law.

$$R = \frac{V_{DD} - V_D}{I_D} = \frac{9.9\,\text{V} - 0.9\,\text{V}}{0.225\,\text{mA}} = 40\,\text{k}\Omega$$

Example 13.1.5

For the circuit shown in Figure 13.25 below, transistors T_1 and T_2 are the same with $V_T = 0.6\,\text{V}$ and $k = 5\,\text{mA}/\text{V}^2$. Find the value of R_2 that will result in T_2 operating at the edge of saturation.

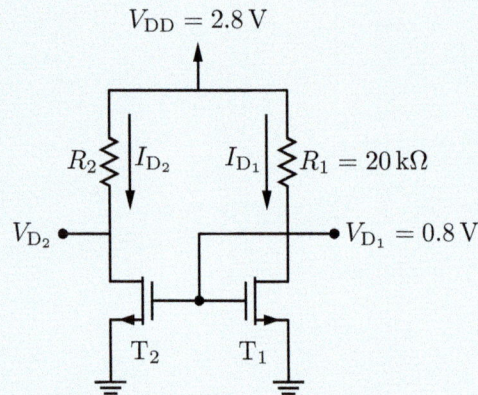

Figure 13.25: Example 13.1.5.

In order to find R_2, we first need to find both V_{D_2} and I_{D_2}. The gate terminals of the transistors are connected to each other as well as to the drain terminal of T_1, so we have

$$V_{G_2} = V_{G_1} = V_{D_1} = 0.8\,\text{V}.$$

Then, to keep T_2 at the edge of saturation,

$$V_{D_2} = V_{G_2} - V_T = 0.8\,\text{V} - 0.6\,\text{V} = 0.2\,\text{V}.$$

The current I_{D_2} can the be found using the transistor's saturation current equation. Using V_{G_2} in place of V_{GS_2} since the source is connected to ground, we have

$$I_{D_2} = \frac{k}{2}(V_{G_2} - V_T)^2 = \frac{5\,\text{mA}}{2\,\text{V}^2}(0.8\,\text{V} - 0.6\,\text{V})^2 = 0.1\,\text{mA}.$$

Finally, R_2 can be found using Ohm's law.

$$R_2 = \frac{V_{DD} - V_{D_2}}{I_{D_2}} = \frac{2.8\,\text{V} - 0.2\,\text{V}}{0.1\,\text{mA}} = 26\,\text{k}\Omega.$$

Example 13.1.6

The transistor in the circuit below (Figure 13.26) has a threshold voltage of $V_T = 0.9\,\text{V}$ and a transconductance parameter of $k = 0.58\,\text{mA/V}^2$. First, find the value of R_D that results in $V_D = 0.2\,\text{V}$. Then, determine the effective resistance between the transistor's source and drain terminals under these conditions.

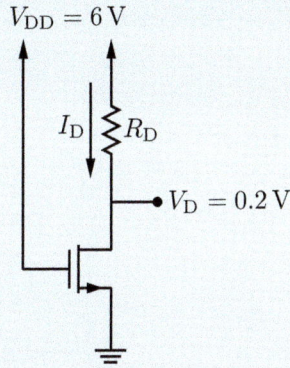

Figure 13.26: Example 13.1.6.

If we knew the current I_D, we could calculate the resistance R_D using Ohm's law as

$$R_D = \frac{V_{DD} - V_D}{I_D}.$$

To find the current, we need to know the transistor's operating mode so we can solve the correct current equation. We can tell that the transistor is not in cutoff because its gate voltage V_{GS} is greater than its threshold voltage V_T. Further, the transistor is not in saturation because its drain voltage V_{DS} (0.2 V) is less than $V_{GS} - V_T$ (5.1 V). Therefore, the transistor must be operating in the triode region. The drain current in the triode region can be calculated by

$$I_D = k \left[(V_{GS} - V_T)V_{DS} - \frac{1}{2}V_{DS}^2 \right]$$

$$= 0.58 \, \text{mA V}^{-2} \left[(6\,\text{V} - 0.9\,\text{V})0.2\,\text{V} - \frac{1}{2}(0.2\,\text{V})^2 \right]$$

$$= 0.58 \, \text{mA}.$$

Now we can find the resistance R_D:

$$R_D = \frac{V_{DD} - V_D}{I_D} = \frac{6\,\text{V} - 0.2\,\text{V}}{0.58\,\text{mA}} = 10.0 \, \text{k}\Omega.$$

Next, we want to determine the effective resistance of the transistor. Recall that when V_D is small relative to $V_G - V_T$, the $i - v$ curve of the transistor is nearly linear (see Figure 13.15b). This means that under these bias conditions, the transistor can be approximated as a resistor with its resistance given by the slope of the transistor's $i - v$ curve at the given bias. Since the $i - v$ curve is almost linear, we approximate its slope at this bias point as

$$\frac{V_D}{I_D} = \frac{0.2\,\text{V}}{0.58\,\text{mA}} = 345 \, \Omega.$$

Example 13.1.7

For the circuit shown below (Figure 13.27), the transistor has the parameters $V_T = 1\,\text{V}$ and $k = \frac{5}{6}\,\text{mA/V}^2$. Find V_G, I_1, I_D, V_D, and V_S.

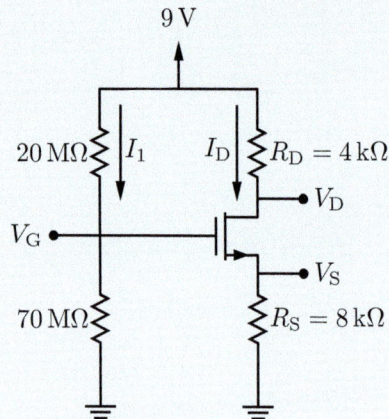

Figure 13.27: Example 13.1.7.

We can find V_G using voltage division. The simplified circuit is shown in Figure 13.28.

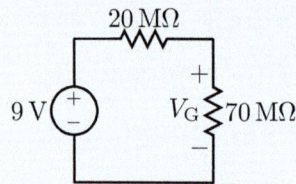

Figure 13.28: Example 13.1.7.

$$V_G = \frac{70\,\text{M}\Omega}{70\,\text{M}\Omega + 20\,\text{M}\Omega}9\,\text{V} = 7\,\text{V}.$$

Next we wish to calculate I_D, but at this point, we don't know in which mode (cutoff, triode, or saturation) the transistor is operating. We can, however, quickly rule out cutoff by noting that if I_D equals $0\,\text{A}$, and then V_S must equal $0\,\text{V}$, which implies $V_{GS} - V_T$ equals $6\,\text{V}$; but a $V_{GS} - V_T$ greater than $0\,\text{V}$ is inconsistent with cutoff. To choose between triode and saturation, we assume saturation, solve the circuit, and then check to see if our assumption is valid. In saturation, the drain current through the transistor is given by

$$I_D = \frac{k}{2}(V_{GS} - V_T)^2,$$

which can also be expressed as

$$I_D = \frac{k}{2}(V_G - V_S - V_T)^2.$$

Both I_D and V_S are unknown, so we need to find another equation relating them in order to complete the solution. Applying Ohm's law at R_S provides

$$I_D = \frac{V_S}{R_S}.$$

Plugging this equation into the one above yields

$$\frac{V_S}{R_S} = \frac{k}{2}(V_G - V_S - V_T)^2,$$

which with a little rearranging can be written as

$$\frac{1}{2}V_S^2 - \left[\frac{1}{kR_S} + (V_G - V_T)\right]V_S. + \frac{1}{2}(V_G - V_T)^2 = 0\,\text{V}.$$

Solving this quadratic equation for V_S provides two solutions:

$$V_S = 7.5\,\text{V and } 4.8\,\text{V},$$

We rule out V_S equal to 7.5 V because it results in a $V_{GS} - V_T$ of less than 0 V, which is inconsistent with saturation. A V_S of 4.8 V, on the other hand, results in a $V_{GS} - V_T$ of 1.2 V. Using this V_S to solve for I_D yields

$$I_D = \frac{V_S}{R_S} = \frac{4.8\,\text{V}}{8\,\text{k}\Omega} = 0.6\,\text{mA}.$$

Now we can use I_D and Ohm's law to calculate V_D as

$$V_D = V_{DD} - I_D R_D = 9\,\text{V} - 0.6\,\text{mA} \cdot 4\,\text{k}\Omega = 6.6\,\text{V}.$$

Finally, we can check our assumption that the transistor is in saturaton by confirming that

$$V_{DS} > V_{GS} - V_T$$
$$V_D - V_S > V_G - V_S - V_T$$
$$1.8\,\text{V} > 1.2\,\text{V}$$

If this had not been true, we would have repeated the solution for V_S, I_D, and V_D with the knowledge that the transistor is in triode mode.

13.2   MOSFET Analog Applications

Amplification is the primary application of transistors in analog circuits. In this section, we will examine a very common transistor configuration known as the **common-source (CS)** amplifier circuit to see how a MOSFET can be used to amplify a signal.

13.2.1   The Common-Source Amplifier Circuit

The CS amplifier circuit is shown in Figure 13.29. In this configuration, the drain of an enhancement mode n-type MOSFET is connected to a DC power source V_{DD} through a load resistor R_{D}. The source terminal is connected directly to ground, hence the name "common-source," and a control voltage V_{G} is connected to the gate.

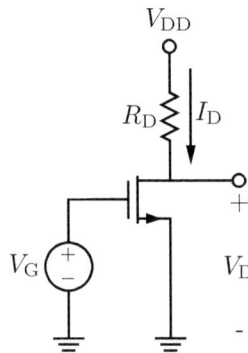

Figure 13.29: The common-source amplifier circuit.

We established in section 13.1.3 that the drain current I_{D} through the transistor depends upon both the gate-to-source voltage V_{GS} and the drain-to-source voltage V_{DS} as given by (13.8) and (13.9). Since the source terminal in the CS amplifier is connected directly to ground (i.e., $V_{\mathrm{S}} = 0$), we can drop the subscript $_{\mathrm{S}}$ from those equations. Also recall that when V_{G} is less than the transistor's threshold voltage V_{T}, the transistor is "off" and the current I_{D} is zero. This operating mode is referred to as **cutoff**. The current equations for each of the three operating modes are summarized below.

$$I_{\mathrm{D}} = \begin{cases} 0 & : \text{cutoff} \\ \dfrac{k}{2}(V_{\mathrm{G}} - V_{\mathrm{T}})^2 & : \text{saturation} \\ k\left[(V_{\mathrm{G}} - V_{\mathrm{T}})V_{\mathrm{D}} - \frac{1}{2}V_{\mathrm{D}}^2\right] & : \text{triode} \end{cases} \qquad (13.15)$$

where

$$k = \mu_n C_{\mathrm{ox}}\left(\frac{W}{L}\right) \qquad (13.16)$$

and is known as the device's **transconductance parameter**.

With the addition of the resistor R_{D} between the fixed DC voltage V_{DD} and the drain terminal, we can write an additional equation relating the drain current to the drain voltage.

$$I_{\mathrm{D}} = \frac{V_{\mathrm{DD}} - V_{\mathrm{D}}}{R_{\mathrm{D}}} \qquad (13.17)$$

Then, given V_{G} and utilizing (13.15) through (13.17), we can, with a bit of effort, solve for both I_{D} and V_{D}. If V_{G} is less than V_{T}, the solution is the trivial solution; I_{D} equals zero and V_{D} equals V_{DD}. However, if V_{G} is greater than V_{T}, we may not be able to guess correctly whether the transistor is in saturation or triode. The solution is to assume that the transistor is in saturation and solve for I_{D} using (13.15). The resulting I_{D} can then be used to solve for V_{D} via (13.17). Finally, we check to see if our initial assumption is correct by confirming (or not) that $V_{\mathrm{D}} > V_{\mathrm{G}} - V_{\mathrm{T}}$. If our initial assumption turns out to be incorrect, we resolve for both I_{D} and V_{D} with the knowledge that the transistor is operating in triode, which will require that we find the solution to a quadratic equation.

13.2.2 Load Line Analysis

To get a visual idea of how V_{G}, V_{D}, and I_{D} are related, we can solve (13.15) and (13.17) graphically by adding the line of (13.17) to the transistor's $I_{\mathrm{D}} - V_{\mathrm{D}}$ characteristic plots (section 13.1.3) as shown in Figure 13.30. The straight line of (13.17), referred to as the **load line**, varies from $V_{\mathrm{DD}}/R_{\mathrm{D}}$ amps when the drain voltage is at zero volts, to zero amps when the drain voltage is at V_{DD} volts. The operating point of the circuit is then the intersection between the load line and the non-linear $I_{\mathrm{D}} - V_{\mathrm{D}}$ curve of the transistor.

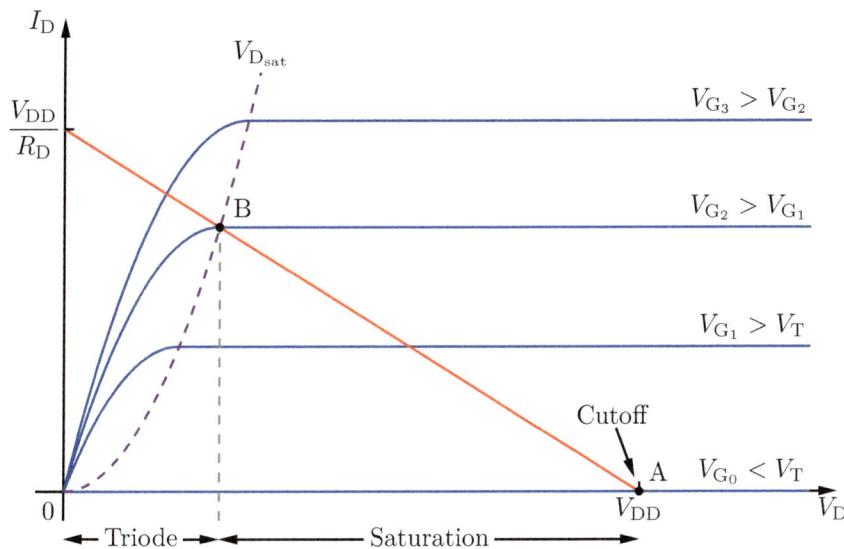

Figure 13.30: Graphical solution for the CS amplifier operating point.

When the gate voltage is less than the voltage threshold ($V_{\mathrm{G}} < V_{\mathrm{T}}$), the transistor is in cutoff at point A of Figure 13.30. The drain current equals zero ($I_{\mathrm{D}} = 0$) and the drain voltage equals the power supply voltage ($V_{\mathrm{D}} = V_{\mathrm{DD}}$).

As the gate voltage is increased slightly above the threshold voltage, the device transitions from cutoff into saturation. A small drain current begins to flow through the transistor, resulting in a small voltage drop across the resistor

R_D and reducing the drain voltage V_D below V_DD. At this point, V_D is only slightly below V_DD and V_G is only slightly above V_T such that $V_\mathrm{G} - V_\mathrm{T}$ is only slightly above zero. Since V_D will be greater than $V_\mathrm{G} - V_\mathrm{T}$, we can confirm that the device is in saturation. Figure 13.30 may be helpful in visualizing this transition from cutoff directly into saturation.

As V_G is increased further, the intersection between the $I_\mathrm{D} - V_\mathrm{D}$ characteristic and the load line slides up and left along the load line from point A toward point B.

Once reaching point B, V_D is no longer larger than $V_\mathrm{G} - V_\mathrm{T}$, so the device transitions from saturation into triode. Recall that in the triode region, an inversion layer connects the transistor's drain and source terminals. This creates a low-resistance connection from the drain to the grounded source. As the gate voltage is increased further, we move beyond point B and the drain current approaches its maximum value, $V_\mathrm{DD}/R_\mathrm{D}$, while the drain voltage approaches zero.

In practice, we cannot reach a drain current of $V_\mathrm{DD}/R_\mathrm{D}$ or a drain voltage of zero because the maximum voltage available in our system is typically used for V_DD. However, even without this constraint, V_G is limited by the breakdown voltage of the oxide layer between the gate and the substrate. Exceeding the oxide's breakdown voltage will result in permanent device failure.

13.2.3   Voltage Transfer Characteristic

The complete relationship between V_G, V_D, and I_D is captured in Figure 13.30, but of particular interest is the relationship between the input voltage, V_G, and the output voltage, V_D. To help visualize their relationship, we can plot V_D as a function of V_G, as shown in Figure 13.31. A plot like this of the output voltage versus the input voltage is known as a **voltage transfer characteristic**.

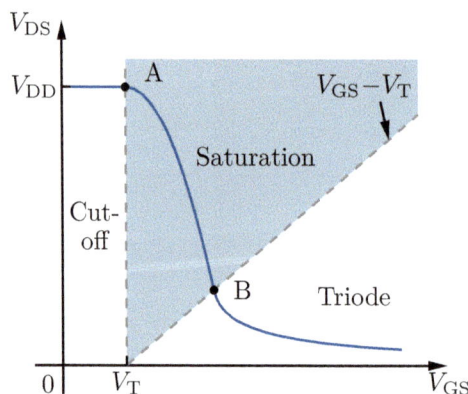

Figure 13.31: The CS amplifier voltage transfer characteristic.

To understand this plot, we will consider what happens to V_D as we sweep V_G up from zero, just as we did with the $I_\mathrm{D} - V_\mathrm{D}$ characteristic in section 13.2.2. The transistor is in cutoff until V_G reaches V_T at point A. With the transistor in cutoff, the current is zero and the drain voltage will equal the power supply voltage V_DD. As V_G increases from point A to point B, the transistor is in saturation because we have $V_\mathrm{D} > V_\mathrm{G} - V_\mathrm{T}$ and $V_\mathrm{G} > V_\mathrm{T}$. Here, the drain voltage, V_D, drops rapidly as the current through the resistor rises. Increasing V_G beyond point B biases the transistor into the triode region with drain voltage V_D asymptotically approaching zero as the drain current I_D approaches V_{DD}/R_D. Note that these modes and transition points correspond exactly to those identified in Figure 13.30.

Example 13.2.1

For the CS amplifier circuit shown below in Figure 13.32, find the gate voltage V_G in terms of V_T, k V_{DD}, and R_D at which the amplifier transitions from saturation mode to triode mode.

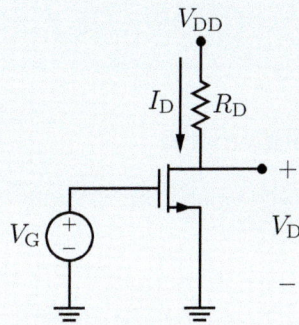

Figure 13.32: Example 13.2.1.

At the transition point between triode and saturation, V_D equals $V_G - V_T$ and the amplifier's current equations for saturation and triode in (13.15) can be written as

$$I_D = \frac{k}{2}V_D^2.$$

Another equation relating I_D and V_D comes from applying Ohm's law at R_D.

$$I_D = \frac{V_{DD} - V_D}{R_D}$$

Combining these two equations, we have

$$\frac{V_{DD} - V_D}{R_D} = \frac{k}{2}V_D^2,$$

which can be rewritten as

$$\frac{1}{2}V_D^2 + \frac{1}{kR_D}V_D - \frac{V_{DD}}{kR_D} = 0 \text{ V}^2.$$

This quadratic equation has solutions:

$$V_D = -\frac{1}{kR_D} \pm \sqrt{\left(\frac{1}{kR_D}\right)^2 + \frac{2V_{DD}}{kR_D}}.$$

We can rule out the negative solution by noting that V_D cannot be less than V_S, which leaves only.

$$V_D = -\frac{1}{kR_D} + \sqrt{\left(\frac{1}{kR_D}\right)^2 + \frac{2V_{DD}}{kR_D}}.$$

Recalling that V_D equals $V_G - V_T$, we can write

$$V_G = V_T + V_D$$

or

$$V_G = V_T - \frac{1}{kR_D} + \sqrt{\left(\frac{1}{kR_D}\right)^2 + \frac{2V_{DD}}{kR_D}}.$$

13.2.4   Signal Amplification

Now suppose that instead of sweeping V_G up from zero, that we apply a time varying input v_G that has both an AC and DC component.

$$v_G(t) = \underbrace{v_g(t)}_{\text{AC}} + \underbrace{V_G}_{\text{DC}} \tag{13.18}$$

We would like to see how the output voltage v_D behaves in response to such a signal. In general, v_g could be any arbitrary AC signal, but to simplify the following analysis, we will assume v_g is a triangle wave, as shown in Figure 13.33a. The complete input voltage v_G is shown in Figure 13.33c. Note that we are using lowercase variables with lowercase subscripts for AC signals, uppercase variables with uppercase subscripts for DC signals, and lowercase variables with uppercase subscripts for complete signals.

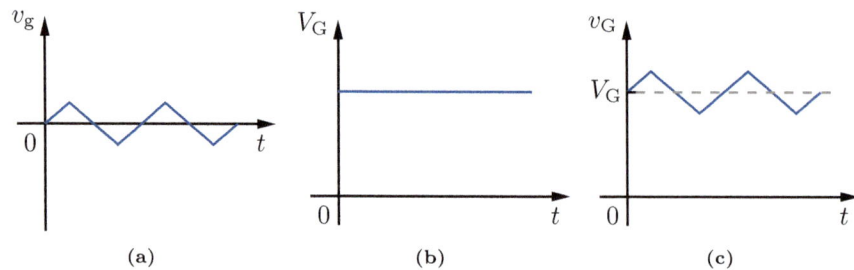

Figure 13.33

To determine what the output voltage v_D looks like we first locate the DC operating point of the transistor, which is called the **quiescent** point or Q point. We have chosen V_G such that the Q point will be in the saturation region between points A and B as shown in Figure 13.34. This establishes the DC output voltage V_D. Next, we slide along the curve of the voltage transfer characteristic plot to see how the output AC signal v_d changes with the applied input AC signal v_g. This allows us to construct the complete output voltage:

$$v_D(t) = v_d(t) + V_D. \tag{13.19}$$

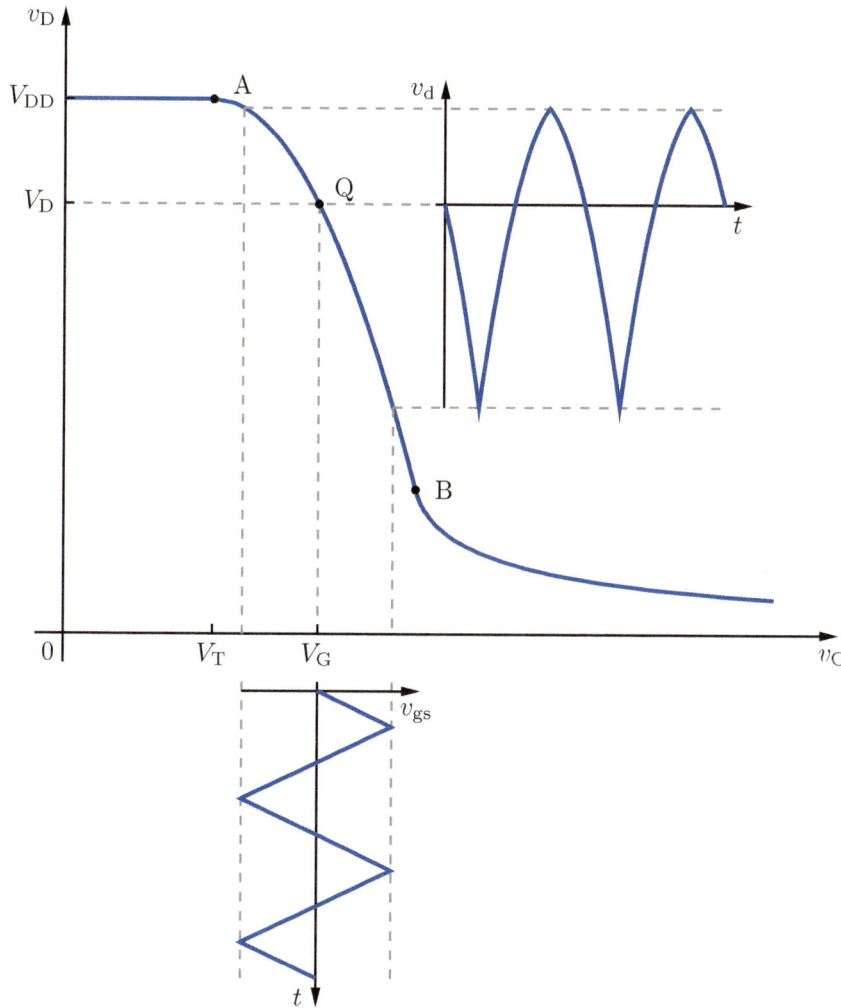

Figure 13.34: Amplification of an AC signal with a CS amplifier. The output signal shows some distortion.

Notice that the amplitude of v_d is greater than the amplitude of v_g and that when v_g is increasing, v_d is decreasing and vice versa. The common source amplifier circuit has amplified and inverted the input signal. Unfortunately, the output signal is somewhat distorted and is not a true triangle wave like the input signal. This distortion occurs because the voltage transfer characteristic is not linear over the region of operation.

Let us also consider what will happen if the Q point is chosen too close to A. When the input signal v_g decreases toward its minimum value, the output signal v_d will increase until reaching V_{DD} and the device will enter cutoff. The peaks of the output signal are effectively clipped off, as shown in Figure 13.35. This type of distortion is appropriately named "cutoff clipping."

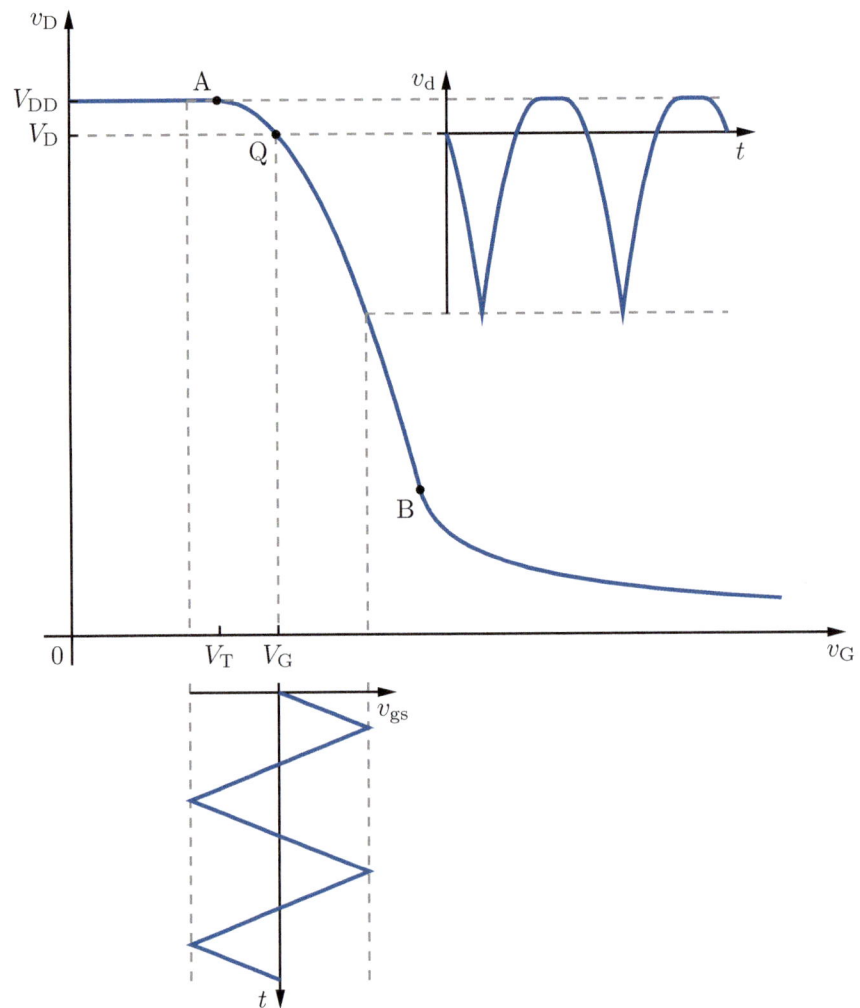

Figure 13.35: Amplification of an AC signal with a CS amplifier. The output signal shows cutoff clipping.

If the Q point is chosen too close to point B, the peaks of the input signal will push the device into the triode region. The magnitude of the slope of the voltage transfer characteristic is much lower in the triode region than the saturation region which results in reduced signal amplification. This produces a flattening of the output signal as shown in Figure 13.36. This type of distortion is named "triode flattening."

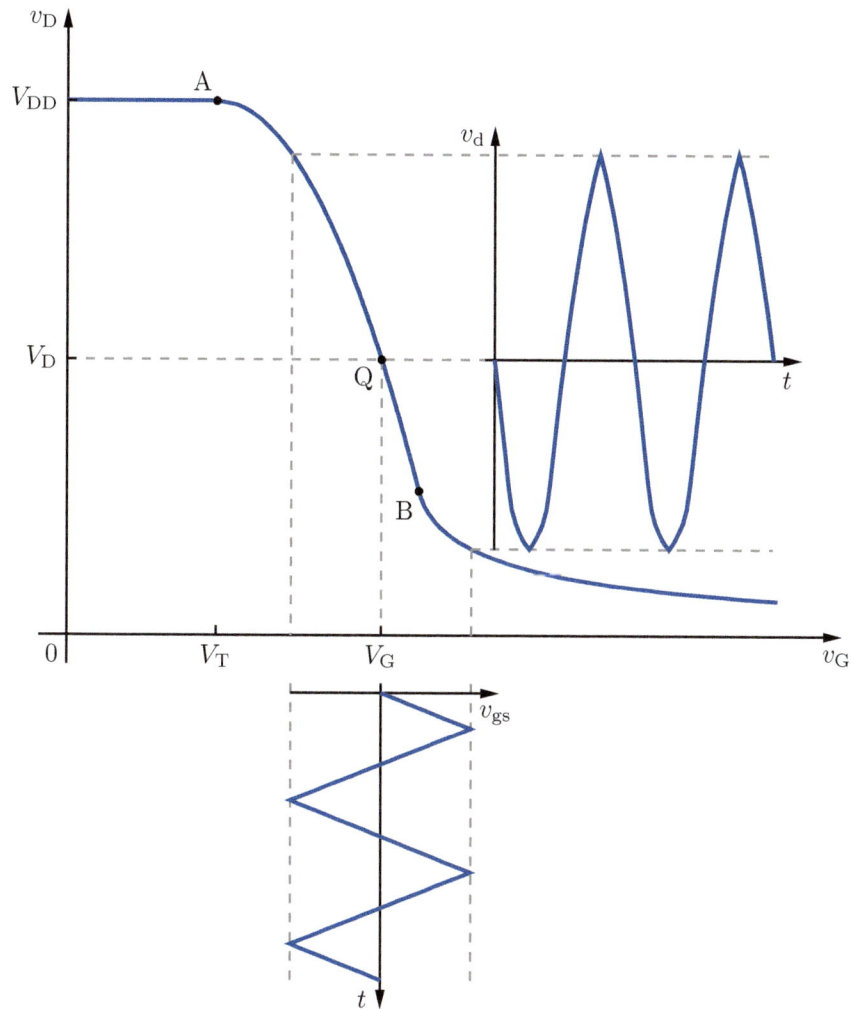

Figure 13.36: Amplification of an AC signal with a CS amplifier. The output signal shows triode flattening.

13.2.5   The Linear Amplifier

Ideally, we would like to have an amplifier without any distortion. This would allow the amplified output signal to accurately reproduce all the features of the input signal. Essentially, what we want is a linear amplifier. To achieve this, we will need to avoid the distortion identified in Figures 13.34, 13.35, and 13.36.

We can avoid cutoff clipping and triode flattening if we can keep the transistor's operating point within the saturation region. This requires that

$$V_G - \hat{v}_g > V_G|_A \tag{13.20}$$

and

$$V_G + \hat{v}_g < V_G|_B \tag{13.21}$$

where $\hat{v}_g$ is the amplitude of the input signal $v_g(t)$. The point $V_G|_A$ is simply V_T, the point where the device transitions from cutoff into saturation. To find the point $V_G|_B$, where the device transitions from saturation to triode, we first solve the load line equation (13.17) for V_D.

$$V_D = V_{DD} - I_D R_D \tag{13.22}$$

and then substitute in the equation for I_D in saturation

$$V_D = V_{DD} - \frac{k}{2} R_D \left(V_G - V_T\right)^2. \tag{13.23}$$

Noting that the transition point B occurs when

$$(V_G - V_T) = V_D, \tag{13.24}$$

we can rewrite (13.23) as

$$V_D|_B = V_{DD} - \frac{k}{2} R_D \left(V_D|_B\right)^2. \tag{13.25}$$

Then, employing the quadratic formula and noting that $V_D|_B$ must be positive, we find.

$$V_D|_B = \frac{\sqrt{2kR_D V_{DD} + 1} - 1}{kR_D} \tag{13.26}$$

Finally, relying on (13.24), we have

$$V_G|_B = V_T + \frac{\sqrt{2kR_D V_{DD} + 1} - 1}{kR_D}. \tag{13.27}$$

The remaining distortion, which was apparent in Figure 13.34, is due to the non-linear response of the transistor's drain current with respect to its gate voltage while in saturation. Let us take a closer look at the source of this distortion to see if there is anything we can do about it. First, we plug the total time varying control voltage from (13.18) into the equation for the drain current in saturation:

$$i_D = \frac{k}{2}(v_G - V_T)^2. \tag{13.28}$$

Breaking v_G into its AC and DC components results in

$$i_D = \frac{k}{2}(v_g + V_G - V_T)^2. \tag{13.29}$$

Then, by expanding the square, we find

$$i_D = \underbrace{\frac{k}{2}v_g{}^2 + k(V_G - V_T)v_g}_{i_d} + \underbrace{\frac{k}{2}(V_G - V_T)^2}_{I_D}. \tag{13.30}$$

We recognize that the third term on the right hand side of (13.30) is the current I_D in response to the DC component of the input voltage V_G. This means the first and second terms must be the response to the AC input signal v_g. The second term is linearly proportional to v_g with a proportionality constant of $k(V_G - V_T)$. This is the term responsible for the desirable linear amplification of the input signal. The first term, however, is a quadratic function of v_g and represents the unwanted non-linear distortion.

We cannot make this non-linear term completely disappear unless k or v_g are equal to zero, in which case we would also have zero amplification. However, if we can choose the device's parameters and DC gate voltage such that the non-linear term is much smaller than the linear term, then we can neglect the non-linear distortion and we will have a viable approximation to a linear amplifier. Hence, we require

$$\frac{k}{2}\hat{v}_g{}^2 \ll k(V_G - V_T)\hat{v}_g. \tag{13.31}$$

Isolating $\hat{v}_g$ on the left, this constraint simplifies to

$$\boxed{\hat{v}_g \ll 2(V_G - V_T)} \tag{13.32}$$

which in known as the **small-signal condition**. In other words, we will have approximately linear amplification if the amplitude of the AC input signal is much less than twice the gate voltage in excess of the threshold voltage.

Knowing that we can neglect distortion if the amplitude of the input signal is "small enough," we naturally wonder: "How small is small enough?" The answer to this question depends on how much distortion can be tolerated in the specific application. The constraint of (13.32) doesn't precisely define the relative values of $\hat{v}_g$ and $2(V_G - V_T)$. Instead, it tells us that because we have design control over those values, we also have design control over the level of distortion and that we can limit it to an acceptable level. Limiting distortion to no more than 10% of the output signal is a commonly used rule of thumb.

Example 13.2.2

The CS amplifier of Figure 13.37 is in saturation. The transistor's parameters are $V_T = 0.7\,\text{V}$ and $k = 2.5\,\text{mA/V}^2$. Given that $V_{DD} = 10\,\text{V}$, $V_G = 2.7\,\text{V}$, and $R_D = 1\,\text{k}\Omega$, find the following:

1. V_D if $v_g = 0\,\text{V}$.

2. v_D and Δv_D if $v_g = +0.04\,\text{V}$.

3. v_D and Δv_D if $v_g = -0.04\,\text{V}$.

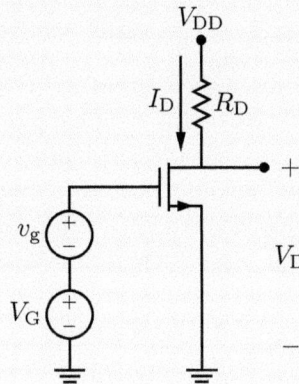

Figure 13.37: A CS amplifier circuit with gate bias voltage V_G and signal voltage v_g.

1. We are given that the amplifier is operating in saturation mode, which allows us to express the transistor's current as

$$I_D = \frac{k}{2}(V_G - V_T)^2 = \frac{2.5}{2}(2.7 - 0.7)^2 = 5\,\text{mA}.$$

V_D can be expressed as

$$V_D = V_{DD} - I_D R_D,$$

which can be solved to yield

$$V_D = 10 - 5(1) = 5\,\text{V}.$$

Finally, we confirm the transistor is in saturation by showing that

$$V_D \geq V_G - V_T,$$

$$5 \geq 2.7 - 0.7.$$

2. We calculate i_D using (13.29) as

$$i_D = \frac{k}{2}(v_g + V_G - V_T)^2 = \frac{2.5}{2}(0.04 + 2.7 - 0.7)^2 = 5.202\,\text{mA}.$$

Then we can find v_D as

$$v_D = V_{DD} - i_D R_D = 10 - 5.202(1) = 4.798\,\text{V}.$$

The change in voltage Δv_D due to v_g can calculated as

$$\Delta v_D = v_D - V_D = -0.202\,\text{V}.$$

3. We will do the same thing again for the new value of v_g.

$$i_D = \frac{k}{2}(v_g + V_G - V_T)^2 = \frac{2.5}{2}(-0.04 + 2.7 - 0.7)^2 = 4.802\,\text{mA}$$

$$v_D = V_{DD} - i_D R_D = 10 - 4.802(1) = 5.198\,\text{V}$$

$$\Delta v_D = v_D - V_D = 0.198\,\text{V}.$$

Example 13.2.3

For the CS amplifier shown in Figure 13.38, find the maximum amplitude value of the input signal ($\hat{v}_g$) to limit the distortion at 10%, and 5%. The transistor's parameters are $V_T = 1\,\text{V}$ and $k = 1\,\text{mA/V}^2$. The transistor is biased with $V_{DD} = 8\,\text{V}$, $V_G = 2\,\text{V}$.

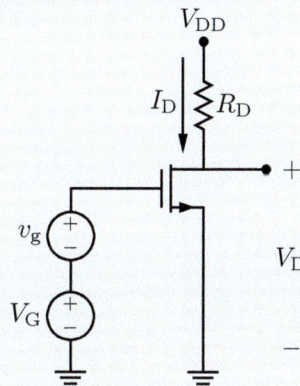

Figure 13.38: Example 13.2.3.

The **small-signal condition** in (13.32) can be used here to determine the value of ($\hat{v}_g$). To limit the distortion to 10%,

$$(\hat{v}_g) < 10\%\,2(V_G - V_T). \tag{13.33}$$

Then

$$(\hat{v}_g) < 0.2\,\text{V}. \tag{13.34}$$

To limit the distortion to 5%,

$$(\hat{v}_g) < 5\%\,2(V_G - V_T). \tag{13.35}$$

Then

$$(\hat{v}_g) < 0.1\,\text{V}. \tag{13.36}$$

13.2.6   Voltage Gain

Now that we have an idea of how to achieve linear amplification, it is worth taking a moment to consider how much amplification, or **gain**, can be achieved with the CS amplifier. The voltage gain is the ratio of the output signal v_d to the input signal v_g:

$$A_v = \frac{v_d}{v_g}. \tag{13.37}$$

To calculate the voltage gain, we start with the total instantaneous output voltage given by

$$v_D = V_{DD} - R_D i_D. \tag{13.38}$$

Then, breaking the total current into its AC and DC components, we have

$$v_D = \underbrace{-R_D i_d}_{v_d} + \underbrace{V_{DD} - R_D I_D}_{V_D}. \tag{13.39}$$

Having identified the AC component of the drain voltage as

$$v_d = -R_D i_d, \tag{13.40}$$

we replace the current i_d with the value found in (13.30) to get

$$v_d = -R_D \left[\frac{k}{2} v_g{}^2 + k(V_G - V_T)v_g \right]. \tag{13.41}$$

Then, under the small-signal condition, we neglect the second-order term, leaving only

$$v_d = -R_D k(V_G - V_T)v_g. \tag{13.42}$$

Finally, the small-signal voltage gain is

$$\boxed{A_v = -R_D k(V_G - V_T).} \tag{13.43}$$

Notice the minus sign in (13.43). This indicates that the output signal is inverted (180° out of phase) with respect to the input signal. This is in agreement with our observations in Figures 13.34, 13.35, and 13.36. Also note that the gain can be increased by increasing the gate voltage V_G as long as

$$V_G + \hat{v_g} < V_G|_B. \tag{13.44}$$

This indicates that, to maximize the amplifier's gain, we should bias the transistor as close to its saturation/triode transition point as possible while still maintaining room within the saturation region for the input signal.

13.2.7 Small-Signal Equivalent Circuit

When designing a linear amplifier circuit, like that shown in Figure 13.39, it is helpful to separate the analysis into AC and DC components. We complete the DC bias analysis first, and then, once the DC operating point has been established, we complete the AC signal analysis. To aid us in this process, recall that the gate and drain voltages can be written as the sum of their AC and DC components, as was shown in (13.18) and (13.19). In those two equations, we dropped the reference to the source terminal voltage because the source terminal was at zero volts in that discussion. Here, we will retain the reference to the source voltage for greater generality.

$$v_{GS} = v_{gs} + V_{GS} \qquad (13.45)$$

$$v_{DS} = v_{ds} + V_{DS} \qquad (13.46)$$

We also know that in order to achieve linear amplification, the device must be operating in saturation, and hence the drain current will be related to the gate voltage, as given by (13.30). Again, we will include the reference to the source voltage in the present discussion for greater generality:

$$i_D = \underbrace{\frac{k}{2}v_{gs}{}^2 + k(V_{GS} - V_T)v_{gs}}_{i_d} + \underbrace{\frac{k}{2}(V_{GS} - V_T)^2}_{I_D}. \qquad (13.47)$$

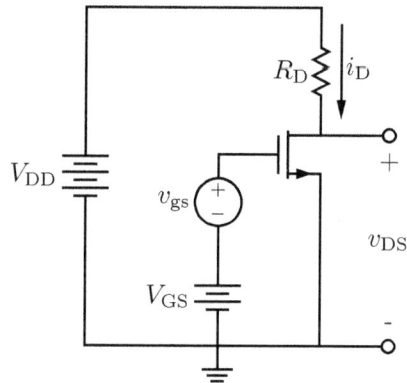

Figure 13.39: CS amplifier circuit with explicit connection to V_{DD}.

In the DC analysis, we first set any AC inputs to zero. If we have them in our circuit, we can also replace any capacitors with an open circuit and any inductors with a short circuit. The results of applying these changes to the circuit of Figure 13.39 are shown in Figure 13.40. Since we know that for linear amplification the transistor must be operated in saturation, the DC drain current and gate voltage are related by

$$I_D = \frac{k}{2}(V_{GS} - V_T)^2. \qquad (13.48)$$

Using this equation, along with the constraints provided by the DC circuit and the amplifier design criteria, we solve for the DC operating point (V_{GS}, I_D).

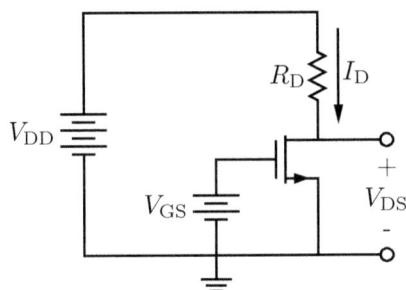

Figure 13.40: CS amplifier DC bias circuit with the AC source shut off.

The AC analysis begins by setting all of the DC sources to zero. This means DC voltage sources become short circuits and DC current sources become open circuits. Also, we will often assume the AC signal frequency is high enough that we can replace any capacitors with short circuits and any inductors with open circuits. The resulting AC circuit is shown in Figure 13.41.

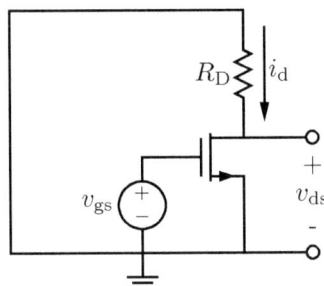

Figure 13.41: CS amplifier AC signal circuit with the DC sources off.

The AC components of the drain current and gate voltage are related by

$$i_\mathrm{d} = \frac{k}{2}v_\mathrm{gs}{}^2 + k(V_\mathrm{GS} - V_\mathrm{T})v_\mathrm{gs}. \tag{13.49}$$

Since the small-signal condition must be satisfied in the design of a linear amplifier, we can neglect the squared term above and simplify the relation to

$$i_\mathrm{d} = k(V_\mathrm{GS} - V_\mathrm{T})v_\mathrm{gs}. \tag{13.50}$$

The term relating v_gs to i_d is known as the MOSFET **transconductance** g_m:

$$i_\mathrm{d} = g_m v_\mathrm{gs}, \tag{13.51}$$

where

$$g_m = k(V_\mathrm{GS} - V_\mathrm{T}). \tag{13.52}$$

Having previously established the DC operating point, g_m will be a fixed constant in the small-signal AC analysis.

The simple relation in (13.51) allows the small-signal AC response of the transistor to be modeled as a linear voltage-controlled current source, as shown in Figure 13.42. By replacing the transistor in the AC circuit with its equivalent circuit model, as shown in Figure 13.43, the resulting circuit can be analyzed with conventional linear circuit analysis techniques.

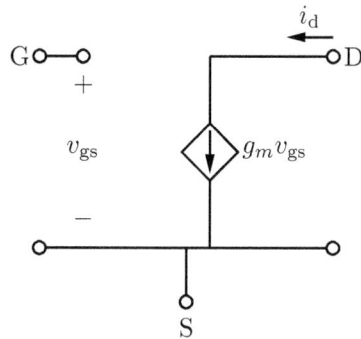

Figure 13.42: The MOSFET small-signal equivalent circuit model.

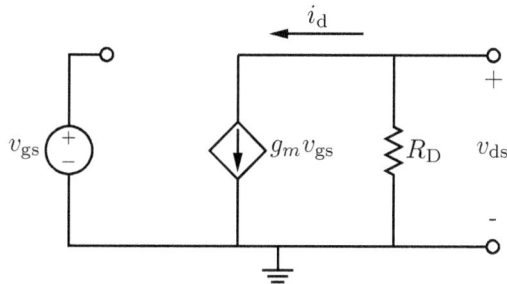

Figure 13.43: CS amplifier small-signal equivalent circuit.

13.3 MOSFET Digital Applications

Digital logic is one of the primary applications of MOSFETs. In this section, we investigate a MOSFET implementation of the digital NOT gate. Also known as an inverter, the NOT gate performs an essential operation in digital logic. While some logic gates such as AND and OR can be constructed from diodes alone, a transistor is necessary to build a NOT gate.

13.3.1 The MOSFET NOT Gate

The CS amplifier circuit presented in section 13.2.1 can also be used in digital circuits to implement a NOT gate. The circuit implementation and the digital circuit symbol for the NOT gate are illustrated in Figures 13.44a and 13.44b, respectively. The digital output of a NOT gate is the opposite of its input. Thus, a low input voltage representing a logical 0 results in a high-output voltage representing a logical 1 and vice versa. The NOT gate's relative input and output voltages are summarized in Table 13.1a, with the associated logical values in positive logic shown in Table 13.1b.

To see how the circuit in Figure 13.44a operates as a NOT gate, we first consider how the circuit behaves with the input voltage set within the low voltage range. In this condition, the gate-to-source voltage V_{GS} will be less than the threshold voltage V_T, which means the transistor will be in cutoff. Since the current I_D is effectively zero in cutoff and the output voltage is given by

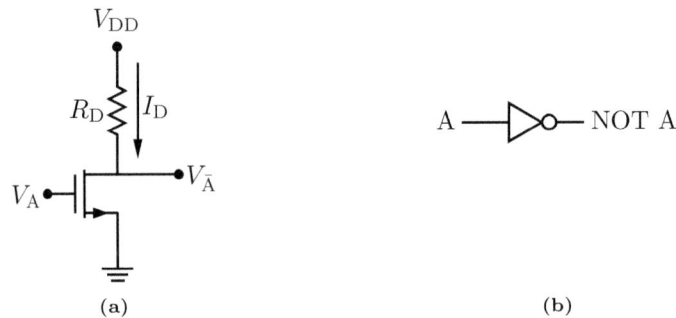

Figure 13.44: (a) An NMOS digital inverter circuit, also called a NOT gate, and (b) its equivalent digital logic circuit symbol.

V_A	$V_{\bar{A}}$
L	H
H	L

(a)

A	NOT A
0	1
1	0

(b)

Table 13.1: Truth tables for a digital NOT gate showing (a) the circuit's relative input and output voltages, and (b) its logical variables in positive logic.

$$V_{\text{high}} - I_D R_D, \tag{13.53}$$

the output voltage will be equal to V_{high}. Thus, a low input voltage results in a high output voltage. Or, stated in positive logic, a logical 0 input results in a logical 1 output.

This confirms the circuit inverts a low input, but what about a high input? When the input voltage is high, it is not immediately apparent whether the transistor is in triode or saturation. To get a clearer idea of how the device is operating, we can plot its voltage transfer characteristics, as shown in Figure 13.45, from which we can see that higher input voltages (V_A) produce lower output voltages ($V_{\bar{A}}$). The overall steepness of the voltage transfer characteristic is strongly influenced by the choice of R_D. Larger values of R_D produce steeper transitions in the output voltage from high to low voltage. Thus, while the output voltage never quite reaches zero, with an appropriate choice for R_D, the inverter can be designed such that any voltage in the high voltage range will result in an output voltage in the low voltage range. Or, stated in positive logic, such that a logical 1 input results in a logical 0 output.

Recall that in digital logic, logical values are assigned to voltage ranges. For example, the 0 V to 0.5 V range is often assigned to a logical zero while voltages in the 2.8 V to 5 V range are assigned to a logical one. This leaves voltages in the 0.5 V to 2.8 V range undefined. In some situations, like when using multiple stages of diode logic or using a logical output to drive a load, the voltage levels can stray outside of the defined ranges. If this happens, we can no longer be sure whether the voltage should be considered a logical 1 or a 0 because the behavior of manufactured logic chips is undefined when the supplied input voltages are outside of their assigned ranges. The following example shows how two NOT gates in sequence can be used as a buffer circuit to help resolve this issue.

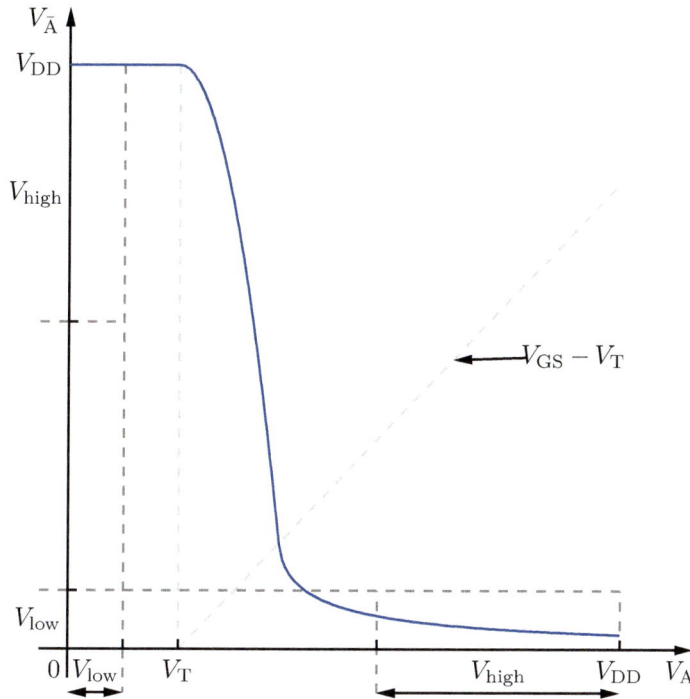

Figure 13.45: Voltage transfer characteristic of a CS amplifier inverter.

Example 13.3.1

Suppose the output from a previous logic stage is only providing $2.8\,\mathrm{V}$. We construct the circuit below to increase this voltage closer to $5\,\mathrm{V}$ before using it as the input to another logic stage. Solve for the buffer circuit's output voltage V_{o2} given $V_{DD} = 5\,\mathrm{V}$, $V_T = 1\,\mathrm{V}$, $k = 1\,\mathrm{mA\,V^{-2}}$, and $R = 10\,\mathrm{k\Omega}$.

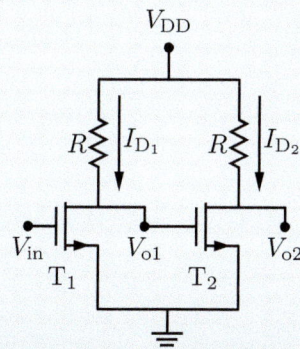

Figure 13.46: A digital buffer constructed from cascaded inverters.

With an input voltage V_{in} in the high range, the transistor should be operating in triode, but we don't know for sure—so we will begin by solving for V_{o1} while assuming saturation. If T_1 is in saturation, the current I_{D_1} is given by

$$I_{D_1} = \frac{k}{2}(V_G - V_T)^2, \tag{13.54}$$

and since

$$\frac{V_{DD} - V_{o1}}{R} = I_{D_1}, \tag{13.55}$$

we have

$$\frac{V_{DD} - V_{o1}}{R} = \frac{k}{2}(V_G - V_T)^2. \tag{13.56}$$

Solving for V_{o1} yields

$$V_{o1} = V_{DD} - \frac{Rk}{2}(V_G - V_T)^2 = 5 - 5(1.8)^2 = -11.2\,\text{V}. \tag{13.57}$$

Since V_{o1} is also the drain voltage of the first transistor and $V_{o1} < V_G - V_T$, this result is not consistent our assumption that the transistor is in saturation, which requires $V_D > V_G - V_T$. So we conclude that the first transistor is operating in the triode region. Replacing the right-hand side of (13.56) with the equation for current through the MOSFET in triode mode, we have

$$\frac{V_{DD} - V_{o1}}{R} = k\left[(V_G - V_T)V_{o1} - \frac{1}{2}V_{o1}^2\right], \tag{13.58}$$

which can be expressed as

$$\frac{1}{2}V_{o1}^2 - \left[(V_G - V_T) + \frac{1}{Rk}\right]V_{o1} + \frac{V_{DD}}{Rk} = 0. \tag{13.59}$$

Solving this quadratic equation for V_{o1} yields

$$V_{o1} = \left[(V_G - V_T) + \frac{1}{Rk}\right] \pm \sqrt{\left[(V_G - V_T) + \frac{1}{Rk}\right]^2 - \frac{2V_{DD}}{Rk}} \tag{13.60}$$

$$= 1.9 \pm \sqrt{1.9^2 - 1} \tag{13.61}$$

$$= 3.52\,\text{V or }0.29\,\text{V}. \tag{13.62}$$

We rule out the higher result because we have already determined that the first transistor must be operating in the triode region, but a V_D of 3.52 V would put it in the saturation region. Thus, the output of the first stage and the input to the second stage is

$$V_{o1} = 0.29\,\text{V}. \tag{13.63}$$

This input to the second transistor is less than the threshold voltage V_T, and so the second transistor is in cutoff. Since no current will flow through the transistor in cutoff, $I_{D_2} = 0$ and $V_{o2} = V_{DD} = 5\,\text{V}$.

Note that the input voltage to the buffer circuit was at the bottom edge of what could safely qualify as a logical 1. If this input voltage fell any further, it would enter the undefined region where we cannot reliably determine if it should be a logical 1 or a logical 0. The output of the buffer, however, provides a clearly defined 5 V logical 1.

In the MOSFET inverter circuit of Figure 13.44a, the magnitude of R_D has a strong impact on the device's voltage transfer characteristics. As the inverter's input voltage increases from zero toward infinity, its output voltage transitions from the supply voltage V_{DD} toward ground. Larger values of R_D result in steeper transitions, as illustrated in Figure 13.47. When designing an inverter, we can ensure that input voltages in the high voltage region produce output voltage in the low voltage region by choosing a sufficiently large value for R_D. In integrated circuits, making resistors with high values of resistance can require large areas of the chip. Since this would limit what can fit on a single chip, we don't want R_D to be any larger than necessary.

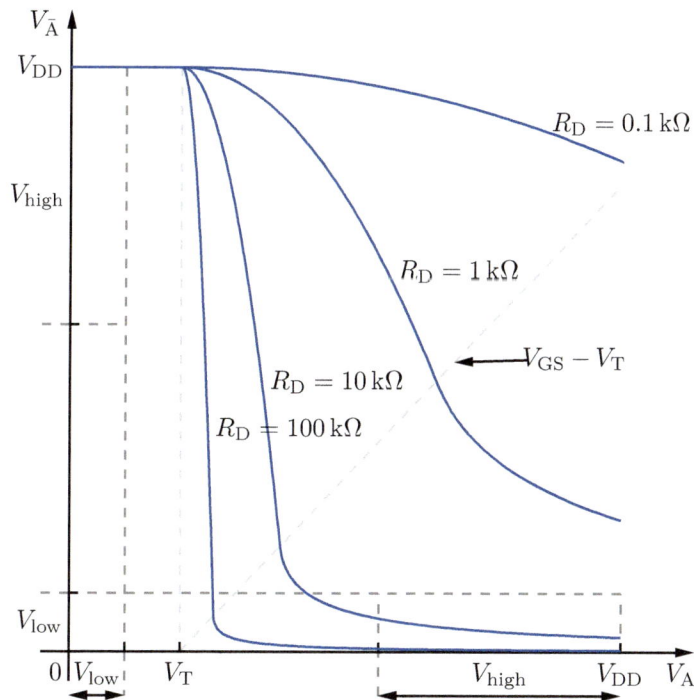

Figure 13.47: Variation of an inverter's voltage transfer characteristics with R.

Example 13.3.2

Find the minimum value of R_D in the MOSFET inverter circuit shown below that will ensure an input voltage in the 2.8 V to 5.0 V range results in an output value in the 0 V to 0.5 V range, given $V_{DD} = 5$ V, $V_T = 1$ V, and $k = 1\,\mathrm{mA\,V^{-2}}$.

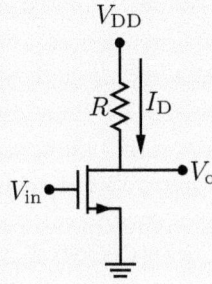

Figure 13.48: MOSFET inverter circuit.

We begin by noting that the highest output voltage corresponds to the lowest input voltage. Hence, we need to ensure that $V_\text{o} = 0.5\,\text{V}$ when $V_\text{in} = 2.8\,\text{V}$. Then, we note that this point, $(V_\text{in}, V_\text{o}) = (2.8\,\text{V}, 0.5\,\text{V})$, is in the triode region because $V_\text{o} = V_\text{D} < (V_\text{G} - V_\text{T}) = (V_\text{in} - V_\text{T})$. The equation for the current I_D in the triode region is given by

$$I_\text{D} = k \left[(V_\text{in} - V_\text{T})V_\text{o} - \frac{1}{2}V_\text{o}^2 \right], \tag{13.64}$$

and from Ohm's law, we have

$$\frac{V_\text{DD} - V_\text{o}}{R} = I_\text{D}. \tag{13.65}$$

By relating these two equations for I_D, we have one equation in which the only unknown quantity is R:

$$\frac{V_\text{DD} - V_\text{o}}{R} = k \left[(V_\text{in} - V_\text{T})V_\text{o} - \frac{1}{2}V_\text{o}^2 \right]. \tag{13.66}$$

Finally, solving for R results in

$$R = \frac{V_\text{DD} - V_\text{o}}{k \left[(V_\text{in} - V_\text{T})V_\text{o} - \frac{1}{2}V_\text{o}^2 \right]} \tag{13.67}$$

$$= \frac{5 - 0.5}{1 \times 10^{-3}\left[(1.8)0.5 - 0.5^2\right]} = 5.8\,\text{k}\Omega. \tag{13.68}$$

The voltage transfer characteristic with R set to $5.8\,\text{k}\Omega$ is shown below.

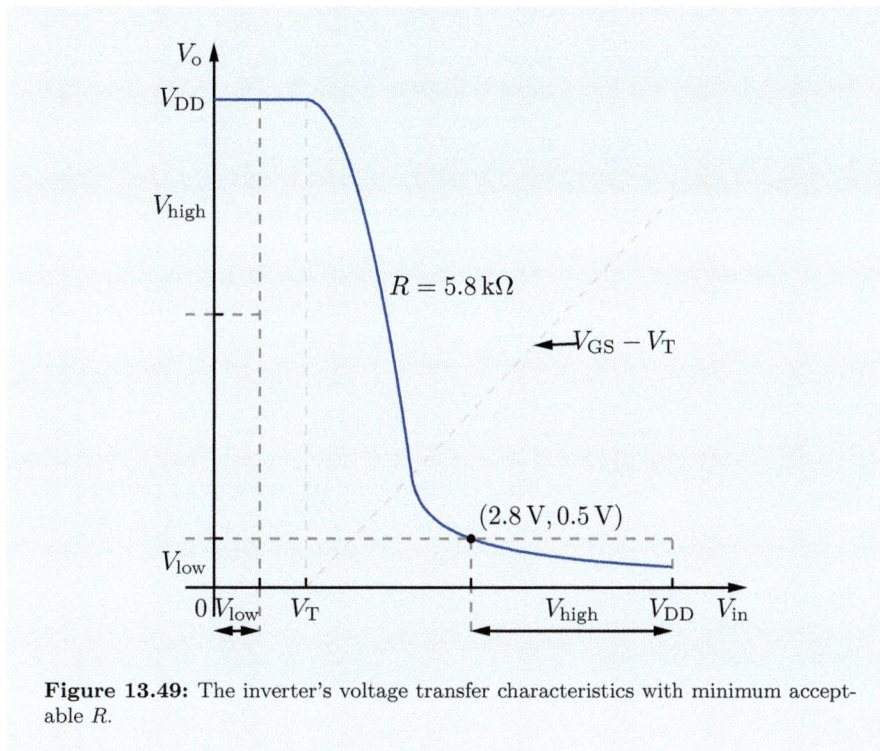

Figure 13.49: The inverter's voltage transfer characteristics with minimum acceptable R.

13.4 Exercises

Subsection 13.1.3

13.1. An NMOS transistor with $V_T = 0.7\,\text{V}$ and $k = 3\,\text{mA}/\text{V}^2$ is operating at a DC current $I_D = 2.5\,\text{mA}$. Find the new DC current I_D if

   a) The transistor's width W was halved.

   b) The transistor's length L was tripled.

13.2. An NMOS transistor with $V_T = 0.65\,\text{V}$ and $k = 4\,\text{mA}/\text{V}^2$ is operating with $V_{GS} = 1.25\,\text{V}$. Find the value of the DC current I_D when the transistor enters the saturation region.

13.3. An NMOS transistor with $V_T = 0.6\,\text{V}$ and $k = 0.4\,\text{mA}/\text{V}^2$ is operating at the edge of saturation with a DC current $I_D = 0.8\,\text{mA}$. Find the biasing voltage V_{GS}. If we want to lower the current at the edge of saturation to $I_D = 0.5\,\text{mA}$, find the new biasing voltage V_{GS}.

Subsection 13.2.1

13.4. The CS amplifier shown in Figure 13.50 is biased using $V_G = 0.5\,\text{V}$ and $V_{DD} = 12\,\text{V}$; determine the operating point V_D, I_D. The transistor parameters are $V_T = 1\,\text{V}$ and $k = 1\,\text{mA}/\text{V}^2$.

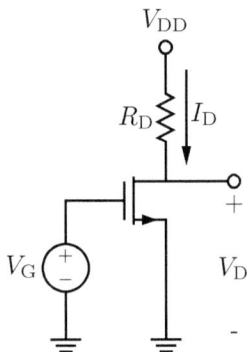

Figure 13.50: The CS amplifier circuit.

13.5. The CS amplifier shown in Figure 13.50 is biased using $V_G = 2.2\,\text{V}$ and $V_{DD} = 10\,\text{V}$; determine the operating point V_D, I_D for $R_D = 5\,\text{k}\Omega$. The transistor parameters are $V_T = 0.7\,\text{V}$ and $k = 1\,\text{mA}/\text{V}^2$.

13.6. The CS amplifier shown in Figure 13.50 is biased using $V_G = 4\,\text{V}$ and $V_{DD} = 15\,\text{V}$; determine the operating point V_D, I_D for $R_D = 30\,\text{k}\Omega$. The transistor parameters are $V_T = 1\,\text{V}$ and $k = 1\,\text{mA}/\text{V}^2$.

13.7. The CS amplifier shown in Figure 13.50 is biased using $V_{DD} = 20\,\text{V}$; determine the gate voltage required to drive the transistor into triode mode for $R_D = 2\,\text{k}\Omega$. The transistor parameters are $V_T = 1\,\text{V}$ and $k = 1\,\text{mA}/\text{V}^2$.

Subsection 13.2.3

13.8. For the CS amplifier shown in Figure 13.51, $V_T = 1\,\text{V}$ and $k = 0.75\,\text{mA}/\text{V}^2$. Find the drain voltage V_D and the drain current I_D.

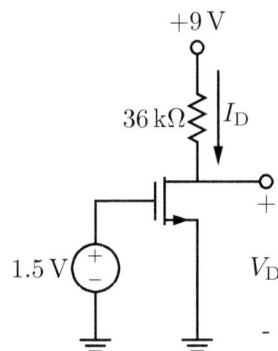

Figure 13.51: The CS amplifier circuit.

13.9. For the CS amplifier shown in Figure 13.52, $V_T = 1\,\text{V}$ and $k = 1\,\text{mA}/\text{V}^2$. Find the drain voltage V_D and the drain current I_D.

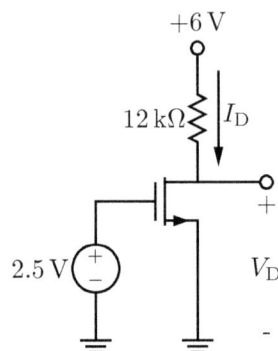

Figure 13.52: The CS amplifier circuit.

13.10. For the CS amplifier shown in Figure 13.53, $V_T = 1\,\text{V}$ and $k = 1\,\text{mA/V}^2$. Find V_G that will make the amplifier operate at the edge of saturation. Also find the drain current I_D.

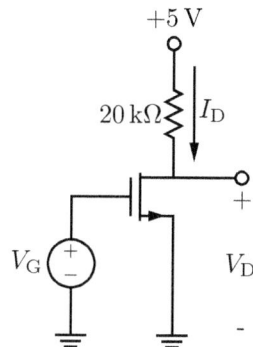

Figure 13.53: The CS amplifier circuit.

Subsection 13.2.5

13.11. The linear amplifier shown in Figure 13.54 has the following transistor parameters: $V_T = 1\,\text{V}$ and $k = 0.75\,\text{mA/V}^2$. The amplitude value of the input signal is $v_g = 0.05\,\text{V}$. Find i_D and v_D.

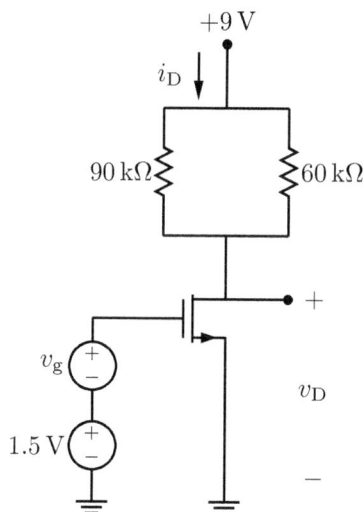

Figure 13.54: The CS amplifier circuit.

13.12. The linear amplifier shown in Figure 13.55 has the following transistor parameters: $V_T = 1\,\text{V}$ and $k = 1\,\text{mA/V}^2$. We have two different values of the amplitude value of the input signal $(\hat{v}_g = 0.2\,\text{V}, 0.5\,\text{V})$. Are these values acceptable

to maintain a distortion $<10\%$? What value of $(\hat{v}_g)$ results in 5% distortion?

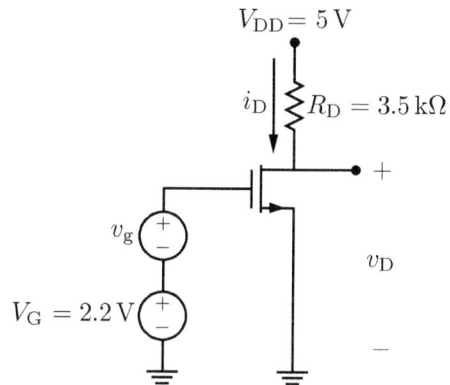

Figure 13.55: The CS amplifier circuit.

Subsection 13.2.6

13.13. The CS amplifier shown in Figure 13.54 has the following transistor parameters: $V_T = 1\,\text{V}$ and $k = 1\,\text{mA/V}^2$. Find the gate bias that produces the maximum linear gain for the input signal amplitudes:

a) $\hat{v}_g = 0.075\,\text{V}$.

b) $\hat{v}_g = 0.1\,\text{V}$.

Is this a good amplifier in your opinion? Why?

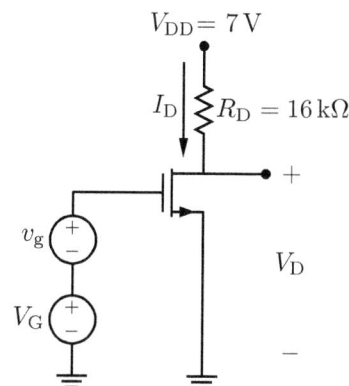

Figure 13.56: The CS amplifier circuit.

Index

Page numbers in italics indicate figures and tables.

About the Authors

Maryam A. Al-Othman received her BSEE degree in electrical engineering from Kuwait University in 2004, and her MS and PhD degrees in electrical engineering from Purdue University in 2006 and 2011, respectively. From 2012 to 2015, she was an assistant professor of electrical engineering at Kuwait University. She has worked for Purdue University since 2015 and is currently a senior lecturer.

John H. Cole received his BSEE and PhD degrees in electrical engineering from Purdue University in 2005 and 2011, respectively. After graduating, he taught at both the American University of Kuwait and the Gulf University of Science and Technology before returning to Purdue University, where he teaches first-year engineering and programming courses.

Dimitrios Peroulis is the senior vice president for Partnerships and Online and the Reilly Professor of Electrical and Computer Engineering at Purdue University. His portfolio includes the offices of industry partnerships, global programs and partnerships, engagement, and online programs. He is an IEEE and IET Fellow and has coauthored over 450 journal and conference papers. He has received numerous research awards and eleven teaching awards, including the 2010 HKN C. Holmes MacDonald Outstanding Teaching Award and the 2010 Charles B. Murphy Award.